About the Author

Dr. Michael Zeilik works as Professor of Astronomy at the University of New Mexico. In his teaching, he specializes in innovative, introductory courses for the novice, non–science major student. His classes include cooperative learning teams to explore key astronomical concepts with hands-on activities. He has been supported by grants from the National Science Foundation, NASA, the Exxon Educational Foundation, and the Slipher Fund of the National Academy of Sciences for innovations in astronomy education, delivery of astronomy to the general public, and astronomy workshops for in-service teachers.

Dr. Zeilik's current research activities focus on two areas: astronomy education and astronomy in the historic and prehistoric Pueblo world in the U.S. Southwest. He has published more than 100 professional articles and four books and has given more than 200 talks to professional and lay audiences.

Dr. Zeilik earned his A.B. in Physics with honors at Princeton University and his M.A. and Ph.D. in Astronomy at Harvard University. He has been a Woodrow Wilson Fellow, a National Science Foundation Fellow, and a Smithsonian Astrophysical Observatory Predoctoral Fellow. At the University of New Mexico, he has been named a Presidential Lecturer, the highest award for all-around performance by a faculty member. In 1998, he was appointed a Research Fellow with the National Institute of Science Education.

In 2000, he became Chair of the Astronomy Education Committee of the American Association of Physics Teachers.

Dr. Zeilik is listed in *American Men and Women of Science*, *The Writers Directory*, *Contemporary Authors*, *Who's Who in the West*, and *Who's Who of Emerging Leaders in America*. He is a member of the Authors Guild and the Text and Academic Authors Association. The 8th edition of *Astronomy: The Evolving Universe* won a 1997 Texty Award from the Text and Academic Authors Association.

Contents

ASTRONOMY

THE EVOLVING UNIVERSE
NINTH EDITION

Michael Zeilik

University of New Mexico

CAMBRIDGE
UNIVERSITY PRESS

PUBLISHED BY THE PRESS SYNDICATE OF THE UNIVERSITY OF CAMBRIDGE
The Pitt Building, Trumpington Street, Cambridge, United Kingdom

CAMBRIDGE UNIVERSITY PRESS
The Edinburgh Building, Cambridge, CB2 2RU, UK
40 West 20th Street, New York, NY 10011-4211, USA
10 Stamford Road, Oakleigh, Melbourne 3166, Australia
Ruiz de Alarcón 13, 28014 Madrid, Spain
Dock House, The Waterfront, Cape Town 8001, South Africa

http://www.cambridge.org

First published 2002

Printed in the United States of America

Typeface Meridien Roman 10/13   *System* QuarkXPress®   [GH]

A catalogue record for this book is available from the British Library.

Library of Congress Cataloging-in-Publication Data

Zeilik, Michael.
    Astronomy : the evolving universe / Michael Zeilik – 9th ed.
      p. cm.
    Includes bibliographical references and index.
    ISBN 0-521-80090-0 (pb.)
    1. Astronomy.   I. Title.

QB45.Z428 2001
520 – dc21   2001035088

ISBN 0 521 80090-0 (paperback)

This project was supported, in part, by the National Science Foundation.

Cover Image
Magnificent Messier 82 (sometimes called the Cigar galaxy) captured by the Subaru Telescope on Mauna Kea, Hawaii,
U.S.A. This irregular galaxy lies about 12 million light years away in the constellation Ursa Major. Messier 82 is the
brightest galaxy in the sky in infrared light. Its core appears to have been disrupted by a close encounter with another
galaxy, triggering a colossal round of starbirth. Winds from these hot, young stars expel supersonic hydrogen gas (red
filaments) from the galaxy. Copyright 2000 National Astronomical Observatory of Japan.

Preface

A *ninth* edition! I really cannot believe that an idea that I had some twenty-five years ago has lived so long and gone so far. Many instructors and students have told me that they've enjoyed using the book, and that feedback has certainly helped me along. I also desire to improve this book so that it innovates and evolves as a better learning tool. Finally, astronomy changes rapidly, especially with the advent of new space and ground-based telescopes. Our outer vision grows, and our inner one intensifies. So what's new for this edition?

One, the obvious **updating of the material.** Constant change drives the excitement of astronomy. That is a main reason why astronomy appeals to me, and, I hope, to you. There's always something new and unexpected for us to discover in the universe. However, I have done the updating with great care to focus on discoveries that I think have long-term value and connect with the major concepts.

Two, a **streamlining of the material** so as to make the descriptions, concepts, and explanations as clear and concrete as possible. I have taken great care to minimize the use of technical terms and the passive voice. I also gave special attention to the Learning Outcomes and Key Concepts to keep them concise and precise.

Three, a **refined art program.** The figures have been reviewed and revised with the goal of better understanding by the novice student (while keeping the science accurate). I have tried to incorporate a human perspective whenever possible. I have also aimed for clarity and simplicity so that careful "viewing" of the figures will illuminate key concepts.

Four, a **research-based pedagogy.** I have targeted concepts in this edition more than ever, based on research in astronomy education carried out by myself and others. But concepts in isolation have little value – they must be *connected*. The right connections of key concepts result in the "big picture" of the cosmos. My goal for my course and this book is to provide a connected understanding of astronomy. To do so, I have introduced a new organizing feature, Celestial Navigators™.

Each chapter contains one **Celestial Navigator™.** These maps provide visual guides of major concepts in the chapter and explicitly show their connections. I have tested them in my classes, and they result in large, robust gains in students' connected understanding of astronomy.

To promote an even higher level of conceptual unification, each part begins with a **Part Concept,** followed by an **Inquiry Focus** – a series of "How do we know . . . ?" questions to link concepts across chapters within a Part. Again, the goal is the "big picture," this time within a thematically-unified Part.

THEMATIC AND TOPICAL STRUCTURE

Previous editions have developed the notion of "how do we know?" in astronomy. I have made that question more explicit and the central theme for the entire book. I believe that answering the "how do we know?" promotes the understanding of any subject, but astronomy more so than many others do. Why? Because once we leave the earth and solar system, astronomy becomes quite abstract, essentially light, physics, and models.

Yet people tend to view science as a disconnected collection of facts, and our culture reinforces this view. Take for instance, the quiz show *Who Wants to Be a Millionaire.* The questions use a multiple-choice format, which is common in schools in the United States. Here is one question for the big prize, $1,000,000:

How long does it take the light from the sun to reach the earth?

A. 42 seconds
B. 3 minutes
C. 8 minutes
D. 1 hour

Do you know the answer? (It is C.) Now, that's OK, but do you know the unasked question? *How do we know that the sun is 8 light-minutes from the earth?* That's a much deeper question and requires a level of understanding far more fundamental than the first (which can just be memorized).

"How do we know?" is the governing theme of this edition. In terms of topics, this book has two main aspects: to describe in narrative form the range of the astronomical universe and its cosmic connections; and to introduce students to how astronomers think about the cosmos so they can gain some understanding about its operation. I hope that students will become so enticed by the contents that they will become intrigued by the concepts linking and illuminating astronomical phenomena.

This edition is designed for a one-semester introduction to astronomy. Note that it is much shorter than previous editions, with fewer and shorter chapters. Yet it retains the previous structure of four coherent parts, each focusing on a key subtheme of cosmic evolution. Like the cosmos, each part connects to the others, so you can really approach the four parts in any order.

Part 1: Changing Conceptions of the Cosmos

This part accents the evolution of cosmological ideas, from the geometric views of the Greeks to the mind-boggling visions of modern astronomy. It leads off with the simplest observations you can make without a telescope from the earth and ends with the farthest reaches of the visible universe. Part 1 acquaints the reader with the idea of *scientific models,* the conceptual core of modern scientific thought. Scientific models are born from our imagination and experience; they mark the essential creative act of the scientific enterprise. As such models evolve, they shape our changing conceptions of the cosmos. The development of scientific models resounds throughout the book; it is *the* fundamental tool to understand the universe.

Essential to the formation of models are new astronomical observations. In the age of microelectronics and space astronomy, we have greatly expanded our vision of the cosmos. Many space telescopes, especially the Hubble Space Telescope, marked great changes in the outer vision of twentieth-century astronomy. In this century, we know that new technology telescopes on the earth will surpass space telescopes in some ways. We can hope that lunar-based telescopes will augment the observational legacy of space telescopes. New generations of computers endow our minds with another way to see the physical processes in the cosmos by the simulation of astronomical systems. Together, computers and telescopes work as the tools to impart innovative ideas to astronomers.

Part 2: The Planets: Past and Present

Flyby spacecraft and gangly landers have provided new insights to our understanding of the planets. This part focuses on the physical properties of the planets to infer their origin and evolution. It first takes a comparative look at our current knowledge of the planets, especially our earth and moon, as well as Mercury, Venus, and Mars. These planets show different degrees of evolution, with the earth being the most evolved. The other planets – Jupiter, Saturn, Uranus, Neptune, and Pluto – have, in contrast, changed little since their birth.

Space missions have disclosed that the moons of the outer planets are really worlds unto themselves, places of rock and ice scarred by violence in the past. These new worlds provide important clues, along with comets and meteorites, of the early history of the solar system. So planetary evolution traces back to origin – the birth of the solar system from an interstellar cloud of gas and dust. By astronomical standards, that birth was quick, violent, and chaotic – shaping the primitive forms of the planets. This picture implies that many other single stars have planetary systems and that these other worlds resemble the local planets in broad ways.

Part 3: The Universe of Stars

Our sun and its planets swing around in a vast island of stars called the Milky Way Galaxy – our home galaxy. As the nearest star, our sun serves as the close-up model for other stars, especially to understand the physical processes within them. The Galaxy contains some hundreds of billions of stars at various stages of their lives. These stars have been born, like the sun, from clouds of interstellar gas and dust. Modern technology supplies us with views deep into the regions of starbirth, showing us the early lives of stars.

Because stars live long by human standards, we cannot directly observe their evolution. We can build models of stars with computers, and these models provide us with the dimension of time to map out the lives of stars. Ordinary stars, powered by fusion reactions, grow old and blow off material before their demise. Many have violent deaths, leaving bizarre corpses such as neutron stars. Massive stars undergo violent deaths, which are signaled by enormous explosions that build heavy elements and propel them into space to seed the next generation of stars. The span of the lives of stars guides us to an understanding of what will happen to our sun in its old age.

Part 4: Galaxies and Cosmic Evolution

The universe contains galaxies, in which the most visible material of the cosmos resides. This part first examines ordinary galaxies, like our Milky Way, and then hyperactive ones. We now realize that galaxies with unusual activity are interacting by gravity with their neighbors, which triggers the celestial fireworks. The show includes long, thin jets of material, confined by magnetic fields, rocketing out of the cores of these galaxies. All these galaxies dwell in clusters with other galaxies, and we have just come to the realization that clusters are laid out in long chains with vast voids in between – the cosmic tapestry. Amid this remarkable layout of visible matter lurks matter that we cannot yet see – the so-called dark matter that shapes the visible universe, both its structure and destiny.

The cosmic design in the large-scale architecture was imprinted in the awesome explosion in which the universe began. That explosion – the Big Bang – linked the smallest pieces of matter to the universe at large. The Big Bang also left relics that we observe today, evidence of the violence of creation. From the Big Bang some 15 billion years ago, cosmic evolution shaped us at where we are and when we are in the universe.

QUICK START

This book is for students, who are probably novices to astronomy. I want them to learn effectively from it. I have designed a four-part structure so that you can investigate each part somewhat independently of the others. Many cross-references, especially to the basic physical and astronomical ideas, should help you to link the parts together. I have made a concerted effort to introduce ideas as concretely as possible. Within parts, I deal with the most familiar first: Chapter 1 (Part 1) with the visible sky; Chapter 8 (Part 2) with the earth; Chapter 12 (Part 3) with the sun; and Chapter 17 (Part 4) with the Milky Way Galaxy. Within chapters, I have tried to present concrete examples before abstract notions.

The **Enrichment Focus** sections furnish another linkage throughout the text. I have set these optional sections off from the main text to enrich ideas by basic mathematics (algebra, trigonometry, and geometry). You will need to decide which of them will be specifically assigned. You will note problems and activities at the end of each chapter. Some of them draw on the material in the Focus sections.

To aid novice science students, I worked to simplify the language as much as I can by using ordinary English rather than technical jargon. My rule when using technical terms is: *Define every term you use, and use every term you define.* I try to avoid a one-time use of a technical term just for the sake of completeness.

If your students are really baffled by a good

strategy to learn astronomy, please tell them to read carefully the special section by Mark Hollabaugh (who teaches at Normandale Community College in the United States) on "How to Study Astronomy." It provides a general strategy for studying plus specific guides to each of the learning features of this book.

> **NOTE.** If you have picked up this book because you are curious about astronomy, you may be interested in taking a college-level course by mail for academic credit. Write to: Independent Study, Continuing Education, The University of New Mexico, 1634 University NE, Albuquerque, NM 87131, USA. The course is called Astronomy 101C; I am the instructor.

INSTRUCTORS' RESOURCES

We have centralized all instructor resources on the Web site for this book. The URL is

http://www.TheEvolvingUniverse.com.

ACKNOWLEDGMENTS

Locally, Sabrina Moore initiated the effort to acquire the new visual images in the book. Louise Shaler provided in-depth editorial work on the manuscript.

She also assisted me in the final round of exploring for astronomical images and acquiring the permissions for them. Artist Boris Starosta and I collaborated on the art and the creation of the Celestial Navigator™ maps.

Other folks include my correspondence students (who have just the book and their brains) and the students in my regular Astronomy 101 course at UNM. I have also received letters and e-mail from students at other colleges and universities, as well as from instructors. Thank you all!

Any errors in the text are my responsibility. It is a little known fact that minor corrections and changes *can* be made in future printings of *this* edition. Please keep this fact in mind and send me any errors you may find. When ordering, please request the *latest printing* of the book so your students will have the most correct version.

Your feedback can improve this book! Please send any comments to me at the Department of Physics and Astronomy, The University of New Mexico, 800 Yale Blvd NE, Albuquerque, NM, 87131–1156, USA. My Internet mail address:

zeilik@la.unm.edu.

Michael Zeilik
Santa Fe, New Mexico, U.S.A., October 2001

Abbreviations

Abbreviations are often used for the names of major observatories and agencies (particularly in the figure captions). These are:

AAO – Anglo-Australian Observatory
AUI – Associated Universities, Inc.
AURA – Association of Universities for Research in Astronomy
CTIO – Cerro Tololo Interamerican Observatory
ESA – European Space Agency
ESO – European Southern Observatory
HST – Hubble Space Telescope
KPNO – Kitt Peak National Observatory
NASA – National Aeronautics and Space Administration
NCAR – National Center of Atmospheric Research

NCSA – National Center for Supercomputer Applications
NOAO – National Optical Astronomy Observatories
NRAO – National Radio Astronomy Observatory
NSO – National Solar Observatory
SAO – Smithsonian Astrophysical Observatory
SOHO – Solar and Heliospheric Observatory
STScI – The Space Telescope Science Institute
TRACE – Transition Region and Coronal Explorer
VLA – Very Large Array radio telescope
VLT – Very Large Telescope, ESO

Concept Clusters

The material in this book falls into four clusters of related concepts; these clusters themselves are interrelated. The Celestial Navigators(tm) include a subset of the most important concepts, also organized by cluster.

1. Cosmic Distances

Angles/angular diameters/positions
Angular speeds/relative motions
Astronomical Unit/Kepler's laws
Heliocentric parallax/triangulation
Inverse-square law for light/flux
Luminosity classes from H-R diagram/
    spectroscopic distances
Doppler shifts of interstellar clouds/rotation
    curve of the Galaxy
Period-luminosity relation for cepheids
Structure of the Milky Way Galaxy/distances
Distance indicators to galaxies (cepheids,
    supernovas)
Clusters and superclusters of galaxies
Hubble law/Hubble constant/red shifts/
    age of universe

2. Heavenly Motions

Angles/angular speeds/relative distances
Motions of sun, moon relative to horizon,
    stars/eclipses

Motions of the planets relative to sun, stars/
    retrogrades/oppositions/elongations
Geocentric/heliocentric
Heliocentric parallax
Kepler's laws/orbits/periodic motion/
    dark matter
Newton's laws of motion and gravitation/
    orbits/mass/weight/freefall/escape speed
Binary stars/masses of stars/center of mass
Tidal forces
Gravitational accretion/contraction
General relativity/spacetime/curved geometry
Conservation of energy/types of energy
Conservation of angular momentum

3. Celestial Light and Spectra

Electromagnetic radiation/spectrum
Emission/absorption
Kirchhoff's rules (emission, absorption, contin-
    uous spectra)
Atomic energy levels/photons/excitation/
    radiative energy

Telescopes: detectors; resolving, light-gathering, magnifying power; interferometers
Planck curve/black body/colors/temperature
Synchrotron emission/magnetic fields
Fusion reactions/nucleosynthesis
Energy transport (radiation, convection, conduction)
Chemical composition/spectra
Doppler shift/radial velocity/blue and red shifts
Stellar spectral and luminosity classes
Hertzsprung-Russell (H-R) diagram/ star clusters
Mass-luminosity (M-L) relation/stellar lifetimes
Spectra of the interstellar medium/ inter-galactic medium
Spectra of galaxies/ red shifts/expansion of cosmos
Cosmic background radiation/Big Bang model

4. Scientific Models

Assumptions, aesthetics/Geometry, physics
Observations (errors)/Predictions, explanations
Quantum theory/photons/energy levels
Age/radioactive dating/half life
Properties of matter (solids, liquids, gases, plasmas)/density
Solar system (geocentric/heliocentric)
Planets/planetary evolution/tectonics/ volcanism/thermal energy
Magnetic fields/dynamo model
Sun/stars/stellar evolution/H-R diagram
Novas/supernovas/nucleosynthesis/mass loss
Stellar corpses /white dwarfs/neutron stars/ black holes
Starbirth/planetary systems/protostars and protoplanets/brown dwarfs
Milky Way Galaxy/normal galaxies
Active galaxies/supermassive black holes
Formation of galaxies/protogalaxies
Big Bang model/critical density/ inflationary models/GUTs

How to Study Astronomy

Mark Hollabaugh
Normandale Community College, Minnesota, U.S.A.

Welcome to astronomy! When I was growing up in Michigan, I would go outside on crisp winter nights to watch the northern lights dance across the sky. In August, I counted meteors during the Perseid meteor shower. I could imagine the terror ancient peoples must have felt as the moon turned blood red during a lunar eclipse. I looked through a real telescope for the first time when I was in seventh grade. A science teacher who spent summers in our town set up his telescope each Friday night in the city park. After a brief peek through the scope, I would run back to the end of the line for another look. I was hooked on astronomy. Later, after studying physics, I did research on comets and asteroids, and on cosmic rays.

Throughout these experiences, I have marveled with fascination at the grandeur and intricacies of the universe. Now, I try to pass that fascination on to my students. That is also what Dr. Zeilik does in *Astronomy: The Evolving Universe*. My association with Mike began with my interest in Native American astronomy. We also share an interest in helping college students like you to enjoy learning astronomy effectively. Your professor has chosen a superb textbook for you to use in your astronomy course.

I'd like to give you some suggestions to assist your learning.

UNDERSTANDING YOUR LEARNING STYLE

Each of us learns in different ways. Some of us learn best by reading. Others find listening to a lecture to be most helpful. Talking with others can help us clarify our own understanding. Drawing diagrams or pictures can help us see the relationships among concepts.

When learning something new, it helps to know our own strengths, weaknesses, and learning style. As you begin your study of astronomy, make a list of the kinds of activities that help you learn best. You will find a variety of learning resources presented in your astronomy course: lectures, textbook readings, laboratory exercises, classroom demonstrations, computer simulations, learning team activities, Web sites, videotapes, or slides. These are all a part of your "tool kit" for understanding astronomy. Concentrate extra effort on those kinds of learning activities where you are the weakest. You may, for example, need to learn how to use your textbook as a resource to clarify ideas you don't understand from a lecture.

Memorization may help you correctly answer questions on an examination, but you will not be able to apply the concepts later. Avoid memorizing and seek rather to understand concepts and relate them to one another. The connected understanding of astronomy will reward you with the grand "big picture" of the cosmos!

Learning to Learn

Ask yourself some questions about your study habits. Do you set aside a specific block of time each day to study a given subject? Do you work in a quiet place, free of distractions, such as the library? Do you study at a time of the day when you are at your best for learning new material? When you prepare for an examination, have you kept up with your reading of the text so all you need to do is review? Do you attend all class lectures and other activities? Do you concentrate on understanding concepts as opposed to memorizing facts or data? If you have answered yes to these questions, you have developed some excellent study habits! If not, set goals for yourself to change your learning habits during the current semester. Find out if your institution has an office that helps students to improve their study skills.

Learning from Lectures

Should I read the book first or go to the lecture first? How you personally answer the question depends on your own learning style. It is usually helpful if you have some basic introduction to a topic before you go to your lecture. Look at the syllabus and make a note of the topics for the day. Then scan the *Learning Outcomes* for the chapter. Read the *Key Concepts* at the end of the appropriate chapter. Then skim the chapter in the textbook. If, for example, the lecture will be on Venus and Mars, page through Chapter 9. Examine the images and make note of any questions you might have about them. Pay special attention to the diagrams; be sure that you understand them. After class you should, of course, read the chapter more carefully.

Learning by Taking Notes

Taking effective notes is a two-step process. First, you will make some notes in class as your professor lectures. Try to keep your notes in outline form. Don't worry about making a formal outline. Rather,

just indent subtopics under major topics. If your professor has provided a course outline for you, *use it*. Concentrate on getting down the "big ideas" and explanations of concepts. Your instructor wants you to *understand* astronomy! You can fill in the details later as you read the text.

The second step in taking good notes is editing them. Check for concepts or learning objectives that you don't understand and look them up in the textbook. Jot down any definitions you may have missed. In short, edit your notes, making sure you understand the principal ideas. This process will keep you well prepared for taking examinations.

Learning with Other Students

Other students are *the* most important resource to your learning. You will understand more about astronomy if you explain, elaborate, or defend ideas to others. Forming small, informal study groups out of class is an important strategy. You can use the resources of the textbook to quiz one another. Come to such study groups prepared: Edit your notes, review the key concepts, and make a list of questions for your group. Learning together means you learn more. Don't wait to work together until the night before an examination. Make time in your schedule for regular meetings of your study group. If your institution has a computer network available to all students, you can contact other students by electronic mail or an electronic bulletin board.

Learning astronomy with a small group of students is an effective and fun way to increase your understanding of this fascinating subject. Your professor may use formal cooperative learning groups or may simply ask you to work together on a task or assignment. In either case following a few simple guidelines will enable your group to work well together. First, and perhaps most important, come to your group *prepared*. Be ready not only by mastering concepts in astronomy but also with questions about things that you do not yet understand.

Actively talk with one another. Make sure everyone participates. Ask questions. No idea or question is too trivial! Don't rush to conclusions, explore all the possibilities, and dig deeper! You will learn more in a learning group if you and the other group members explain or elaborate upon ideas. This process often happens in response to the simple question, "Why?" So don't be hesitant to challenge or to be skeptical about a group member's idea. Finally, ask yourselves, "What did we do well as a group and what can we do better next time?" Working together will give you a better understanding of the evolving universe.

Learning from the Textbook

Dr. Zeilik has incorporated many useful learning tools into your textbook to help you discover the excitement of astronomy. You should be aware of what your instructor expects of you. He or she probably will omit some topics, even whole sections of the textbook, perhaps even entire chapters! You may or may not be assigned specific Learning Outcomes from each chapter. It may help you to view your textbook as your primary reference. It supplements the lecture and other activities in the class. The textbook is your teacher away from the classroom. Your professor will expect you to read carefully and understand the assigned sections in *Astronomy: The Evolving Universe*.

Take notes as you read. Simply highlighting text is not particularly effective in placing concepts into our long-term memory. You will learn more, and recall it later, if the process of jotting down notes stimulates your brain as you read. Make notations in the margins of your textbook. Later, transcribe these notes into outline form to your notebook. Try to focus on the big ideas and reasons the universe works the way it does.

Take a few moments to become familiar with the design of the book by looking at the various sections of a chapter.

Learning Outcomes. Your instructor may indicate to you the learning outcomes for which you are responsible. These outcomes will help you know what is important. Before beginning to study the chapter, read the objectives. Then, as you read the chapter, think of the outcomes. *After* you have studied the chapter, write out a brief (one or two sentences) response to *each* outcome. If the outcome says "explain" or "describe," do just that. Although this may seem time-consuming, it is an excellent way to summarize and apply what you have learned. You will find review questions at the end of each chapter keyed directly to the outcomes.

Central Concept. Each chapter has a theme. Keeping this theme in mind as you read will help you to understand the unifying idea of each chapter. Try to relate concepts back to the central question.

Celestial Navigators™. To help you see the "big picture," each chapter contains one Celestial Navigator™ map. These serve as visual guides to the main concepts of a chapter and their connections with other concepts. You can copy these maps and use them as an organizing guides to the entire chapter. Write in your annotations on them as you read and study the material.

Key Concepts. When you read a mystery novel, you probably wouldn't read the ending first. However, in reading *Astronomy: The Evolving Universe*, begin a chapter by studying the *Key Concepts* at the end of the chapter. Along with the Concept Clusters and Celestial Navigators™, this will give you an overview of the principal ideas in the chapter. The *Key Concepts* also will be very useful when reviewing for an examination.

Enrichment Focus. For some astronomy students, encountering a mathematical equation in a text can be both distracting and discomforting. You will notice that mathematical topics are discussed in short, compact essays. If your instructor covers these topics, you will want to review carefully these focus boxes. In most cases you should actually work through the calculations with your calculator. By using some of the data tables in the text, you can make up other examples to work.

Study Exercises. At the end of each chapter are a series of review exercises. They require you to explain concepts you have encountered in the chapter. Use these in your study group. These review exercises are keyed back to the Learning Outcomes. Test your understanding of specific outcomes with specific questions.

The most effective way to use these Study Exercises is to work on them with other students. You, and your friends, will learn more in the process. If you do not answer a question to your satisfaction, check the objective and discuss the material again. Then attempt the question a second time. It is not particularly beneficial for you simply to look up the answers to the multiple-choice questions. Try to give explanations of *why* your answer is correct, or *why* other answers to multiple-choice questions are incorrect.

Problems and Activities. These typically require some knowledge of simple geometry or algebra. Or they may involve graphing or some observations. Your instructor will inform you about them.

Key Terms. Astronomy, and all of science, is rich with specific, technical terms. Each chapter includes boldface key terms. Write a brief definition of each term as you encounter it and then check it with the *Expanded Glossary.* Simply being able to define a term may not mean you understand it. Always give explanations. When you read, you may even encounter nontechnical words you don't fully understand. If you can't determine the meaning from the context, look the word up in a dictionary. A small, paperback dictionary in your book bag is as essential as a pen or calculator. If you forget the meaning of an astronomical term, look in the Expanded Glossary first. The Expanded Glossary contains additional terms that may not be used in the text itself but which you may encounter in other astronomical readings.

Appendices. At the end of the textbook, these contain many tables of useful data and information. If it has been some time since you have taken a science class, or you are unfamiliar with the units of measurement in astronomy, you should carefully review Appendix A.

We hope you will enjoy your study of astronomy. Approach your class with a sense of awe and curiosity. Try to understand how the universe works. And, as another famous "science officer" has said, "Live long and prosper!"

PART ONE
Changing Conceptions of the Cosmos

PART OUTCOME

Describe how astronomical observations and physical ideas interact to generate a scientific model of the cosmos.

INQUIRY FOCUS

How do we know:

The patterns of the motions of celestial objects we can see with the naked eye?

If the solar system is centered on the earth or centered on the sun?

When a scientific model is successful?

The effects and consequences of gravitation on celestial bodies?

The nature and messages of light we receive from celestial objects?

The impact of Einstein's ideas on the astronomical universe?

From Chaos to Cosmos

LEARNING OUTCOMES

After studying this chapter, you should be able to:

1-1 Describe the daily motions of the sun, moon, stars, and planets relative to the horizon from a midnorthern or midsouthern latitude.

1-2 Describe the seasonal positions of the sun – at sunrise, noon, and sunset – relative to the horizon from a midnorthern or midsouthern latitude.

1-3 Describe the motions of the sun and the moon, as seen from the earth, relative to the stars of the zodiac.

1-4 Describe the motions of the planets, as seen from the earth, relative to the sun and the stars of the zodiac, with special attention to retrograde motions.

1-5 Tell what astronomical events or cycles set the following time intervals: day, month, and year.

1-6 Describe the astronomical conditions necessary for the occurrence of a total solar eclipse and a total lunar eclipse.

1-7 Describe the phases of the moon in terms of the moon's angular position in the sky relative to the sun.

1-8 Define the ecliptic and tell how to find its approximate position in the sky.

1-9 Argue, from naked-eye observations and simple geometry, an order of the sun, moon, and planets from the earth.

1-10 Make use of angular measure to find positions of celestial objects relative to the horizon and relative to one another.

1-11 Define angular speed and angular size and relate angular speed to actual speed and distance and angular size to actual size and distance in verbal, graphical, and analytical form; apply these concepts to astronomical and everyday situations.

CENTRAL CONCEPT

The motions of astronomical objects you can see by eye follow distinctive patterns and cycles in the sky over both short and long periods of time. These repeated motions suggest an underlying design to the heavens.

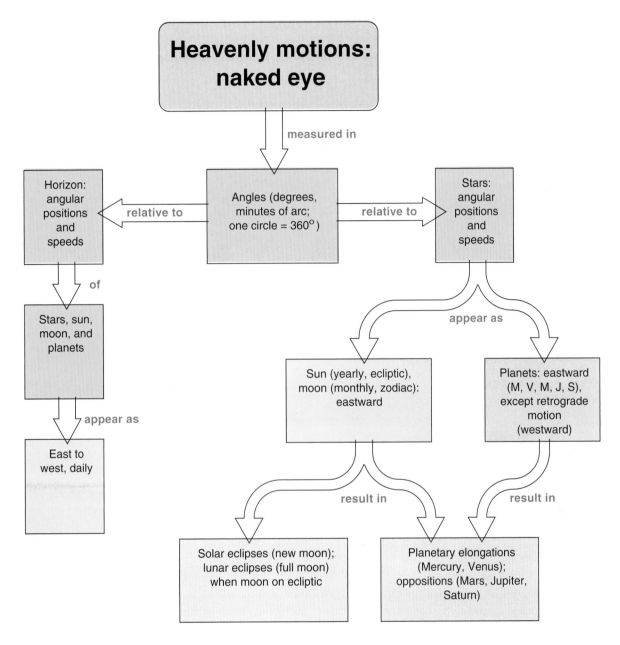

Chapter 1
Celestial Navigator™

Do the heavens have an order? Can we make sense of the events in the sky? Pure wonder, which provoked the observations of the early skywatchers, gave way to a desire to find harmony in the seeming chaos of the heavens. This quest drove people to develop concepts of space and time. The arching heavens, far removed from the earth, displayed cycles of celestial motions that skywatch-ers read as cosmic clocks. Early astronomers found an order in the heavens, a structure in space, and a sequence in time. A cosmos emerged, shaped by the patterns in the sky.

This chapter deals with naked-eye observations of the sky, like those made by early astronomers – observations you can make without optical equip-ment. You will be able to sense the regular cycles of motions in the heavens. The chapter does *not* explain these motions. (That explanation comes in Chapters 2, 3, and 4.) Long-term, naked-eye observations do establish the periods of celestial cycles with amazing precision. Your recognition of these rhythms will help you appreciate the first steps taken by ancient astronomers in the development of astronomy and of early concepts of the cosmos.

1.1 THE VISIBLE SKY

Have you ever looked carefully and curiously at the night sky far away from city lights? If so, at first glance, you may have found no pattern in the stars, and no way to judge their distances, beyond sensing that the stars are far away from the earth. With patience, you can do better than these first impressions. (See the Naked Eye Celestial Navigator™.)

Constellations

Take the time to study the stars for a while; you will find that they fall into patterns – designs imposed by your mind (Fig. 1.1). Ancient skywatchers noticed stellar patterns, named them, and passed the names on. Such collective patterns are called **constellations.** Early constellations marked groups of stars, with ill-defined boundaries that typically were seen as outlining mythological or realistic figures. Constellations today (88 official ones in all) are established by international agreement of astronomers.

If you observe the sky nightly, you will see that the shapes of the constellations do not change. In fact, if you watched them for your whole life, you would not notice any change. The stars appear to hold fixed positions relative to one another.

Angular Measurement
(Learning Outcomes 1-10, 1-11)

How do you measure how far apart stars appear in the sky? You need a convenient sighting device such as your hand, held at arm's length (Fig. 1.2). You can then measure the **angular distance** between the stars, that is, the angle between one star and another. Angular measurement is based on counting by 60: a circle is divided into 360 **degrees** (°), each degree into 60 minutes (**arcmin** or ') of arc, and each minute into 60 seconds (**arcsec** or ") of arc. (An *arc* is any part of the circumference of a circle.)

At arm's length, your fist covers about 10° of sky; each fingertip masks roughly 1°. You can use your hand to measure the angular size of the sun and moon (both about 0.5°, or half a fingertip) and the separation of the pointer stars in the Big Dipper (about 5°, or half a fist: Fig. 1.3).

The sun and the moon each span a certain **angular size** of the sky. Because the sun and moon appear as disks, their **angular diameter** is the angular distance between opposite edges. The sun and moon are the only celestial objects whose angular diameter is clearly visible to our eyes.

Now, angular distance and angular size are related. Suppose you are looking at two objects some distance apart. The farther they are away from you, the smaller the angle between them will appear. When you look at the sky, you see the angular distances between celestial objects. That is all you know – angles – unless you can measure physical distances.

Angles are crucial in astronomy. If you look at the sky, you observe many angles: between stars, between the moon and stars, between planets and stars, between the sun and the horizon. They provide information about positions of objects in the sky and their motions, as their positions change.

Motions of the Stars
(Learning Outcomes 1-1, 1-11)

Stay out one night and watch the stars from dusk to dawn. They move relative to your horizon – rising in the east, slowly traveling in arcs against the sky, and setting in the west. The **horizon** is an imaginary line at which the earth and sky would meet if no terrain were in the way. The horizon you would view at sea comes closest to this ideal.

If you live in the continental United States and face south in December, you will see Orion east of south early in the evening (Fig. 1.4a). After a few hours, Orion will appear due south at about midnight (Fig. 1.4b). If you face south, west is to your right, east to your left in the Northern Hemisphere.

Using your fist, you can measure how far stars move, in angle relative to the horizon, in an hour. It comes out to about 1.5 fists, or some 15° per hour. An angular amount of movement in a certain time is **angular speed.** It is the rate with which an object covers a certain angular distance. Because we only measure angles directly on the sky, we see motions as angular speeds.

Figure 1.1 The constellation Orion. The three bright stars that make a diagonal line form Orion's Belt. Below the center star is the fuzzy patch of the Orion Nebula. The very bright star at bottom right is bluish Rigel; to the upper left is reddish Betelgeuse.

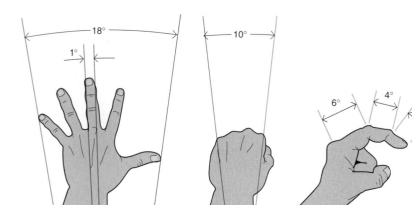

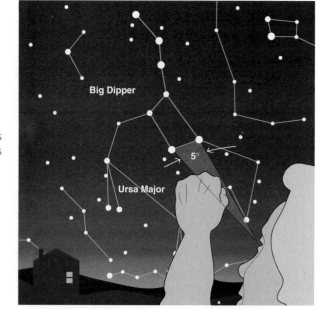

Figure 1.2 Angular measurements made with a hand extended at arm's length. The angles are approximate and typical for an adult. You will need to calibrate your own hand and fingers to find their angular extents.

Figure 1.3 Measuring angles on the sky with a hand at arm's length. The angle between the two pointer stars in the Big Dipper is about 5°, or half a fist at arm's length.

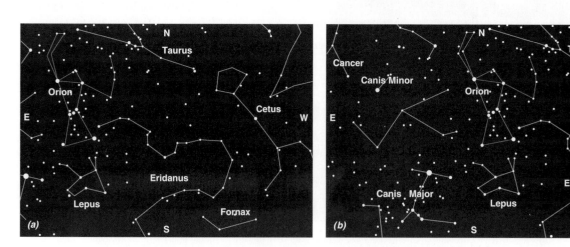

Figure 1.4 Motions of the stars on the same night, relative to the horizon. (a) Facing south at a midnorthern latitude on December 21 at about 22:00 local time. Note that Orion is still east of due south. (b) The same view about two hours later, at local midnight. Now Orion and all other stars have moved to the *west* relative to the horizon. The view is 80° by 60°.

Figure 1.5 Motions of the stars around the north celestial pole. This time exposure shows the circumpolar motions above Star Hill Inn, New Mexico, U.S.A. As the stars move around the north celestial pole, they trace out arcs that are progressively larger the farther a star lies from the pole. Polaris, the brightest star near the center of the stellar circles, has traced a small arc, which indicates that it is not exactly at the north celestial pole.

If you acquire a regular stargazing habit, you will see that the visible constellations change with the seasons. In winter, consider Orion again in the south at midnight. Look south on the following nights at the same time. Orion moves slowly to the west toward the sun. In summer, you cannot see Orion at all because it is next to the sun and is up during the day. In winter, a year later, Orion again lies south at night. All constellations take one year to return to their initial places in the sky relative to the sun.

CAUTION. Do not confuse this gradual yearly change relative to the sun with the much faster daily motion relative to your horizon (from east to west)! The two angular speeds here are very different: *1° per day* relative to the sun, and *15° per hour* relative to the horizon. Why do these angular speeds differ? They use different references for the motions.

Here we have a good example of **relative motion.** Objects, such as stars, can have different motions simultaneously. How that motion appears depends on what you use as a reference. If you watch a short sprint, the runners appear to move much faster relative to the track than relative to each other. Stars appear fixed relative to each other, to move slowly relative to the sun, and to move quickly relative to the horizon. You must understand this concept of relative motion to keep the motions in this chapter clear. If something moves, always ask: relative to what?

If you live in the Northern Hemisphere and look north, you will find that some stars never dive below your horizon but trace complete circles above it (Fig. 1.5). These are the **circumpolar stars.** The center of these circles marks the **celestial pole,** the point about which the stars seem to pivot. A modestly bright star called Polaris lies close to the north celestial pole. Polaris is now the North Pole star. (No bright star now falls close to the south celestial pole, so we do not have an obvious South Pole star.)

1.2 THE MOTIONS OF THE SUN

Like the stars, the sun moves relative to the horizon. It also moves relative to the stars!

Motions Relative to the Horizon
(Learning Outcomes 1-1, 1-2, 1-5)

The sun's daily motion – rising along the eastern horizon, arching across the sky, and setting along the western horizon – defines the most basic time cycle: day and night. Midway between sunrise and sunset, the sun reaches its highest point relative to the horizon, which defines **noon.** The interval from one noon to the next sets the length of the **solar day.**

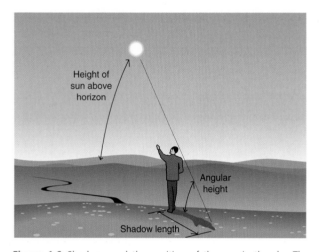

Figure 1.6 Shadows and the position of the sun in the sky. The higher the sun relative to the horizon, the shorter the shadow. The shortest shadow of the day occurs at noon. The sun is then due south. The angle between the horizon and the sun is the angular height of the sun, which has its maximum value for any day at noon.

Consider your shadow cast on a flat ground. You can use the length of your shadow to study the sun's daily and seasonal motions in the sky relative to the horizon. The tip of the shadow marks the end of an imaginary line that connects the tip itself, the top of your head, and the sun (Fig. 1.6). The shadow points in the direction opposite to the sun's location in the sky. The length of the shadow tells the height of the sun relative to the horizon. When the sun hangs low in the sky, the shadow is longer. At noon the shadow reaches its shortest length for that day. Also at solar noon, in the northern latitudes, the shadow points due north – so the sun lies due south.

Observe your shadow throughout a year. You will find that the height of the sun in the sky at noon varies with the season. The shadow is shortest at noon on the **summer solstice** (around June 21), the day with the greatest number of daylight hours. At the summer solstice, the noon sun hits its highest point in the sky for the year (Fig. 1.7a). In winter at noon, the shadow stretches longest on the **winter solstice** (around December 21), the day with the fewest daylight hours. The noon sun has dropped to its lowest point in the sky for the year (Fig. 1.7c). On the first day of spring and autumn, you cast a shadow with a length between its summer minimum and winter maximum (Fig. 1.7b) – these days are the **spring** and **fall equinoxes:** around March 21 for spring (vernal) and September 21 for fall (autumnal). The cycle registered by your shadow defines the second basic unit of time: the year of seasons.

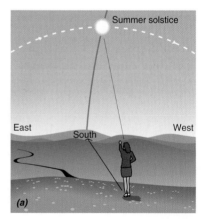
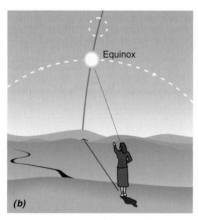
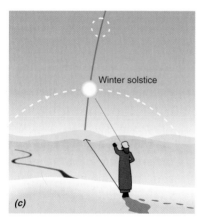

Figure 1.7 Seasonal position of the sun at noon, when it hits its highest point in the sky, relative to the horizon, for the day. The angular height varies with the seasons. Measuring with a fist, an observer can find that for a midnorthern latitude, the sun is highest at the summer solstice (a), lower at the fall equinox (b), and lowest at the winter solstice (c).

The seasonal change in the sun's noon position is tied to a cyclical change in the sunrise and sunset points. Consider sunrise. At the summer solstice, for midnorthern latitudes, the sun rises at its farthest point north of east for the year (Fig. 1.8). For about four days, the sun seems to rise at the same place on the horizon – the sunrise point appears to stand still. After the summer solstice, the sunrise position moves southward. At the winter solstice, the rising point reaches as far south of east as it will get for the year. The sun again stands still for a few days, and then moves northward. Halfway between the solstice points lies the point for sunrise of the equinoxes. *Only at the equinoxes does the sun rise east and set due west.* At these dates, the sunrise and sunset points show the greatest angular speeds for the year. Sunset points show a symmetrical pattern around due west.

Motions Relative to the Stars (Learning Outcomes 1-3, 1-8)

The sun also moves with respect to the stars – a motion tough to observe, for you cannot see the stars during the day. Try this: pick out a bright constellation, such as Gemini, visible in spring just above your western horizon right after sunset (Fig. 1.9a). (We are using the horizon here to fix the sun's position and to block out the sun's direct light.) Look again at the same time about two weeks later. Because the sun has moved east relative to the stars, the constellation will appear to have shifted to the west (Fig. 1.9b). About two weeks later, again at the same time, the sun will have moved out of Gemini and into Cancer (Fig. 1.9c). Relative to the stars, the sun appears to move to the *east.* In one year the sun returns to the same position relative to the stars, so it circuits at an angular speed of 360° in a year, or about 1° a day. (In contrast, the sun moves 360° per day westward relative to the horizon.)

Imagine that you recorded the sun's position among the stars for a year. If you drew an imaginary line through these points, you would trace out a complete circle around the sky: this is called the **ecliptic** (Fig. 1.10). The traditional 12 constellations through which the sun moves define the **zodiac.** (Note that the sun also travels through a part of the constellation Ophiuchus, which is not a zodiacal constellation.)

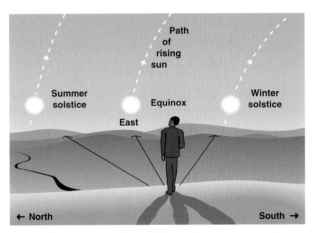

Figure 1.8 Seasonal positions of the sunrise points along the horizon (from the summer solstice to the winter solstice for an observer at midnorthern latitudes). Note north is to the left, south to the right.

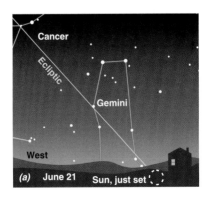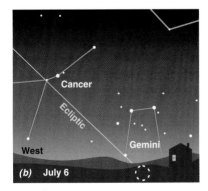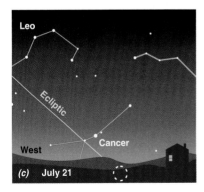

Figure 1.9 Motion of the sun relative to the stars. Find a constellation (such as Gemini, shown here) near the western horizon at sunset (a). About a week later, look for the same stars just after sunset (b). They will have moved down the western horizon. Eventually, they will set before the sun goes down (c). The stars seem to move to the west with respect to the sun, or the sun seems to move *east* with respect to the stars.

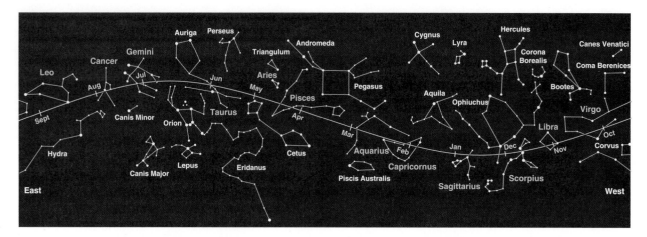

Figure 1.10 Ecliptic and zodiacal stars. If there were no atmosphere, you could see the sun's changing position with respect to background stars. The sun's path is called the *ecliptic* (yellow line). This diagram indicates the sun's position by date along the ecliptic and shows the major constellations. The zodiacal constellations are indicated in blue. Note that the sun moves west (right) to east (left) among the stars and returns to the same position on the ecliptic in one year.

Table 1.1 The Zodiacal Constellations

Constellation	Traditional Pattern	Approximate Date for Sun's Location within Constellation
Aries	Ram	April 30
Taurus	Bull	May 30
Gemini	Twins	July 5
Cancer	Crab	July 30
Leo	Lion	September 1
Virgo	Virgin	October 11
Libra	Balance	November 9
Scorpius	Scorpion	December 3
Sagittarius	Archer	January 7
Capricornus	Goat	February 8
Aquarius	Water Bearer	February 25
Pisces	Fishes	March 27

CAUTION. Do not confuse the *eastward* motion of the sun with respect to the stars in a year with the much faster *westward* motion of the sun relative to the horizon in a day! Again, it's relative motion. Consider watching a bus going westward along a street with a passenger moving to the rear. Relative to the bus, the passenger is moving eastward.

Relative to the street, though, the passenger moves westward with the bus.

The sun's position along the ecliptic is associated with the constellations of the zodiac (Table 1.1). For instance, to say that the sun is "in Taurus" specifies a place along the ecliptic. The zodiac probably arose from a desire to mark the sun's position with respect to the stars. (You cannot see the sun in a constellation during the day, but you can get a rough idea near sunrise and sunset.)

The sun's location in the zodiac also roughly indicates the time of year. In spring and fall, the equinoxes occur. The summer and winter solstices mark two other key times. Nowadays the sun lies in Aries on the spring equinox, Gemini on the summer solstice, Virgo on the fall equinox, and Sagittarius on the winter solstice (see the dates along the ecliptic in Fig. 1.10). These seasonal locations change slowly with time (Enrichment Focus 1.1).

NOTE. It is standard practice to say that the sun is "in" a constellation. But the stars that form the constellation are far from the sun and are not necessarily close together in space; they appear close together to us because they are located roughly in the same direction along our line of sight. The statement "the sun is in Pisces" means that the sun lies in the same direction in the sky as the pattern of stars called Pisces.

ENRICHMENT FOCUS 1.1
Precession of the Equinoxes

In 3000 B.C. the sun appeared in Taurus at the vernal equinox, the first day of spring (Fig. F.1a). Today the sun lies in Pisces at the start of spring. Over the passage of 5000 years, the position of the vernal equinox has moved to the west out of Taurus, through Aries (Fig. F.1b), and into Pisces (Fig. F.1c). That is, in 5000 years the position of the vernal equinox has moved through two constellations. To circuit the whole zodiac will take six times as long, or about $6 \times 5000 = 30,000$ years. (A more precise calculation gives 25,780 years.) This slow westward drift of the equinoxes with respect to the stars is called the *precession of the equinoxes.*

The most dramatic effect of the precession of the equinoxes is to change the zodiacal location of the sun at the equinoxes and solstices. Precession has another less obvious effect: the celestial poles move in the sky, with the result that the North Pole star changes (Fig. F.2). The north celestial pole now lies near the star Polaris. About 5000 years ago, the north celestial pole was near the star Thuban (in Draco). About 12,000 years from now, precession will have carried the north celestial pole near the bright star Vega in Lyra.

Precession is hard to observe without a telescope because it takes place so slowly. But if a culture kept astronomical records for a few centuries, its astronomers could notice the shift in the equinoxes and solstices with respect to rising stars. Keep in mind, though, that precession does *not* change the rising and setting positions of the sun on the horizon.

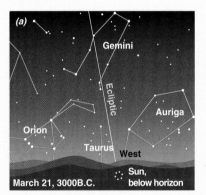 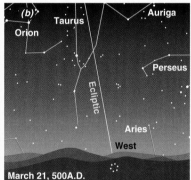 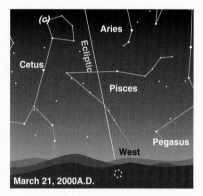

Figure F.1 Precession and the change of position of the vernal equinox with respect to the zodiacal stars. (a) Sunset scene in 3000 B.C. on March 21. (b) Note that the sun lies just below the horizon in Taurus. (c) Sunset scene in A.D. 2000, also on March 21. Now the sun lies in Pisces. Note that the equinox point moves to the west along the ecliptic.

Figure F.2 Precession and the change of position of the north celestial pole with respect to the stars. The pole's motion completes a circle counterclockwise in the sky in about 26,000 years. About 12,000 years from now, the pole will be near the bright star Vega.

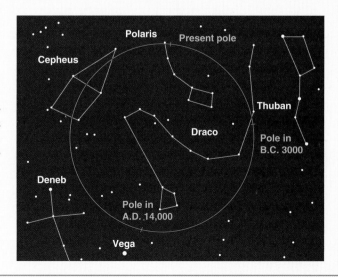

Figure 1.11 Motion of the crescent moon relative to Venus and the bright star Aldebaran (in Taurus). East is to the left. The field of view is 9° by 7°; the moon covers an angle of about 0.5°. (a) The scene at 3:00. (b) One hour later. (c) Another hour later.

View from the Southern Hemisphere (Learning Outcomes 1-1, 1-2, 1-3, 1-8)

How does the sky look from the earth's Southern Hemisphere? If you view the earth as a sphere, you should be able to rework a Northern Hemisphere outlook to southern latitudes. Be careful! When you face south, east is to your left, west to your right.

Start with the stars. You will see new ones, invisible from the north – a stunning sky endowed with bright stars. Those that fall in the middle of the sky, like the zodiacal ones, will look upside down (compared with the view from the Northern Hemisphere)! You still have the same zodiac in which the sun, moon, and planets travel. Direction? *Eastward,* except during retrograde motion for the planets! (Section 1.4 presents retrograde motion.) If you face south, you will see circumpolar stars but no bright star near the south celestial pole. The difference: stars rotate *clockwise* around the south pole. They still rise in the east and set in the west.

Where is the sun at noon? North, at its highest elevation. Winter and summer solstice still mean the same for the sun's noon altitude – lowest in the winter, highest in the summer, but the dates are interchanged. The summer solstice occurs in December, the winter in June. The equinoxes occur around March 21 for autumn and September 21 for spring. The sun's rising points still vary in a seasonal cycle, but the north–south directions are reversed compared to the Northern Hemisphere. For instance, in Figure 1.7, the labels change so that south becomes north, east changes to west, and west shifts to east.

1.3 THE MOTIONS OF THE MOON

If you carefully watch the moon for a few hours, you can spot two of its motions. First, like the sun and the stars, the moon rises in the east and sets in the west relative to the horizon. Second, the moon also journeys eastward against the backdrop of the zodiacal stars.

Motions Relative to the Stars (Learning Outcome 1-3)

How can you observe the eastward motion? Wait until the moon appears close to a bright planet or star (Fig. 1.11a). On the same night, observe the moon and star again an hour later (Fig. 1.11b). Repeat the observation after another hour has passed (Fig. 1.11c). The moon will have moved to the east with respect to the star (both will have moved westward in the sky with respect to the horizon). If you measure the moon's rate of motion, you will find an angular speed of about 0.5° per hour. At this rate, the moon circuits the zodiac in about 27 days. (Although the moon's path does not fall right on the ecliptic, it lies close to it, so the moon stays within the zodiac.) This eastward motion of the moon relative to the stars causes it to rise later every day (by about 50 minutes).

Note that the moon's angular speed eastward relative to the stars is *much* faster than that of the sun's. In a year, the sun circuits the zodiac only once. The moon laps the zodiac about 13 times during the same interval. On the basis of angular speed, which object might you judge as closer to the earth? Right, the moon, because of its greater angular speed.

Phases (Learning Outcomes 1-3, 1-5, 1-7)

Watch the moon for a few nights; you will note that the amount of its surface that is illuminated – its **phase** – follows a regular sequence (Fig. 1.12). When the moon rises at sunset and its face is completely illuminated, it's a full moon. About a week later, the moon is at last quarter, some 90° west of the sun. A week after that phase, the moon is new and is not visible in the sky. A week later, the moon appears in the west at sunset, partially illuminated – a first quarter moon. Another week passes; the moon rises in the east as a full moon at just about the time the sun sets. A complete cycle of phases – say, from one full moon to the next – takes 29.5 days. Here we have a third basic unit of time: the month of phases.

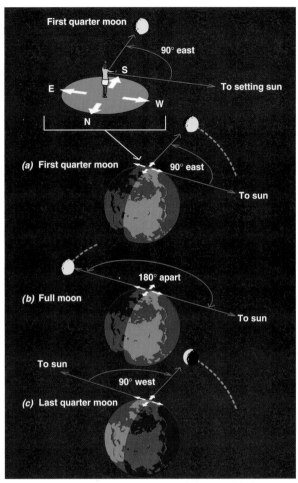

Figure 1.13 Orientation of the sun and moon in the sky as viewed by an observer on the earth for the phases of the moon in Fig. 1.12. At first quarter (a), the moon is 90° east of the sun. At full (b), it is 180° away (in opposition). At last quarter (c), the moon is 90° west of the sun (due south). The view here is facing south from a midnorthern latitude.

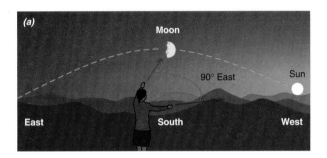

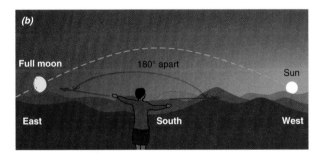

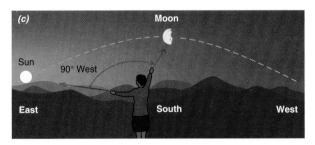

Figure 1.12 Views of sun and moon relative to the horizon, facing south from a midnorthern latitude. At first quarter (a), the moon is 90° east of the sun (due south as the sun sets). At full (b), it is 180° away (in opposition, so the full moon rises as the sun sets). At last quarter (c), the moon is 90° west of the sun (due south as the sun rises in the east).

The different phases of the moon relate to specific alignments of the sun and the moon in the sky. At new moon, the angular distance of the sun and moon is small – less than a few degrees. At first quarter, the moon lies 90° east of the sun (Fig. 1.13a). If you were to point to the setting sun with one arm and to the moon with the other, the angle between your arms would be 90°. At full, the moon is 180° from the sun (Fig. 1.13b); at last quarter, it is 90° west of the sun (Fig. 1.13c). First and last quarters refer to the position of the moon in the sky – one-quarter of a full circle away from the sun – not to the amount of illumination. The moon at quarter phase looks half full

1.4 THE MOTIONS OF THE PLANETS

If you observe the sky often, you can quickly pick out heavenly bodies that do not belong to the constellations. Five such objects wander in regular ways through the stars of the zodiac: the **planets** Mercury, Venus, Mars, Jupiter, and Saturn. (Uranus, Neptune, and Pluto cannot be seen without a telescope.) Viewed by naked eye, the planets look pretty much like stars, though they twinkle less than stars do. At times, some planets are brighter than the brightest stars; all planets vary in brightness. It is their motions, however, that really separate the planets from the stars.

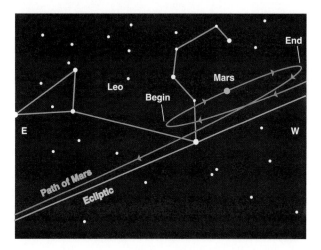

Figure 1.14 Retrograde motion of Mars, relative to the stars. The line across the center marks the ecliptic; the outline of the zodiacal constellation Leo is shown. "Begin" and "End" mark the points on Mars' path in the sky when the retrograde motion begins and ends. Note that during its retrograde motion, Mars moves from east to west; otherwise it moves from west to east.

Retrograde Motion
(Learning Outcome 1-4)

The planets display a peculiar motion that sets them apart from all other objects in the sky. For example, suppose you observe Mars every night for a few months near the time when Mars appears brightest in the sky. (This occurs about every two years.) At first, Mars moves slowly eastward with respect to the stars through the zodiac close to the ecliptic (Fig. 1.14). In this regard, it moves like the sun and the moon. But later, Mars falters in its eastward motion with respect to the stars and stops for a short time.

Then for about three months, Mars moves *westward* – the opposite of its normal direction relative to the stars. After that, its westward motion slows down and stops. Finally, Mars resumes its normal eastward track. The planet's backward motion to the west is called **retrograde motion.** In the middle of its retrograde motion, Mars shines at its brightest.

The retrograde begins when a planet stops moving relative to the stars. Later it reaches a midpoint in its westward motions, stops again, and then returns to its normal eastward motion. All planets progress along or near the ecliptic with periodic retrograde motion – but generally they are not all in retrograde motion at the same time or for the same duration. For instance, Mars takes about 72 days to undergo its complete retrograde cycle; Saturn takes 138 days (Table 1.2). The sun and the moon never move retrograde, and, in general, the planets travel eastward along or near the ecliptic.

Table 1.2 Main Motions of the Visible Planets

Planet	Typical Duration of Retrograde (days)	Average Time between Retrogrades (days)	Period around Ecliptic (eastward)	Average Angular Speed (degrees per day)
Mercury	22	116	≈ 1 year	≈ 1
Venus	42	584	≈ 1 year	≈ 1
Mars	72	780	≈ 2 years	0.5
Jupiter	120	399	≈ 12 years	0.08
Saturn	138	378	≈ 30 years	0.03

Note: The duration of retrograde varies somewhat from the values above from retrograde to retrograde. The fourth column gives a time interval for the planet to circuit the ecliptic once, covering 360°.

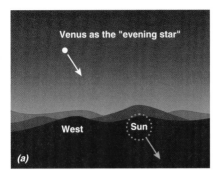
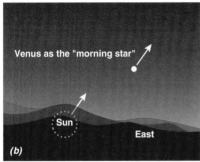

Figure 1.15 Venus goes through retrograde motion when it switches from being an evening "star" (a) to a morning "star" (b): that is, from being on the east side of the sun to being on the west side. It moves retrograde *after* its greatest eastern elongation.

Elongations, Conjunctions, and Oppositions (Learning Outcome 1-4)

The alignment of the sun and a planet in the sky at the time of retrograde motion divides the visible planets into two groups. Mercury and Venus make up one; they never stray very far in angular distance from the sun. Because they stick close to the sun, they are visible only as morning and evening "stars" (Fig. 1.15). That means that Mercury and Venus gleam over the western horizon after sunset or burst above the eastern horizon before sunrise (but keep in mind that they are *not* stars).

You can use your fist to measure the maximum angular separation of Mercury or Venus from the sun. For Mercury, the average maximum separation is 23° (about 2½ fists); for Venus, about 46° (about 4½ fists). When either planet is at its greatest angular separation from the sun, it has reached **maximum elongation** (Fig. 1.16). (Mercury's maximum elongation varies a lot; its maximum possible elongation is 28°.)

Mercury and Venus begin their retrograde motions after they hit maximum elongation east of the sun as evening "stars." They then move westward, pass the sun, and reappear as morning "stars" west of the sun. When Mercury or Venus lies very close to the sun in the sky, the planet and the sun are aligned in **conjunction.** (Whenever two celestial objects appear to come close together in the sky, they are said to be "in conjunction," usually for a short time.)

> **CAUTION.** Note that Mercury and Venus are evening "stars" when they are *east* of the sun, so they set *after* the sun does. They are morning "stars" when *west* of the sun and rise *before* it does.

The second group of planets visible to the naked

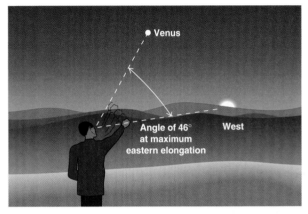

Figure 1.16 Measuring the greatest eastern elongation of Venus at sunset. At this time, the angle between Venus and the sun is about 46°. The same observation for Mercury gives a maximum angle of about 23°.

eye comprises Mars, Jupiter, and Saturn. In contrast to Mercury and Venus, these planets can lie at any angle with respect to the sun. During their retrograde motion, they reach a point when they are in **opposition** to the sun: opposite the sun in the sky. At opposition, the sun and the planet are separated by 180° (similar to when the moon is full). Then the planet rises as the sun sets. If you were to point one arm to Mars at opposition and the other at the sun, your arms would be 180° apart (they would make a straight line). A planet at opposition hits the middle of its retrograde loop and shines its brightest since it is closest to the earth.

In summary, Mercury and Venus can *never* be in opposition to the sun; they retrograde after passing their greatest eastern elongation and continue until reaching their greatest western elongation. Mars, Jupiter, and Saturn can be in opposition or conjunction (or anywhere in between), but they retrograde *only* at opposition.

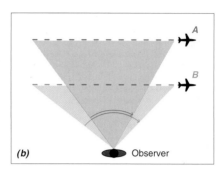

Figure 1.17 Judging relative distances from angular speeds. (a) The scene is at night, so all you can see are the lights from the airplanes. In the same time, airplane A covers a smaller angle than airplane B. (b) Airplane B is closer if both airplanes fly at the same speed.

Angular Speed and Relative Distances
(Learning Outcomes 1-9, 1-10, 1-11)

The time each planet takes to circle the ecliptic provides clues to the relative distances of the planets from the earth (Table 1.2). Assume for a moment that all planets move at the same speed. The one that appears to move at a faster angular rate is closer than another that moves at a slower angular rate. The slowest-moving planet is the most distant from the earth; the swiftest is the nearest.

Here is an analogy (Fig. 1.17). Suppose you are watching the lights of two airplanes at night and want to estimate the planes' relative distances from you. Assume that both planes are flying at the same speed. The one that appears to move faster – at a greater angular speed – is the closer of the two. If both were at the same distance, you would observe the same angular speed for both. The angular speed depends on both the actual speed and the distance. How? *Angular speed is proportional to the ratio of the actual speed to the distance.* If two airliners travel at the same speed, then the closer one will appear to have a greater angular speed. If it is twice as much, the nearer is at half the distance as the farther one.

Apply this argument to the moon, sun, and planets: the fastest (the moon) is closest to the earth; the slowest (Saturn) is the farthest. (The assumption that the speeds are the same turns out not to be correct, but the basic conclusion is still OK.) The angular speeds give an inkling of the relative distances from the earth.

But what about retrograde motion of the planets? The evolution of ideas about the motions of the planets – including retrograde – marks a central question of the next three chapters: how do the planets move?

1.5 ECLIPSES OF THE SUN AND MOON
(Learning Outcomes 1-6, 1-7)

An eclipse of the sun – a **solar eclipse** – occurs when the moon passes in front of the sun (Fig. 1.18). Although the moon is smaller in diameter than the sun, it is closer to the earth by an amount that makes the angular diameter of the sun and moon almost the same – about 0.5°. So the moon may just cover the sun's disk when it passes directly between the sun and the earth, as it may do at new moon.

Why do eclipses *not* happen each and every month? Mainly because the moon's path in the sky relative to the stars does not usually coincide exactly with the ecliptic; it is tilted at an angle of about 5°. The ecliptic and the moon's path cross at two points. Only at or near these points will the sun and the moon be so close that an eclipse occurs. If the moon is more than 0.5° above or below the sun, it will pass by without blocking out the sun's disk. No eclipse will occur.

When the sun and moon are exactly lined up, the new moon completely covers the sun, and we have a total solar eclipse (Fig. 1.19). During a total solar eclipse, the moon's shadow is a few hundred kilometers (abbreviated km: see Appendix A on units) wide. Only people in this narrow band on the earth see a total eclipse as the shadow sweeps by. Those just outside the central band see a partial solar eclipse, in which the moon does not completely cover the sun.

An eclipse of the moon – a **lunar eclipse** – occurs when the moon passes through the shadow cast by the earth (Fig. 1.20). Then the sun's direct light is cut off from the moon (Fig. 1.21). A total eclipse of the moon takes place only when the moon is full – when the earth lies directly between the sun and the moon. Also, the moon must be close to the ecliptic; otherwise it will miss the earth's shadow.

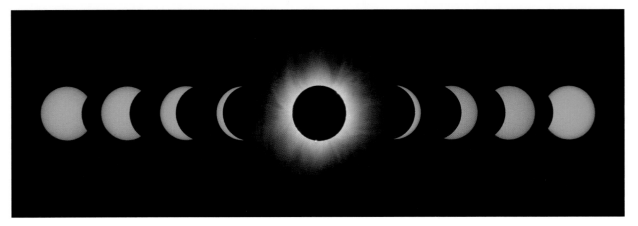

Figure 1.18 A panoramic view of a total eclipse of the sun on August 11, 1999. These nine images capture the entire two-hour process. As viewed from Lake Hazar in Turkey, the moon covers the sun's visible disk (starting at left) until the sun's outer atmosphere (its corona) is visible (central image). The moon then moves out of line with the sun (to the right).

Figure 1.19 Alignment of moon, sun, and earth to scale for a total solar eclipse. The moon must be new and on (or very close to) the ecliptic as viewed from the earth. The moon's central shadow is usually long enough to hit the earth; a total eclipse is seen in the path made as the central shadow moves along the earth.

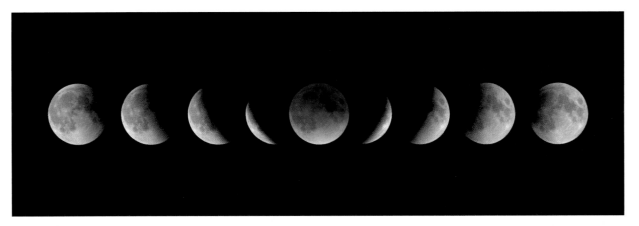

Figure 1.20 A panoramic view of a total eclipse of the moon, January 20, 2000. These nine images show a full moon gradually entering the earth's shadow (starting at left). It reaches the center of the earth's shadow in the middle image, then slowly moves out of the central part of the shadow (to the right). The reddish color comes from dust in the earth's atmosphere, which scatters blue light but allows red light to pass.

Figure 1.21 Alignment of sun, moon, and earth to scale for a total lunar eclipse. The moon must be full and on (or close to) the ecliptic and pass through the earth's inner shadow. A total lunar eclipse is visible to everyone on the night side of the earth.

ENRICHMENT FOCUS 1.2
Angular Size and Speed

Angles are the basis of simple observational astronomy. Some angles are large, such as the circle around the horizon that covers 360° or the 90° from the horizon to a point overhead. Most angles in astronomy are fairly small, such as the 5° between the pointer stars of the Big Dipper and the 0.5° angular diameter of the sun and moon.

Go out on a night with a full moon and hold your hand out at arm's length, with your little finger in front of the moon. Its width covers the moon! The angular size of your little finger at arm's length is about the angular size of the moon. However, this comparison tells you nothing about the actual size of the moon – or the width of your little finger! Luckily, your finger is close enough to permit you to measure its width; do so with a centimeter ruler. Hold your little finger at arm's length again and measure with the ruler the distance from your eye to your little finger, still in centimeters. Divide the width of your finger (its size) by the eye–finger distance. You have then determined the size-to-distance ratio of your little finger at arm's length. What value do you get for this ratio?

The size-to-distance ratio is related to the angular size of an object. For small angles (about 10° or less), the *size-to-distance ratio is directly proportional to the angular size*. Here, "directly proportional" means that the smaller the angular size, the smaller the size-to-distance ratio. Any object with a size-to-distance ratio of 1/57 has an angular size of 1°. What did you find for your little finger at arm's length? You should have found a result between 1/50 and 1/60. Your little finger covers an angle of about 1°, or about twice the angular size of the moon. For convenience, you may use 1/60 as the size-to-distance ratio that corresponds to 1° (Table F.1).

Now apply this relation. A tennis ball has a diameter (size) of about 6 cm. For a tennis ball to have an angular diameter of 1°, how far away from you must it be placed? Well, since a size-to-distance ratio of 1/60 is required, the ball must be at a distance of 60 × 6 cm = 360 cm. It then covers the same angle as

TABLE F.1 Angular Diameter and Distance

If the angular diameter in *degrees* is	Then the distance, in *diameters*, is roughly
4	15
2	30
1	60
1/2	120
1/4	240
1/8	480

your little finger at arm's length and twice the angle covered by the moon.

If you knew the distance to the moon, you could find out its actual diameter from its angular diameter. How? You know that the moon has an angular diameter of some 0.5°, so its diameter-to-distance ratio is about 1/120. The problem is finding that distance!

We can summarize these concepts analytically as follows:

size ≈ (angular size × distance)/60

Note that the units for size are the same as those for distance.

Angular speeds are analogous. For instance, the moon moves eastward relative to the stars at about 1/2° per hour, so its speed-to-distance ratio is about 1/120 per hour. Since the moon's angular size is 1/2°, we then know that the moon has an angular speed of one diameter per hour. But we do not know the moon's diameter – and so its speed is unknown – unless we can measure or estimate its distance. That is the key issue!

We can summarize these concepts analytically as follows for small angles:

actual speed ≈ (angular speed × distance)/60

This relationship is the same as the statement that angular speed is directly proportional to the ratio of the actual speed to distance. Note that the distance units of actual speed are the same as the distance, and the time units of actual speed are the same as those of angular speed.

Lunar eclipses play to a large potential audience: all the people on the night side of the earth (because that whole side faces the moon). During your lifetime, you can expect to see roughly 50 lunar eclipses, about half of them total. In contrast, since solar eclipses are visible only along a narrow band of the earth, a total eclipse of the sun occurs rarely for any one location.

Eclipses take place less frequently than other celestial events described here. Their spectacular nature motivated people to study them, however, and we have reliable records dating as far back as 750 B.C. The Greeks, for example, found that eclipses were predictable from the motions of the sun and the moon. They also noted that solar eclipses prove the moon is closer to the earth than the sun because the moon blots out the sun. In addition, they decided that the sun must be larger than the moon (it is!).

Now you can see how the *ecliptic* got its name. Only when the moon lies on or close to the ecliptic can *eclipses* occur.

Eclipses bring us back to the concept of angular diameter. The angular size of the moon gives us a good idea of the size of the moon compared to its distance (Enrichment Focus 1.2). The same is true for the sun. That diameter-to-distance ratio is the same for the sun and the moon, so both objects have distances from the earth such that the same number of diameters of each fit into that distance. Which is farther away, and the larger? Because the moon eclipses the sun, it must be closer. Since the angular diameters are the same, that's how we know that the sun must be farther than the moon in distance and larger than the moon in physical diameter.

KEY CONCEPTS

1. Celestial bodies participate in cyclic motions; some are short term (hours, days) and others long term (months, years: see Table 1.3). Most of these motions are periodic; that is, they repeat after a given period of time has elapsed. These repetitions give the impression that the sky has an order.

2. Relative to the horizon, the stars rise in the east and set in the west every day. The stars make up fixed patterns, called *constellations,* which do not change over a human lifetime. Different constellations are visible in the night sky during different seasons – except for the circumpolar groupings, which are always visible. The stars act as the background for the motions of other celestial objects.

3. Relative to the horizon, the sun rises roughly in the east and sets roughly in the west daily, with the exact position changing with the seasons: its highest point in the sky defines noon. The height of the noon sun varies with season – it is highest at the summer solstice, lowest at the winter solstice, and midway at the equinoxes. Relative to the stars, the sun moves eastward along a path called the *ecliptic,* which cuts through the constellations of the zodiac; the sun completes one eastward circuit of the zodiac in a year.

4. Relative to the horizon, the moon rises in the east and sets in the west; relative to the stars, the moon moves eastward (completing one circuit in about a month). As the moon moves around the sky relative to the sun, it goes through a cycle of phases, each of which depends on the angle on the sky between the sun and the moon.

5. Relative to the horizon, the planets rise in the east and set in the west daily: generally, the planets move on or close to the ecliptic eastward with respect to the stars. Occasionally, the planets loop westward relative to the stars in retrograde motion for weeks or months. Of the five planets visible without a telescope, Mars, Jupiter, and Saturn reach opposition to the sun during their retrograde motion; Mercury and Venus retrograde when moving from evening "star" to morning "star" (which also means that Mercury and Venus never move far from the sun).

6. A lunar eclipse occurs when the full moon passes through the earth's shadow; a solar eclipse occurs when the new moon passes between the sun and the earth. Solar eclipses show that the moon is closer to the earth than the sun is.

7. The relative angular speeds of the planets with respect to the stars allow us to estimate their relative distances from the earth: the slower planets are further away.

8. The measurement of angles is basic to astronomy. We only measure directly angular positions and speeds on the sky. We can use relative angular speeds to estimate the relative distances of the sun, moon, and planets from the earth, based on the assumption that they all move at the same speeds. We cannot find the actual speeds of an object unless we know its distance. By eye, we can see just the angular diameters of the sun and the moon. If we knew their distances, we could directly find their actual diameters. Otherwise, we just know their diameter-to-distance ratios.

Table 1.3 Naked-Eye Celestial Motions

Object	Daily Motion	Long-Term Motion
Sun	E to W in about 12 hours from sunrise to sunset. Length of day varies from season to season.	W to E along ecliptic 1° per day. Height of sun in sky at noon is maximum in summer, minimum in winter.
Moon	E to W in about 12 hours 25 minutes from moonrise to moonset. Moonrise is about 50 minutes later each day.	W to E within 5° of ecliptic. It takes 27.3 days to travel 360° relative to the stars. Phases repeat in cycles of 29.5 days.
Planets	E to W in about 12 hours from rising to setting; interval varies depending on the rate of a planet's motion with respect to the stars.	W to E within 7° of ecliptic. Average speed along ecliptic varies, fastest for Mercury and slowest for Saturn. Retrograde motion from E to W at a time specific to each planet.
Stars	E to W in about 12 hours from star rise to star set. Circumpolar stars never set; their motions center on the celestial pole.	In fixed positions with respect to each other. Relative to the sun, a constellation returns to the same position in 1 year. Position of the celestial pole changes slowly, returning to its initial position in about 26,000 years.

STUDY EXERCISES

1. How can you find the ecliptic in the sky? The zodiac? (Learning Outcomes 1-4 and 1-8)
2. Draw a schematic diagram of the retrograde motion of a planet relative to the stars. Indicate the east and west directions clearly. (Learning Outcome 1-4)
3. What celestial bodies *never* show retrograde motion? (Learning Outcome 1-3)
4. Into what two groups can the planets be divided on the basis of retrograde motion? (Learning Outcome 1-4)
5. When Mars is at opposition, approximately when will it rise? Set? (Learning Outcome 1-4)
6. For what two reasons did ancient astronomers believe (correctly) that the moon is closer to the earth than the sun is? (Learning Outcomes 1-6 and 1-9)
7. a. You go outside at about 9 P.M. and face south. The moon is up and off to your right, near the horizon. Is it rising or setting? What is its phase?
   b. The next night you go out again at 9 P.M. Where is the moon? Is it higher, lower, or not up at all? Did it move east or west with respect to the stars? Has its phase changed? If so, how? (Learning Outcomes 1-3, 1-5, and 1-7, 1-10)
8. Describe the changing position of the rising sun on the eastern horizon throughout a year, with special emphasis on the solstices and equinoxes. (Learning Outcomes 1-2 and 1-10)
9. What phase must the moon be in for a solar eclipse to occur? A lunar eclipse? (Learning Outcome 1-6)
10. On what observational basis can you argue that Mars is closer to the earth than Saturn? (Learning Outcomes 1-9 and 1-11)
11. At what time of year would the daily angular speed of the sun along the horizon (at rising or setting) be the greatest? The smallest? (Learning Outcome 1-2)
12. Consider observing the sun at sunset from a midnorthern latitude. How does the setting point change from winter to summer solstice? What would you observe from a *midsouthern* latitude? (Learning Outcome 1-2)
13. Is the sun always overhead at noon when observed from a midnorthern latitude? If not, when is it overhead? (Learning Outcome 1-2)
14. Imagine that, from a midnorthern latitude, you measured the length of a shadow cast by a telephone pole at noon on the day of the spring equinox. A week later, you measure the shadow again. Is it shorter or longer? What happens at a midsouthern latitude? (Learning Outcomes 1-1 and 1-2)
15. What observation defines the constellations of the zodiac? (Learning Outcome 1-3)

PROBLEMS AND ACTIVITIES

1. The moon circuits the zodiac once in about 27 days. What is its average angular speed relative to the stars per day? per hour? Explain why this differs from the month of phases (29.5 days).
2. What is the diameter-to-distance ratio of the sun? How many diameters of the sun would fit in the earth–sun distance?
3. How does the diameter-to-distance ratio of the sun compare to the diameter-to-distance ratio of the moon? Which object is farther? Can you make any confident statement about the actual size of the moon relative to the sun?
4. A stick that's 10 cm high casts a shadow that's 3.5 cm long at noon. What is the sun's altitude? *Hint:* Draw a diagram to scale. Or you can divide the height of the stick *l* by the shadow length *s*. This ratio is the tangent of the angle of the sun's height h above the horizon,

$$\tan h = \frac{1}{s}$$

Calculate the ratio and use a pocket calculator with trigonometric functions to find *h*.
5. The sun circuits the zodiac in about a year. What is its average angular speed relative to the stars per month? per day? per hour? How does it compare to the moon's average angular speed?
6. A Boeing 747 jumbo jet has a length of about 70 meters. If you saw one flying in the sky and measured (with your hand) an angular length of 1°, how far is the airplane from you in airplane lengths? In meters?
7. Use the information in the last column of Table 1.2 to estimate the relative distances of Saturn and Jupiter from the earth. What assumption will you make?
8. Plot the information given in Table F.1. How are the angular diameter and distance (in actual diameters) related? *Hint:* Direct, inverse, or inverse square?

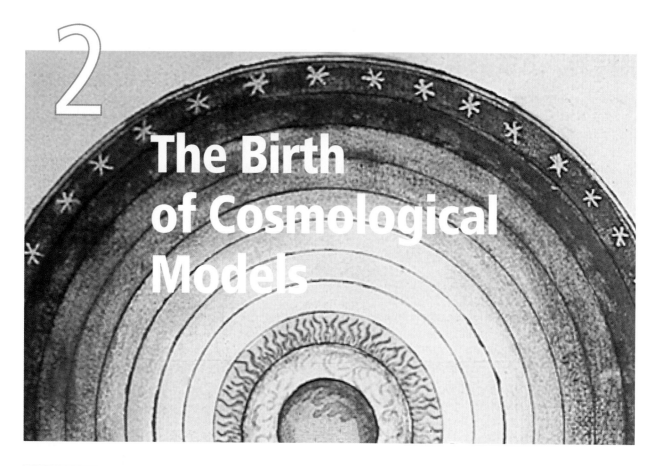

2
The Birth of Cosmological Models

LEARNING OUTCOMES

After studying this chapter, you should be able to:

2-1 Describe and explain the essential aspects of a scientific model.

2-2 Summarize the observations that must be explained in a cosmological/solar system model and evaluate such models in the context of scientific model making.

2-3 Describe the physical basis of Aristotle's geocentric model and tell how it influenced his picture of the cosmos.

2-4 Describe the geometric devices that Aristotle used to explain basic astronomical observations in the context of a geocentric model.

2-5 State the assumptions and physical basis for Ptolemy's geocentric model.

2-6 Sketch a basic geocentric model for the major motions of the moon, the sun, and the planets, and show how motions on circles were used to explain retrograde motions and planetary elongations.

2-7 Indicate the geometric devices and physical ideas that Ptolemy used to explain the main celestial motions and any variations in the major cycles.

2-8 Evaluate the essential assets of Ptolemy's geocentric model that led to its wide, long-term acceptance; as part of this appraisal, be able to construct a simplified version of the model.

2-9 Use at least one specific case to show how geometric and aesthetic concepts influenced Greek ideas about the cosmos.

2-10 Describe the difference between a sun-centered model and an earth-centered one with respect to an annual stellar parallax.

2-11 Explain sky observations witha geocentric model.

CENTRAL CONCEPT

Scientific models of the cosmos can explain and predict the motions of celestial bodies, especially those of the planets. Early models of the cosmos were centered on the earth.

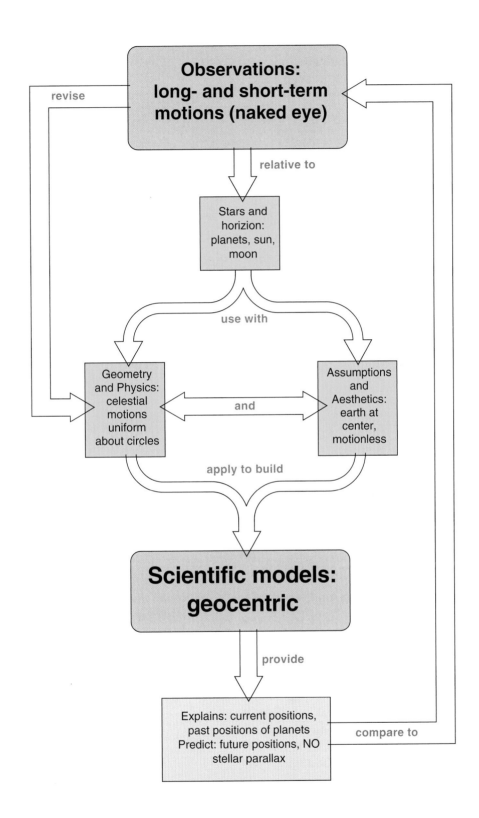

Observations:
long- and short-term
motions (naked eye)

revise

relative to

Stars and
horizion:
planets, sun,
moon

use with

Geometry
and Physics:
celestial
motions
uniform
about circles

and

Assumptions
and
Aesthetics:
earth at
center,
motionless

apply to build

Scientific models:
geocentric

provide

Explains: current positions,
past positions of planets
Predict: future positions, NO
stellar parallax

compare to

How have people pictured the cosmos? Older cosmologies generally paid little attention to the details of the celestial motions even if the cycles were carefully observed. Most ancient cultures viewed the cosmos as finite and geocentric, closed off by a shell of stars. Order in the cosmos tended to be explained in terms of myths.

In the fourth century B.C., the Greeks first attempted to take cosmological ideas beyond myths. Grappling with the vexing problem of planetary motion, Greek thinkers refined their cosmological picture with geometric devices intended to account for the celestial cycles. These designs marked the first earnest scientific models: mental images with features to explain those observed in nature. Early cosmological thought culminated in the geocentric model of Claudius Ptolemy in the first century A.D. So well did Ptolemy succeed that no one soundly challenged his model for more than 1,400 years. That challenge (to come in Chapter 3) eventually seeded the birth of models used today.

2.1 SCIENTIFIC MODELS

Chapter 1 described naked-eye astronomical observations. You probably felt the urge to place these observations into a model for the operation of the heavens, to answer "how" and "why." People in general want to make sense of what they see in the world. This natural drive gave birth to **scientific models** – conceptual plans to explain the workings of nature. Scientific models lie at the heart of the workings of modern science. (See the Scientific Models Celestial Navigator™.)

Building Models
(Learning Outcome 2-1)

A scientific model evokes a mental picture that is based on geometric ideas, physical concepts, aesthetic notions, and basic assumptions. It draws on analogy and metaphor of what is seen in nature. In this process, a model comes to grips with a seemingly chaotic world by casting it in familiar terms.

Certain elements always enter into a scientific model. First, our observations of the world provide raw information and stimulate our curiosity. Then we try to interpret these impressions by geometry,

physics, and aesthetics in astronomical models. Our choices are filtered by whatever assumptions – both explicit and implicit – we apply to the situation. In astronomy, geometry outlines a visual framework for the model and the shapes of objects within it. Physics deals with the motions and interactions of various components within the model. And aesthetic ideas – gut judgments of what seems beautiful – compel us to select the simplest, most pleasing models from those we can imagine. Simplicity also forces us to include only the essential aspects of explanation. Scientific models are mental, essential, simple, and beautiful.

Finally comes the crucial test: how well do the features of the model correspond to actual observations? If they agree well (within the known errors of the observations), the model is confirmed as workable. If not, various aspects of the model are modified to get a better match.

A scientific model has two key features: it *explains* what is seen and it *predicts* what will happen accurately. A model's predictions of future events or descriptions of past events must relate directly to observations and do so with ample accuracy to be convincing. Although all good scientific models contain both explanation and prediction, one feature may overshadow the other in a particular model.

A mature model's predictions must be *numerical* to permit direct comparison with measurements. But keep in mind that every observation has an error attached to it. For instance, at best the human eye can see an angle as small as 1 arcminute. So all unaided observations of the sky include an error of at least this amount, which is about one-thirtieth the angular size of the moon. Instruments such as telescopes can improve on this natural limitation; even today's telescopes spawn errors in every observation. A model is deemed sufficient if its numerical predictions and explanations fall within the recognized range of intrinsic errors in observations – as judged by the standards of the day.

Evaluation of Models
(Learning Outcomes 2-1, 2-2)

To confirm a model, we look at how well its predictions fit our best observations. The search for confirmation can make or break a model. They are necessarily tentative, containing the seeds of their own destruction. For scientific models must change when new or more accurate observations become available. As models evolve, we acquire a deeper understanding of the cosmos, the order in the physical world. The conception of models is the creative act of science.

As models change, the "facts" of science also change. For example, this chapter highlights a model of the cosmos centered on the earth. People who accepted this model regarded the earth's central position as a "fact," the proper interpretation of naked-eye observations as presented in Chapter 1. We have a different model today. Beware that "facts" in science may depend on specific models – many are *model dependent.*

To understand scientific models as completely as possible, you should be able to:

1. State the aesthetic, geometric, and physical bases of the model.
2. State clearly the assumptions behind the model and evaluate how well they are supported.
3. State the key observations the model attempts to explain and evaluate how well it succeeds.
4. Describe the relative importance of various aspects of the model, making clear the connections between parts.
5. Indicate how the model deals with new observations and suggest predictions it can make for new situations.

Some models prove to have a great deal of power by tying together disparate observations. Others direct scientists into new areas never imagined. When a model proves very robust and potent, it may be elevated to the status of *scientific theory.* Some theories – like Einstein's theory of relativity – prove so powerful at conception that they quickly attain this status.

People sometimes belittle the word "theory," such as in "It's just a theory," with the emphasis on "just." Such people use "theory" not in its scientific sense but more in the sense of "hypothesis." Detective stories frequently use "theory" to mean "hypothesis," which is a tentative idea vulnerable to ready falsification. Scientific theories verify themselves by surviving repeated experimental tests, unifying physical phenomena, and creating opportunities to explore new aspects of nature.

2.2 GREEK MODELS OF THE COSMOS

A good model must tackle the major and minor motions in the sky (review Tables 1.2 and 1.3 in Chapter 1) so that the model predicts the angular positions and angular speeds that are observed. Early Greek philosophers devised the first geometrical and physical models of the cosmos. A brief look at Greek astronomy and cosmology shows how scientific model making arose in Western culture.

Harmony and Geometry
(Learning Outcomes 2-2, 2-9)

The ideas of Pythagoras (who lived in the sixth century B.C.) forged the aesthetic basis of Greek cosmology. The Pythagorean cosmos first incorporated some aspects of a scientific model. In it, the earth has a spherical shape for reasons of symmetry and perfection – a sphere is the most perfect of all solids. Here geometry and beauty intertwine and add power to a mental model.

One spherical shell enclosed the cosmos and held the stars. Smaller spheres within it carried the planets around. The stellar sphere, driving all other spheres, rotated daily east to west – an explanation of the westward daily motion of the sky. At the same time, the planetary spheres rotated slowly at different rates from west to east. Their rotational periods matched the time for each planet to circuit the zodiac. These rotations explained the eastward motion of the planets relative to the stars as seen from the earth and their average angular speeds. This model is **geocentric,** meaning that the earth lay motionless in the center of the cosmos.

The Pythagorean model relied strongly on the notions of harmony and symmetry. Such aesthetic ideas play an essential role in scientific models. Though this early model contained forceful aesthetic and geometric elements, it lacked physical ideas. And its correspondence to actual observations was awkward. For example, the basic model failed to account for the pesky retrograde motions of the planets.

Roughly a century later, Plato (427–347 B.C.) proposed ideas derived from Pythagorean concepts. Plato, too, saw the perfection of the cosmos in the form of a sphere. So he assumed – demanded! – that all heavenly bodies move at a uniform rate around circles. (A uniform rate is expected from the rotation of a sphere.) Plato also succinctly stated the problem of the planets: the goal of an astronomical model was "to save the appearances" – to devise a model that explained the observed motions. This goal preoccupied astronomers for centuries. It also burdened them with an aesthetic ideal that turned out to be wrong!

A Physical Geocentric Model
(Learning Outcomes 2-2, 2-3, 2-4, 2-9)

Aristotle (384–322 B.C.), the most famous of Plato's pupils, also based his model of the cosmos on the idea of uniform, circular motions (Fig. 2.1). He even tried to incorporate the retrograde motions of the planets. Nevertheless, the results were a mess – a model with 56 spheres centered on the earth. Even with this complexity, the model did not match observations very well. Geometry by itself was not enough.

Aristotle's model was more scientific than earlier efforts, however, because it did incorporate – for the first time! – *physical* ideas of motion. What were these physical concepts? Aristotle viewed the cosmos as divided into two distinct realms: a region of change near the earth and an eternal region in the heavens. The realm of change contained bodies made of four basic elements – earth, air, fire, and water. Each element had its own natural motion toward its place of rest in the cosmos: earth to the center, fire to the greatest heights reaching toward the moon, air below fire, and water between the earth and the air. In contrast, the heavens were made of an immutable, crystalline material, which did not obey terrestrial laws.

The ancient Greeks believed that each realm had different versions of **natural motion:** motion without forces. In the heavens, the celestial spheres rotated naturally, so *no* forces were needed to move the planets around the earth. In the terrestrial realm, earth, air, fire, and water each had its natural motion. For example, the natural motion of earthy material was toward the center of the cosmos. Motion is natural if an object returns to where it "belongs." No more needs to be said.

Other motions required a cause. These motions, Aristotle reasoned, needed a **force,** *a push or a pull,* to keep them going. That is the crucial physical distinction between forced and natural motion. These ideas seem common sense. For instance, to keep a bike moving, you must continue to pedal it. When you stop this effort, the bike rolls a bit but will stop – naturally. No force, no motion!

These physical ideas shaped Aristotle's image of the cosmos. He argued that the earth must be stationary and in the center of the universe. How so?

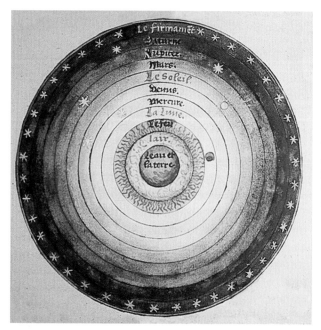

Figure 2.1 Geocentric model of the cosmos in the tradition of Aristotle: picture of the system of geocentric spheres, drawn by French astronomer Oronce Fine in *Le Sphère du Monde* (1542). The earth lies in the center; above it are the natural realms of water, air, and fire below the sphere of the moon. Beyond the moon are the heavenly spheres to which the planets and the stars are attached.

First, the natural motion of earthy material to seek the center of the cosmos was taken to explain the location of the earth there. Second, Aristotle reasoned that if the earth rotated, bodies thrown upward would not drop back to their point of departure. Yet, he noted, heavy objects thrown upward do indeed return to their starting place. Conclusion: the earth does not rotate. (Note how these ideas of natural and forced motion influenced Aristotle's incorrect model of the cosmos. Concepts of motion profoundly affect cosmological models, even today.)

A model gains credibility when it explains observations in a natural way. In this respect, Aristotle's model was successful, for it explained two important observations: the lack of a stellar parallax and the spherical shape of the earth. Aristotle noted that if the earth moved around the sun, the stars must display an annual shift in their positions, called a *stellar parallax* (described later; see Fig. 2.2). No one had observed this change, so Aristotle concluded that the earth did not move around the sun, a conclusion that reinforced his geocentric view. (Stellar parallax does occur, but it is too small to detect with the unaided eye.)

Aristotle followed Pythagorean ideas and added observations of his own (such as the curve of the earth's shadow on the moon during a lunar eclipse) to conclude that the earth was a sphere. This idea led to a calculation of the earth's size by Aristotle, who cited a diameter of 5100 km, and by a later Greek, Eratosthenes, who found a diameter of 13,400 km (Enrichment Focus 2.1).

A Contrary View: A Sun-Centered Model
(Learning Outcomes 2-1, 2-2, 2-9)

After the time of Alexander the Great (356–323 B.C.), the Greek scientific tradition centered on the great library at Alexandria in Egypt. Here worked astronomers such as Eratosthenes, who calculated the earth's size, and Aristarchus, who proposed a sun-centered, or **heliocentric,** model of the cosmos. Aristarchus lived in the third century B.C. His heliocentric model had the earth rotating on its axis once a day to explain the daily motion of the sky. The earth also moved around the sun in one year, which explained the annual motion of the sun through the zodiac.

Unfortunately, when the library at Alexandria burned, all the major writings of Aristarchus were destroyed. We know of his ideas from comments by others and from one fragment of a work. In it, Aristarchus figured out the earth–sun distance relative to the earth–moon distance and inferred that the sun was a body much larger than the earth. We have no evidence that Aristarchus worked out planetary motions in any detail using his heliocentric model.

One person who may have read Aristarchus was Hypatia (A.D. 370–415), the last astronomer to work in the great library. Her work is lost to us, too. A mob brutally murdered her, burned her remains, and all but obliterated her work and life – and any continuity with Aristarchus' ideas.

This early heliocentric model was attacked on two fronts: it contradicted Aristotle's physics by stating that the earth moved, and it required a stellar parallax, which could not be observed with the naked eye. For these reasons, and also because of the dominant influence of Aristotle's ideas, Aristarchus' model languished.

Stellar Parallax in a Finite Cosmos
(Learning Outcome 2-10)

Parallax is an apparent shift in the positions of a body (or bodies) because of the motion of the observer. When the body in question is a star, we use the term **stellar parallax.** In a heliocentric model, stellar parallax arises from the earth's motion around the sun; it is called **heliocentric stellar parallax.** How it works depends on whether the stars in space are confined to a thin shell (as in the Greek picture) or spread throughout space (as in modern concepts; see Chapter 13).

Parallax occurs when observations take place from two distinct locations. Consider a merry-go-round. As it turns, a rider on it sees nearby objects shifting in their positions relative to more distant ones. That shift is parallax. The greatest angular shift of any single nearby object occurs from the two positions separated by the diameter of the merry-go-round. And the shift recurs in a periodic cycle as the rider moves around.

Now imagine that the stars are stuck in a thin shell that closes off the cosmos, as in the model of Aristarchus (Fig. 2.2). Pick out two stars close together on the celestial sphere (A and B in Fig. 2.2). Observe these stars when they are in the southern sky at midnight (position 2 in Fig. 2.2); they will appear some angular distance apart (angle A2B).

ENRICHMENT FOCUS 2.1
Surveying the Earth

Around 200 B.C., the Greek astronomer Eratosthenes, believing the earth to be round, calculated its circumference. Here's how Eratosthenes reasoned. While on vacation from his studies at the library in Alexandria, he visited Syene (now Aswan), Egypt. He noted that at the summer solstice, sunlight fell directly down a well at noon (Fig. F.3), indicating that the sun was directly overhead. At noon on the same date the following year, back in Alexandria (located directly north of Syene), Eratosthenes observed that a sundial shadow indicated the sun to be about 7° south of directly overhead. Because the earth's circumference totals 360°, the astronomer concluded that the distance from Syene to Alexandria must be 7/360 of the circumference (Fig. F.4).

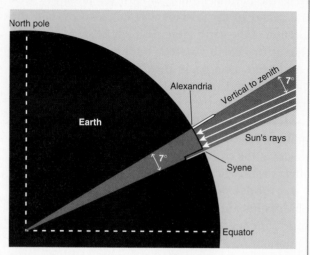

Figure F.4 Geometrical basis for calculating the earth's circumference. If the sun is so far away from the earth that its incoming rays are parallel, only a curved earth could simply account for the difference in the sun's noon angle at the two locations: in this case, the 7° between Alexandria and Syene at noon.

To find the length of the circumference, Eratosthenes needed the distance from Alexandria to Syene. The trip by camel between these two cities took about 50 days. The average camel then traveled about 100 stadia per day, so the trip covered 5000 stadia. The stadium (plural stadia) was a unit of length measuring about ⅙ km; its exact value varied in the ancient world. Eratosthenes calculated that the earth's circumference was 360/7 × 5000 = 250,000 stadia. Take the length of a stadium as ⅙ km; then the earth's circumference comes out to 42,000 km – surprisingly close to the modern value of 40,030 km. (This may be coincidence. Although Eratosthenes measured the angular separation of Syene and Alexandria very accurately, we do not know how well he knew the distance between the cities, nor do we know the exact length of a stadium in kilometers. In any case, he got the *ratio* right.) Divide the circumference by 2π and you have the radius: about 6700 km. Double the radius to get the diameter – 13,400 km.

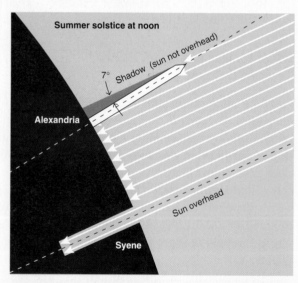

Figure F.3 Eratosthenes' solar observations at noon at Alexandria and Syene (Aswan) in Egypt. He noted that at Syene on the summer solstice, the noon sun's light came down from directly overhead. In another year, he saw that in Alexandria on the summer solstice, the noon sunlight fell at an angle of 7° from the vertical.

Just after sunset three months later, observe the stars again (position 3 in Fig. 2.2). They will appear closer together (angle A3B is less than angle A1B). Observe them again six months later (position 1 in Fig. 2.2); their angular separation is A1B, which is the same size as angle A3B.

Imagine viewing these stars from positions 1, 2, and 3. You'd see the stars close together (1), farther apart (2), and then closer together again (3). This cyclical shift in angular position because of the earth's motion around the sun is heliocentric stellar parallax. Note that the size of the earth's path compared to the size of the shell of stars determines the size of the shift: the smaller the ratio, the smaller

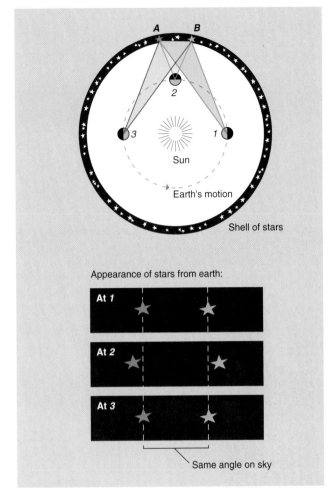

Figure 2.2 Stellar parallax in a finite, heliocentric model. Top: A and B are two stars fixed on the nearby stellar sphere. The earth moves around the sun in a year, causing a change in the observed angle between the stars. The largest angle, A2B, occurs at position 2; there are intermediate angles, A1B and A3B, at positions 1 and 3, respectively. Bottom: View of stellar parallax on the sky from earth.

the shift. If the earth were in the center of the stellar shell and did not move, no shift would occur, as expected in a geocentric model.

Greek astronomers did not observe this parallax. The heliocentric model disagreed with such observations and was rejected partly on this basis. A geocentric model predicts no such parallax. (We now know that heliocentric parallax is too small to detect without a telescope, because the stars are very far away compared to the earth–sun distance.)

Expanding the Geocentric Model (Learning Outcomes 2-2, 2-9)

The distinguished astronomer Hipparchus lived and worked on the Greek island of Rhodes from 160 to 127 B.C. To the basic geocentric model, Hipparchus added the geometric devices of **eccentrics, epicycles**, and **deferents** to explain aspects of planetary motions that had been shrugged aside by earlier observers. Each of these devices accounted for observed features: the epicycle and deferent together explained the retrograde motions; the eccentric accounted for the variable motion of the planets and sun through the zodiac. Hipparchus used geometry "to save the appearances" and stuck to the assumption of uniform, circular motions. How?

Hipparchus assumed that the planets move at a uniform speed on circular paths. Yet he knew from observations that the planets' motions are not uniform: The average angular speed of a planet is faster in one region of the zodiac and slower in the opposite region. Hipparchus explained this variation with an eccentric (Fig. 2.3a). The earth was displaced from the center of the pla-net's circular motion. Then the planet under consideration appeared to go faster through the zodiac when it was closer to the earth and slower when farther away (Fig. 2.3b). But

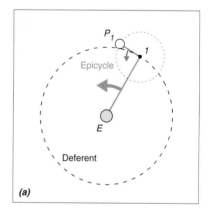
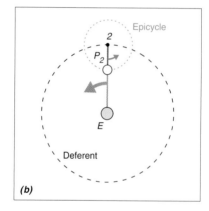
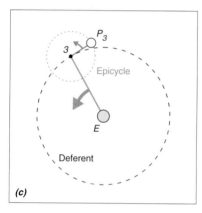

Figure 2.3 An eccentric (a). A planet (P) revolves with uniform circular motion about the center (C) of its path. As observed from C, the planet moves at a constant angular speed. (b) The effect of the eccentric for modeling a planet's motion. As seen from C, the planet moves through equal angles in equal amounts of time – a constant angular speed – but as seen from the earth (E), the planet covers different angles in the same times. The planet has the fastest angular speed when closest to the earth (at point 10) and the slowest when farthest away (at point 4). So the planet's motion as seen from the earth is *not* uniform.

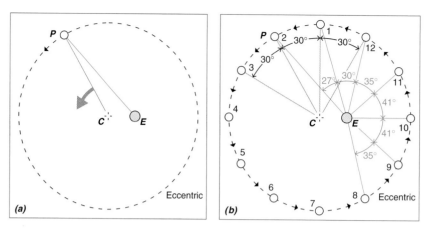

Figure 2.4 Epicycle and the deferent. A planet (P) is attached to a small circle (epicycle) whose center rides on a larger circle (deferent). The earth (E) lies in the center of the deferent. The radius of the epicycle turns in the same direction as the radius of the deferent. At P_1, the planet begins its retrograde motion. At P_2, it is in the middle of its retrograde motion. At P_3, retrograde motion ends. Since the planet is closest to the earth in the middle of its retrograde motion, it appears at its brightest then.

its actual motion along the circle remained at a constant angular rate as seen from its center rather than from the earth.

The eccentric did *not* explain retrograde motion. For that, Hipparchus used the combination of an epicycle and deferent (Fig. 2.4). The deferent was a large circle, sometimes centered on the earth, sometimes offset (then it was also an eccentric). The center of the epicycle, a smaller circle, moved along the circumference of the deferent. The planet was fixed to the epicycle, so its motion comprised its circuit around the epicycle plus the epicycle's circuit around the deferent. As the planet moved on the part of the epicycle within the deferent, it moved in the opposite direction from the deferent motion and so imitated the backward swing of retrograde motion. As a bonus, while the planet is in the middle of its retrograde motion, it was closest to the earth and shone the brightest. Note that all motions were uniform as viewed from the center of the epicycle or deferent.

Figure 2.5 A fifteenth-century Italian portrait of Claudius Ptolemy from *Cosmografia,* from the Biblioteca Estense in Modena, Italy: "we shall only report what was rigorously proved by the ancients."

The motion of the deferent represented the planet's general motion from west to east through the zodiac. The model matched up with observations because the periods of the epicycle and the deferent were set from observations of the time each planet took to circuit the zodiac once (the deferent's period) and the time between retrograde motions (the epicycle's period).

2.3 CLAUDIUS PTOLEMY: A TOTAL GEOCENTRIC MODEL

Two-and-a-half centuries after Hipparchus, Claudius Ptolemy (Fig. 2.5) worked in Alexandria and so had access to records in the great library. There, he molded the astronomical tradition into a comprehensive model that would endure for centuries. (See the Geocentric Model Celestial Navigator™.)

We know little of Ptolemy's life, not even the dates of his birth and death. His observations indicate that he worked around A.D. 125. Ptolemy's most influential astronomical work – the *Almagest* – was the first professional astronomy textbook. Many of the geometric devices that Ptolemy used in the *Almagest* did not originate with him, but he was the first to design a complete system that accurately predicted planetary motions, with errors of usually not more than 5° and often less. His careful application of geometry nicely described the celestial motions.

A Geocentric Model Refined (Learning Outcomes 2-2, 2-5)

At the start of the *Almagest*, Ptolemy defines the problem: use geometry to describe astronomical observations, especially the major cycles and their variations, known since Babylonian times. Note that Ptolemy, like Hipparchus, follows Plato's dictum "to save the appearances."

Ptolemy assumes that the earth is spherical and in the center of the cosmos. The earth has no motions. It is much smaller than the outer sphere of stars. In addition, he assumes that uniform motion around the centers of circles is the aesthetically pleasing motion for the celestial spheres. Ptolemy's model has Aristotle's physical basis.

In practice – and this is a key point – Ptolemy ends up *violating* this uniform motion precept. Recall that there are two main variations in the motions of the planets: retrograde motion and the variable motion through the zodiac. Also, older records showed that retrograde motions varied in size, shape, and duration (Fig. 2.6). How can uniform circular motion account for these many variations?

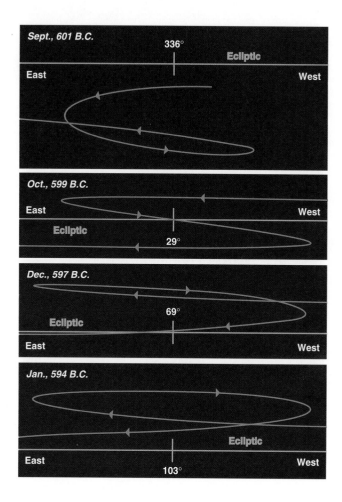

Figure 2.6 Retrograde motions of Mars in Babylonian times. Mars appears at different positions along the ecliptic (indicated in degrees from the vernal equinox, which is 0°) for different times of retrograde motion indicated by the dates. The dates are for the middle of the retrograde motion. Note how the shape and size (and so the duration) of the retrograde loop vary. Babylonian astronomers knew of such variations. (Adapted from a diagram by Owen Gingerich.)

Nonuniform Motion
(Learning Outcomes 2-6, 2-7)

Think of the problem of the motion of the planets as a model-building puzzle that needs a different device for each observed aspect. The two main aspects are (1) eastward motion through the zodiac – with variations, and (2) retrograde motion – also with variations! Like Hipparchus, Ptolemy modeled the first with a planet moving around the earth at uniform speed on a circle, but he offset the earth from the center of the circle – the eccentric – to account for variations. For the second, again like Hipparchus, Ptolemy used smaller circles (epicycles) moving on the larger ones (deferents). Then, for variations in the retrograde motions, Ptolemy invented a new geometrical device: the *equant.*

To make an equant, Ptolemy started with an eccentric. Here the earth lies away from the center

of the circle of a planet's deferent. Ptolemy then imagined another point, not at the circle's center, from which the motion would appear uniform. This imaginary point, the **equant** point (Fig. 2.7a), lay opposite the center of the deferent from the earth (which is at an eccentric point). If you stood there, you would see the planet move around the sky at a uniform angular speed relative to the stars. From the earth and from the circle's center, though, the motion is not uniform (Fig. 2.7b). The observed angular speed varies inversely as the distance from the earth.

Imagine a planet's epicycle as lying on the outside of the face of a clock with one hand, which drives the epicycle around. The ideal is to have the hand at the center of the clock. The equant, though, positions the hand off-center. The hand turns uniformly and so moves the epicycle around as viewed from the point of attachment (Fig. 2.8). As seen from the center, however, the motion is nonuniform.

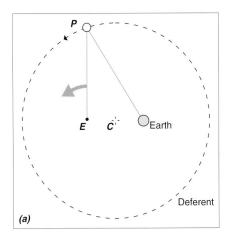
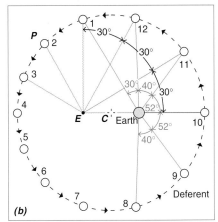

Figure 2.7 Ptolemy's equant. For clarity, the epicycle is not shown. (a) Here the earth is at the eccentric point. The equant point is an imaginary point (E) on the opposite side of the center from the earth. (b) A planet moves on the deferent circle so that, as viewed from the equant (E), it would appear to cover equal angles in equal times and so move at a uniform angular speed. However, as seen from C or the earth, the planet moves through different angles in equal times, which means that its angular speed varies. The motion illustrated here ignores the motion around the epicycle, which generates retrograde motion.

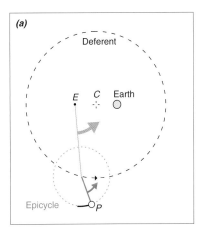
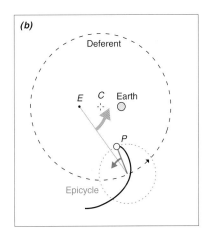
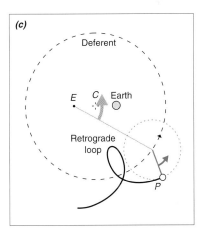

Figure 2.8 The equant combined with the epicycle and the deferent to model the actual retrograde motions of a planet. Note that if the planet retrogrades at a different position along the deferent, the shape and size of its retrograde loop will differ from that shown here. (Adapted from a diagram by Owen Gingerich.)

You do not need to understand the nitty-gritty details of the equant to appreciate Ptolemy's contribution to the evolution of cosmological models. The main point is that with the equant, *celestial motions no longer had to be uniform around the centers of circles*. The equant was a nonphysical, totally geometric device that broke the fundamental assumption that planetary motion had to be uniform along circles.

Why did Ptolemy violate this aesthetic ideal? Probably because he demanded that his model fit observations reasonably well. With the equant, he could better match the model's predictions to the observations, especially the annoying variations of the planets in their overall retrograde cycles. This desire for accuracy struck Ptolemy as more important than the traditional precept of uniform, circular motion – a crucial step to a robust scientific model.

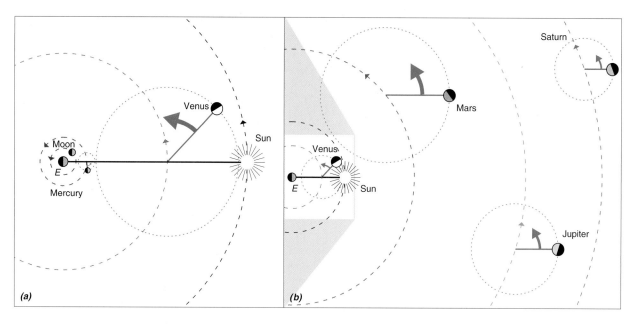

Figure 2.9 Overall Ptolemaic model. (a) For the moon, Mercury, and Venus. The centers of the epicycles of Venus and Mercury are fixed along a line with the sun; this explains why they are never very far from the sun. The size of each epicycle accounts for the measured size of the maximum elongation angles. (b) For Mars, Jupiter, and Saturn. To account for the observed sizes of the retrograde loops, the epicycles of these planets decrease in size, so that Mars has the largest and Saturn the smallest. For the retrograde motions to occur at opposition, the radii of these epicycles must align with the earth–sun radius.

Ptolemy's Complete Model
(Learning Outcomes 2-2, 2-8)

When completed, Ptolemy's model (Fig. 2.9) treated Mercury and Venus differently from the rest of the planets. Why? Recall (Section 1.4) that Mercury, on average, stays within about 23° of the sun, and Venus averages within 46°. To explain these observations, Ptolemy made Mercury's epicycle smaller than Venus' and demanded that the centers of these epicycles always lie on the line connecting the earth and the sun, so Mercury and Venus were constrained near the sun (Fig. 2.9a).

In contrast, Mars, Jupiter, and Saturn can be anywhere along the zodiac relative to the sun, so the centers of their epicycles could be anywhere on the perimeter of the deferents. To ensure that these planets retrograde only at opposition, Ptolemy set the radii of their epicycles parallel to the earth–sun radius (Fig. 2.9b).

In Ptolemy's model, the heavenly spheres are actual, physical spheres. The motion of these spheres – rotation – was assumed to be natural and

to drive all the heavenly motions. No force was required.

Although Ptolemy's model lost acceptance centuries ago, do not berate it! This model remained in use for some 1400 years mainly because it worked – it predicted planetary positions to the accuracy needed (a few degrees) by astronomers who had no telescopes. It agreed with early Greek physics and aesthetics. It survived because no other satisfactory model was advanced to compete with it. More than any other astronomical book, the *Almagest* showed that the motions in the heavens, though complex in appearance, could be described in simple terms by geometry.

The Size of the Cosmos

Ptolemy's cosmos was finite. How large was it? Ptolemy gives the distances to the sun and the moon in terms of earth radii, as worked out by Aristotle and Hipparchus. Going farther out, he assumes that no space is wasted between the heavenly spheres.

Figure 2.10 An evening scene, facing south. The moon is at first quarter; the sun is setting as Mars is rising. Venus lies between the sun and moon; Jupiter between Mars and the moon.

Then, with the earth–moon distance set at about 60 earth radii, the distances to Saturn can be laid out with the celestial spheres nesting tightly together. The sphere of the stars then lay 20,000 earth radii distant. It was a small cosmos – only about the distance we know today between the earth and the sun! But it did mark a rational attempt to establish the distance scale of the known universe.

Model Evaluation
(Learning Outcomes 2-1, 2-2, 2-8)

Now to evaluate Ptolemy's model based on the criteria of a scientific model in Section 2.1. First assumptions. These are: (1) the earth is spherical, lies in the center of the cosmos, and has no motions, and (2) celestial motions are uniform about the centers of circles. These relate to aesthetics: spheres and circles are simple and pleasing, and so is uniform motion. They also come from basic geometry.

Ptolemy then places the moon in a circular path moving around the earth once a month. The sun has a slower angular speed than the moon, so it is farther away but also in a circular path. The planets are placed on circles with centers on the earth to explain their eastward motion relative to the stars. The epicycle combines with the deferent to produce retrograde motions. No forces are needed. According to the accepted physics, these are all natural motions.

The variations in these motions are explained using eccentrics and equants. Now the model is more complex but able to predict positions within a few degrees or less – acceptable by the standards of the time. The equant violates one of the basic assumptions, however.

2.4 DESCRIBING BASIC OBSERVATIONS WITH A GEOCENTRIC MODEL

To help grasp the main features, let's connect it with observations of the sky. Suppose you go out at night and face south (Fig. 2.10). You observe Mars rising in the east just as the sun sets in the west, so Mars is in opposition to the sun, and the angle between them is 180°. Jupiter is 45° west of Mars. You also observe Venus, in the western sky, 45° east of the sun (greatest eastern elongation). The moon is at first quarter phase and due south, so it is 90° east from the sun.

Let's draw a geocentric model (Fig. 2.11). Put the earth at the center and draw five circles around it for the paths of the moon, Venus, the sun, Mars, and Jupiter. Imagine that you are looking at the earth from above the North Pole. Put the sun off to the right, because (if you face south) that's where it is setting. Next, mark the center of Venus' small circle (which must always be on the earth–sun line). Now put Venus on its small circle at the point where a line from the earth is tangent to it, because this orientation gives the greatest angle of Venus away from the sun (greatest eastern elongation). As Mars is just rising at sunset, it must be opposite to the sun (at opposition). Because the radius of its small circle must be parallel to the earth–sun line, the center of its small circle must be on the same direct line as that from the earth to Mars.

The moon is observed halfway around the sky between the sun and Mars, and so must be on its circle 90° around from the earth–sun line. What about Jupiter? Draw a line 45° clockwise from the earth–Mars line. Now place the center of Jupiter's small circle on its larger one such that the radius to Jupiter is parallel to the earth–sun line.

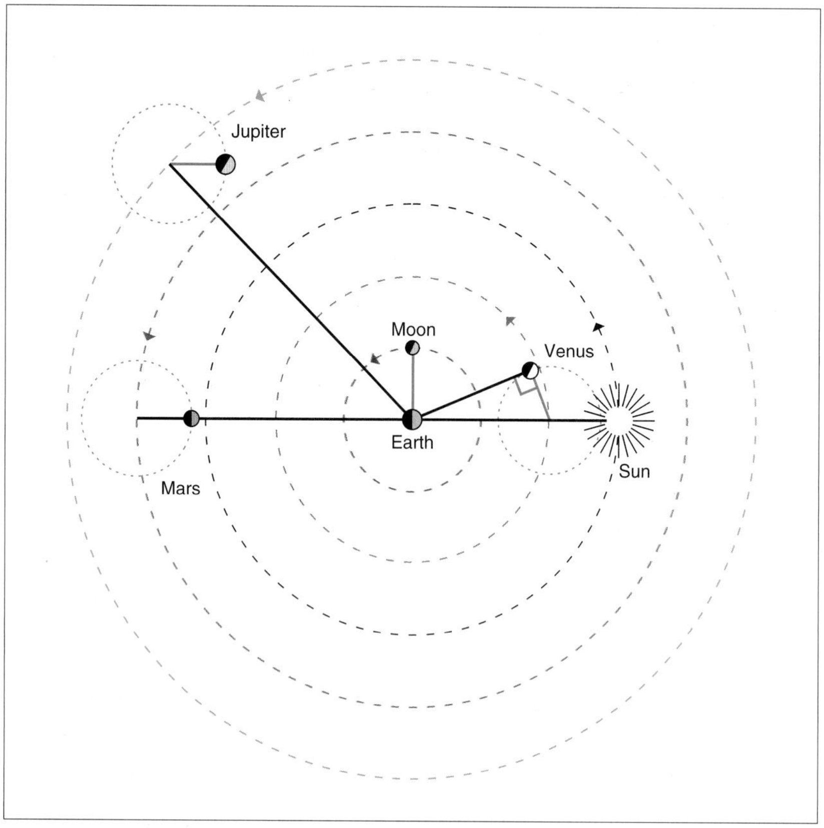

Figure 2.11 A geocentric model of the observations in Figure 2.10. Note all the angular positions of the moon and planets must match those observed.

Note that nothing fixes the relative distances. We could make Jupiter's circles, for example, twice as big and double Jupiter's speeds around the circles. Then the angular speeds and relations would still turn out the same. In fact, this model had no *fixed* distance scale because of its focus on angular speeds. An angle gives a size-to-distance ratio; an angular speed gives a rate-to-distance ratio. But we cannot know actual sizes or speeds unless we know distances! For any of the planets, we can adjust its speed around its circles to fit the observations for any reasonable range of distances. The whole system can grow or shrink in an optional way.

KEY CONCEPTS

1. Scientific models are the cores of creativity and understanding in the scientific process. Models are designed to explain and predict what is observed; astronomical models contain geometric, physical, and aesthetic elements. Scientific models are never complete or final – they evolve.

2. The retrograde motions of the planets vary from planet to planet and also for the same planet at different times. Constructing an explanation of retrograde motion was a most vexing puzzle for the construction of early cosmological models.

3. The Greeks made the first scientific models, which were based on aesthetic notions and on a desire to explain observations. These early models included the central concepts of harmony, symmetry, and simplicity, as well as the use of geometry "to save the appearances." The celestial motions were locked into the concept that required them to occur at a uniform rate along circles.

4. Aristotle developed a complete cosmological model based on physical ideas. Different concepts of motion applied to the terrestrial and celestial realms (no forces were believed to be needed to turn the celestial spheres). The model

Table 2.1 Summary of Ptolemy's Model

Observation	Explanation
SKY	
Motion of entire sky E to W in about 24 hours	Daily motion E to W of the sphere of stars, carrying all other spheres with it.
SUN	
Motion yearly W to E along ecliptic	Motion of sun's sphere W to E in a year.
Nonuniform rate along ecliptic	Eccentric (earth displaced from center of sun's circle).
MOON	
Monthly motion W to E compared to stars	W to E motion of the moon's sphere in a month.
PLANETS	
General motion W to E through zodiac	Motion of deferent W to E. Period is set by observation of period of planet to go around ecliptic.
Retrograde motion	Motion of epicycle in same direction as deferent. Period is the time between retrograde motions.
Variations in speed through zodiac, in retrograde motions	Eccentrics, equants.
MERCURY, VENUS	
Average greatest elongation of 23° and 46°	Size of epicycles set by those angles.
Appear near to sun	Centers of epicycles on earth–sun line.
MARS, JUPITER, SATURN	
Retrograde at opposition, when brightest	Radii of epicycles aligned with earth–sun radius.

Note: Error of predictions was usually 5° or less, occasionally larger.

was geocentric. It did not predict the celestial motions very accurately but did account for the major cycles.

5. Aristarchus proposed a heliocentric model that never gained headway because it violated Aristotle's physics and predicted an annual stellar parallax, which could not be observed without optical aid. In this sun-centered model, the earth moved around the sun, as did the other planets, while the moon moved around the earth.

6. Ptolemy built up a comprehensive cosmological model based on the physics of Aristotle and on the geometric ideas of the Greeks (Table 2.1). To make it complete, he invented the equant, which violated the precept that celestial motions had to be uniform and circular. His model predicted planetary positions with acceptable accuracy (5° or less in general, usually about 1° or better).

7. The assumed or observed nature of natural and forced motions strongly influences the development of cosmological models. For instance, geocentric models place the earth at the center in part because the "natural" motion of earthy material is to fall toward the center of the cosmos. Celestial motions were viewed as natural, because spheres rotate.

8. The basic aspects of a geocentric model explain in a reasonably simple way most of the major and minor motions of the sky by various geometric devices. The model was strong enough to survive for a long time – more than a thousand years. However, because such a model focuses on angular speeds of celestial objects, it does not have a fixed distance scale. Later observation of an annual stellar parallax would disprove the model.

STUDY EXERCISES

1. How did Aristotle's model explain (a) the apparent lack of motion by the earth, (b) the daily motion of the stars, and (c) the annual motion of the sun through the zodiac? (Learning Outcomes 2-2 and 2-4)

2. How did Aristotle argue against a heliocentric model? (Learning Outcomes 2-3 and 2-4)

3. In the Ptolemaic model, how did long-term observations set the period of the epicycle? The deferent? (Learning Outcomes 2-2 and 2-6)

4. How did Ptolemy treat Mercury and Venus differently from the other planets? (Learning Outcomes 2-2 and 2-6)

5. How did Ptolemy violate his own precept of uniform, circular motion? (Learning Outcomes 2-5, 2-6, 2-7, and 2-9)

6. What observation was Ptolemy trying to explain with each of the following geometric devices: (a) epicycle, (b) eccentric, and (c) equant? (Learning Outcomes 2-2, 2-6, and 2-7)

7. State one strength of Ptolemy's model; one weakness. (Learning Outcomes 2-1, 2-8, and 2-9)

8. How does a geocentric model differ from a heliocentric one in its prediction of an annual stellar parallax? (Learning Outcome 2-10)

9. What is the crucial observational test for a heliocentric model? (Learning Outcomes 2-1 and 2-10)

10. What is the function of geometry in building scientific models, especially in astronomy? (Learning Outcome 2-1)

11. What is the critical test of *any* scientific model? (Learning Outcome 2-1)

12. Ptolemy's model was accepted in scientific circles for more than 1,400 years. Why? (Learning Outcomes 2-1 and 2-8)

13. Judge whether the following statements are model dependent or model independent (Learning Outcome 2-1):
    a. The shape of the retrograde loop of a planet varies from one retrograde to another.
    b. The earth lies in the center of the cosmos.
    c. Mars moves on a small circle (epicycle) for its retrograde motion.

PROBLEMS AND ACTIVITIES

1. Suppose you tried to duplicate Eratosthenes' observations (Enrichment Focus 2.1) and came up with value of 8°. What would you calculate for the earth's circumference and radius?

2. With naked-eye observations, the smallest angle you can measure is about 1 arcminute. Assume that the stars are fixed on a sphere and consider two stars 1° apart as seen from the sun in a heliocentric model. How far must the sphere of stars be placed from the sun so that their parallax is 1 arcmin or less? *Hint:* What is the size-to-distance ratio of 1 arcmin?

3. The sun does not move at a uniform speed along the ecliptic. Over the half-year centered on the summer, it covers 180° in 186 days; over the winter half, 180° in 179 days. Using the eccentric to model this motion in a geocentric picture, calculate how far away the earth must be located from the center of the circle of the sun's motion. *Hint:* Draw a circle with a radius of 10 cm. Use half the circle to show the summer interval, and half the winter. Assume that the sun's motion along the circle is uniform. Then how far would the earth have to be offset from the center of the circle to explain the sun's observed motion?

4. Suppose that in a heliocentric model, such as that of Aristarchus, the shell of stars had a radius of 10 earth–sun distances. If two stars located in the plane of the earth's orbit are observed to be 6° apart when the earth is closest to them, how far apart will they be in angle when the earth is on

the side of its orbit farthest from them? Do you think this difference would be observable with the naked eye?

5. What is the size-to-distance ratio of the moon? How could you estimate the moon's distance from the earth if you knew (Enrichment Focus 2.1) the diameter of the earth in units like stadia or kilometers? What, then, is the moon's speed along a circular path?

6. Aristarchus is reported to have concluded that the moon has a diameter one-fourth that of the earth. Using Eratosthenes' diameter for the earth, at what distance does that place the moon?

7. What is the size-to-distance ratio for the sun? How many solar diameters fit in the earth–sun distance? If you assume that the moon is the same size as the earth, what size does that make the sun? What, then, is the sun's speed along a circular path?

3

The New Cosmic Order

LEARNING OUTCOMES

After studying this chapter, you should be able to:

3-1 List the assumptions and arguments that Copernicus used to support his model and refute the Ptolemaic one.

3-2 Explain why Copernicus disliked Ptolemy's use of nonuniform motion and how this bias influenced the development of his heliocentric model.

3-3 Show how the Copernican model explains the major motions in the sky; in particular, use simple diagrams to explain retrograde motions and elongations in the Copernican model.

3-4 Outline how the Copernican model's geometry plus simple observations determine the periods and relative distances of the planets from the sun, and so set a distance scale for the heliocentric model.

3-5 Evaluate the strengths and weaknesses of the Copernican model compared to the Ptolemaic one, including stellar parallax.

3-6 Describe how Copernican ideas influenced the astronomical work of Kepler.

3-7 Argue that Kepler, not Copernicus, properly deserves the title of the "first astrophysicist."

3-8 Describe the important geometric properties of ellipses and apply these to planetary orbits.

3-9 Compare and contrast the Copernican model and the Keplerian one in terms of physics, simplicity, geometry, and prediction.

3-10 State Kepler's three laws of planetary motion and apply them to appropriate astronomical situations.

3-11 Explain sky observations in a heliocentric model.

CENTRAL CONCEPT

A heliocentric model of the cosmos was reinvented during the sixteenth century in Europe, but this break with the geocentric tradition required new physical laws and a revolution of the cosmological views of the time.

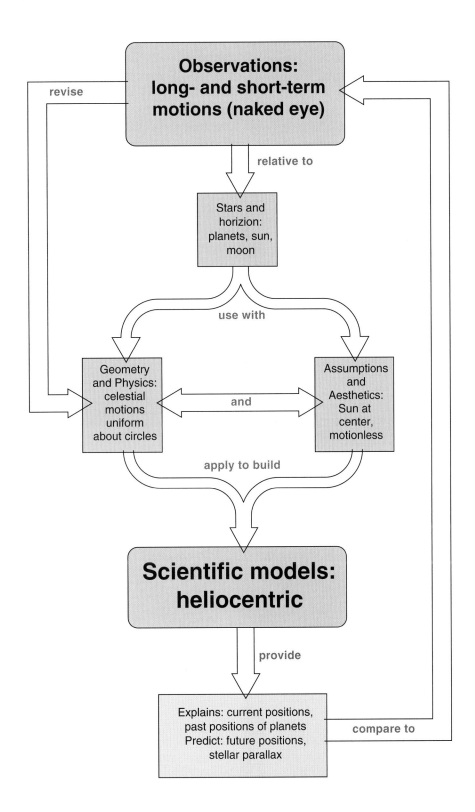

Astronomy languished with the decline of Greek civilization in the first few centuries after Ptolemy. Most of the Greek astronomical works were unknown in Europe until the twelfth century, when Arabic manuscripts of Greek thought were translated into Latin. In the thirteenth century, Alfonso the Great of Spain sponsored the publication of the lists that became known as *Alfonsine Tables,* handy compilations of planetary positions based on the Ptolemaic model. The essential Ptolemaic model still worked well enough. Practicing astronomers did not discard it.

One astronomer was unhappy: Nicolaus Copernicus, whose great work *De revolutionibus orbium coelestium* (On the revolutions of the heavenly spheres) was published in the year of his death. In *De revolutionibus,* the earth was shaken from a static place at the center of the cosmos into a path around the sun – just another planet with the rest. Copernicus' model was heliocentric.

Copernicus' view did not immediately wrench the minds of astronomers from their geocentric notions. Some, like the Danish observer Tycho Brahe, considered Copernican claims but rejected them. Others, like Johannes Kepler, were struck by essential harmonies in Copernicus' model. The revolution in astronomy after Copernicus marked the first major shift in our concept of the earth's place in the cosmos – from geocentric to heliocentric. This revolution also injected a new concept into the models – that of a physical force working in the heavens. Ultimately, these shifts shaped our modern views of the solar system and the cosmos.

3.1 COPERNICUS THE CONSERVATIVE

Nicolaus Copernicus (1473–1543) initiated an intellectual revolution (Fig. 3.1) by developing a new heliocentric model of the cosmos when the old one seemed adequate to most astronomers of the time.

Figure 3.1 Portrait of Nicolaus Copernicus, 1575, from the Collegium Maius in Cracow, Poland: "In the center of all rests the sun."

This idea sparked a firestorm during the Reformation and Renaissance years and eventually transformed the earth's place in the cosmos. See the Heliocentric Model Celestial Navigator™.

The Heliocentric Concept
(Learning Outcomes 3-1, 3-2)

While studying medicine and law in Italy, Copernicus read works of Aristotle, Pythagoras, and Plato. An offshoot of Plato's philosophy asserted that the sun is the source of the godhead and of all knowledge. These ideas singled out the sun as a body quite different from the planets and encouraged Copernicus to ponder a heliocentric model.

In about 1514, Copernicus wrote a summary of his new model and circulated it to some friends. In this general outline of the heliocentric model (Fig. 3.2), the sun replaced the earth at the center of the cosmos, the earth revolved around the sun yearly and also rotated on its axis daily. (These ideas had been outlined earlier by Aristarchus; review Section 2.3.)

Copernicus still assumed that all celestial motions must be uniform, circular motions. He viewed Ptolemy's equant as a violation of this ideal and so attacked the Ptolemaic model as "not sufficiently pleasing to the mind." Recall (Section 2.4) that the equant required that a planet move non-uniformly as seen from the center of its deferent.

Figure 3.2 Diagram of the Copernican model from *De revolutionibus orbium coelestium*, Book 6 (1543), emphasizing one main departure from the Ptolemaic model: the sun is placed in the center of the cosmos, with the planets, including the earth, revolving around it. Like Ptolemy, Copernicus placed the sphere of stars beyond the sphere of Saturn.

The new model would reinstate the uniform circular motion that Ptolemy, in practice, gave up.

Just what compelled Copernicus to offer a new model is still a puzzle. The basic Ptolemaic model worked well enough, by the demands of the day, in its planetary predictions. Contrary to some tales, the Ptolemaic model did not require the continual addition of circles to match observations and so had not become a monstrosity by medieval times. Copernicus took 20 years to develop his model, yet his predictions came out no better than those based on the Ptolemaic model.

However, Copernicus asserted that his model surpassed the Ptolemaic one on aesthetic grounds – the notions of harmony and order. What were the aesthetically pleasing aspects of his model?

The Plan of *De Revolutionibus*
(Learning Outcome 3-3)

Long after Copernicus had privately distributed his ideas, his work became known to two astronomers at the University of Wittenberg in Germany: Georg Rheticus and Erasmus Reinhold. Fascinated by the new model, Rheticus visited the aging Copernicus, who showed him a final manuscript copy of *De revolutionibus*. Rheticus begged Copernicus to allow its publication. (Not until Copernicus' lifetime had printed texts, rather than handwritten copies, become common.) Copernicus finally agreed. The book was published in April 1543, after much trouble with the printer. Meanwhile Copernicus had suffered a stroke and was confined to his bed. Although the book was delivered to him before he died in June, he probably did not read it.

How fortunate that he did not, for the final version of his life's work had been renamed! The original one was simply *De revolutionibus;* the published title sported two more words: *De revolutionibus orbium coelestium.* An unsigned preface, added without the astronomer's knowledge, stated that the work contained a new hypothesis about a heliocentric cosmos, for use in computing planetary positions – but not to be taken as reality.

For some years the preface was attributed to Copernicus and was taken to imply that he had not believed in the reality of his model. Later, Johannes Kepler (Section 3.4) discovered that the author of the preface was Andreas Osiander, a Lutheran clergyman who oversaw the completion of the book's publication. Osiander probably felt that a disclaimer was necessary to protect the book from criticism by those who believed that the Bible taught that the earth was motionless. He may have also changed the title to emphasize the motions of the heavens rather than those of the earth.

De revolutionibus took after the *Almagest* in outline and basic intention. It explained planetary motions using only uniform circular motions – the traditional precept of the Greeks. Copernicus thought that although Ptolemy's model matched observations, it clashed with the Greek ideal because it used equants. Copernicus was greatly offended by the equant – his major objection to the Ptolemaic model. In a basically conservative mood, he wished to devise a model that was faithful to uniform circular motion and to eliminate the equant. Copernicus decided to discover a natural explanation for retrograde motion and a new harmony for the celestial spheres that would relate the planets' distances to their periods around the sun.

3.2 THE HELIOCENTRIC MODEL OF COPERNICUS

In the introduction to *De revolutionibus*, Copernicus lays out his assumptions. First, he requires that the planets move in circular paths around the sun at uniform speeds. Second, he assumes that the closer a planet is to the sun, the greater is the speed of its revolution. For instance, because Mercury is closer to the sun than the earth is, Mercury travels around the sun faster than the earth does. Except for the position of the sun, these ideas are identical to Ptolemy's assumptions.

Details of the Model (Learning Outcome 3-3)

Then Copernicus treads on different ground. Here's a short summary of his ideas:

1. All the heavenly spheres revolve around the sun, and the sun is at the center of the cosmos.
2. The distance from the earth to the sphere of stars is much greater than the distance from the earth to the sun.
3. The daily motion of the heavens relative to the horizon results from the earth's motion on its axis.

4. The apparent motion of the sun relative to the stars results from the annual revolution of the earth around the sun.
5. The planets' retrograde motions occur from the motion of the earth relative to the other planets.

Let's expand on these points to emphasize key differences between the Copernican and Ptolemaic models.

In point 1, Copernicus asserts, for philosophical and aesthetic reasons, that the cosmos is heliocentric rather than geocentric. But as the Greeks had realized from Aristarchus' model, a heliocentric scheme demands a stellar parallax (Section 2.3) – which was not observed! Point 2 addresses this issue: if the stars are very far away, the parallax would be too small to be detectable by naked-eye observations. (Later, telescopes would show that Copernicus' intuition about the vast distances to stars was correct.)

In point 3, Copernicus declares that the daily westward motion of the skies results from the earth's rotation (from west to east) rather than from the rotations (from east to west) of celestial spheres. This idea has aesthetic appeal, for it replaces the rotations of many celestial spheres with the rotation of just one sphere, the earth. (You may wonder how the earth rotates without objects flying off its surface. This physical objection was voiced against the sixteenth-century Copernican model, and Copernicus had no good answer for it. Now we explain it by gravity and inertia: see Chapter 4.) Copernicus suspected that the cosmos had more than one center of motion but could not prove it.

Point 4 explains how, in a heliocentric model, the sun appears to move around the earth. Here's an analogy. Imagine that you are walking slowly counterclockwise around a lamppost. Look in the direction of the lamppost and note the background behind it. You'll see the background slowly change: the lamppost appears to move counterclockwise (the same direction as you're going) with respect to background objects.

Now imagine that the lamppost is the sun, you are the earth revolving around the sun, and the background consists of the stars of the zodiac (Fig. 3.3). From the earth, you see the sun in a constellation, say Leo. As the earth revolves counterclockwise, the stars behind the sun change. After one

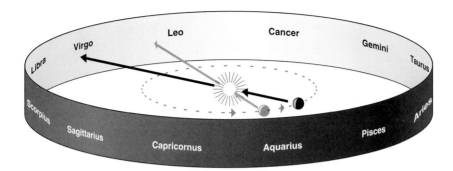

Figure 3.3 The sun's apparent motion through the zodiac in a heliocentric model. As the earth travels around the sun (counterclockwise, west to east), the line of sight to the sun and toward the background stars moves in the same direction (eastward). For example, start with the sun in Leo. A month later, the earth will have moved eastward far enough to put Virgo behind the sun. The sun seems to move eastward through the zodiacal stars at the rate of about one circuit a year.

month the sun appears in Virgo, one constellation to the east. The sun seems to have moved, relative to the stars, counterclockwise. Actually, the earth has moved; it's the line of sight from the earth to the sun that has changed.

Point 5 makes retrograde motion a natural result of the planets' revolutions. Retrograde motion arises from the planets chasing one another around, the faster (inner) planets regularly passing the slower (outer) ones. An analogy: when you pass a car on a highway, the slower car appears to move backward with respect to background objects. Similarly, as the earth speeds past another planet, that planet seems to move westward (backward) against the backdrop of stars.

Retrograde Motion Explained Naturally (Learning Outcome 3-3)

In the Ptolemaic model, epicycles moving on deferents generated the retrograde motions. Since retrograde motion arose naturally in his model, Copernicus eliminated five Ptolemaic epicycles and made a simpler model in which the relative motions of the earth and planets produced retrograde motions. When the earth passed any of the outer planets or when the inner ones passed the earth, retrograde motion occurred. The *passing* is the key to retrograde motion in this model.

Take Mars as an example (Fig. 3.4). Its retrograde motion occurs at opposition. In a heliocentric model, opposition happens when the earth is in line between the planet and the sun. As the earth approaches, the line of sight from the earth to Mars moves eastward. But as the earth passes Mars, the line of sight swings westward relative to the stars. As the earth moves on, the line of sight eventually moves eastward again. Mars undergoes the illusion of retrograde motion as the earth passes it. Note (Fig. 3.4) that the earth and Mars are closest together in the middle of the retrograde motion, so Mars should be brightest in the sky then – and it is! The same explanation applies to the motions of Jupiter and Saturn.

The identical chase-and-pass scenario results in the retrograde motions of Venus and Mercury. These inner planets pass the earth because they move faster around the sun. Consider Venus. When Venus moves around the back of the sun (from greatest western elongation), it appears to move eastward. But as Venus catches up to the earth (from the east side of the sun) and passes it, you observe Venus moving westward with respect to the stars (toward the west side of the sun). For these two inner planets, retrograde also arises from relative motion, but the role of the earth is reversed. Note that the passing of any two planets produces the illusion of retrograde motion of one as seen from the other. If you stood on Venus and viewed the earth, it would appear to undergo retrograde motion at opposition!

Copernicus aimed for simplicity in his model. This aesthetic goal is common to scientific models. It is an essential standard by which to judge the best among competing ideas.

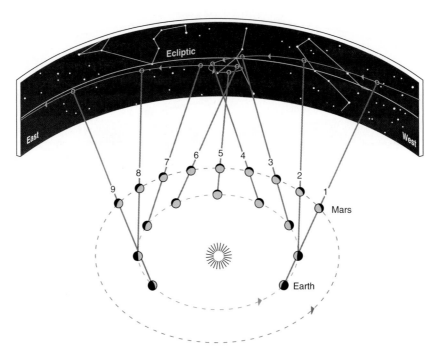

Figure 3.4 Retrograde motion of Mars in the Copernican model. The numbered positions show one-month intervals. As the earth comes around the same side of the sun as Mars, it is moving faster along its path and so passes Mars at 4, 5, and 6. During this passing interval, Mars appears to move backward (to the west) with respect to the stars. Since Mars appears in the middle of its retrograde motion just as the earth passes it, Mars is in opposition to the sun as viewed from the earth. Also, it is closest to the earth and so appears brightest in the sky. The illusion of retrograde motion results from the passing situation.

Planetary Distances
(Learning Outcome 3-4)

Using the planets' observed synodic periods, Copernicus achieved another basic goal: the establishment of the order and distances of the planets from the sun. (The **synodic period** is the time required for a planet in some alignment with the sun as seen from the earth – such as opposition – to move around the sky and return to that same alignment.) From the synodic periods, Copernicus calculated the **sidereal periods:** the periods of revolution of the planets with respect to the stars as seen from an observer at the sun (Enrichment Focus 3.1). Examples of synodic periods are: (1) Venus from one greatest western elongation to the next (584 days) and (2) Mars from one opposition to the next (780 days). Their sidereal periods are 225 days and 687 days.

Copernicus found that the planetary order from the sun falls into a natural sequence when based on sidereal periods: Mercury, with the shortest period, is closest to the sun; Saturn, with the longest, is the farthest away. He savored the harmony in this sequence, which later scientists – such as Kepler and Galileo – considered the essential elegance of the heliocentric model.

Copernicus also calculated the distances of the planets from the sun relative to the earth–sun distance. (Our term **astronomical unit,** abbreviated AU, is one earth–sun distance; the value was not well determined until the nineteenth century.) The distances Copernicus used came from direct observations and simple geometric arguments (Table 3.1). He was very pleased to relate the motions of the planets to their distances from the sun. Let's see how this is done, for it forged a crucial difference from the Ptolemaic model.

Relative Distances of the Planets
(Learning Outcomes 3-4, 3-5)

The planets' distances from the sun, relative to the earth–sun distance (the AU) can be found directly from observations in the Copernican model. The method, which assumes circular orbits, differs for planets interior and exterior to the earth.

ENRICHMENT FOCUS 3.1
Sidereal and Synodic Periods in a Heliocentric Model

From the earth you can see a planet in a particular alignment with the sun – opposition, for example. The time interval for that planet to return to the same alignment with respect to the sun in the sky is called the *synodic period*. In contrast, consider the time it takes for a planet to return to a specific location in the zodiac as seen from the sun. That's the planet's *sidereal period*.

The synodic period is a geocentric property, while the sidereal period is a heliocentric one. How can you transform synodic periods into sidereal ones? Assume that the planets move in circles and that the closer a planet's path is to the sun, the faster the planet moves. Its sidereal period is shorter. You then have a chase situation, in which an inner planet always catches up to an outer one.

Imagine that two adjacent planets start out lined up (Fig. F.5), with the outer planet in opposition as seen from the interior one. How long until they are aligned again? Note that the inner planet will make one circuit around the sun and then catch up to the outer one to reform the alignment. In other words,

the inner planet laps the outer one. So as seen from the inner planet, the synodic period will be longer than its sidereal one.

The inner planet moves at $360/P_i$ degrees per day, where P_i is the sidereal period of the inner planet. The outer planet, moving at the slower rate of $360/P_o$ degrees per day, lags behind the inner planet more and more each day, where P_o is the sidereal period of the outer planet. At the end of one day, the inner planet has gained an angle of $(360/P_i - 360/P_o)$ degrees on the outer one. This day's gain is equal to $360/S$ degrees, where S is the synodic period of either planet seen from the other. (These periods are the same.) In algebraic terms,

$$\frac{360°}{S} = \frac{360°}{P_i} - \frac{360°}{P_o}$$

or, by dividing by 360°,

$$\frac{1}{S} = \frac{1}{P_i} - \frac{1}{P_o}$$

Take the earth as the inner planet and any planet exterior to it as the outer one. Since the earth's sidereal period is one year, $P_i = 1$; then

$$\frac{1}{S} = 1 - \frac{1}{P_o}$$

where P_o is the sidereal period of the outer planet, and S, its synodic period as seen from the earth, is in years.

Now consider the earth to be the outer planet and choose any inner one. Then $P_o = 1$ is the earth's sidereal period, and

$$\frac{1}{S} = \frac{1}{P_i} - 1$$

where all the quantities are in years. Use Venus as an example. The synodic period of Venus is 585 days, or $585/365 = 1.60$ years. So

$$\frac{1}{P_{Venus}} = 1 + \frac{1}{1.60} = 1.62$$
$$P_{Venus} = 0.62$$

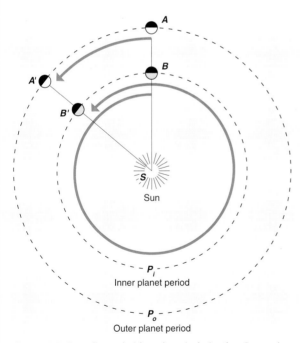

Figure F.5 Synodic and sidereal periods in the Copernican model. As well as traveling a shorter distance, the inner planet moves faster along its path and so must have a shorter sidereal period (P_i) than the outer planet (P_o).

This transformation from synodic to sidereal periods played a key role in the Copernican model, because it places the planets in a natural order from the sun, from the shortest (Mercury) to the longest (Saturn) periods.

Table 3.1 Copernicus' Relative Distances of the Planets

Planet	Copernicus' Value (AU)	Modern Value (AU)
Mercury	0.38	0.387
Venus	0.72	0.723
Earth	1.00	1.00
Mars	1.52	1.52
Jupiter	5.22	5.20
Saturn	9.17	9.54

For Mercury and Venus, the inner planets, the method rests on the observed maximum elongation angle of the planet from the sun (discussed in Section 1.4). Remember that this angle averages about 23° for Mercury and 46° for Venus.

Here's the procedure for Venus. In a heliocentric model, maximum elongation occurs when the line of sight from the earth to Venus hits tangent to Venus' orbit (Fig. 3.5). The line from the sun to the line of sight makes a right angle. Draw a triangle with one side the sun–Venus distance (*SV*), another side the sun–earth distance (*SE* equals 1 AU), and the third side the earth–Venus line (*EV*). You've drawn a right triangle.

Now observe the elongation angle, *SEV.* Let *SE* be one unit long. Then draw a triangle to scale with *SEV,* and Venus' distance from the sun (*SV*) will be 0.72 the earth's (*SE*). Since *SE* is 1 AU, Venus is 0.72 AU from the sun. Similarly, use 23° as the maximum elongation angle for Mercury to find its distance as 0.39 AU. Note that just one observation establishes the distances of the inner planets in a heliocentric model.

A different method applies to the outer planets. Take Mars as an example. Start with some alignment of the earth and Mars (*A* in Fig. 3.6a). Wait one sidereal period of Mars. By the time the earth has returned to the same position in its orbit relative to the stars as seen from the sun (Enrichment Focus 3.1), our planet – moving faster – will have gone around more than once (*B* in Fig. 3.6b). Find the position of Mars from these two observations from the earth by extending the lines of sight for both. The point at which the lines intersect is the position of Mars in its orbit (*M* in Fig. 3.6c).

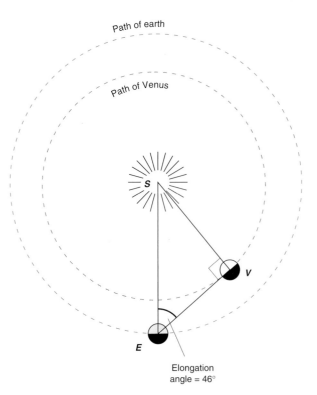

Figure 3.5 The distance of Venus from the sun. As seen from the earth at its time of greatest elongation, Venus makes an angle (*SEV*) of about 46° from the sun. The line of sight from the earth then is tangent to the orbit of Venus, so the radius of Venus' orbit drawn to this point makes a right triangle with one side the sun–Venus distance (*SV*) and the other the earth–sun distance (*ES*).

Suppose you have made an accurate scale drawing of these observations. Calling the earth–sun distance 1 AU, you can also measure the sun–Mars distance in AU. Or you can solve the triangle *MSB* for side *SM* using trigonometry, because you know all the angles. A similar technique works for the other outer planets. These distances (Table 3.1) are *fixed* by the planets' sidereal periods and the geometry of the heliocentric model.

Problems with the Heliocentric Model (Learning Outcome 3-5)

Copernicus eliminated epicycles to explain retrograde motion and expelled the equant, yet the heliocentric model did *not* predict planetary positions any better than the Ptolemaic model. In fact, the Copernican model was not even simpler, judged on the basis of a count of circles. Copernicus eliminated five, but he was forced to add many smaller ones to accommo-

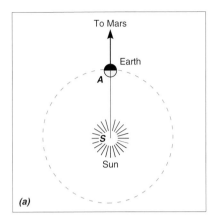
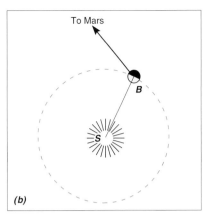
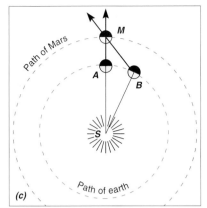

Figure 3.6 The distance of Mars from the sun determined in a heliocentric model. (a) Start with the earth at *A* and Mars at opposition. We know the angle of Mars relative to the sun, but we don't know the distance. (b) Wait one sidereal period of Mars. Then Mars returns to the same place in its orbit (as seen from the sun), but the earth has gone around almost twice (to *B*). (c) The line of sight from the earth (at *B*) to Mars crosses the Martian orbit at the same spot (*M*) that the earlier one did (from *A*). Set the earth–sun distance at 1 unit. The distance from the sun to Mars is known on this scale because all the angles are known.

date variations in planetary motions. (Ptolemy had accounted for these variations with equants.) By a simple count of the total of circles involved, the complete Copernican model is somewhat more complicated than Ptolemy's original model.

Copernicus also sidestepped the fact that his model violated Aristotelian physics. And he did not offer new physical ideas to support his model; those still needed development. Copernicus had created a new *geometric* model, not a *physical* one.

> **CAUTION.** Don't fall into the trap of thinking that the Copernican model included a force between the sun and the planets. It did *not!* Planetary motions were believed to arise from the natural motions of the celestial spheres – uniform rotation. As in the Ptolemaic model, this was understood as natural motion without forces.

The Impact of the Heliocentric Model (Learning Outcomes 3-4, 3-5)

Although Copernicus' claims clashed with tradition, a few astronomers in the middle of the sixteenth century tested the new setup. They knew that the venerable *Alfonsine Tables* (based on the Ptolemaic model) contained some inaccurate positions.

Foremost among those willing to try the new model was Erasmus Reinhold, who had encouraged Rheticus, his fellow professor at Wittenberg, to visit Copernicus. Reinhold made up planetary position tables based on the new model; these were adopted by almanac makers as the *Prutenic Tables*. Reinhold did not swing completely to the Copernican model, however. He saw it as more geometrical than physical. Because Copernicus clung to uniform circular motions, his model was geometrically similar to Ptolemy's (except for the different placement of earth and sun). What astronomers would switch if the new model offered no advantages over the old?

Why did Copernicus bother to develop his model at all? Certainly he was motivated by aesthetic reasons: he saw a new harmony in the "design of the universe and fixed symmetry in its parts" that was "pleasing to the mind." The "fixed symmetry" referred to the spacing of the planets as determined by observations. In the Ptolemaic model, each planet's circle is independent of the others. Overall, Copernicus' model had a unity lacking in the Ptolemaic one, so it was simpler in grand design. Aesthetics mattered – a lot!

In 1616 the Catholic Church placed *De revolutionibus* on the Index of Forbidden Books because of specific passages, which were supposed to be changed by the censors. Very few of the copies were changed, however, with most of the censored books being found in Italy. So the printing of Copernicus' ideas resulted in their rapid spread in Europe, intellectual sparks that kindled new ideas.

3.3 TYCHO BRAHE: FIRST MASTER OF ASTRONOMICAL MEASUREMENT

Tycho Brahe (1546–1601) was born in Denmark into a family of noble standing. While he was a child, his uncle – who had no children – stole him from his parents, supported the boy, and decided that his adopted son should become a lawyer. Young Tycho was sent off to study law, but secretly he worked on astronomy, spent his allowance on astronomy books, and sneaked out at night to make observations.

Later in his career, Tycho (Fig. 3.7) rejected the heliocentric model on both physical and observational grounds. He proposed an alternative model that mixed the classic geocentric model with the Copernican one. His work is infused with the attitudes typical of a professional astronomer just after the publication of *De revolutionibus*. It revealed that the reality of the Copernican model had escaped most astronomers, who saw the heliocentric model as merely a geometric device for calculating planetary positions. Copernicus invited this stance because his model violated Aristotelian physics and offered no alternative physical basis.

The New Star of 1572

In November 1572 a new star, called then a **nova**, burst into view in the constellation of Cassiopeia. At its brightest, it could be seen during the day. (Today we know that a nova is a star normally too faint to be seen that suddenly increases many times in brightness. Such an apparently new star is called a *nova* from the Latin *stella nova*, "new star.")

Tycho observed this new star for more than two years and collected observations from all over Europe of the star's position in the sky. Using others' data as well as his own, Tycho showed that the nova was in the same place in Cassiopeia regardless of where it was observed. That is, the nova did not have an observable angular shift from different places on the earth, so it was at a great distance from the earth. Tycho concluded that the nova had to lie in the sphere of stars, thus rebutting the Aristotelian tenet that the heavens never changed.

Tycho's work on the new star of 1572 catapulted him to fame as an astronomer, and King Frederick II of Denmark offered Tycho an observatory on the island of Hven. With royal funds – estimated by Tycho to be more than a ton of gold – Tycho built on Hven the first modern astronomical observatory, Uraniborg, Castle of the Heavens. Here he worked in grand style with the finest observing equipment (designed by himself) and a crew of assistants. His instruments had an accuracy of 1 arcmin (as good an angular precision as can be had with the naked eye).

At Uraniborg, Tycho began the first comprehensive observational program. The planets were observed not only at times of importance, such as during retrogrades, but also in between notable events. No one had tried for this continuity before. Tycho eventually amassed records of precise planetary positions covering the years 1576 to 1591.

Tycho's Hybrid Model

Tycho concluded that the nova of 1572 made the classic geocentric model untenable, so he devised his own. He had read about the Copernican model, but he could not accept the idea of the earth revolving around the sun. In Tycho's model, the moon and sun revolved around the earth, but all the other planets revolved around the sun (Fig. 3.8). This model was geometrically the same as the Copernican one – only the one center of motion was changed.

Also, Tycho had tried to measure a stellar parallax (Section 2.2) but he failed to detect it. The lack of a detectable parallax fortified Tycho's belief that the Copernican model was invalid. (We now know that the largest heliocentric parallaxes are almost 100 times smaller than Tycho could have detected with his best instruments.)

When Frederick II died, Tycho fell from royal favor because of his autocratic ways. He moved his equipment to Prague, where he worked for Emperor Rudolf of Bohemia and took on the young Johannes Kepler as an assistant. When Tycho died in 1601, an era in astronomy ended. Another was born when Kepler created a new astronomy based on Tycho's observations.

Figure 3.7 Engraving of Tycho Brahe at work, surrounded by some of his wonderful "machines for the reform of astronomy." From Joan Blaeu's *Atlas Major,* published in Amsterdam in 1662.

3.4 JOHANNES KEPLER AND THE COSMIC HARMONIES

Ptolemy's model, because it kept the earth in the center of the cosmos, upheld traditional physical ideas. In contrast, the Copernican model violated concepts of Aristotelian physics. The heliocentric model lacked a physical basis, judged by the standards of the time.

Johannes Kepler (1571–1630) forged new ideas about planetary motion that built the foundation of modern cosmological concepts. Kepler (Fig. 3.9) reshaped the Copernican model into a truly physical one in which the sun determined the planetary orbits by a force between the sun and the planets.

Kepler also simplified the Copernican model, perfected its usefulness for the prediction of planetary positions, and moved it to the realm of modern ideas about the solar system.

The Harmonies of the Spheres (Learning Outcome 3-6)

Born on December 27, 1571, in the small town of Weil der Stadt in southwestern Germany, Kepler in 1589 entered the University of Tübingen. He was an outstanding student. In his last year at Tübingen, he was selected to replace a teacher of mathematics at a Protestant high school in Graz, Austria, where he then made his first key astronomical discoveries.

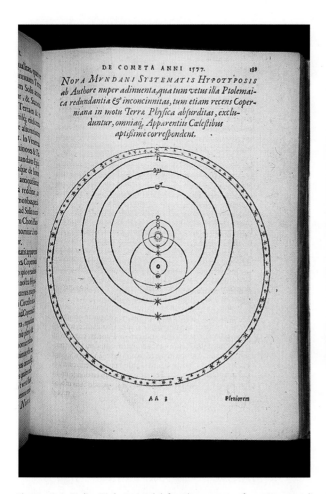

Figure 3.8 Tycho Brahe's model for the cosmos, from *De mundi aetherei recentioribus phaenomenis*, printed by Brahe himself in 1588. It is a geocentric modification of the Copernican one. The sun moves around the earth and all planets circle around the sun.

Figure 3.9 Johannes Kepler as a young man (artist and date unknown): "Astronomy has two ends, to save the appearances and to contemplate the true form of the edifice of the world."

One day while teaching, Kepler was struck by the most compelling insight of his life. As he was telling his class about conjunctions of Jupiter and Saturn, which occur at regular intervals, he saw that the ratio of the radii of two circles – one drawn around a triangle and one drawn inside it – is about 2 to 1. That is almost the same as the ratio of the distances of Saturn and Jupiter from the sun in the Copernican model, 1.8 to 1. Kepler found this result exciting, for Saturn was the first planet in from the "sphere of stars," and the triangle was the simplest plane figure. He felt he had found a geometrical design for the solar system.

While still at Tübingen, Kepler had been delighted by the harmonies he found in the Copernican model, which he adopted. One such harmony appears in the spacing of the planets and their sidereal periods around the sun. Copernicus, however, had not explained this relationship. Kepler thought he could: perhaps a geometric layout determined the planets' spacings.

Kepler approached this idea by first considering plane geometric figures. Then he realized that the situation was actually three-dimensional and required solid figures. From classical geometry, Kepler knew that only five regular solids existed, those with faces having the same kinds of regular plane figure. The known planets numbered six, separated by five spaces, so Kepler felt that these five regular solids might establish the spacing of the planetary spheres. He worked out a nesting scheme of solids and spheres that gives distances from the sun within a few percent of those given by Copernicus. Kepler's intuition later proved to be wrong – the discovery of any new planet shoots it down – but it energized his later study of planetary motions. It led to radical results.

With great excitement, Kepler wrote to his former astronomy teacher, Michael Maestlin, about this idea and another: the sun must have some power or force to propel the planets in their orbits. This conviction seemed reasonable because Mercury, for example, moves faster in its orbit than Venus. A force from the sun would push Mercury harder than Venus, because Mercury is closer to the sun. Kepler failed to work this idea out at the time, but it was firmly implanted in his mind.

In 1594 Kepler published the results of his work in the *Mysterium cosmographicum* (*The Cosmic Mystery*), which announced the details of his nesting scheme and declared his adherence to the Copernican model. It was the first such treatise by a professional astronomer. His intuition that the sun physically directs planetary motions set the stage for the next advance of astronomy in terms of physical laws.

3.5 KEPLER'S NEW ASTRONOMY

Kepler was not yet satisfied with his geometric model. He needed better observational data to calculate the thickness of the nesting shells. To obtain them, Kepler turned to Tycho Brahe, who had read the *Mysterium cosmographicum* but did not like its mystical approach to astronomy. When Tycho received Kepler's request for observations, he invited Kepler to come to Prague to discuss the matter personally. Tycho and Kepler met on February 4, 1600, an encounter fateful for them both and for the progress of astronomy. Their personalities clashed so strongly that they could not work on astronomy together.

Later, Tycho was at a party with a baron and drank too much. Because it was impolite to leave during the affair, Tycho did not relieve himself and as a result suffered a urinary infection that led to his death. Just before he died, however, Tycho relented in his hostility to Kepler. He consented to Kepler's use of his observations but urged Kepler to support his cosmological model with these data. After Tycho's death, Kepler was promoted to Tycho's position and allowed access to his unpublished records. With them, Kepler first attacked the problem of the orbit of Mars.

The Battle with Mars (Learning Outcome 3-7)

Kepler first tried to match Tycho's observations with a heliocentric scheme that had a circular orbit, an eccentric, and an equant. With this model, he could fit Tycho's data to within 2 arcmin along the ecliptic, but his predictions above and below the ecliptic were wide of the mark. This discrepancy struck Kepler as important – he realized that he must treat the orbit in three dimensions. He then moved the center of the earth's orbit and fitted positions above and below the ecliptic correctly but was 8 arcmin off along the ecliptic. Kepler rejected this result, for he knew that Tycho's observations were accurate to 1 arcmin. This demand – that his model's predictions be as good as the observations – marked the first time anyone seriously set this tough standard in astronomy. It qualifies Kepler as the first *modern* physicist.

Frustrated, Kepler pondered the notion that the sun drove the planetary motions. But by what force? Influenced by William Gilbert's book *De magnete*, Kepler imagined the sun showering out a magnetic force that propelled the motion of the planets. This idea required that a planet move faster in its orbit when it was closer to the sun.

Kepler found that a magnetic force idea worked out if the orbit of Mars was *elliptical* rather than circular. He wrote, "With reasoning derived from physical principles agreeing with experience, there is no figure left for the orbit of the planet except a perfect ellipse." Kepler had found one of his three famous laws of planetary motion, which rested on a physical explanation of Tycho's precise observations.

Properties of Ellipses
(Learning Outcomes 3-8, 3-9)

To understand Kepler's laws, you need to have in mind the basics of the geometrical properties of ellipses. (Enrichment Focus 3.2 provides the details.) Ellipses are a lot like circles, so their basic properties are fairly easy to grasp. You can make an ellipse by taking a loop of string and holding it down on a board with two tacks. Keeping the string taut with a pencil held against it, draw a curve around the tacks (Fig. 3.10). That curve is an ellipse. The two tacks mark the two foci of the ellipse. Each point on the ellipse has the property that the sum of its distances from the two foci is the same as the sum of those distances for every other point. The line through the foci to both sides of the ellipse is called the **major axis.**

ENRICHMENT FOCUS 3.2
Geometry of Ellipses

Ellipses are related to circles. If you draw a curve as shown in Figure 3.10, the two tacks mark the two foci (F_1 and F_2) of the ellipse. Each point on the ellipse has the property that the sum of its distances from the two foci (F_1 to P and F_2 to P in Fig. F.6) is the same. The line through the foci to both sides of the ellipse (R_a to R_p) is called the **major axis.** Half this length is the **semimajor axis,** usually designated by a; the major axis has a length of $2a$. The distance from the center of the ellipse to a focus is designated c, so the distance between the two foci is $2c$.

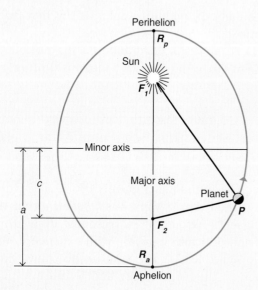

Figure F.6 Geometrical properties of an ellipse.

The most essential property of an **ellipse** is its eccentricity. To define it exactly, the eccentricity, e, is

$$e = \frac{c}{a}$$

If the sun is at one **focus** (F_1) of a planet's orbit (Fig. F.6), then R_p, the closest point to the sun, is the **perihelion** point in the orbit. The perihelion distance R_pF_1 is

$$\overline{R_pF_1} = a - c = a - ae = a(1 - e)$$

In comparison, R_a, the farthest point from the sun, is the **aphelion** point. The aphelion distance R_aF_1 is

$$\overline{R_aF_1} = a + c = a + ae = a(1 + e)$$

Note that a is the average distance from one of the foci of a point moving on the ellipse. So for planetary orbits, a is the average distance of a planet from the sun.

As an example, the orbital eccentricity of Mercury is 0.21, and its semimajor axis is 0.39 AU. Then Mercury's perihelion distance is

$$R_p = a(1 - e)$$
$$= 0.39(1 - 0.21)$$
$$= 0.39(0.79)$$
$$= 0.31\,\text{AU}$$

Similarly, the aphelion distance is

$$R_a = a(1 + e)$$
$$= 0.39(1 + 0.21)$$
$$= 0.39(0.21)$$
$$= 0.47\,\text{AU}$$

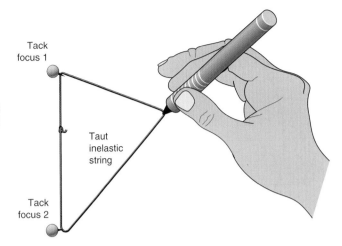

Figure 3.10 Drawing an ellipse with a loop of string around two tacks on a board. The pencil must hold the loop taut at all times to draw the proper curve.

Tack
focus 1

Taut
inelastic
string

Tack
focus 2

The amount by which an ellipse differs from a circle is defined as an ellipse's **eccentricity.** Imagine that you moved the two tacks at the foci closer together. The ellipse would become more circular. In fact, when the two tacks coincided, the ellipse would become a circle, with an eccentricity of zero. Moving the two tacks farther apart would increase the eccentricity until the tacks reached the limits of the string. The eccentricity would then be 1. Note that the eccentricity increases as the distance between the foci does. The most essential property of an ellipse is its eccentricity.

Kepler's Laws of Planetary Motion (Learning Outcomes 3-8, 3-9, 3-10)

Today, Kepler's fame rightly rests on his three laws, which broke the tradition of uniform motion on circular orbits. Yet Kepler's laws of planetary motion were small fragments of his wider vision for harmonies in the physics of celestial motions. Let's look briefly at each law in modern terminology.

Law 1. Law of Ellipses (1609): The orbit of each planet is an ellipse, with the sun at one focus (Fig. 3.11). The other focus is located in space. The distance from the sun to the planet varies as the planet moves along its elliptical orbit.

Law 2. Law of Equal Areas (1609): A line drawn from a planet to the sun sweeps out equal areas in equal times. Physically, this law holds that the orbital speeds are nonuniform but vary in a regular way: the farther a planet is from the sun, the more slowly it moves (Fig. 3.12).

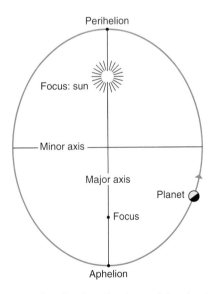

Figure 3.11 Kepler's first law. The shape of the planetary orbits is elliptical (exaggerated here), with the sun located at one focus of the ellipse. The closest point along the orbit to the sun is called the *perihelion;* the farthest, the *aphelion.* The sum of the distances of any point on the ellipse from the two foci is a *constant.*

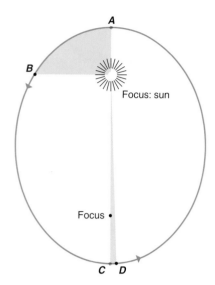

Figure 3.12 Kepler's second law. Consider two equal periods of time: in one the planet is closest to the sun (*AB*) and in one it is farther away (*CD*). At AB, the planet–sun distance is shortest and the planet moves fastest in its orbit. At *CD*, the planet moves more slowly because it is farther from the sun. In both cases, the areas (shaded areas *AB* to the sun and *CD* to the sun) covered by the line drawn from the planet to the sun are equal.

Law 3. Harmonic Law (1618): The square of the orbital period of a planet is directly proportional to the cube of its average distance from the sun (Fig. 3.13). This law reveals that the planets with larger orbits move more slowly around the sun and so implies that the sun–planet force decreases with distance from the sun.

What do these laws mean?

Well, for the first time we can talk about physical **orbits** of the planets and the patterns related to them. Law 1 points out that the shape of the orbits is elliptical, so a planet's distance from the sun varies. (Note that the sun does not lie in the center of the ellipse but at one focus.) It also shows that the planetary orbits follow a special and regular geometrical shape. Note that, as observed from the sun, a planet's angular speed would vary inversely with distance (even if its orbital speed were constant).

Law 2 notes that as a planet's distance from the sun varies, so does its orbital speed: the closer a planet is to the sun, the faster it goes. These regular variations in orbital speeds, along with the changes in distance, explain the variations observed in angular speeds. Gone is the equant, or any device like it! You also get the sense that the sun pulls a planet toward it; then the planet whips around the sun and slows down as it moves away – a force from the sun acts upon it. This law also allows us to find the planet in another part of the orbit once it has been observed at some position.

Law 3 states that the more distant a planet's orbit is from the sun, the slower its average orbital motion will be. Look at Table 3.2. You can see that a greater distance from the sun means that a planet takes much longer to orbit the sun once. You can also note that if you square the orbital period (in years) and divide by the cube of the distance (in AU), the ratio comes out the same for all the planets – unity, in these units. The fact that this ratio is the same for such different orbits – a common pattern – suggests a similar physical cause.

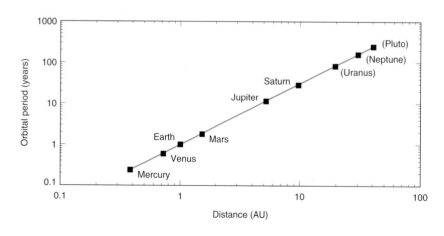

Figure 3.13 Kepler's third law. A planet's orbital period (once around the sun) is related to its average distance from the sun. For each planet now known, this diagram plots the orbital period (in years) squared to its distance (in AU) cubed; note the powers of ten axes. The line indicates the values expected from the modern formulation of Kepler's third law. The points for the planets fall along the same straight line, even Uranus, Neptune, and Pluto, which were not known in Kepler's time.

Table 3.2 Kepler's Third Law

Planet	Period (P, in years)	P^2	Distance (a, in AU)	a^3	P^2/a^3
Mercury	0.24	0.058	0.39	0.059	0.97
Venus	0.62	0.38	0.72	0.37	1.0
Earth	1.0	1.0	1.0	1.0	1.0
Mars	1.9	3.6	1.5	3.4	1.1
Jupiter	12	140	5.2	140	1.0
Saturn	29	840	9.5	860	0.98

Both laws 2 and 3 imply that a force between the sun and the planets weakens with increasing distance. Hidden in law 3 is an exact description of how the sun–planet force decreases with greater distance, but Kepler was not able to figure it out. You can catch a glimpse of this by examining Figure 3.13, which plots the data from Table 3.2. Note that all the points fall along the same line – even Uranus, Neptune, and Pluto, which were not known at the time. (Later, Isaac Newton figured out the force: see Section 4.4.)

Law 3 can be written algebraically as

$$P^2 a^3 = k$$

where P is the orbital period, a the average distance from the sun, and k a constant. If P is in years and a in AU, then k equals 1. For any body orbiting the sun (even spacecraft!), law 3 can be used to find the average distance from the orbital period or the orbital period from the average distance. For example, if $a = 4$ AU, then $P = 8$ years. This result depends only on a; no amount of eccentricity of the orbit can change the orbital period. If the orbiting mass is much smaller than the mass it orbits, the result does not depend on the mass of the orbiting object.

Note what Kepler's laws did to the planetary orbits. No circles, nor equants, nor uniform orbital speeds! Yet Ptolemy and Copernicus did not miss by much. The actual orbits have small eccentricities (Fig. 3.14) – they are very close to circles. In fact, offset circles mimic these ellipses fairly well. Kepler's planets toe the line of ellipses exactly, just enough to differ from circles. Circles are not a bad first guess. The eccentricity of an ellipse does them in.

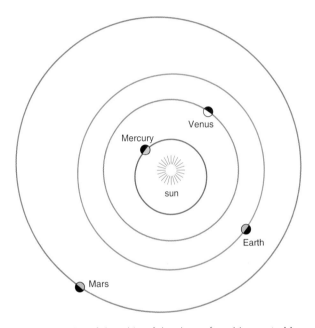

Figure 3.14 Plot of the orbits of the planets from Mercury to Mars, to scale. Note how they appear circular, with centers offset from the sun. Mars is the most noticeable with the largest eccentricity.

The New Astronomy (Learning Outcomes 3-7, 3-10)

As he announced in *Astronomia nova* in 1609, Kepler had developed astronomy based on geometry and on physics (a force between the sun and the planets). Elliptical orbits satisfied both his intention and the observations of Tycho Brahe. Using ellipses, Kepler calculated the *Rudolphine Tables* (1627), which supplanted both the *Prutenic Tables* based on the Copernican model and the *Alfonsine Tables* based on the Ptolemaic one.

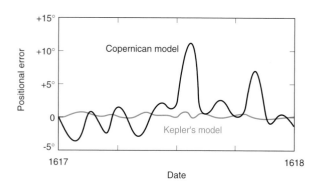

Figure 3.15 Accuracy of predictions of planetary positions: difference between predicted and actual positions of Mercury in 1617 and 1618 in Kepler's model (heliocentric, elliptical orbits, nonuniform speeds) and in the Copernican one (heliocentric, circular orbits, uniform speeds). In the Copernican model, the error is as large as 10°. For Kepler's scheme, the maximum error is 10 times smaller. (Adapted from a diagram by Owen Gingerich and Barbara Welther.)

Kepler's *Rudolphine Tables* completely revised the heretofore mediocre science of astronomy. Earlier, a few degrees of error between predictions and observations was considered to be "good enough." Kepler's calculations were some 10 times more accurate than those based on either the Copernican or the Ptolemaic model (Fig. 3.15). The reason was that the older models assumed uniform, circular motions for the planets, whereas, the motions are neither circular nor uniform.

Kepler broke the ancient spell of perfect circles and uniform motion that had mesmerized astronomers for centuries. He rewove the fabric of the heavens into a pattern that proclaimed the end of an ancient era and the birth of astronomy as a physical science.

KEY CONCEPTS

1. Copernicus proposed his heliocentric model mainly for aesthetic reasons; he especially disliked the equant in the Ptolemaic model, which violated the traditional precept of uniform heavenly motions about circles. Copernicus believed that, overall, his model was more pleasing than a geocentric one.

2. The Copernican model (Table 3.3) views the daily motion of the sky as arising from the earth's rotation, the sun's eastward motion along the ecliptic as a reflection of the earth's revolution around the sun, and retrograde motion as an illusion when one planet passes another. The order of the planets is set by their periods of revolution about the sun; the planets' distances are set by observations and simple geometrical layouts.

3. Copernicus' model did not predict planetary positions any better than the Ptolemaic

model did, nor was it simpler in terms of the total of circles used. It also violated the accepted physics of the day (that of Aristotle), and it required a stellar parallax, which was not observed. All these points made the Copernican model quite unattractive in principle to contemporary astronomers.

4. Tycho Brahe typified the reaction of practicing astronomers toward the Copernican model. He thought it an interesting geometric idea but could not accept a moving earth. Tycho tried to observe stellar parallax but didn't detect it. He eventually developed a hybrid model with the earth at the center of the cosmos. Tycho did, however, compile a years-long series of accurate and continuous observations that formed the basis of Kepler's incisive work.

5. Kepler used Tycho's observations and tried to match them with the Copernican model. His failure to do so within the errors of Tycho's fine observations drove him to decide that planetary orbits are elliptical (first law); that the planets move around the sun nonuniformly but predictably (second law); and that the farther a planet is from the sun, the longer it takes to complete one orbit (third law). Kepler held that the sun exerts a force that keeps the planets in their orbits.

6. Kepler's model predicted planetary positions much more accurately than either the Ptolemaic or the Copernican model, mainly because Kepler broke with the long tradition of uniform, circular motions for celestial objects. He was driven to this radical approach by his faith in the reliability and accuracy of Tycho's observations. Here we see how good observations can work to change a scientific model.

7. Kepler's notion that astronomy should be based on physics as well as on geometry produced the crucial concept that planetary motions and the layout of the solar system arose from a force between the sun and the planets. The planets actually moved on orbital paths, and the sun compelled and controlled their motions.

Table 3.3 Comparison of Ptolemaic (P), Copernican (C), and Keplerian (K) Models

Observation	Explanation
Motion of entire heavens daily from E to W	P: Motion of all heavenly spheres E to W C: Reflection of rotation of earth from W to E K: Same as C
Annual motion of sun W to E through zodiac	P: Rotation of sun's sphere W to E in a year C: Reflection of annual revolution of the earth about the sun K: Same as C
Nonuniform motion of sun through zodiac	P: Circular path of sun eccentric, with uniform speed C: Same as P K: Elliptical orbit of the earth, with nonuniform speed
Retrograde motions of the planets	P: Epicycles and deferents C: Relative motions of planets, including earth, around sun K: Same as C
Variations in retrograde motions	P: Equant, eccentrics C: Small epicycles, eccentrics K: Nonuniform orbital motion and elliptical shape of orbits
Distances of planets	P: Arbitrary as long as angular relationships are correct C: Relative distances set by observations K: Distances related to periods
Cause of planetary motions	P: Natural motion of celestial spheres; no force C: Same K: Magnetic force from sun
Accuracy of prediction	P: Error typically 5° or less, sometimes 10° C: Same as P K: Error generally about 10', sometimes as large as 1°

STUDY EXERCISES

1. What was Copernicus' primary objection to the Ptolemaic model? In what essential way were *De revolutionibus* and the *Almagest* similar? (Learning Outcomes 3-1 and 3-2)
2. How did Copernicus account for retrograde motion in his model? (Learning Outcome 3-3)
3. If you were on Mars, when would you see Jupiter retrograde? The earth? (Learning Outcome 3-3)
4. What is the difference between a planet's synodic and sidereal periods? How can sidereal periods be found from synodic ones? (Learning Outcome 3-4)
5. In your opinion, what was the major advantage, if any, of the Copernican model over the Ptolemaic one? (Learning Outcomes 3-1 and 3-5)
6. What force keeps the planets moving around the sun in the Copernican cosmos? (Learning Outcomes 3-6 and 3-8)
7. How did Copernicus' model deal with the not-yet-observed stellar parallax? (Learning Outcome 3-5)
8. Give at least two key differences between the models of Copernicus and Kepler. (Learning Outcome 3-9)
9. What two main advantages did Kepler's model have over that of Copernicus? (Learning Outcome 3-9)
10. According to Kepler's second law, a planet travels slowest at what point in its orbit? (Learning Outcome 3-10)
11. In what sense did Kepler develop a "new" astronomy? (Learning Outcome 3-7)
12. Keeping in mind that the orbit of Mercury is an ellipse with a large eccentricity, explain the fact that the maximum elongation angle of Mercury can vary over a few degrees. (Learning Outcomes 3-8 and 3-10)
13. Imagine that a new planet is discovered between Mars and Jupiter. Using Table 3.2, what would you predict that this planet would have for the ratio of P^2/a^3? (Learning Outcome 3-10)
14. Judge whether each of the following statements is model dependent or model independent:
    a. The sun lies in the center of the cosmos.
    b. A force moves the planets in their paths around the sun.
    c. The shape of the planetary paths is circular.
    (Learning Outcomes 3-3, 3-5, and 3-9)

PROBLEMS AND ACTIVITIES

1. The maximum elongation angle of Mercury averages 23°. Calculate the average Mercury–sun distance in AU. *Hint:* sin *SEM* = *SM/SE*.
2. As an alternative procedure for Problem 1, draw on graph paper a circle with a radius of 10 cm. Mark the earth at some position on the circumference and the sun at the circle's center. Now follow Figure 3.5. Draw a line between the sun (*S*) and the earth (*E*). With a protractor, find an angle of 23° outward from the earth–sun line and draw a line marking it. Find on this line the position at which another line from the sun strikes it at 90°. Measure the distance from the sun to this point. What is your result for the sun–Mercury distance? It should be close to 4 cm, or 0.4 AU, on this scale.
3. Take the sun–Mercury distance you calculated in Problem 1 or 2 and use one of Kepler's laws to find Mercury's orbital period.
4. Mercury's synodic period is 116 days. What is its sidereal period? Compare your answer here with your answer in Problem 2.
5. The semimajor axis of the orbit of Mars has a size of 1.52 AU and an eccentricity of 0.09. Calculate

the perihelion and aphelion distances of Mars. Assume that the earth's orbital eccentricity is zero; then what are the closest and farthest possible distances of Mars at opposition?

6. Jupiter's average distance from the sun is 5.2 AU. What is its sidereal period?

7. Suppose you lived on Mars. What would be the maximum elongation of the earth from the sun?

8. The synodic period of Saturn as seen from the earth is about 378 days. (a) What is the synodic period of the earth as seen from Saturn? (b) What is the earth's maximum elongation angle as seen from Saturn? (c) What is Saturn's sidereal period around the sun?

9. Compare Mars, Jupiter, and Saturn on the basis of their respective speeds around the sun. Consider the orbits to be circular. Calculate for each an orbital speed in units of AU per year. What is the trend that you see?

4 The Clockwork Universe

After studying this chapter, you should be able to:

4-1 Describe Galileo's important telescopic discoveries and their impact on the controversy over the Copernican and Ptolemaic models.

4-2 Indicate Galileo's purpose in developing a new science of terrestrial motions, and contrast Galileo's ideas about motion to those of Aristotle.

4-3 Contrast Galileo's astronomy and cosmology to those of Copernicus and Kepler.

4-4 Describe the difference between speed and velocity and between accelerated and unaccelerated motion, giving everyday and astronomical examples of each.

4-5 Cite Newton's three laws of motion, describe each in simple terms, provide concrete examples, and apply them to astronomical and everyday cases.

4-6 Contrast Newton's concept of natural motion to that of Aristotle, especially with regard to celestial motions.

4-7 Describe Newton's law of gravitation in simple physical terms, and apply this law to the concept of weight.

4-8 Define and describe the concept of centripetal force and acceleration, and use it in the moon–apple test to support Newton's law of gravitation.

4-9 Contrast Newton's astronomy and cosmology with those of Copernicus and Kepler.

4-10 Use Newton's physical ideas to support the Copernican model, and apply these ideas to appropriate astronomical situations, especially to those involving orbits.

4-11 Apply Newton's law of gravitation and Newton's revision of Kepler's third law in verbal, graphical, or analytical form.

4-12 Describe the physical concept of escape speed and apply it to astronomical situations in verbal, graphical, or analytical form.

CENTRAL CONCEPT

Newton's laws of motion and gravitation explain, predict, and unify the motions of the bodies in the solar system. These laws are universal and apply to objects outside of the solar system.

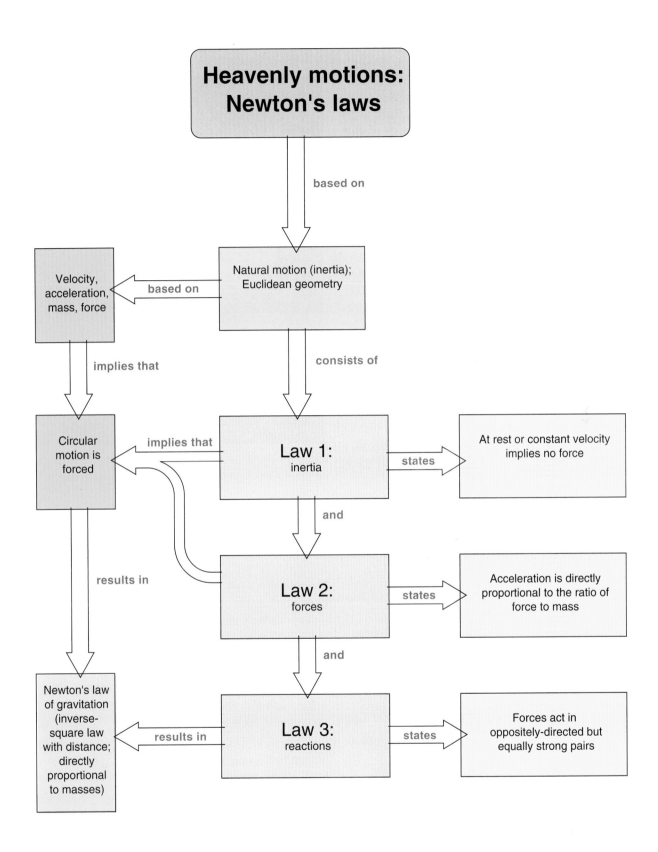

Heavenly motions:
Newton's laws

based on

Natural motion (inertia);
Euclidean geometry

based on

Velocity,
acceleration,
mass, force

implies that

Circular
motion is
forced

implies that

consists of

Law 1:
inertia

states

At rest or constant velocity
implies no force

and

Law 2:
forces

states

Acceleration is directly
proportional to the ratio of
force to mass

and

results in

Newton's law
of gravitation
(inverse-
square law
with distance;
directly
proportional
to masses)

results in

Law 3:
reactions

states

Forces act in
oppositely-directed but
equally strong pairs

What makes the world go 'round? For the planets, according to Isaac Newton, it's gravitation. By linking the cosmos with gravity, Newton achieved the goal that had eluded Kepler: a model of the cosmos run by a single force.

Newton arrived at this goal after crucial groundwork had been laid by Copernicus, Kepler, and Galileo. Copernicus violated the physics of the day and offered no substitute. This gap was an embarrassment for the Copernican model. Kepler sought to establish a clockwork cosmos driven by a single force. His idea of magnetic force finally failed, but his concept of a force directly influenced Newton's vision. Galileo's concern with terrestrial motions also shaped Newton's ideas. Galileo's use of the telescope supported the heliocentric model.

Newton erased the old physical separation between the earth and the heavens. He linked gravity (terrestrial physics) to the orbital motion of the planets (celestial physics). The links he forged were his laws of motion and gravitation, which sprang from his concept of natural motion. With Newton's ideas, the universe took on a new aspect. No more would there be the closed, finite universe of Ptolemy, Copernicus, and tradition. In Newton's grand vision, the universe embraced an infinite expanse, tethered by a single force: gravitation.

4.1 GALILEO: ADVOCATE OF THE HELIOCENTRIC MODEL

The Italian scientist Galileo Galilei (1564–1642) ventured to establish celestial physics on a firm experimental and mathematical basis. In contrast to Kepler, Galileo (Fig. 4.1) concerned himself almost exclusively with terrestrial motions, especially those of falling bodies. He felt that to establish the laws of terrestrial motions would do more to cement the structure of the Copernican cosmos than any observations, including those made with a telescope. He did make use of telescopic observations, but he valued physics more highly. He failed to achieve his goal, but his discoveries guided the work of Newton.

The Magical Telescope (Learning Outcome 4-1)

Galileo used his telescope to bolster the Copernican model. He did not invent optical lenses, nor was he the first to use them in a telescope. Rather, he pro-

Figure 4.1 Portrait of Galileo Galilei by Ottario Leoni, c. 1624: "By denying scientific principles, one may maintain any paradox."

moted his astronomical ideas by exploiting the novelty and shock value of telescopic observations. Galileo's report of his observations in his book *Sidereus nuncius* (*The Starry Messenger*) was widely circulated and brought him fame.

By Galileo's time, glass lenses had been in use for about 300 years. Their origins are not clear, but with them, eyeglass makers could correct defects of vision. In 1609 Galileo heard that a Dutchman had constructed a spyglass that made distant objects appear to be nearby. Although Galileo had little experience with optics, he immediately worked to duplicate the instrument and succeeded in constructing an optical device that made objects appear thirty times closer than when viewed with the naked eye (more on telescopes in Chapter 6). He put the device to astronomical use at once. Within a few weeks in 1609 and 1610, he made a series of astronomical discoveries that marked a new era in astronomy. Although Galileo was not the first person to build a telescope, he first recognized that the telescope increased our power to perceive reality.

When Galileo turned his telescope to the moon,

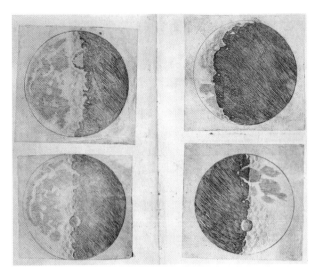

Figure 4.2 One of Galileo's drawings of the moon as seen through his telescope, from *Sidereus nuncius,* published in 1610 (from Biblioteca Nazionale in Florence, Italy). Note the craters and mountains, especially along the boundary between the illuminated and dark regions.

he saw that the moon's surface does not conform to the Aristotelian ideal for a perfect heavenly body. Rather than being smooth and spherical, it is rough, with chains of mountains and valleys and many craters (Fig. 4.2). Galileo then determined the height of a lunar mountain from the length of its shadow: roughly 6 km was the astonishing (and correct) result.

Galileo next peered at the stars. His instrument fragmented the faint band of the Milky Way into innumerable stars, more than can be distinguished with the unaided eye. This observation refuted the Aristotelian idea that the sky contained a set number of stars that cannot change.

Then Galileo hit on what he considered to be his most important discovery – four new "planets." Actually they were not new planets but rather were the four brightest moons of Jupiter. (At least seventeen moons are now known: see Chapter 10.) To his amazement, Galileo found that these four bodies revolved around Jupiter (Fig. 4.3). From continuous nightly observations, he estimated the orbital periods of the Jovian satellites. Here he found another argument against tradition, for Jupiter and its satellites resembled a miniature solar system. This fact required a second center of revolution in the cosmos and contradicted the Aristotelian doctrine that only the center of the universe – the earth – could be the center of revolution.

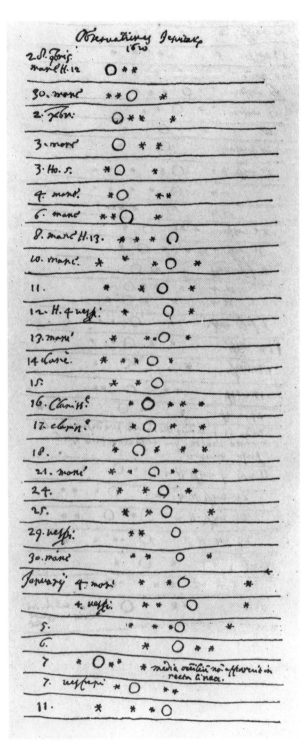

Figure 4.3 Drawings of early telescopic observations of the moons of Jupiter. The large circle represents Jupiter; the "stars" are the moons. Their changing positions next to the planet indicated that they were revolving around Jupiter, rather than being just background stars. These appear to be copies of Jesuit observations made in 1620.

Historically, Galileo may not have been the first to report on moons orbiting Jupiter. An old Chinese record notes that Kan Te, an astronomer in China in the fourth century B.C., made many observations of Jupiter. In one of his books, Kan Te states that Jupiter looked as if it had "a small reddish star attached to it." This may record the brightest of Jupiter's moons, which can, under ideal circumstances, be visible to the unaided eye. If so, Kan Te recorded this moon some 2000 years before Galileo observed it with a telescope!

The Starry Messenger

Galileo's *Sidereus nuncius* was published on March 12, 1610. Written in Italian rather than Latin (except for the title), the book was accessible to the general public, and it sold as fast as it could be printed. These telescopic discoveries, especially the news of Jupiter's moons, not only brought fame to Galileo but also created a demand for the telescopes produced in his workshop. He was still cautious in print: although he used his observations to demolish the Aristotelian cosmos, he did not openly advocate the Copernican scheme or support it with his observations.

Critics quickly spurned Galileo's work by invoking atmospheric phenomena or flaws in the lenses. Such arguments seem strange to us today, but they appeared reasonable enough to seventeenth-century people. The art of lens-making was just beginning, and many lenses produced ghost images. (Galileo's telescopes produced images far inferior to those in a cheap modern telescope.) Opponents conceded that Galileo might have honestly reported what he had seen but did not believe that his observations had a direct connection with reality.

Galileo convinced himself by the regularity of the telescopic phenomena that he was viewing reality and not illusion. In 1611 Kepler, who was well regarded in scientific circles, backed up Galileo with observations of his own and also with his theory of optics, which supported the validity of telescopic observations. He also showed that the moons of Jupiter obeyed his third law and so demonstrated that their orbits were like those of the planets around the sun.

Galileo's Discoveries and the Copernican Model (Learning Outcome 4-1)

Goaded by the opposition, Galileo continued to scan the skies for new marvels. He observed *sunspots* – dark regions on the sun. Defects on the supposedly perfect sun, the sunspots dealt another blow to the tenets of Aristotelian cosmology. In 1613 Galileo published his results in *The Letters on Sunspots* and made his first direct, printed declaration of his belief in the Copernican model.

Note that none of these observations directly confirmed the Copernican model. Rather, they worked as evidence *against* the old Aristotelian–Ptolemaic model. Galileo's observations of the phases of Venus (Fig. 4.4) could support the Copernican model. Half of Venus, which has a spherical shape, is illuminated by the sun. In the Copernican model, we observe Venus from outside its orbit, and Venus moves around the sun faster than the earth does. So we view Venus from different angles relative to the sun, and – like our moon – Venus shows different phases, from crescent to full. (The full phase is very hard to observe because it occurs when Venus lies on the other side of the sun when viewed from the earth.) Galileo observed part of this phase cycle of Venus.

Now in the Ptolemaic model, Venus does show limited phases. Because Venus must stay along an epicycle whose center must lie fixed along the earth–sun line (review Section 2.4), the range of phases is less than in the Copernican model. For instance, Venus can never be full because it cannot lie opposite the sun.

> **CAUTION.** The observations of the phases of Venus *do* favor the Copernican model over the Ptolemaic one, but they do *not* show that a heliocentric model is better than a geocentric one. For example, Tycho Brahe's geocentric model permits the same cycle of phases as the model of Copernicus.

The Crime of Galileo

Galileo publicly declared himself a Copernican in 1613. His intellectual vehemence and enthusiasm irritated many of his opponents, some of whom enjoyed substantial power in the Church. In 1616

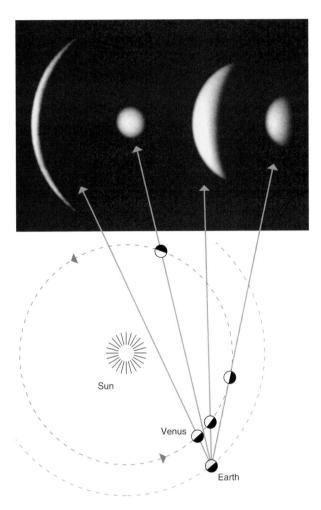

Figure 4.4 Phases of Venus in the Copernican model. The range of phases is restricted from our view because Venus' orbit is interior to the earth's. In particular, we can never observe a "full" Venus. The actual view through a telescope at each relative position of the earth and Venus is shown above each position of Venus.

In 1623 Cardinal Barberini, an erstwhile friend of Galileo and a patron of the sciences, was elected pope and took the name Urban VIII. Galileo thought he saw his chance and had a long audience with the new pontiff, during which Urban discussed the decree forbidding the heliocentric teachings. Feeling that he had the support of the pope and also of the Jesuit astronomers, Galileo wrote the *Dialogue on the Two Chief World Systems.* With the approval of the Roman Catholic censors in Florence, the book was published in 1632, after a few minor corrections had been made. The book supposedly relates an objective debate about the relative merits of the Ptolemaic and Copernican systems, with the judgment being rendered in favor of the traditional view. In reality, the text is a thinly veiled polemic favoring the Copernican cosmos.

Instantly Galileo found himself in trouble, for his opponents swiftly countered his claims with theological arguments. The pope himself fumed because his favorite arguments had been put in the mouth of Simplicio, the supporter of the geocentric viewpoint in the *Dialogue.* (Galileo had not drawn Simplicio kindly, and although there had been a sixth-century commentator on Aristotle named Simplicio, Galileo's readers knew the word also meant "simpleton.") Some copies of the book were seized before they left the printers, and the Inquisition summoned Galileo to Rome and forced him to recant his scientific beliefs. His friends were afraid to come to his support. As punishment he was placed under perpetual house arrest. The *Dialogue* restricted until 1835, when the works of Galileo, Copernicus, and Kepler were finally removed from the Index.

Galileo's trial by the Inquisition, although motivated as much by internal Church politics and Galileo's abrasive personality as by conviction that the Copernican system was wrong, frightened intellectuals in regions where the Church exerted power. The confrontation with the Church surprised Galileo, who thought he had divorced theology from science.

In October 1992, the Church decided to clear Galileo of his official condemnation of 1633. The Vatican finally admitted to error in its action over 300 years ago.

an official of the Inquisition apparently warned Galileo to stop teaching the Copernican model as truth rather than as a hypothesis. Galileo's position was held to be contrary to Holy Scripture. At the same time, *De revolutionibus* was placed on the Index of Forbidden Books until "corrections" had been made. (Only a few of the copies were actually changed.) These two events initiated a complex chain: Galileo, an obedient Catholic, avoided teaching the Copernican model as true; he also entered into arguments about scientific truth versus revealed truth. His argumentative spirit promoted his downfall.

4.2 GALILEO AND A NEW PHYSICS OF MOTION

Despite his observational triumphs, Galileo still felt the Copernican system lacked the anchor of a physical understanding of motion like the one Aristotle had provided for his model. Kepler had taken an important step to a new physics by his mathematical description of celestial motions. Galileo took the next step by devising a mathematical description of terrestrial motions, particularly motions of falling bodies. Galileo aimed to justify the Copernican model by his systematic, experimental search for physical ideas of motion. In unraveling the details of motion produced by applied forces, he was inventing the study of **mechanics.**

To set up the new science of mechanics, Galileo needed an important tool: a description of the motion of falling bodies, that is, the motion of masses under the influence of gravity. To understand Galileo's contribution, you need a precise physical description about such motion. Galileo's views on motion became a pivot on which modern physics turned.

Acceleration, Velocity, and Speed
(Learning Outcomes 4-2, 4-4)

When you step on a car's accelerator, the vehicle does just that: it **accelerates.** If you start out at rest, the car goes from zero speed to faster speeds, as you can tell from the speedometer. **Speed** is the rate of your motion with time. If you are cruising on the highway, you can pass a slower car by stepping on the accelerator to reach a faster speed (Fig. 4.5a). In both instances your speed changes with time; that's one aspect of the meaning of acceleration.

When you travel in a car, however, you do not simply race along at a given speed. You must also go in specific directions; otherwise, you'd never make it to your destination. Suppose, for example, that you travel from Albuquerque to Santa Fe, New Mexico; you go from one place to another at some speed in some direction. **Velocity** involves the direction as well as rate of motion. If you take one hour to drive from Albuquerque to Santa Fe, a distance of about 100 km, your speed is about 100 km per hour (km/h). When you drive back at the same speed, your velocity is different because you're heading in a different direction. In the first case, your velocity is 100 km/h north; in the second, 100 km/h south. So velocity is the rate of change of position with time in a certain direction. It's not simply speed, which tells only how fast you're going not where you're heading!

Imagine now that you're driving around a circular race track at a constant speed. Are you accelerating, even though your speed is constant? Yes! Why? Because your car is constantly changing direction as you negotiate the circle of the track (Fig. 4.5b). Your speed does not change, but your direction does, so your velocity changes and you accelerate. In fact, since your direction is constantly changing, you are undergoing a constant acceleration (Enrichment Focus 4.1).

> **TO SUM UP.** You are accelerating if your speed changes while your direction remains the same. You accelerate if *either or both* your speed and your direction change.

Note that speed and velocity have *different* and *precise* physical meanings. Speed is the average rate of travel and is measured in distance per unit of time – such as miles or kilometers per hour (mph or km/h) or even just meters per second (m/s). Velocity is speed with something added: direction. Acceleration is the rate at which velocity (not speed!)

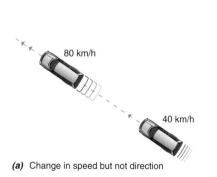

(a) Change in speed but not direction   *(b)* Change in direction but not speed

Figure 4.5 Acceleration. (a) A car accelerates if its speed changes while its direction remains the same. (b) A car also accelerates if it moves at a constant speed around a circular track, because the direction of its motion is always changing.

ENRICHMENT FOCUS 4.1

Speed, Velocity, and Acceleration

You must understand the differences among these terms, because they are physically real, not just semantics. You may see the distinctions more clearly from simple equations.

If you travel over some distance in a certain amount of time, then your speed is

$$\text{speed} = \frac{\text{distance covered}}{\text{time taken}}$$

where we really mean the *average* speed over the time interval. For now, let's assume a constant speed. Then you are probably more familiar with this relationship in the form

$$\text{distance} = \text{speed} \times \text{time}$$

Now specify a direction to the motion. Then we are talking about velocity rather than speed, but the relations are the same for a constant velocity:

$$\text{velocity} = \frac{\text{distance}}{\text{time}}$$

and

$$\text{distance} = \text{velocity} \times \text{time}$$

We can change velocity by changing speed, direction, or both. The rate of change in the velocity with time is called *acceleration*, which is defined by

$$\text{acceleration} = \frac{\text{change in velocity}}{\text{time interval}}$$

In some special cases, like bodies falling near the earth's surface, the acceleration is *constant*. Then we have

$$\text{velocity} = \text{acceleration} \times \text{time}$$

Now imagine an object, say a mass at the end of a string, moving in a circle at a constant speed. The direction of the motion is constantly changing, so the mass is accelerating. The magnitude of the centripetal acceleration is

$$\text{centripetal acceleration} = \frac{(\text{circular speed})^2}{\text{radius}}$$

and its direction is toward the center of the circle. This is not an intuitive result, but it was correctly obtain by Newton.

---

changes. It is measured in distance per unit of time per unit of time, such as meters per second per second: (m/s)/s. What happens when you hit the brakes of a car? You accelerate opposite the direction of motion. We commonly call this *deceleration.*

A car with a fairly powerful engine can accelerate from zero to 100 km/h in about 10 seconds. What is its acceleration? It averages 10 (km/h)/s (kilometers per hour per second) in the forward direction. If it has excellent brakes, the car can stop in about 4 seconds from 100 km/h. What's its deceleration? It averages about 25 (km/h)/s in the backward direction.

Natural Motion Revisited
(Learning Outcome 4-2)

Aristotle divided motions into two types: *forced* and *natural* (as in Section 2.2). An example of Aristotelian natural motion is the fall of a rock dropped to the earth. The rock follows the supposedly natural tendency of earthy material to seek the central point of the cosmos; its motion requires no force. What about forced motion? Throw a rock at a right angle to the natural downward motion! Aristotle would require a continuously acting force to keep up this unnatural, forced motion horizontal to the ground. As you know, however, after you have thrown an object horizontally, its motion continues for a while after the object leaves your hand. This tendency for the motion to continue, even when the force has stopped, is called **inertia.** It presented a nagging problem to Aristotelian physics, because inertia contradicts the basic premise about forced motion.

Galileo inverted Aristotle's ideas of natural and forced motion. He concluded that the downward motion of objects results from an attractive force: gravity. In addition, he viewed the horizontal motion of objects through the air as from their inertia. This inertial motion, he argued, was a natural one and would continue if no forces, such as air resistance, countered it. So for Galileo, the downward motion of dropped objects was a forced motion, not a natural one.

To cast this revision, Galileo first had to arrive at a new version of inertia. He explained his idea this way (Fig. 4.6): if a perfectly smooth ball were placed on a frictionless flat surface that sloped, the ball would roll down the slope forever (assuming the surface were infinitely long). The greater the slope, the faster the ball would roll. By the same argument, a ball traveling up a slope would eventually stop. If, however, the surface had no slope and the ball were placed on the level with no horizontal velocity, the ball would remain at rest. A ball on a level surface, if pushed, would move straight ahead at a constant velocity for-ever, since it is not slowed down by an ascent or speeded up by a descent. He defined inertia as a natural tendency for a body in motion to remain in motion.

Forced Motion: Gravity
(Learning Outcome 4-2)

Using this concept of inertia, Galileo dealt with the problem of falling bodies. He viewed falling not as natural motion (as had Aristotle) but rather as motion due to a force: gravity. Motion under the influence of gravity only is called **free-fall.** Drawing on both his experiments and intuition, Galileo concluded that free-falling motion took place at a constant acceleration, with the object's velocity changing at a constant rate downward as it fell. (The acceleration of gravity at the earth's surface has a value of 9.8 (m/s)/s and is usually denoted by g. This means that for every second of fall, an object gains 9.8 (m/s) of speed. You can recall this number by its approximate value, 10 (m/s)/s.)

Galileo also concluded that, in the absence of air resistance, all falling masses have the same acceleration. When dropped, they reach the same velocity after the same elapsed time. They also fall the same distance in the same time. His conclusion – that all masses fall with the same acceleration near the earth – directly contradicted the traditional teaching that a massive body falls faster than a less massive one and so moves a greater distance in a given time.

What do you think from your own experience? Try dropping a lightweight ball and a piece of paper from the same height. The ball hits the ground first as the paper drifts down through the air. Now crumple the paper so that it has about the same shape and size and as the ball. Drop both again. What's the result? Both reach the ground in about the same time!

In the case of constant acceleration, such as that of falling bodies, the velocity increases as time passes. That is, a dropped object falls with greater and greater speed, the closer it gets to the ground. When you throw a ball near the earth's surface, its path describes an arc (Fig. 4.7). This pattern results from a combination of the ball's falling motion (which occurs with a constant acceleration) and its horizontal motion (which occurs at pretty much a constant speed, because air friction has little effect on the ball during its time in the air). The curved path results from a force in one direction only: that pulling the ball to the ground, toward the center of the earth. So the ball's path results from a combination of two types of motion: a natural, inertial one in the horizontal direction and a forced one vertically.

Planets and Pendulums

Galileo recognized that the motion of a pendulum corresponded to that of a falling body constrained by a cord. We can apply this analogy to the motion of the planets – a leap that Galileo himself did not make.

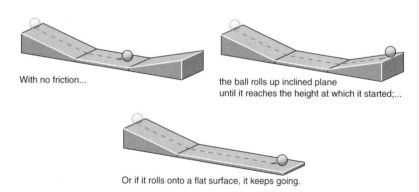

With no friction...

the ball rolls up inclined plane until it reaches the height at which it started;...

Or if it rolls onto a flat surface, it keeps going.

Figure 4.6 Galileo's concept of inertia (natural motion). The trick is to ignore friction in this thought experiment of rolling a ball down an inclined plane. It will roll up an equal distance onto a plane with a similar incline. If the ball rolls down the plane onto a flat surface, it will keep moving along with no change in speed.

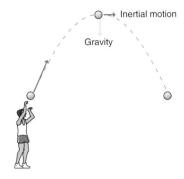

Figure 4.7 Path of a thrown ball. Gravity pulls it straight down and its inertia keeps it moving horizontally. Both motions take place simultaneously. The result is a curved path called a *parabola*, if air friction can be ignored.

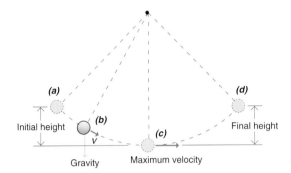

Figure 4.8 Swing of a pendulum. When at its maximum heights, its velocity is zero. The velocity is greatest at the middle of the swing.

Imagine a pendulum with a mass of 1 kg hanging at the end of a string 1 meter in length (Fig. 4.8). Move the mass and let it go. As you watch the pendulum swing, note two things. One, the motion is *periodic*. It repeats over a certain interval of time. Consider the motions of the planets, as described by Kepler's laws (Section 3.5). Their orbits are also periodic. That's one analogy between the motions of a pendulum and that of the planets. Two, the pendulum moves the fastest in the middle of its swing, when it is closest to the ground. What about planets? They move at their fastest orbital speeds when they are closest to the sun, at perihelion.

Now take two pendulums, side-by-side, each with the same mass and length. Using both hands, raise them to the same height. What do you expect to find about their periods? Right, they are the same. They swing side by side. Now increase the mass on one, keeping the length the same (Fig. 4.9a). Your prediction about their periods? Perhaps a surprise. They will again be the same! The swings stay synchronized. Does this equality seem strange? Think about Kepler's third law. It relates a planet's orbital period only to its distance from the sun.

Return to pendulums with the same mass. Now make one longer than the other (Fig. 4.9b). Use two hands to raise them so that the strings line up.

Let them go at the same time. Your prediction about the periods? The longer one will have the longer period. The swings get out of synchronization quickly. Consider again the planets. Those with larger semimajor axes have longer orbital periods than those with shorter ones.

The motions of pendulums mimic the motions of the planets, as described by Kepler's laws. A coincidence? No! Both are forced motions, and the force is gravity.

Galileo's Cosmology (Learning Outcome 4-3)

In his *Dialogue on the Two Chief World Systems*, published in 1632, Galileo compared the traditional model to the Copernican one. In fact, he displayed the planetary system as a pure Copernican scheme with no evidence at all of the ideas of Kepler (such as elliptical orbits). Nor did Galileo attempt to apply his own terrestrial mechanics to the motions of the planets. The physics of the heavens and earth remained separate.

Galileo paid little attention to the use of the Copernican model in predicting the planetary positions. What struck Galileo was the Copernican order, rather than the detailed astronomy of planetary motions. He apparently felt, with the same conviction as Copernicus, the harmony of the planetary order

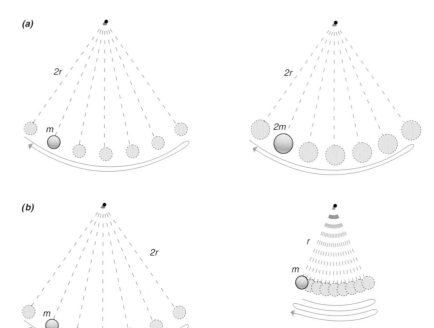

Figure 4.9 Comparison of motions of pendulums. (a) If they have two different masses and the same lengths, their periods are the *same*. (b) If they have two different lengths and the same masses, their periods *differ*. The shorter one has the shorter period.

established from observation. Of course, Galileo relied heavily on his telescopic observations along with the general arguments of Copernicus (Section 3.1).

Galileo placed the stars in a thick shell far beyond the planets. This stellar wall, a holdover from Greek ideas, was included in the original Copernican model. However, the 1632 *Dialogue* keeps open the possibility that the universe is not spherical and finite but is open and infinite, with the stars sprinkled "through the immense abyss of the universe." (In fact, as early as 1563, some accounts of the Copernican model already had stars scattered through unlimited space.) Newton would later support the idea of the cosmos as an infinite universe rather than a closed space.

4.3 NEWTON: A PHYSICAL MODEL OF THE COSMOS

Despite his interest in – and insight into – terrestrial motions, Galileo did not worry about the details of celestial motions. Clearly, he believed that the model of Copernicus was on the right track. But until the general concepts of motion were understood, Galileo thought it premature to consider explanations of planetary motion that would provide a physical foundation for the Copernican model.

Sir Isaac Newton (1642–1727) emerged as the genius destined to fuse the terrestrial and celestial realms and so to end the long-standing separation first assumed by the Greeks (Fig. 4.10). The publication, in 1687, of Newton's *Principia*, containing his new physics of motion and the concept of gravitation, provided a unified physical view of the universe. At last, Newton's discovery of gravitation solved the old puzzle of the motions of the planets. (See the Newton's Laws Celestial Navigator™.)

The Prodigious Young Newton

In the small English village of Woolsthorpe, on December 25, 1642, Newton's widowed mother gave birth to a sickly child. The fragile baby was so small at birth that it is said he could have fitted into a quart mug!

At age 18, Newton enrolled in Trinity College at Cambridge University. He first intended to study mathematics as applied to astrology, but decided to study physics. Then, in 1665, the bubonic plague overwhelmed England, the university shut down, and Newton returned to Woolsthorpe. There, in quiet isolation, he made discoveries in mathematics, optics, and the science of mechanics. He wrote, "In those days I was in the prime of my invention, and minded mathematics and philosophy more than at any other time since."

Figure 4.10 Portrait of Sir Isaac Newton from the Uffizi Gallery in Florence, Italy. "I have laid down the principles of philosophy; principles not philosophical but mathematical."

This fertile period in Woolsthorpe generated the legend of the falling apple. As is common in a creative flash of genius, Newton – whether he in fact saw an apple fall or not – linked two seemingly unrelated phenomena: the fall of an apple and the orbit of the moon. He was puzzled by the fall of objects, such as apples, and wondered about the force that attracts masses, such as the moon, to the earth's center.

On returning to Trinity, Newton showed his work to Barrow, who soon actually resigned his position so that Newton could have it. Newton, pursuing his interest in optics, invented one version of the reflecting telescope (Chapter 6), which uses a mirror as the primary light gatherer. His design was communicated to the Royal Society of London, for which he constructed a small reflector. This application of his ideas resulted in Newton's election as a Fellow of the society. The association was not completely happy, however, for when his work on light was also communicated to the society, such bitter controversy arose that Newton resolved never to publish again.

The Magnificent *Principia*

Newton broke his vow about 10 years later, when Edmond Halley (1656–1742) requested his advice on the problem of elliptical orbits described by Kepler's laws. Halley queried Newton on the nature of the force between the sun and the planets that produced such orbits and was surprised to hear that Newton had solved the problem in exact detail. Newton, however, had misplaced the solution and could not find it at the moment; he promised Halley that he would send it along later.

Recognizing the importance of Newton's discovery, Halley cajoled his introverted friend to make the studies public, promising to oversee and finance their publication. Stimulated intellectually by Halley's support, Newton labored for nearly two years to complete his *Philosophiae naturalis principia mathematica* (*The Mathematical Principles of Natural Philosophy*). Published in 1687, the *Principia* solved the vexing problem of planetary motions.

Forces and Motions
(Learning Outcomes 4-4, 4-5, 4-6)

Before dealing with Newton's analysis of orbits, let's reflect on our own experiences with forces. Imagine that you have a toy car that moves along a flat surface with little friction. Consider the car at rest. Then push the car quickly, applying a force for a brief time. What happens? The car accelerates from zero velocity to some velocity. The direction of the acceleration (and the resulting velocity) is the same as the direction of the initial force. This example shows that a force can produce accelerations (a rate of change in velocity, remember!).

Imagine that you push the toy car, again at rest, with a force greater than the one you applied the first time. What happens? Right – the car achieves a greater acceleration and reaches a greater velocity than before. In fact, if you apply twice as much force as the first time, the car will accelerate twice as much. (The acceleration is still in the same direction as before.) If you use three times the force, you get three times the acceleration, and so on. Hence, the acceleration of any object is *directly proportional* to the force exerted on it. "Directly proportional" means here that the acceleration increases in the same way the force increases. (What would happen if you applied four times the force?)

Add weights to the car until it has twice its original mass and reapply our initial degree of force. What occurs when we apply one unit of force? The acceleration is still in the same direction as the applied force, but it is less, because there is more mass to be moved. In fact, if you apply the same amount of force as you did the first time, the car will have half the acceleration for twice the mass. With three times the mass and the same force, the resulting acceleration would be one-third as much, and so on. The acceleration of any object is *inversely proportional* to its mass – the greater the mass, the less the acceleration (for the same amount of force). "Inversely proportional" means here that the acceleration decreases as the mass increases.

TO SUM UP. The acceleration of an object depends directly on the amount of force applied to it and inversely on the mass of the object.

"Amount of force" means the net force on the object. Why *net* force? Because several forces may act on an object, and they may balance and cancel out if they act in opposite directions. Consider a tug-of-war between matched teams. Both exert the same amounts of force on the rope, but in opposite directions. The net force is zero because the opposite, equal forces balance. They have achieved an **equilibrium.** This example emphasizes that forces have direction and can work either together or against each other.

Newton's Laws of Motion

In the *Principia,* Newton defines mass, velocity, and acceleration. He then expounds his three axioms, or **laws of motion.** In modern terms, these famous laws are as follows.

Law 1. The Inertial Law: A body at rest or in motion at a constant speed along a straight line remains in that state of rest or motion unless acted on by a net outside force (Fig. 4.11).

Law 2. The Force Law: The rate of change in a body's velocity due to an applied net force is in the same direction as the force and proportional to it but is

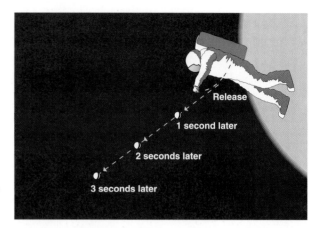

Figure 4.11 Newton's first law. An object thrown by an astronaut moves at a constant speed along a straight line (no friction slows it down in space), so it covers equal distances in equal intervals of time.

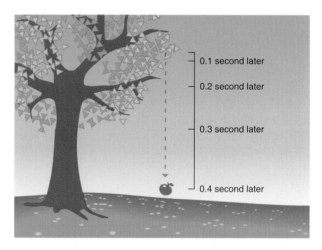

Figure 4.12 Newton's second law. A net force applied to a mass changes its speed, direction of motion, or both in the direction of the applied force. Here, gravity pulls on an apple and accelerates it to the ground. As this apple accelerates, it falls faster so that it covers a greater distance in the same amount of time (intervals of 0.1 second).

inversely proportional to the body's mass (Fig. 4.12).

Law 3. The Reaction Law: For every applied force, a force of equal size but opposite direction arises (Fig. 4.13).

Newton's first law takes a logical step beyond Galileo's concept of inertia by postulating that constant, uniform motion is the natural state of moving

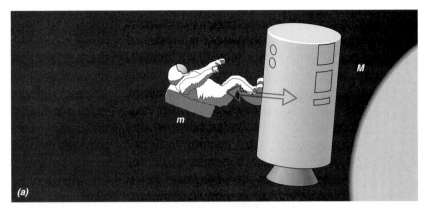

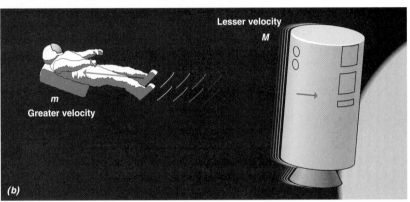

Figure 4.13 Newton's third law. Imagine that you are in space next to a spacecraft that has a larger mass. (a) Push off it quickly; it pushes back on you with a force equal to the force you apply to it, but in the opposite direction. You and the spacecraft move apart. (b) If you push off from a spacecraft more massive than you, you end up moving with a greater velocity than the spacecraft because the same force is accelerating your lesser mass applied to the more massive object.

mass anywhere in the universe. The first law tells you whether a net force is acting on an object: look for a change either in an object's speed or in the direction of its motion, or in both speed and direction.

The second law extends the recognition of a force to its results. The direction of the change in motion is the same as the direction of the applied force. Also, the amount of acceleration depends directly on the size of the force. Exerting a force (simply as a push or a pull) accelerates an object; that is, it slows it down, speeds it up, or changes the direction of its motion.

For example, suppose you are drifting in space next to two different objects. You push one. It accelerates momentarily and moves away, traveling in the direction of your push. You measure its acceleration, its change in velocity. Now push the other mass with the same force. You measure and find that its acceleration is half the acceleration of the first. (Its resulting velocity is half as much.) You have applied the same force to both objects, but the accelerations differ, so the second must have

twice the mass of the first. Newton's second law provides a means to measure mass.

In algebraic form, Newton's second law is

$$a = \frac{F}{m}$$

where F is the net applied force, m the object's mass, and a the acceleration resulting from the force. Note that, like velocities, forces have directions, and so do accelerations. An object will accelerate in the same direction as an applied force. The unit of mass in the International System of Units (Appendix A) is the *kilogram*. If we apply enough force to accelerate one kilogram to 1 (m/s)/s, we've applied a force of 1 *newton* (N) – named in honor of Isaac Newton! It takes about 10,000 N to brake a car.

A word about equations in this book: you will find very few equations in the main body of this text. Their primary function is to serve as a mathematical shorthand of selected, crucial physical and astronomical concepts. In this context, I expect you to grasp them conceptually, with an understanding at least in words of their essential meaning. The Enrichment Focus sections show additional equations and how to manipulate them, which your instructor may require.

Newton's third law recognizes that forces interact in simultaneous pairs. If you were at rest in space and pushed against a massive spacecraft, it would react to your applied force with an equal but oppositely directed force, pushing you away (Fig. 4.13). Now the forces applied to you and to the spacecraft were the same, but the resulting accelerations were different. According to Newton's second law, the acceleration is greater for you than for the spacecraft because your mass is less. As a result, you would move quickly away from the spacecraft, which would hardly budge. Also, you and the spacecraft would move in opposite directions.

This third law marks Newton's most original contribution to the understanding of motion in general and of gravitation in particular. Here Newton reasoned that gravitation is an interaction among the sun and the planets, and among all the planets – in fact, of all masses in the cosmos!

4.4 NEWTON AND GRAVITATION

From the platform of his ideas about force and motion, Newton attacked the problem of the planets by devising the law of gravitation. Two questions needed to be answered: in what direction does the force of gravity act? What is the amount of the force?

The first question involves a recognition of the general nature of the force, and the second involves a recognition of the physical properties that determine the force's strength. To answer these two questions, Newton combined his laws of motion with Kepler's planetary laws to arrive at a law of universal gravitation.

Newton demonstrated that the type of force that causes the elliptical orbits of Kepler's first law is a **central force,** one directed to the center of the motion. Also, he showed that planets moving in orbits under the influence of a central force follow Kepler's second law: the law of areas. Finally, Newton showed from the geometric properties of ellipses that a specific type of force law describes the force and then he rederived Kepler's third law with this force law. Thus, Newton ensured that his procedure fell in line with planetary laws known in his time.

Centripetal Acceleration
(Learning Outcomes 4-4, 4-5, 4-6, 4-8)

How do the moon and the famous apple enter into this scheme? Newton recognized that gravity caused the apple's fall. Might it not be that the earth's gravity, pulling on the moon, also keeps the moon in its orbit? For simplicity, assume that the moon's orbit is circular. The direction of the moon's orbital motion changes constantly. The moon stays on a curved path around – rather than moving along a straight line away from – the earth (Fig. 4.14). Newton's first

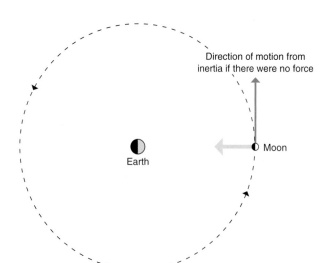

Figure 4.14 Central force and the moon's orbit. Since the moon moves along a circular orbit, a force must be acting on it. This force pulls the moon toward the earth in a direction toward the center of the moon's orbit.

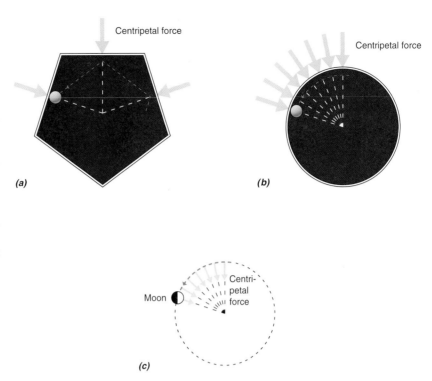

Figure 4.15 Centripetal force and circular motion. (a) A ball moving on a table with five sides strikes the center of each side. The force at each impact is centrally directed, that is, centripetal. (b) On a circular table, a centripetal force is exerted on the ball at every point of contact in its motion around the circumference. (c) As for the ball, the moon orbiting the earth requires a centripetal force acting on it at each and every point along the orbit.

law tells us that a force acts on the moon; moreover, according to the second law, this force results in an acceleration toward the earth's center. Such a centrally directed force is called **centripetal force.** (*Centripetal* means "directed toward the center.") The resulting acceleration toward the center is called **centripetal acceleration.**

Let's try to get a feel for centripetal force and acceleration. You experience centripetal force when you turn a corner in a car. The faster you go around the corner, the greater the acceleration. The radius of the turn also affects the acceleration. For instance, if you make a long, gentle curve – even at 90 km/h – you turn the steering wheel gently to change the car's direction. Here the turn has a large radius. But if you take a sharp turn at 90 km/h, you must turn the wheel sharply. Here the turn has a smaller radius. Centripetal force depends on both the speed of an object around a circle and the radius of the circle (Enrichment Focus 4.1).

Now apply the concept of centripetal force to the moon. Consider first a billiard ball hitting the middle of each of the side cushions of an imaginary, square billiard table. At each collision, the direction of the ball's velocity changes, so it must accelerate from an applied force. The direction of that force is toward the center of the table for each collision. Consider next a pentagonal table (Fig. 4.15a); the force of collision at each side still points to the center, but the angle of strike and rebound is smaller than for a square.

Finally, imagine a circular table with its infinite number of sides (Fig. 4.15b). The ball always touches the cushion (the angle between the cushion and the path is zero), and at every point, the force on the ball points to the center of the circle, and so its acceleration is also in that direction. Now, imagine the lunar orbit as circular. By analogy to the circular table, at every point in this orbit, a centripetal force acts on the moon (Fig. 4.15c). That force keeps the moon bound to the earth.

What causes the centripetal force that holds the moon in its orbit? Newton generalized from the apple to the moon: "I began to think of gravity extending to the orb of the moon." In this statement he expressed the new insight that every body in the universe attracts every other body with a gravitational force. This statement became the first *universal* physical law.

Newton's Law of Gravitation
(Learning Outcomes 4-7, 4-11)

What, in Newtonian terms, does *gravitation* mean? First, it means that all masses in the universe attract all other masses. (The gravitational force can only attract; it does not repel, despite some claims to the contrary in science fiction stories.)

Second, Newton's law of gravitation says that – if you consider just two masses as an example – the amount of the gravitational force depends directly on the amount of material each mass has. If you doubled the mass of one and kept the distance between the two the same, the force would also double. If you doubled the other mass, the force would double again, so doubling both masses results in four times the initial force. (The kind of material that makes up the masses does not matter.)

Third, masses at greater distances from each other have less gravitational force than those closer together. The drop-off of force with distance happens in a special way – as the *inverse square* of the distance. Consider, for example, a mass at different distances from the earth. The mass at 1 unit of distance from the earth experiences one unit of force (Fig. 4.16). Now move the mass away by two units of distance. The force is less. How much less? By ½ squared, which is $1/(2 \times 2)$, or ¼ as much as when the mass and the earth were 1 unit apart. Now move the mass to half the unit distance. The force is greater, by 1 over ½ squared, or 4 times as much.

Now consider any two masses 1 meter apart. A certain amount of gravitational force attracts them. Move the masses until they are 2 meters apart. The force is less by ½ squared, or $1/(2 \times 2)$, ¼ as much as when the masses were 1 meter apart (Fig. 4.16). At 3 units apart, the force is ⅑ as much; at 4 units, ¹⁄₁₆ as much. Note in all these examples that the masses remain the same. (What if the distance is 5 meters?)

You'll run across a number of relations in this book. Some are direct, such as that between the gravitational force and the masses of the objects involved. If you double one mass or the other, then the force doubles. For this kind of direct relation, I'll use the phrase "directly proportional to." In contrast, the relationship between the force and the distance between the two objects is an inverse one. As the distance increases (or decreases), the force decreases (or increases). For such inverse relations, I will use the phrase "inversely proportional to."

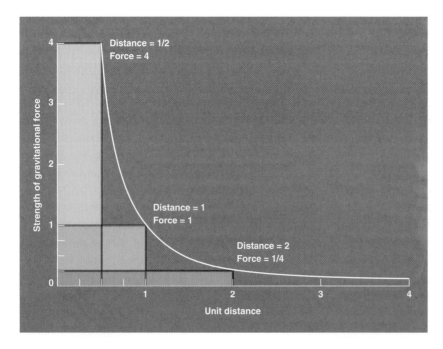

Figure 4.16 Gravitation and distance. This graph shows how the strength of the gravitational force changes with the distance between the earth and another mass: at a separation of 1 unit of distance, the force has 1 unit of strength. When the masses are twice as far apart, a distance of 2, the force has only ¼ the original strength.

Figure 4.17 The moon's orbit. In one second, the moon moves about 1 km. If it followed just its inertial path, this single kilometer would be a straight line. During the second, the earth's gravity causes the moon to fall about 0.14 cm toward the earth. The combination of the inertial motion and the accelerated one keeps the moon on an elliptical orbit around the earth.

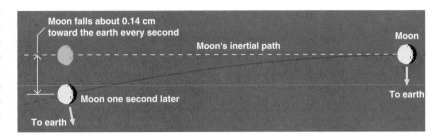

CAUTION. Gravity cannot be understood in terms of a simple inverse relation. Rather, it's expressed by an *inverse-square relation.* Here, the force is *inversely proportional to the square of the distance,* not just the distance by itself! So the gravitational force between two masses decreases much faster than you'd expect intuitively, probably on the basis of a simple inverse relation.

In modern algebraic form, Newton's law of gravitation is

$$F = \frac{Gm_1m_2}{R^2}$$

where F is the gravitational attraction between two spherical bodies m_1 and m_2 whose centers are separated by a distance R. The symbol G is a constant, a number whose value is assumed not to vary with time and location in the universe; its value relates the size of the force to the sizes of the other quantities. The value for G in the International System of units (SI units, see Appendix A) is 6.67×10^{-11}. (Here I am expressing the constant in powers-of-ten notation, also in Appendix A.) In this system, a unit of gravitational force is the newton (N) – the same as for forces in general. Recall that one newton is the amount of force needed to accelerate a mass of one kilogram (kg) at a rate of one meter per second per second ([m/s]/s). This force is small – between two 1-kg masses 1 meter apart, the amount of gravitational force is a mere 6.67×10^{-11} N. That's 0.0000000000667 newton!

Gravitation has a far reach. Newton used the moon and the apple to test the validity of this law. He knew that the earth's gravity at its surface (1 earth radius from the center) caused the apple to fall. The earth–moon distance is about 60 earth radii, so if an inverse-square law correctly describes gravitational forces, the acceleration of the moon toward the earth must be $1/60^2$ or $1/3600$ as much as the acceleration of the apple (Enrichment Focus 4.2).

Newton compared his predicted centripetal acceleration with the centripetal acceleration derived from observations of the moon's orbit. As he put it, the predicted and the observed accelerations were "pretty nearly" the same. Newton concluded that the cause of the moon's centripetal acceleration is the same as that of the apple's: the earth's gravity. This force, extended out to the distance of the moon, keeps the moon in its orbit. In other words, the moon's motion results from two independent actions – its natural, inertial motion (along a straight line) and its centripetal acceleration from the earth's gravity. In fact, the moon moves about one kilometer in a second along its orbit while it falls 0.14 cm toward the earth in the same time (Fig. 4.17). That 0.14 cm is about the height of the capital letters in this book! In the same second, an apple near the earth's surface falls about 5 meters.

CAUTION. An apple orbiting the earth at the same distance as the moon would have the same centripetal acceleration as the moon (even though the forces would differ)! So how far would the apple fall in one second?

Consider Galileo's law of free-fall from Newton's standpoint. Imagine dropping a cannonball and a tennis ball. The earth exerts a much greater gravitational force on the cannonball than on the tennis ball because of the cannonball's greater mass. However, when the two are dropped, they fall side by side and land at the same time (in a vacuum). Gravity's effect is the same on both objects; more precisely, the acceleration of each is the same because the two spheres speed up at the same rate. Although the forces are different, the accelerations turn out to be the same. Why? Because the extra

ENRICHMENT FOCUS 4.2
Newton, the Apple, and the Moon

How did Newton connect the apple's fall to the moon? He knew that the distance to the moon was approximately 60 earth radii. He also surmised that if the earth's influence extended to the moon, it grew weaker as the distance increased. How much did the force weaken? Newton guessed that it weakened as the inverse square of the distance (Fig. F.7). An apple dropping from a tree is 1 earth radius from the center of the earth. So, comparing the accelerations of the moon and the apple, Newton predicted that the moon's acceleration must be $1/60^2$ less, or $1/3600$ as large. The acceleration due to gravity at the earth's surface is 9.8 (m/s)/s. For the moon, then, the predicted acceleration is 9.8 (m/s)/s divided by 3600, or about 2.7×10^{-3} (m/s)/s. This acceleration of the moon toward the earth is a centripetal acceleration, for the necessary force, gravity, pulls the moon toward the earth's center.

Note that Newton did not need to know the mass of the apple or the mass of the moon, because the accelerations of these bodies do not depend on their masses. Galileo (Section 4.1) had reached the same conclusion from his own experiments. Newton made crucial use of this conclusion in the moon–apple test. How did Newton's predicted value compare with the actual rate of the moon's fall? Newton had found that for a circular orbit, the centripetal acceleration has a value of

$$a = \frac{V^2}{R}$$

where a is the centripetal acceleration, R the radius of the orbit, and V the orbital speed. For the moon, R equals 3.84×10^8 m. Its speed is the distance it travels in one orbit divided by the period for one orbit, or

$$V = \frac{2\pi R}{P}$$

One day contains $24 \times 60 \times 60 = 8.64 \times 10^4$ s, so the moon's orbital speed in a sidereal month of 27.3 days is

$$V = \frac{2\pi(3.84 \times 10^8 \text{ m})}{(27.3 \text{ days})(8.64 \times 10^4 \text{ s/day})} = 1.02 \times 10^3 \text{ m/s}$$

Then the moon's orbital acceleration is

$$a = \frac{V^2}{R} = \frac{(1.02 \times 10^3 \text{ m/s})^2}{3.84 \times 10^8 \text{ m}} = 2.7 \times 10^{-3} \text{ m/s}^2$$

This result comes close to the prediction made from an inverse-square law.

Newton did not have the modern values used in these calculations for the period and size of the moon's orbit and the acceleration by gravity. But he chose a value of the moon's distance that made his results match. Newton felt that his approach was correct even though he fudged the figures slightly.

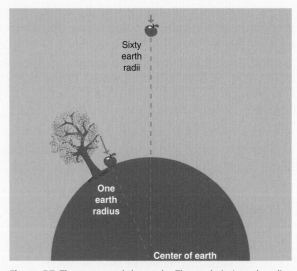

Figure F.7 The moon and the apple. The apple is 1 earth radius from the earth's center and it falls with an acceleration of 9.8 (m/s)/s. If an inverse-square law for gravity is correct, then the moon accelerates at a rate of 1/3600 less than the apple.

force on the cannonball is exactly offset by its greater inertia to changes in motion. (This result influenced Einstein in the development of his theory of gravitation: see Chapter 7.)

This concept applies to other bodies. On the moon, for example, the acceleration of gravity is about one-sixth less than at the surface of the earth. If you drop a hammer and a feather from the same height, they will hit the lunar ground at the same time. Not only do both objects fall with the same acceleration but, without any atmosphere, air friction does not complicate the result. Apollo astronauts did the experiment and proved Galileo's prediction.

Finally, Newton's third law requires that the gravitational force of the earth on the moon be *exactly equal* to that of the moon on the earth (even though the masses differ). They are interacting bodies.

4.5 COSMIC CONSEQUENCES OF UNIVERSAL LAWS

The gravitational force of the earth causes the centripetal acceleration of the moon. This was Newton's central discovery: the earth's gravity keeps the moon swinging around. Newton's vision resulted in a new understanding of the planetary orbits. The most obvious: the sun's gravity locks the planets in their elliptical orbits. Newton had found the physical interaction between the sun and the planets first sought by Kepler. He derived a new version of Kepler's third law that included the gravitational constant and the masses of the interacting bodies (Enrichment Focus 4.3). Related to basic physical laws, the revised third law became a potent tool in determining the mass of the sun, as well as the masses of the planets that have at least one satellite – quantities never known before! In fact, mass turns out to be a key physical property of astronomical objects. Newton's ideas provide us with the masses of stars – and even galaxies!

Newton answered in detail the ancient question of how the planets moved, and he answered it precisely. His predictions of planetary positions were far more accurate than those of his predecessors. Newton's ideas provided the physical support sorely needed for the Copernican model. The sun tethers the planets in a heliocentric system.

The Earth's Rotation (Learning Outcome 4-10)

One objection to the Copernican model: objects not tied down to the earth should fly off because of its rotation. Newton noted that gravity holds things down. Another objection was that dropped objects should not land at their starting position; rather, they should be left behind by the turning earth. Newtonian physics explained that objects on the earth have inertia, which they retain when falling. That is, unattached objects do not lose their forward inertial motion but continue to move with the ground beneath them. When you throw an object straight up, it is moving in the direction of the earth's rotation while it is in the air; moreover, it retains the inertia it had while in your hand. The object aloft keeps up with the turning earth because no force is acting on it to change its eastward velocity. When it drops, therefore, it falls right back into your hand.

Precession of the Earth's Axis (Learning Outcome 4-10)

Section 1.2 described the precession of the equinoxes; here I'll explain it briefly using Newton's law of gravitation. You can understand this effect if you picture the earth as spinning like a top. If you place a spinning top on a table, its axis of rotation moves in a circle – an effect called **preces-**

ENRICHMENT FOCUS 4.3

The Mass of the Sun

By using his laws of motion and gravitation, Newton reworked Kepler's third law so that it had the form

$$P^2 = \frac{4\pi^2}{G(M_{sun} + m_{planet})} a^3$$

where m_{planet} is the mass of the planet and M_{sun} the mass of the sun. Compare this to Kepler's third law in the form

$$P^2 = ka^3$$

with

$$k = \frac{4\pi^2}{G(M_{sun} + m_{planet})}.$$

You can use the earth's orbit to find the sun's mass. The earth–sun distance, a, is 1.50×10^{11} m. The earth's period, P, is 365.25 days, or 3.16×10^7 s. Because the mass of the sun is much larger than that of the earth, we can approximate $m_{earth} + M_{sun}$ by M_{sun}. That is, $m_{earth} + M_{sun} \approx M_{sun}$. Then Newton's revision gives

$$M_{sun} = \frac{4\pi^2}{G} \frac{a^3}{P^2}$$

And, substituting in the values, we have

$$M_{sun} = \frac{4\pi^2}{6.673 \times 10^{11}} \frac{(1.496 \times 10^{11})^3}{(3.156 \times 10^7)^2} = 1.99 \times 10^{30} \text{ kg}$$

Thus, the mass of the sun is about 3.3×10^5 times that of the earth.

sion. The precessional motion occurs in response to gravity trying to pull the top down so it falls over.

The earth, because it is spinning, is not a perfect sphere but oblate – it has an equatorial bulge. (Newton actually predicted roughly the amount of the oblateness.) Now, the moon's orbit does not lie in the same plane as the earth's equator, but near the ecliptic. In general, the moon lies above or below the earth's equator, so the moon's gravitational force attracts the earth unevenly – it tends to pull the bulge closest to it more than the part of the bulge on the opposite side of the earth. The net force tries to tip the earth's axis of rotation from its orientation in space. The spinning earth responds by a precession of its axis. This motion makes the earth's poles describe a circle in the sky, so over 26,000 years what stars are closest to the poles change. The earth's equator projected into space also moves, which we see as the precession of the equinoxes. (The sun's gravity also takes part, but its effect is about ⅜ that of the moon.)

The Earth's Revolution and the Sun's Mass (Learning Outcomes 4-10, 4-11)

You can use the earth's orbit to find the sun's mass. The earth–sun distance, a, is 1.50×10^{11} m. The earth's period P is 365.25 days. With this information, you can use Newton's second law and the law of gravitation to work out the mass of the sun (Enrichment Focus 4.3).

The knowledge of the sun's mass bolstered the Copernican model in the framework of Newtonian physics. Newton's laws show that the sun has roughly 3.3×10^5 times the mass of the earth. Newton's third law of motion requires equal gravitational forces between the sun and the earth – the force of the sun on the earth equals that of the earth on the sun.

However, the second law demands that the earth's acceleration be much greater than the sun's (in fact, 3.3×10^5 times greater!). This law is fulfilled only because the sun has a small acceleration in one year. The earth orbits the sun rather than the other way around. That is, sun and earth interact gravitationally, and because of their mutual attraction, both orbit around a common point called the center of gravity or the center of mass. As seen from this point, the earth's centripetal acceleration is 3.3×10^5 times greater than the sun's, so the earth has greater change of velocity than the sun. But the sun also moves!

The *center of mass* is the balance point of two connected objects. If the masses are equal, the center of mass lies at the midpoint of two masses connected by a rod (Fig. 4.18) With two unequal masses, the balance point is closer to the more massive of the two. If you threw the spinning rod in the air, the two masses would spin around the center of mass (Fig. 4.19). Gravity works to bind two masses together, as the rod does here, and two masses linked by gravity also have a balance point, a center of mass. For the earth and sun, the center of mass is very close to the sun because the sun is much more massive than the earth.

Gravity and Orbits (Learning Outcomes 4-9, 4-10)

Newton's laws allow us to picture planetary orbits (and Kepler's laws) in simple physical concepts. Each planet's path is an ellipse, with the sun at one focus, and each planet has a unique semimajor axis, eccentricity, and orbital period. Imagine a planet at aphelion, when it moves most slowly in its orbit (Fig.

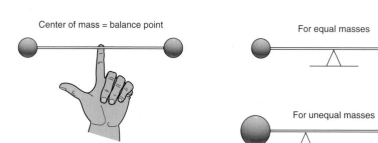

Center of mass = balance point

For equal masses

For unequal masses

Figure 4.18 The center of mass is the balance point for two masses connected by a rod.

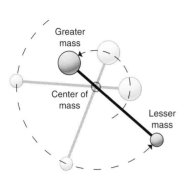

Figure 4.19 Motions of two unequal masses at the end of a rod as the rod is thrown in the air. Both masses revolve around the center of mass in the *same* period of time.

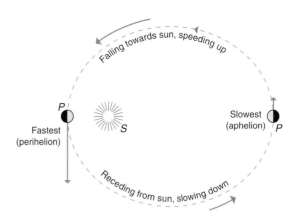

Figure 4.20 Elliptical path of a planet around the sun and the variation of its orbital speed over one orbit.

4.20). As the planet moves toward the sun, its speed increases in response to the sun's gravitational force. The planet whips around the sun at perihelion at its fastest speed. As it moves away from the sun, the gravitational pull of the sun retards its speed. So it slows down, until it turns the corner at aphelion and falls in again. This natural rhythm is a consequence of Kepler's second law. The motion is just like that of a pendulum (Section 4.2).

Newton's physics also correctly describes the orbits of comets around the sun. Through the Middle Ages, astronomers believed that comets were objects confined to the earth's atmosphere. Later, Tycho showed by parallax observations that comets are not atmospheric phenomena. Later still, Newton and Halley decisively demonstrated that comets orbit the sun in accordance with the law of gravitation. In fact, Halley correctly predicted the return of the comet that bears his name, but he did not live to see the comet return in accordance with his prediction. (More on comets in Chapter 11.)

Newton's ideas also led to the discovery of Neptune in 1846, long after the first publication of the *Principia*. Neptune was the first planet to be found because of its gravitational effects on another. Newton's laws predicted the existence of Neptune before the planet was observed.

The discovery of Neptune rested on observed irregularities in the orbit of Uranus (a planet discovered with a telescope in 1781). Astronomers had noted small discrepancies between the observed positions of Uranus and those predicted from Newton's laws. Such irregularities occur because all planets attract one another. Jupiter's tug, for instance, causes the orbit of Uranus to differ from what we would expect if Uranus and the sun were the only attracting bodies. The existence of Neptune was revealed before the planet was discovered when Uranus' motion was seen to deviate from the pertubations caused by known planets.

Applying Newton's laws to explain the deviations of Uranus as attributable to a planet beyond Uranus, the Englishman John C. Adams (1819–1892), in 1845, estimated where the unknown planet should be in the sky. He communicated his result to the Astronomer Royal, Sir George Airy (1801–1892), hoping that a search for the new planet would begin immediately. But Airy did not believe the result, partly because he thought the gravitational force deviated from an inverse-square law at large distances.

At almost the same time in Paris, Urbain J. J. Leverrier (1811–1877) made similar calculations and transmitted them to Johann Galle (1812–1910) in Berlin. Galle found the planet on September 23, 1846 – the very night he received the predicted position! Newton's laws triumphed. In the twentieth century, discrepancies in Neptune's orbit stimulated the search for other planets. The search resulted in the discovery of Pluto (Chapter 11), even though we now know that the mass of Pluto is too small to affect Neptune.

An equally dramatic discovery, which extended the validity of Newton's laws beyond the bounds of the solar system, was the observation of *binary star systems* in the late eighteenth century. (A binary star system consists of two stars, held together by their mutual gravity, orbiting around their center of mass.) William Herschel (1738–1822) and his sister, Caroline Herschel (1750–1848), observed many pairs of stars over a long time, looking for heliocentric parallax. Unexpectedly, the Herschels observed the orbital motions of some pairs. These observations showed that some pairs of stars lie close together in the sky not simply because they both happened to be in the same direction but because they are actually linked by gravitational forces.

Calculations of the orbital periods of binary star systems and the separation between the stars confirmed that the motions of these stellar pairs follow Kepler's laws. These laws can be derived from Newton's laws of motion and gravitation. Since binary star systems obey Kepler's laws, they indirectly confirm Newton's laws. (Chapter 15 has more details on binary star systems. In the same way that we found the mass of the sun, we can use binary systems to find the masses of stars – a powerful extension of Newton's laws.)

Orbits and Escape Speed
(Learning Outcome 4-12)

From Newton's laws of motion and gravitation arises the concept of **escape speed,** the minimum speed an object needs to attain to be free of the gravitational bonds of another. Let's see what that means.

Consider a ball at the edge of a table (Fig. 4.21a). Let it drop, and it falls straight to the ground. Now give the ball a push (Fig. 4.21b). As it passes the end of the table, it falls down, but the push makes it travel some distance horizontally before it hits the ground. Push it harder and it goes farther horizontally (Fig. 4.21c).

Now imagine, as Newton did, a giant cannon placed on the top of a very high mountain and aimed parallel to the ground (Fig. 4.22). Fire a cannonball. It travels some distance, then falls to the ground. If you use more powder in the cannon, the ball travels farther along the earth's curve before it hits the ground. If you put a large enough charge in the cannon, the ball goes completely around the earth in a circular orbit, returning to the cannon (orbit *C* in Fig. 4.22). The inertial motion of the ball exactly offsets the falling due to gravity.

The ball's circular orbit defines a crucial path around the earth, for the ball (or any mass) requires

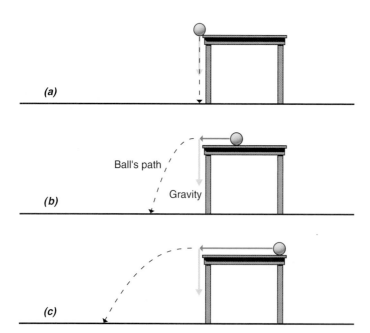

Figure 4.21 Motion of a ball near the surface of the earth. (a) Dropped from the edge of a table, the ball falls straight down. (b) When pushed along the table before it reaches the edge, the ball moves horizontally (along a parabola) before it hits the ground. (c) When pushed harder, it moves a greater distance before it strikes the ground.

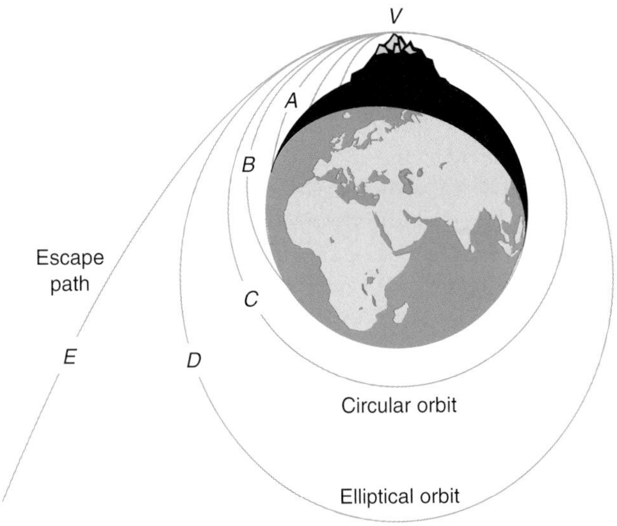

Figure 4.22 Launching earth satellites, according to Newton. From the top of a high mountain (*V*), cannonballs fired at a low speed hit the earth after traveling some distance (*A* and *B*). A certain minimum speed places them in a circular orbit (*C*); a somewhat greater speed, an elliptical orbit (*D*); and above the escape speed, an escape orbit (*E*).

a specific speed – no more, no less – for the circular orbit. A somewhat lower speed results in an elliptical orbit. The position of the top of the mountain then marks the farthest point from the earth in such an orbit – the *apogee* point.

Now dump an even larger charge in the cannon. With a starting speed greater than that needed for a circular orbit, the cannonball also travels around the earth in an elliptical orbit, but increasing the starting speed makes the orbit more eccentric. The point at the top of the mountain will now be the point in the orbit closest to the earth – the *perigee* point (orbit *D* in Fig. 4.22). Eventually the orbit becomes so eccentric that the semimajor axis is infinitely long, and it would take the ball an infinite time to return (orbit *E* in Fig. 4.22). So it never does. The speed that is necessary to achieve this result is called the *escape speed*. (Note from Kepler's second law that when the ball is an infinite distance away, its speed relative to the object it left is zero.) Any speed larger than the escape speed produces the same effect: the ball leaves the gravitational grip of the earth, never to return.

What determines the escape speed from the surface of an object? Consider the earth. Its escape speed is 11 km/s. Suppose that you kept the earth at its present radius but increased its mass. Its escape speed would increase. Now suppose that you kept its mass the same but decreased its radius. The escape speed from the surface would again increase. Basically, an object's mass and radius determine the escape speed from its surface. A greater mass or a smaller radius (for the same mass) results in a larger escape speed. A smaller mass or a greater radius (for the same mass) results in a smaller one (Enrichment Focus 4.4). Note that the mass of the escaping object does not matter. The escape speed is the *same*, whether for 100 kg or one million kilograms!

Today we use rockets rather than Newton's imaginary cannon to place satellites in orbit around the earth. We also send them off with greater than escape speed to journey to other planets – and even out of the solar system!

In terms of a scientific model, the predictive success of Newton's laws was astounding. Time after time, the numbers – not just the general ideas – came out right. Planetary positions could be predicted precisely, as could the motions of planetary satellites. Newton's ideas were universal. No longer were separate laws needed for falling bodies, the moon, and binary stars. One simple force binds them all: gravitation!

Newton's Cosmology (Learning Outcome 4-9)

For the Greeks, Ptolemy, and Copernicus, the cosmos – enclosed by the sphere of fixed stars – appeared to be finite and bounded. Galileo had left open the possibility that the universe might be infinite but did not strongly push his case. Newton held that his laws required that the universe be infinite in extent.

ENRICHMENT FOCUS 4.4

Escape Speed

From Newton's laws of motion and gravitation arises the concept of escape speed, the minimum speed an object needs to be free of the gravitational bonds of another. Algebraically, the formula is

$$V_{esc} = \sqrt{\frac{2\,GM}{R}}$$

where V_{esc} is the escape speed, M the mass of the object you want to escape, and R its radius. (Assume that the mass is spherical.) Note the mass of the escaping object does *not* enter into the calculation of escape speed. (Why not?)

Let's calculate the earth's escape speed. The mass of the earth is 6.0×10^{24} kg, its radius is 6.4×10^6 m, and $G = 6.7 \times 10^{-11}$ in SI units (Appendix A). So

$$V_{esc} = \sqrt{\frac{2\,GM}{R}}$$

$$= \sqrt{\frac{2(6.7 \times 10^{-11})(6.0 \times 10^{24})}{6.4 \times 10^6}}$$

$$= 1.1 \times 10^4 \text{ m/s} = 11 \text{ km/s}$$

For objects in the solar system, the radius and mass of the earth are convenient units to use. With these units, the escape speed in kilometers per second is

$$V_{esc} = 11\sqrt{\frac{M}{R}}$$

This form of the equation emphasizes that escape speed *increases* as mass *increases* or as radius *decreases*. For example, increasing the mass by 4 with the same radius doubles the escape speed. So will decreasing the radius by 4, if the mass stays the same. What if you did both? Then the escape speed increases to four times the original amount. *The escape speed is directly proportional to the square root of the mass and inversely proportional to the square root of the radius.*

His argument went like this: if the universe were finite, gravitation would eventually pull all matter to the center. As a result, only one large mass would exist. However, we see other masses – stars, for example. In an infinite universe, the matter would be pulled into an infinite number of small condensations. Newton concluded that the universe is infinite.

Not all things in the heavens and on earth sat happily in Newton's infinite universe. The innermost planet, Mercury, posed an annoying problem: the axis of its orbit rotates in space, and this motion cannot be explained completely by the attraction of existing planets. Some astronomers thought that the excess rotation was caused by a planet between Mercury and the sun. Although observations of this hypothetical body – called Vulcan – have been reported, they have so far proved to be mistaken. The supposed Vulcan has never been seen, even when modern observation techniques have attempted to catch its swift flight. (Einstein later solved the problem of Mercury's orbit: see Chapter 7.)

Newton was sorely disturbed by the mutual forces among the planets, which he thought must eventually lead to the disintegration of the solar system. To avoid this awful event, Newton envisioned the hand of God occasionally descending to reset the clockwork mechanisms of planetary motions, like a conscientious artisan making adjustments. The order of the mechanical universe – ordained by the Divine Being – was maintained by His intervention, and the expanses of Newtonian space were benevolently watched by the distant God.

Table 4.1 Evolution of Cosmological Models

Aspect of Model	Ptolemy	Copernicus	Kepler	Newton
Planetary motions	Uniform, circular	Uniform, circular	Nonuniform, elliptical	Nonuniform, elliptical
Force on planets?	No, natural motion	No, natural motion	Yes, magnetic	Yes, gravitation
Kind of cosmos	Geocentric finite	Heliocentric finite	Heliocentric finite	Centerless, infinite

KEY CONCEPTS

1. Galileo used his telescopic observations to promote the Copernican model (which he believed in for aesthetic reasons), even though none of his data gave direct proof of its validity. Galileo recognized that the Copernican model lacked a physical basis and sought one by analyzing the motions of free-falling objects near the earth. To this end, he developed a new concept of inertia. He concluded that falling objects have accelerated motion and so are subject to a force: gravitation.

2. With his laws of motion and gravitation, Newton abolished the ancient physical distinction of motions in the terrestrial and celestial realms. The result was a physically unified cosmos (Table 4.1), linked by gravity and powerful laws of explanation and prediction of motions.

3. Newton's laws of motion described natural motion and forced motion in a new way. He viewed natural motion as that of a body at rest or moving at a constant speed along a straight line. Forces result in accelerations (changes in velocities); forces act in mutual pairs; and a net force results in acceleration – forced motion.

4. From these ideas about motion, Newton concluded that planetary orbits result from a centrally directed or centripetal force; that force is the gravitational force between the sun and the planetary masses. Newton used the moon's orbit to confirm that the force of gravity varies with the inverse square of the distance.

5. Newton's law of gravitation describes the strength of the gravitational force as (a) depending directly on the product of the masses of the two objects and (b) inversely as the square of the distance between their centers. The direction of the force pulls the masses together. This description of gravitation resulted in elliptical orbits for the planets.

6. Newton's physics justified the Copernican model as revised by Kepler, with the correct force, gravity; the earth's rotation and revolution were naturally explained. Newton's work also resulted in very accurate predictions of planetary motions along elliptical orbits, as well as a revised version of Kepler's third law. The final result was a pleasing and accurate physical foundation for a heliocentric system.

7. The discovery of Neptune confirmed the predictive validity of Newton's law of gravitation within the solar system. The discovery of binary stars whose motions follow Kepler's laws verified the validity of Newton's law of gravitation beyond the solar system.

8. Newton's laws provide the mathematical tools for placing satellites in orbit around the earth, for setting the trajectories of spacecraft in the solar system, and for defining the escape speed of an object from a planet or even the solar system itself.

9. Newton's cosmology demanded that the universe be infinite in extent; if it weren't, gravity would bring all the material together, which had not been observed and did not appear to be remotely likely.

STUDY EXERCISES

1. Use one of Galileo's major telescopic discoveries to support the Copernican model and refute the traditional Aristotelian–Ptolemaic one. (Learning Outcome 4-1)
2. What important discovery of Kepler's about planetary orbits was ignored by Galileo? (Learning Outcome 4-3)
3. Describe Galileo's concept of inertia and contrast it to the Aristotelian one. (Learning Outcomes 4-2 and 4-4)
4. You have two balls of the same size and shape. One is lead, the other wood. You drop them together from the same height. What happens? Explain. (Learning Outcomes 4-2, 4-4, 4-5, and 4-6)
5. Imagine that you are out in space and push away from you an object having a mass identical to your own. What happens? Explain. (Learning Outcomes 4-4 and 4-5)
6. Describe two ways in which Newton's model of the cosmos differs from that of Copernicus. (Learning Outcome 4-9)
7. To answer the main physical objections to the Copernican model, use Newtonian physics to argue that the earth rotates on its axis and revolves around the sun. How do we know the conclusions are correct? (Learning Outcome 4-10)
8. Describe one way in which Newton and Kepler had similar ideas about the cosmos and one way in which their ideas differed. (Learning Outcome 4-9)
9. Give a simple example, different from those in the text, of each of Newton's laws of motion. (Learning Outcome 4-5)
10. Imagine that you hold a ball in your hand with your arm out to your side. You walk along quickly toward a target on the floor. You release the ball just as it reaches a point above the target. Use Newton's concept of inertia to predict where the ball will strike: behind, on, or in front of the target. (Learning Outcomes 4-5 and 4-6)
11. Consider an object shot upward from the earth with *less* than escape speed. Describe its motion. (Learning Outcome 4-12)
12. Consider a special satellite sent into space to orbit the earth at the same distance as the moon. How would the centripetal acceleration of such a satellite compare to the centripetal acceleration of the moon? (Learning Outcomes 4-7 and 4-8)
13. When you see Mars at opposition to the sun in the sky, the earth and Mars lie closest together on the same line from the sun. Consider the gravitational attraction of Mars on the earth at that time. How does it compare with the gravitational attraction of the earth on Mars at the same time? (Learning Outcomes 4-5 and 4-7)
14. In words, explain the physical meaning of Kepler's third law of planetary motion. (Learning Outcome 4-11)
15. How does the gravitational force of the earth on the apple compare to that of the apple on the earth? (Learning Outcomes 4-7, 4-8, and 4-9)

PROBLEMS AND ACTIVITIES

1. Refer to Appendix B to find out the distances of the planets from the sun (in AU) and the masses of the planets (relative to the mass of the earth, when the earth's mass = 1.0). Place these numbers in a table from Mercury outward, and then calculate the force of gravity on each planet from the sun.
2. From Newton's second law, $F = ma$, you can see that the acceleration of a mass is $a = F/m$. Hence, to find the centripetal acceleration of the force of gravity, divide both sides of Newton's law of

gravitation by *m*, the mass of the accelerated object (say, the earth moving around the sun). As in Problem 1, refer to Appendix B to find the relative distances of the planets and calculate the relative centripetal acceleration for each.

3. Use information from Appendix B to calculate the escape speed from the moon.

4. Imagine that you have a force of 10 newtons applied to a mass of 1 kg. What happens when you apply it to a mass of 5 kg? 10 kg?

5. What is the gravitational force between you and another person standing 10 meters apart? 20 meters? 100 meters? *Hint:* Estimate your mass!

6. A satellite near the surface of the earth completes its orbit once in 90 minutes. What is its centripetal acceleration? Compare your result to the acceleration of gravity near the earth's surface.

7. Use the information about the satellite in Problem 6 to find the mass of the earth. *Hint:* Apply Kepler's third law.

8. What is the approximate distance between the earth and Jupiter at opposition? Estimate this distance in AU and then convert to kilometers using $1 \text{ AU} = 1.5 \times 10^8$ km. Jupiter's brightest moon is called Ganymede. Its distance from the center of Jupiter is about 1.1×10^6 km. Given the smallest angle observable by the eye (about 1 arcmin), could Ganymede be seen as an object separate from Jupiter during Ganymede's maximum elongation from Jupiter?

9. The earth orbits the sun in an elliptical orbit. At aphelion it is farthest from the sun; at perihelion, closest. How does the amount of gravitational force between the earth and sun at aphelion compare to that at perihelion? Give a conceptual answer rather than a numerical one.

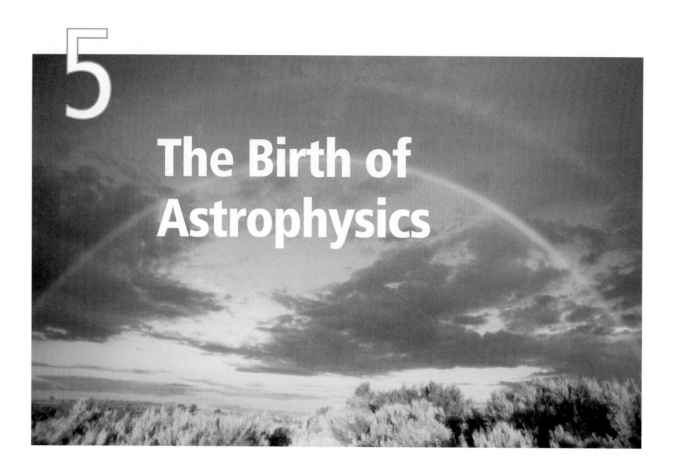

5 The Birth of Astrophysics

LEARNING OUTCOMES

After studying this chapter, you should be able to:

5-1 Describe the differences in the appearance of continuous, absorption, and emission spectra as seen through a spectroscope.

5-2 Explain how an understanding of spectra made it possible for astronomers to determine the chemical compositions of stars and the physical conditions inside their atmospheres.

5-3 Use Kirchhoff's rules to relate the three basic spectral types to the physical conditions of their production.

5-4 Describe how wavelength, frequency, and speed of light are related.

5-5 Describe the relationship between energy and frequency or wavelength for light and apply it to the spectrum.

5-6 Briefly describe the electromagnetic spectrum with examples from each major region.

5-7 Use an energy-level diagram to explain in general how atoms absorb and emit light.

5-8 Use the energy-level diagram of a hydrogen atom to explain how the Balmer series is produced, both as emission and absorption lines.

5-9 Explain in simple physical terms how absorption lines occur in spectra.

5-10 Describe the concept of the conservation of energy and apply it to ordinary and astrophysical situations.

5-11 Describe the concept of energy and the various types of energy; apply these to light and astrophysical situations.

5-12 Describe, sketch, and explain the three major types of spectra in graphical form.

CENTRAL CONCEPT

Matter produces light, and this light carries physical information about the sun, stars, and other celestial objects that emit it. Light comes in discrete units that are emitted and absorbed by atoms.

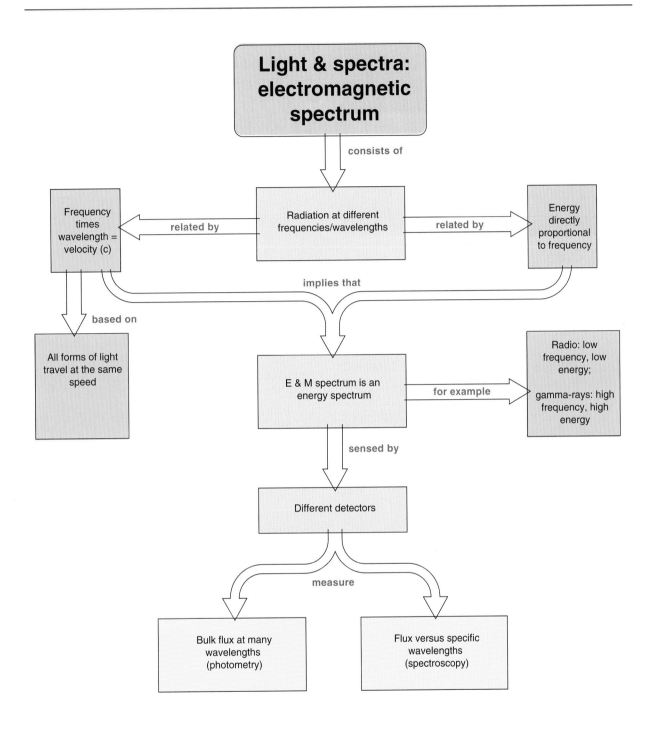

Chapter 5
Celestial Navigator™

Astronomers had for centuries charted the positions of the stars and planets. Their attention focused on the question of how the planets moved, a problem solved by Isaac Newton. In contrast, the stars played a passive role, forming a mere backdrop to the complex motions of the planets.

How could they find the physical makeup of the stars and planets? Astronomers had no means to bring a piece of a star or planet into the lab for

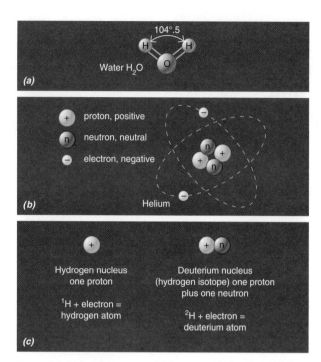

Figure 5.1 Molecules, atoms, and isotopes. (a) The chemical structure of a molecule of water, H_2O. (b) The structure of a helium atom. (c) The difference between a hydrogen atom and deuterium, an isotope of hydrogen, lies in the neutron.

examination; all they had at hand was the light funneled into their telescopes.

The birth of astrophysics at the end of the nineteenth century gave astronomers physical insight about light. Experimental attempts to understand light unexpectedly led to a study of atoms. The atoms in stars (and the sun) emit starlight (and sunlight), so an analysis of light reveals information about the physical conditions and chemical compositions of stars. Astronomers could finally penetrate the environments of the distant stars and the sun. Astrophysics transformed our conception of the cosmos. No more were the stars regarded as simple points of light. Now they appeared as other suns made of the same stuff found in our sun and the earth.

5.1 SUNLIGHT AND SPECTROSCOPY

Matter makes up the sun and stars – matter in the form of **atoms.** Atoms give off and absorb light; the sunlight we see originates from atoms. The structure of atoms can be investigated by analyzing the light

they emit and absorb. We also investigate the physical environment of atoms by analyzing light. To understand more of the sun (and stars), you need to understand how atoms and light interact.

Atoms and Matter

In the eighteenth and nineteenth centuries, chemists discovered that substances fall into two classes: chemical **elements** and **compounds.** Elements cannot be broken by chemical reactions into simpler substances. (See Appendix F for a table of the elements.) They are the most basic substances, such as hydrogen (H), helium (He), carbon (C), and oxygen (O). Although 92 elements occur in nature (and more have been created in the laboratory), many are rare; only a few dozen are common. Matter is mostly compounds, made of two or more elements. For example, water (H_2O) is a chemical compound composed of the elements hydrogen and oxygen (Fig. 5.1a).

A **molecule** is the smallest unit of a compound that still has the chemical properties of the compound. However, a compound can be broken down into elements, so a unit of matter smaller than a molecule must exist. An atom is the smallest unit of an element that displays the chemical properties of the element. A compound is created when atoms join together to form molecules.

A Model of the Atom

Physicists developed a useful model of the atom in the twentieth century, one that pictures a tiny, dense **nucleus** surrounded by rapidly moving electrons (Fig. 5.1b). The study of electricity revealed that matter has two opposite charges, positive and negative. Like charges repel each other and opposite charges attract. **Electrons,** which carry a negative charge, are low-mass particles (a mere 10^{-30} kg; see Appendix C). **Protons** and **neutrons,** which are about 2000 times as massive as electrons, make up an atom's nucleus. The protons are positively charged and the neutrons have no electrical charge. The nucleus of an atom, which contains protons, is positively charged and attracts the negatively charged electrons.

This attractive electrical force binds the electrons to the nucleus and holds the atom together. (A *strong*

Figure 5.2 A rainbow. Water droplets disperse the white light of the sun to make the continuous band of rainbow colors.

nuclear force binds the protons and neutrons together in the nucleus, despite the mutual repulsion of the positively charged protons.) The electrons whiz around the nucleus in orbits, but their distance is very great in terms of the size of the nucleus. Most of an atom is empty space! If the proton of a hydrogen atom were the size of a marble, the electron would be about a kilometer away from the nucleus.

Elements differ from one another because their nuclei differ. The nucleus makes the difference; the nuclei of different elements contain different numbers of protons. The nucleus of a hydrogen atom, for example, contains one proton; helium, two (Fig. 5.1b); and carbon, six. (See Appendix F.) The number of electrons orbiting the nucleus of a normal atom is the same as the number of protons in the nucleus, so the atom carries no net electric charge. The electrons determine the chemical properties of atoms.

Nuclei that contain the same number of protons but a different number of neutrons are called **isotopes** of the element. For example, heavy hydrogen, sometimes called *deuterium* (Fig. 5.1c), has one proton and one neutron. Ordinary hydrogen has one proton and no neutrons. They are the same element, but each atom in a deuterium molecule has more mass than the atoms of ordinary hydrogen, because of the extra neutron. (Appendix F gives only the most common isotope of each element.)

Most elements contain the same number of protons and neutrons, or a few more neutrons. For example, atoms of the most common isotope of carbon contain six protons and six neutrons.

Although an atom is neutral electrically, taking away or adding electrons can change this state. If the number of electrons is less than or greater than the number of protons (so the net charge is positive or negative), the atom becomes an **ion.** In most astronomical situations, we find positive ions, which have resulted from the loss of one or more electrons.

Simple Spectroscopy (Learning Outcome 5-1)

Breaking up light into its component colors to detect how atoms and light interact is called **spectroscopy.** You do a little spectroscopy when you look at a rainbow. Next time you see one, look carefully at the colors, which run from red to violet. What's happening? Raindrops are dispersing sunlight into a continuous band of colors – the visible rainbow (Fig. 5.2).

A prism does the same. Use a slotted piece of cardboard to pass a beam of sunlight through a prism. Let the light coming out of the prism fall on a white sheet. You'll find the sunlight spread into an array of colors, with red light bent the least and violet the most. This sequence of colors is called a **spectrum.** A spectrum with no breaks in it is called a **continuous spectrum;** it has a continuum of

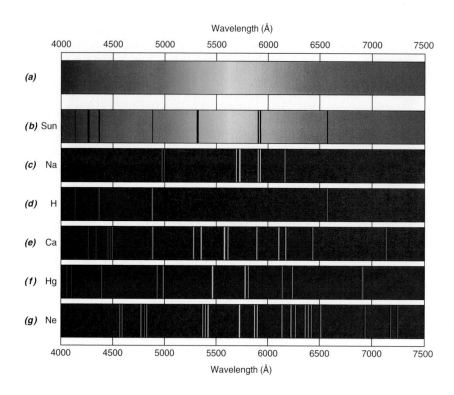

Figure 5.3 Major spectral types: the shorter wavelengths lie to the left, with wavelength increasing to the right. At top (a) is a continuous spectrum such as that from an incandescent lightbulb. Below it (b) is the absorption-line spectrum of the sun; only the most prominent dark lines are shown from the following elements: H, hydrogen (the Balmer series); Ca, calcium; Fe, iron; and Na, sodium. Below the sun's spectrum are the emission-line spectra of selected elements: sodium (c), hydrogen (d), calcium (e), mercury (f), and neon (g). Note how some bright lines of sodium, calcium, and hydrogen line up with the dark lines in the sun's spectrum, which indicates the presence of these elements in the sun's atmosphere.

shades of color. White light contains all these colors. You can think of white light as a mixture of all colors at equal intensity.

An inexpensive device to disperse white light into its colors is called a *diffraction grating*. If you have one, you can examine various light sources with it. An ordinary incandescent lightbulb shows a continuous spectrum (Fig. 5.3a). If you look at reflected sunlight, a continuous spectrum with some well-defined dark lines appears (Fig. 5.3b). A neon sign shows distinct, bright lines (Fig. 5.3g), many bunched in the red region of the spectrum. That's why such a sign appears red to our eyes.

With identifications of local sources, you can examine distant ones, such as streetlights at night (Fig. 5.4). Some may show continuous spectra; they are probably ordinary lightbulbs. Others may display strong emission in the blue and green; they have a mercury gas light. These lines combine in the greenish color of mercury street lamps. And yet others may emit strong lines in the yellow; these contain low-pressure sodium gas. They cast a ghastly yellow glow.

Going on a night spectra quest with a diffraction grating replicates an astronomer's work with spectra. The main difference: astronomical sources are

much, much farther away, so we need telescopes to scoop up their light.

Suppose you put another slotted piece of cardboard (or slit) behind your prism or grating to let just a single color pass through. You'd then see a bright line of that color, not a continuous spectrum. This line is called a **spectral line.** If you let sunlight pass through a prism and slit, and then magnify the spectrum, you'll see a continuous spectrum filled with dark lines (Fig. 5.3b). Light is missing from some of the colors. The dark lines, where these colors have been removed, or absorbed, are called **absorption lines,** and a spectrum exhibiting them is called an **absorption-line spectrum.** When light is emitted only at certain colors, as in the case of a neon sign, we speak of **emission lines** or **emission-line spectra** (Fig. 5.3c–5.3g).

Straight lines appear in a spectrum because the slit admitting the light is a straight-line source of light. Each line is an image of the slit in the light of one color.

5.2 ANALYZING SUNLIGHT

To unravel the message of sunlight, consider a few experiments with a **spectroscope,** an instrument for observing fine details in a spectrum.

Figure 5.4 Streetlights and their spectra. This night scene was shot through a diffraction grating, so the lights appear with their spectra above them. At far right are two lights with continuous spectra. The small light to the left of these two and the bright one on the bridge have strong emission in the blue and green. Near the center is a light with a spectrum with strong yellow lines. Turn the page horizontally to view the spectra with red on the right and violet on the left.

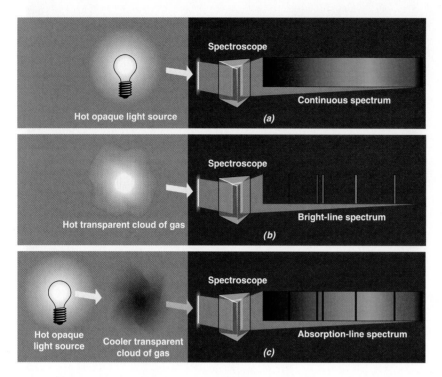

Figure 5.5 Basic spectral types as viewed through a spectroscope. (a) Continuous spectrum with no lines from a hot, opaque solid (the light filament). (b) Emission-line spectrum with a few lines from a hot, transparent gas. (c) Absorption-line spectrum with the same spectral lines as in (b); in this case, the continuous spectrum emitted by the light has passed through a transparent, cooler gas with the same chemical composition as in (b).

First, put sodium (in the compound form of table salt) in a hot, colorless flame. Visually, you'll see the flame turn yellow. Through the spectroscope, you'd see a series of bright lines, the brightest, a pair, in the yellow region (Fig. 5.3c). With the spectroscope, you can prove that different chemical elements give off different patterns of bright lines. Each chemical element displays a unique arrangement of bright lines when excited in the gaseous state. A particular spectral pattern demonstrates the presence of a particular element. Each pattern is unique to an element, as fingerprints are to people.

The atoms in the gas produce the bright lines. Different gases composed of different atoms produce different patterns of spectral lines. The bright-line spectrum of a hot gas is characteristic of the atoms in the gas. The pattern of the lines reveals which elements the gas contains.

Next, look at sunlight with a spectroscope. You'll see dark lines against a bright, continuous background of color (Fig. 5.3b). A pair of dark lines in the yellow region of the spectrum particularly stands out. Astronomers call this pair the *D lines.* Could these dark lines be related to the bright lines of sodium observed when salt is put in a flame? Could the dark lines in the sun's spectrum come from sodium in the sun?

To test this idea, pass light from a glowing solid (which gives off a continuous spectrum) through the sodium flame. A pair of dark lines appears at exactly the same position in the yellow region as the pair in the sun's spectrum. It must be sodium making these dark lines, removing yellow light from the solar spectrum. Now pass sunlight through a sodium flame into the spectroscope. What do you expect? That the added light of the sodium flame will make the lines less dark? Wrong! The pair of lines becomes even darker! The sodium in the flame absorbs even more of the sun's yellow light. That result proves that sodium made the lines. But why did the lines get *darker?*

Kirchhoff's Rules (Learning Outcomes 5-1, 5-3)

From experiments such as those described above, the German physicist Gustav Kirchhoff (1824–1887) formulated empirical rules of spectroscopic analysis, the determination of the physical state and the composition of an unknown mixture of elements. **Kirchhoff's rules** are as follows:

Rule 1. A hot and opaque solid, liquid, or highly compressed gas emits a continuous spectrum (Fig. 5.5a). Example: filament of a lightbulb.

Rule 2. A hot, transparent gas produces a spectrum of bright lines (emission lines). The number and colors of these lines depend on which elements are present in the gas (Fig. 5.5b). Example: a neon sign.

Rule 3. If a continuous spectrum (from a hot, opaque solid, liquid, or gas) passes through a transparent gas at a lower temperature, the cooler gas will cause the appearance of dark lines (absorption lines), whose colors and number will depend on the elements in the cool gas (Fig. 5.5c). Example: light from the sun.

The first rule says that an opaque, hot material produces a continuous spectrum. Recall that atoms, instead, usually emit and absorb at discrete colors. However, when the atoms are jammed together so densely that the material becomes opaque to light, a continuous spectrum appears. For instance, the sun is so hot that it is gaseous, but it is still opaque enough to produce a continuous spectrum.

A solar spectrum shows dark lines against a continuous background. What causes these dark lines? By Kirchhoff's third rule, you expect a cooler and less dense gas to lie between the visible surface of the sun and us. Such a gas could be in the atmosphere of the earth or in the atmosphere of the sun. Although gases in the earth's atmosphere (water vapor in particular) produce same lines in the solar spectrum, most dark lines are produced in the solar atmosphere.

Light from the continuous spectrum passes through a cooler atmospheric layer, which absorbs it to produce the dark lines. The positions of the lines in the spectrum, their colors, tell which elements are present. The solar composition determined from the dark lines relates only to the region of the sun that produces the absorption spectrum (its atmosphere).

With these ideas in mind, you should now understand what happens when sunlight passes through the sodium vapor to make darker D lines. The sodium in the flame is cooler than the sun's visible surface, so the sodium absorbs the light in the D lines. Though hot, the sodium gas removes some specific wavelengths from the beam of sunlight. That's why the sodium D lines got *darker* when sunlight passed through the sodium flame. This experiment also proves that the atoms in the cool gas of the solar atmosphere absorbed the colors that are missing from the continuous solar spectrum.

Whether the atoms in the gas emit or absorb depends on the physical conditions in the gas. Emission requires high temperatures in a transparent gas; absorption occurs when a continuous spectrum passes through a *cooler* transparent gas. In either case, the pattern of emission lines is the same as the pattern of absorption lines for a gas with the same chemical composition.

Here's the fundamental point concerning emission and absorption lines: *the lines absorbed by a gas from a continuous spectrum are the same lines emitted by the gas when energy is put into it.*

The statement above is one way to describe the conservation of energy. **Energy** allows events to happen in the universe. The concept of energy has fundamental importance because, although energy comes in many forms, the total amount of energy that an object has can be described and assigned a number. If you have a bunch of objects, the sum of their individual energies is the total energy of the group. Moreover, the law of **conservation of energy** says that the total energy of a group of objects, isolated from the rest of the world, has a constant value.

The Conservation of Energy
(Learning Outcome 5-11)

Energy comes in many different forms. Three common ones are **kinetic energy,** which is from motion; **potential energy,** which is stored up by position under an applied force; and **radiative energy,** which is carried by light. This chapter deals mostly with radiative energy.

All forms of energy are measured in the same unit: the **joule** (J). You do a joule of work when you exert a force of one newton over a distance of one meter. (Table 5.1 gives the energy outputs of some familiar objects and events.) If you lift an apple from the ground to over your head, you've done about one joule's worth of work. You have also changed the apple's energy. If you release it, it will fall to the ground. The apple has acquired potential energy (from your work on it), which is transformed to kinetic energy when the apple falls. Though the forms of energy have changed, the total is still the same – it has been conserved.

Table 5.1 Some Energy Outputs

Energy Source	Total Energy (J)
Sun's radiation (1 year)	10^{34}
Volcanic eruption	10^{19}
H-bomb	10^{17}
Thunderstorm	10^{15}
Lightning flash	10^{10}
Barrel of oil	10^{9}
One medium pizza	10^{7}
Battery (D cell)	10^{4}
Baseball pitch	10^{2}
Typing (per key)	10^{-2}
Flea hop	10^{-7}

Kinetic Energy (Learning Outcome 5-11)

Kinetic energy is associated with motion. You can understand it in terms of work. To stop a moving object requires an obvious effort. The more massive the object or the faster its motion, the greater the effort (work!) needed to stop it. For instance, it is easier to grab and stop a bicycle traveling at 5 km/h than one moving at 10 km/h. You'd not want to try that with a massive car, even at 5 km/h!

The kinetic energy of a moving object depends on its mass and its speed: it is directly proportional to the mass and to the square of the speed (Enrichment Focus 5.1). So a bicycle moving at 10 km/h has four times the kinetic energy of one that travels at 5 km/h. At 15 km/h, it has nine times the kinetic energy of one that goes at 5 km/h. Investigators use this concept when unraveling an automobile accident. With the brakes locked, a car that has been clocked at 100 km/h will skid four times as far as one moving at 50 km/h, if the cars have equal mass. The faster car has four times the kinetic energy, and so stopping it takes four times the work.

You must distinguish between the kinetic energy of a single object, such as a baseball, and the average kinetic energy of the particles that make it up. A baseball's kinetic energy allows it to go somewhere, but that property doesn't affect the microscopic motions of its atoms, which move small distances at random within the ball. The motion of the atoms makes up the **thermal energy** of the ball – a measure of the total kinetic energy of a large collection of particles. When you add energy to a gas, the average value of the random speeds of its constituent molecules increases. What have you done? You have used a hot flame, say, to raise the thermal energy of the gas. **Heat** is the thermal energy that is transferred from one body to another. *Heat always flows from a hotter body to a colder one.*

Temperature measures the average kinetic energy of particles in, say, a gas. That, in turn, depends on the average speed of the particles. The temperature scale that indicates this average speed most directly is called the **kelvin** (K) scale; at 0 K, all microscopic motions have ceased. (See Appendix

---

ENRICHMENT FOCUS 5.1

Kinetic Energy

The kinetic energy of an object depends on both its mass and its speed. In algebraic form, the expression for kinetic energy is

$$KE = \frac{1}{2}mv^2$$

where KE is the kinetic energy (in joules), m the mass of the moving object (in kilograms), and v its speed, which is squared. Note that moving at a higher speed results in a *large* increase in kinetic energy. Many highways in the United States have speed limits of 55 mph.

If a car travels at 75 mph instead, its kinetic energy is roughly doubled: $(75/55)^2 \approx 2$.

Let's do an example. In its orbit, the earth moves at about 30 km/s. What is its kinetic energy? The earth's mass (Appendix C) is about 6×10^{24} kg, so

$$KE = \frac{1}{2}(6 \times 10^{24} \text{ kg})(30 \times 10^3 \text{ m/s})^2$$
$$= \frac{1}{2}(6 \times 10^{24} \times 9 \times 10^8)$$
$$= 2.7 \times 10^{33} \text{ J}$$

That amount of kinetic energy is equivalent to the amount of radiative energy that the sun emits in a month!

A for temperature units.) On this scale, your body temperature is about 310 K.

> **TO SUM UP.** Heat is the flow of thermal energy from a hotter body to a cooler one; temperature measures the average kinetic energy of the random motions of these particles. These are all manifestations of energy, or the ability to do work.

You should now be able to tell the difference – physically – between a fast baseball and a hot baseball (and a fast, hot baseball!). What is it?

Potential Energy (Learning Outcome 5-11)

Potential energy is the stored energy of position, as in a stretched rubber band. Fuels have potential energy, which is released when they are burned. On the chemical scale, changes within and between the electrical charges of molecules result in the release of the energy. These, too, are related to positions and are a form of potential energy. Why? Basically because electrical charges create forces among themselves. Likewise, in an atom, the electrons orbiting the nucleus have specific kinetic energies that are due to their motion. They also have specific potential energies that correspond to their distances from the nucleus and their electrical attraction to the protons there.

One significant kind of potential energy relates to gravitational attraction. On the earth, all masses feel the earth's pull, although some, such as an apple held in your hand, do not immediately respond to the force. When you drop the apple, however, it gains kinetic energy. The higher the point from which you drop it, the greater the kinetic energy it has attained by the end of its fall. The position above the earth's surface is related to the total amount of potential energy the object has. Energy is transformed from one form to another (potential to kinetic), yet the total energy of the apple remains constant. As it loses potential energy, it gains an equal amount of kinetic energy.

Energy is naturally transformed from one form to another, but as long as *all* the energy is carefully tallied, the total amount remains the same – it is, again, conserved. One way to phrase the concept of the conservation of energy is: *energy cannot be created or destroyed; it may be transformed, but the total does not change.* The absorption and emission of light by atoms obey this conservation law. You will see the conservation of energy applied throughout this book. (This principle is modified somewhat because of Einstein's discovery that matter is actually a form of energy: to come in Section 7.2.)

5.3 SPECTRA AND ATOMS

How can the physical conditions in stars be discovered from their spectra? The spectral code of the patterns of lines was cracked in a great revolution of physics in the twentieth century: the **quantum theory** and its explanation of the nature of the atom and of light. To understand the power of spectroscopy, you must first know something about light.

Light and Electromagnetic Radiation (Learning Outcomes 5-5, 5-6)

Today we view light as having *both* particle and wave properties. Light sometimes behaves as waves and sometimes as particles, depending on how it is observed. Both models are needed to describe the properties of light completely. In either case, light carries energy – radiant energy. I will use both models; let me emphasize first some wavelike features of light.

Waves (Learning Outcome 5-4)

You probably have some experience with waves. Imagine that you're at a beach with waves arriving at regular intervals. You time them with your watch. The number of waves that arrive in a given period is the **frequency.** For instance, one wave hitting the shore every minute is a frequency of one per minute. The distance between the crests (peaks) of these waves is their **wavelength.** For example, if the crests are 10 meters apart, the wavelength is 10 meters. The wave velocity indicates how fast and in what direction the waves are traveling. Waves that move 10 meters in a minute have a speed of 10 m/min.

Note that waves do *not* involve the mass motion of material over long distances. If you are in a boat on a lake with no currents, waves from passing motorboats cause your boat to bob up and down, but it stays in the same location. A wave carries energy of motion from one place to another but does not transport material.

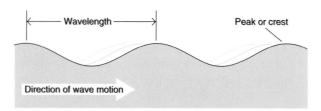

Figure 5.6 Some properties of waves. Every wave has a peak and a trough: the distance from one peak of a wave to the next is its wavelength. The number of waves that pass by each second (per second is one hertz) is the frequency.

TO SUM UP. Waves have three fundamental properties (Fig. 5.6): wavelength, frequency, and velocity. The wavelength is the distance between two successive crests of a wave, the frequency is the number of waves that pass you each second, and the velocity is the distance covered in one second by a crest traveling in a certain direction.

The units of velocity are length per time. When you multiply frequency (number per time) and wavelength (length), the result equals wave velocity (length per time): frequency times wavelength equals velocity. Or, as an equation,

$$f \times \lambda = v$$

where f is the wave's frequency, λ the wavelength (in meters, say), and v the velocity (in m/s). This fundamental rule applies to waves of all kinds. It tells you that for waves traveling through some material at constant velocity, a change in frequency requires a change in the wavelength.

For light in a vacuum, the velocity is the same for all wavelengths: 299,793 km/s. This value is usually designated by a lowercase c and you can recall it by its approximation, 300,000 km/s (3×10^8 m/s). So

$$f \times \lambda = c$$

where again λ is the wavelength and f the frequency. Because c is constant for all kinds of electromagnetic radiation, *different wavelengths must have different frequencies.*

The Electromagnetic Spectrum (Learning Outcome 5-6)

Visible light is one type of wave produced whenever electric charges are accelerated. *Radio waves* constitute another type. A transmitter uses electricity to move electrons rapidly back and forth in an antenna. When the electrons are accelerated in a periodic fashion, they produce radio waves at a certain wavelength. Such radiated energy, generated by accelerated electrical charges, is termed **electromagnetic radiation.** The range of all different wavelengths of electromagnetic radiation makes up the **electromagnetic spectrum** (Fig. 5.7); visible light covers only a small part of the electromagnetic spectrum. All forms of electromagnetic radiation have the same physical nature. They make up the cosmic rainbow. (See the Electromagnetic Spectrum Celestial Navigator™.)

Although the wavelength of electromagnetic radiation is sometimes measured in meters or centimeters, scientists often use special wavelength units for different regions of the electromagnetic spectrum. For example, visible light is usually measured by astronomers in *angstroms* (abbreviated Å), where 1 Å is 10^{-10} m. Physics people tend to use *nanometers* (abbreviated nm), where 1 nm is 10^{-9} m, so 1 Å = 0.1 nm. In the infrared region, the unit of length commonly used is the *micrometer* (abbreviated μ), with 1 μm equal to 10^{-6} m.

In the radio region, astronomers commonly talk in terms of frequency rather than wavelength. The unit of frequency is the **hertz** (abbreviated Hz), which is one cycle (or vibration) per second. When you see a wave go by, from peak to peak, it has gone through one cycle. Even in the radio region, many cycles go by in a second. For example, the AM band (your car radio) covers the range from 540 to 1650 kHz (kHz is the abbreviation for kilohertz, or 1000 Hz). The FM band ranges from 88 to 108 megahertz; 1 megahertz (abbreviated MHz) is 1 million hertz. Radio astronomers work at such high frequencies that they use the unit called the *gigahertz* (abbreviated GHz), 1 billion hertz. Police radar used in the United States to trap speeding motorists typically operates at a frequency of about 10 GHz.

In a sense, a spectroscope works like a radio tuner. Both devices arrange a part of the electromagnetic spectrum by spreading it out from shorter

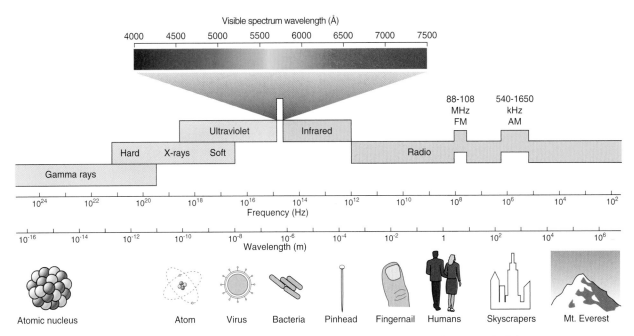

Figure 5.7 The electromagnetic spectrum from the shortest (gamma rays) to the longest (radio) wavelengths. Note that visible light makes up just a tiny piece of the complete spectrum. (Adapted from a diagram by E. J. Chaisson.)

to longer wavelengths. A certain wavelength can then be selected for examination. You make a selection on a radio by tuning it to a specific station that broadcasts at a certain wavelength. A spectroscope also can be "tuned in" to look at a particular line of some region of wavelengths.

Atoms, Light, and Radiation
(Learning Outcome 5-2)

Although a wave model for light was championed during the nineteenth century, it did not explain how atoms produced light. Astronomers were puzzled by spectral lines from stars. Why should stars emit light at certain wavelengths and not at others? The pattern of lines of hydrogen appeared in a simple series in the visible region of the electromagnetic spectrum. This series of hydrogen lines was named the **Balmer series** (look back at Fig. 5.3d), after the Swiss mathematician Johannes J. Balmer (1825–1898), who, in 1885, devised an empirical formula describing the regular sequence of lines. Because the Balmer series is a patterned sequence, the inference was that emission and absorption

were related to a simple structure of atoms. But what was that structure?

In 1911 the New Zealand–born physicist Ernest Rutherford (1871–1937) proposed a model for the atom in which all the positive charge is concentrated in a tiny nucleus, with a surrounding negatively charged cloud of electrons. Rutherford imagined that the mutual attraction of unlike charges binds the atom. If an atom loses one or more of the electrons, it becomes an ion. Physicists had noticed that an element's spectrum changes when it is ionized. For example, when hydrogen is ionized, it no longer produces the Balmer series of lines. The physicists inferred that the arrangement of electrons somehow determines its light-emitting properties.

In 1900 the German scientist Max Planck (1858–1947) announced a revolutionary idea about atoms and light. Perplexed by experimental work on spectra, Planck developed a startling new model for light: radiating matter emits light in discrete chunks of energy that he called **quanta.** From this basis Planck explained some of the observed properties of radiation.

ENRICHMENT FOCUS 5.2
Energy and Light

The energy-level diagram of an atom describes a situation analogous to that of objects pulled by gravity. Instead of gravitational force, however, we have the force of electrical attraction between the positively charged nucleus and the negatively charged electrons. When an electron falls from one energy level to a lower one, it loses potential energy, which must manifest itself in some other form, in this case a photon (radiative energy). To reverse the process requires the electron to gain energy, again produced from the absorption of a photon having exactly the energy required to raise the electron to the next level.

The amount of energy carried by light depends on its frequency (or wavelength). The fundamental relation is

$$E_{photon} = hf$$

where E is the energy in joules, h a constant (called *Planck's constant*) with a value of 6.63×10^{-34} J · s, and f the frequency in hertz. Note that light with twice the frequency carries twice the radiative energy; three times the frequency, three times the energy; and so on.

Since wavelength and frequency are related by the wave equation

$$f = \frac{c}{\lambda}$$

we can rewrite the energy equation as

$$E = \frac{hc}{\lambda}$$

with the wavelength λ in meters.

A few examples. Consider police radar operating at a microwave frequency of 10 GHz. What is the energy of a single photon? Since 1 GHz = 10^9 Hz, then

$$E = hf$$
$$= (6.63 \times 10^{-34} \text{ J·s})(10 \times 10^9 \text{ Hz})$$
$$= 6.63 \times 10^{-24} \text{ J}$$

Contrast this to light at 5000 Å in the visible part of the spectrum. Since 1 Å = 10^{-10} m, then

$$E_{photon} = \frac{hc}{\lambda}$$
$$= \frac{(6.6 \times 10^{-34} \text{ J·s})(3 \times 10^8 \text{ m/s})}{(5000 \text{ Å}) (10^{-10} \text{ m/Å})}$$
$$= 4.0 \times 10^{-19} \text{ J}$$

so that the visible photon has about 10^5 times more energy than the microwave one. However, an intense beam of microwaves (for example, in a microwave oven) can carry more energy than a beam of visible light, say from a flashlight. Each microwave photon carries less energy than each visible light photon, but the oven generates many photons.

Taking up this quantum idea, Albert Einstein (1879–1955) showed that each quantum, usually called a **photon** when talking about light, carries an amount of energy that is directly related to its frequency (Enrichment Focus 5.2). Light of higher frequency (shorter wavelength) transports more energy than light of lower frequency. For example, an x-ray photon carries much more energy than a photon of visible light. This relationship between energy and frequency is a direct one: a photon with twice the frequency carries double the radiative energy. Similarly, a photon with half the frequency transports half as much energy. For example, a photon with a wavelength of 5000 Å carries 4×10^{-19} J; light at 2500 Å has 8×10^{-19} J.

With this idea in mind, you should now realize that the sun's continuous spectrum is, in fact, an energy spectrum, from lower (infrared) to higher (ultraviolet) energies. You have seen that atoms emit and absorb certain colors at specific wavelengths. For each wavelength you can imagine a photon carrying off a certain amount of energy. Atoms must absorb and emit energy in the form of whole photons. This quantum model emphasizes that in the emission or absorption of light, an atom loses or gains energy in discrete amounts, as the energies of individual photons.

Solving the Puzzle of Atomic Spectra (Learning Outcome 5-7)

The Danish scientist Niels Bohr (1885–1962) meshed Planck's quantum and Rutherford's atomic pictures. The resulting scheme, known as the **Bohr model**

of the atom, explained and predicted the absorption and emission of photons.

Bohr pictured atoms as having many possible arrangements of electrons. He imagined that the emission or absorption of light arises from a **transition** between electron arrangements. In line with Planck's theory and observed spectra, Bohr realized that only certain electron arrangements are permitted. Similarly, electron transitions can occur only between these special arrangements. You can visualize this atomic model by using concepts of energy. For a given nuclear charge (which is positive and varies from element to element), electrons have available a large number of possible states with specific energy values or **energy levels.** Each level corresponds to a certain orbit the electrons can have – a certain total energy, which is the sum of its potential and kinetic energies.

Consider the hydrogen atom: it has one proton for a nucleus and one electron attached by the electric force between itself and the proton. (Other atoms have more protons and electrons but are held together in the same way.) The electron has a certain total energy. The essence of quantum theory is that electrons remain in stable states of specific energies, and each has a particular orbit (Fig. 5.8a). For example, the electron in the lowest energy level (called the **ground state**) securely orbits the nucleus. The electron must gain energy to move out to larger orbits, but not just any amount of energy. The orbits and energy levels follow strict spacing rules, and an electron can only gain an exact amount of energy as it moves from one orbit to the next.

Energy can be added to the atom either by collision with another particle or by absorption of a photon with sufficient energy. The electron jumps up one or more energy levels (Fig. 5.8b). The atom is then agitated and the process of reaching this state is called **excitation.** The excited condition does not last long; the electron drops to a lower level in about 10^{-8} s. To descend to a lower energy level, however, the electron must lose some energy (Fig. 5.8c). The electron achieves such a loss by emitting a photon with an amount of energy equal to the amount it needs to lose. This energy is directly related to the photon's frequency.

If an atom gains enough energy, the electron flies away from the nucleus. The atom is then ion-

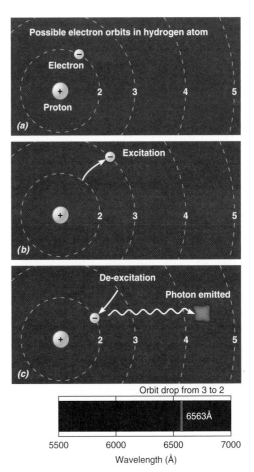

Figure 5.8 Energy orbits for hydrogen. (a) The first five orbits for an electron; energy increases outward from the nucleus. (b) The addition of energy can excite an electron to a higher orbit, such as from 1 to 3. (c) When an electron drops an orbit, it loses energy, usually as a photon; here in the drop from 3 to 2, a photon at 6563 Å is emitted.

ized, and the process is called **ionization.** The loss of an electron changes the energy arrangements available and also changes the atom's spectrum. For a given atom and electron in an atom, a certain minimum energy is necessary to break the electron loose from the electrical grip of the nucleus – the *ionization energy.* For hydrogen, the ionization energy is a mere 2.2×10^{-18} J.

Energy Levels (Learning Outcomes 5-7, 5-8)

So far, I've been describing energy levels in terms of orbits of an electron. This description is generally valid, but it may mislead you into imagining that the electron orbits are sharp circles. They are not.

The electrons actually orbit in fuzzy clouds, many of which are not even spherical. A better way to talk about how atoms emit and absorb light is to conceive of energy levels themselves without images of orbits of any kind. This picture works even for atoms more complex than hydrogen.

As an analogy, imagine jumping up and down the stairs. Each step is an incremental change in the ball's energy, and only full-step changes are permitted. If you try to add less than one step's worth of energy, you remain at its initial level. If you add exactly one step of energy, you move up one step. When you lose energy, you descend in steps until you hit the floor at the bottom of the staircase; this is the ground level, the state of lowest energy. The natural tendency of an atom is to attain this ground level. The key point is this: since electron transitions can occur only between certain energy levels (stairs), an atom must produce a particular pattern of lines (colors of light).

Let's apply the energy-level concept to hydrogen with an *energy-level diagram.* Every upward step of an electron requires the absorption of energy (Fig. 5.9a) and every downward one the emission of energy (Fig 5.9b). The greater the energy difference needed for a transition, the higher the frequency (hence the shorter the wavelength) of the photon produced from that transition. All transitions to and from the lowest energy level (level 1) involve large energy changes (short-wavelength photons), so they correspond to wavelengths in the ultraviolet range of the spectrum. The eye cannot see this set of lines, called the Lyman series, because it lies in the ultraviolet. The set of transitions down to and up from the second energy level (level 2) is the Balmer series (Fig. 5.9c); it lies mainly in the visible region of the spectrum. The hydrogen absorption lines in the sun's spectrum are those in the Balmer series.

Keep in mind that the Balmer series involves the set of spectral lines arising from transitions between the second energy level of hydrogen and the levels above the second. The transitions are upward from level 2 for absorption and downward to level 2 for emission. Note again that the larger the transition, the higher the energy of photon emitted, and the shorter its wavelength. The H-beta photon, for example, which comes from a downward transition from level 4 to level 2, has a wavelength lower than that of H-alpha (4861 versus 6563 Å: Fig. 5.9c).

Also, even though one atom can emit (or absorb) only one line at a time, many atoms can undergo different transitions at the same time, with the result that many lines are visible at once. Note that in all these cases the atom acts to conserve energy by transforming it among various forms: radiative, potential, and kinetic.

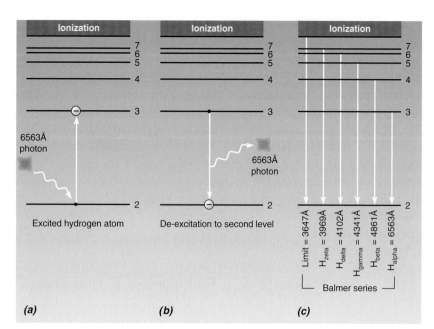

Figure 5.9 Energy-level diagrams for the series of hydrogen. The energy-level differences are drawn to scale. (a) Absorption of a 6563-Å photon by an excited atom boosts an electron upward to level 3 from level 2. (b) When the electron drops from level 3 to 2, the atom emits a 6563-Å photon. (c) The Balmer series from 3647 to 6563 Å involves transitions to and from the second energy level. (Note that level 1 is not shown.)

Other Atoms (Learning Outcome 5-2)

I have concentrated on the hydrogen atom because it has the simplest characteristics of all the elements and plays a dominant role in the cosmos. Also, the Bohr model correctly describes the emission and absorption of light by hydrogen atoms.

The simple Bohr quantum model needs drastic modification to work with other elements. Despite complications, one essential point endures: each atom and each ion (even of the same atom, for more than one electron can be lost from bound states) has its own unique set of electron energy levels, so each has its unique energy-level diagram. Because spectral lines are produced by electronic transitions between energy levels, each element or ion has a unique set of spectral lines. Hence, the study of spectra reveals key information about the internal structure of atoms.

Let me give schematic examples with two hypothetical elements, A and B (Fig. 5.10). Note the different spacings of the energy levels of each element. Also note the correspondence between energy differences of each transition and color (or wavelength). Specifically, transitions with the smaller amounts of energy (smaller differences between energy levels) result in longer wavelengths of light (in the red region of the spectrum).

In a sense, atoms are tuned. They respond to energy input by resonating at their natural frequencies. The energy levels are like the keys on a piano, which, when struck, cause their respective wires to resonate. You can go up or down the keyboard in whole steps or in half steps or multiples of these, but no interval smaller than a half step can be played. The natural resonators of atoms are their energy levels, so that the light emitted and absorbed by atoms provides information about their natural resonances, much as the playing of the keys of a piano tells you how the instrument is tuned.

CAUTION. Do not confuse the energy of photons with the brightness of emission lines! The energy determines the wavelength (color). The total number of photons emitted at a given wavelength sets the brightness or intensity of the line.

In summary, the electrons in atoms are stable only at certain energies. Add sufficient energy and an electron moves to a higher energy level. Shortly it drops to a lower energy level and emits a photon. The bigger the downward jump, the more energy the photon has. Any jump, however, produces or absorbs a photon with a specific energy and wavelength. Atoms absorb and emit photons, producing absorption and emission lines, by means of this discrete jumping of electrons. (See the Spectra Celestial Navigator™.)

5.4 SPECTRA FROM ATOMS

So far I have represented spectra by a picture or diagram with the spread-out colors so that lines are visible (as in Fig. 5.3). Astronomers also show spectra by using graphs that plot brightness or intensity against wavelength (Fig. 5.11). You will see pictorial and graphic representations in this book, so you should know how to use both. A continuous spectrum (Fig. 5.11a) has no sharp dips or peaks, even though the brightness or intensity may vary.

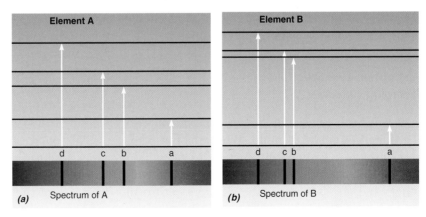

Figure 5.10 Different spectra from hypothetical element A (a) and element B (b), which have different sets of energy levels. Note how energy differences and wavelength correlate: the larger the difference, the shorter the wavelength of the absorbed photons.

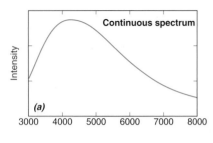 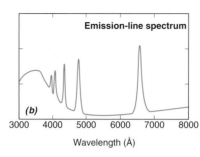 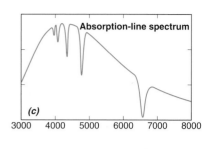

Figure 5.11 Graphical representations of the intensity or brightness profiles of the main spectral types: (a) continuous, (b) emission-line, and (c) absorption-line. The intensity is plotted versus wavelength, so that emission lines rise above the level of the continuous spectrum and absorption lines fall below it. Sharp peaks correspond to emission lines, sharp dips to absorption lines. The emission lines in (b) are those of the Balmer series of hydrogen. The same lines are shown in absorption in (c), against the background of the continuous spectrum of (a).

Emission lines appear as sharp peaks (Fig. 5.11b); absorption lines as sharp dips (Fig. 5.11c). Note that Fig. 5.11a represents the spectrum of Fig. 5.3a (a continuous spectrum of a hot, opaque source); 5.11b that of 5.3d; and 5.11c that of 5.3b with only the Balmer absorption lines.

What excites atoms for emission or absorption? Imagine a box filled with a gas. The atoms of the gas collide with one another (and with the sides of the box). Heat up the gas. The atoms then move faster and knock together harder. Collisions can bump electrons into higher energy levels; the harder the collision, the higher the level. This process is called *collisional excitation*. Some of an atom's kinetic energy is transferred to an electron of another atom. Photons can also excite atoms when absorbed, but only certain photons, namely those with energies corresponding to the difference in energies of two energy levels of the atom. This process is called *photon excitation*. Whether excited by collisions or by photons, the excited atom usually radiates quickly, sending off a photon in any direction. In some cases, a particle strikes an excited atom and carries off its energy before the excited atom has emitted a photon.

Absorption-Line Spectra
(Learning Outcomes 5-2, 5-7, 5-9)

Now return to the sodium vapor experiment with the quantum model in mind. When sodium chloride burns in a flame, individual atoms of sodium are released from the salt. Some atoms collide with others and are excited. As the electrons return to lower levels, they emit photons of yellow light at wavelengths of 5890 and 5896 Å – the sodium D lines.

Let's follow a sodium D-line photon (5890 Å) on its way from the sun to the spectroscope through the sodium vapor cloud. If a photon of just this energy encounters an unexcited sodium atom, the sodium atom will absorb the photon. The gas contains enough sodium atoms to absorb almost all the photons that try to pass through. The sodium atoms don't absorb any other wavelength in the visual region, so other photons with different wavelengths will pass right through the gas, preserving the continuous spectrum. A dark line appears at 5890 Å because the sodium gas absorbed these photons.

What happens to the sodium atom that has been excited by the photon absorption? An excited atom quickly emits a new 5890-Å photon. The original 5890-Å photon was headed directly for the spectroscope slit before a sodium atom absorbed it. The brand-new 5890-Å photon can be reemitted in any direction (Fig. 5.12). Very rarely will it be emitted in the same direction in which the original photon was traveling; only very few of these new photons enter the spectroscope slit. That's why the sodium line in the sun's spectrum becomes *darker* when passed through the sodium vapor. If you observe the sodium vapor from the side without the sun behind it, you would see only emission lines from the reemitted 5890-Å photons.

This basic idea of how atoms in transparent gases produce line spectra applies to the sun and also to stars, because their spectra resemble the spectrum of the sun. The absorption lines arise as follows. Low down in the sun's atmosphere, the

Figure 5.12 Paths of photons from the sun through a hot sodium flame and into a spectroscope. Photons with wavelengths of 5890 Å are scattered out of the beam, with the result that an absorption line is observed at this wavelength. If you observed the flame from the side, you would see the 5890 Å line in emission.

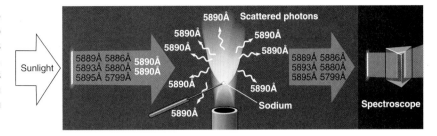

solar gas is compressed and opaque; here the continuous spectrum is produced. Above this region, the gas is transparent and cooler; these layers of the sun's atmosphere are the last barriers through which the continuous radiation must pass before it escapes. Atoms and ions at this level in the atmosphere absorb light at their characteristic wavelengths from the continuous spectrum and create the absorption lines in the sun's spectrum. The atmospheres of stars produce absorption-line spectra by similar atomic processes.

Most stars have dark-line spectra. Figure 5.13 is one plotted as intensity against wavelength for a star that has a higher temperature than the sun at its surface. The overall trend is the star's continuous spectrum; note that it peaks around 4000 Å. The sharp dips in the spectrum are absorption lines; the strongest ones here are those from the Balmer series of hydrogen. Compared to the sun's spectrum, this star has much more intense Balmer lines. (Chapter 14 will explain this difference.)

By matching the spectral lines of stars to those made in labs from known elements, astronomers inferred that stars contained almost all elements found on the earth. One important difference emerged, however: in contrast to the earth, stars and the sun contain mostly hydrogen. (Details will be found in Chapter 12 for the sun and Chapter 13 for other stars.) The similarity of spectra and chemical composition gave astronomers a new insight about the nature of the stars: stars are other suns.

Figure 5.13 Spectrum of a star hotter than the sun (called HD116608; it has a surface temperature of about 9400K). The general shape of this intensity graph shows the continuous part of the spectrum; the sharp dips are absorption lines. The strongest of these are from the Balmer series. (From *A Display Atlas of Stellar Spectra* by B. Margon, Department of Astronomy, University of Washington.)

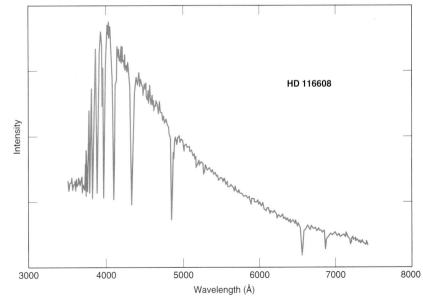

KEY CONCEPTS

1. A spectroscope spreads light into its component colors, and this display is called a *spectrum.* Three general types of spectra appear through a spectroscope (Table 5.2): continuous, with a smooth spread of colors; emission, which shows discrete, bright lines; and absorption, in which dark lines appear in a continuous spectrum. Spectra can be represented in graphical form, as a plot of intensity versus wavelength (such as in Fig. 5.11).

2. The sun and most other stars have absorption-line spectra. By matching the dark lines in a star's spectrum with those produced by elements in the lab, the chemical composition of a faraway object can be inferred; this technique works because each element has a unique set of spectral lines. These lines are determined by the arrangement of the electrons bound to the nucleus of the atom.

3. Kirchhoff's rules of spectral analysis describe the physical conditions under which each type of spectrum is produced (Table 5.2): (a) continuous spectrum from a hot opaque gas, liquid, or solid; (b) emission-line spectrum from a hot transparent gas; (c) absorption-line spectrum from a continuous spectrum passing through a cooler, transparent material (usually a gas).

4. Light has wave properties; red light has longer wavelengths than blue light. Light also carries radiative energy: the shorter the wavelength, the greater the energy. So blue light has more energy per photon than red light. The total range of wavelengths of light is called the *electromagnetic spectrum,* which goes from very long radio waves to very short gamma rays. All forms of electromagnetic radiation travel at the same speed, about 300,000 km/s.

5. In an energy-level model for the hydrogen atom, Balmer lines arise from an electron starting in the second level and jumping to any higher level (for absorption lines) or dropping from a higher level to the second level (for emission lines).

6. The quantum model for the emission and absorption of light by atoms views the electrons as playing a critical role; electrons can inhabit only stable regions called *energy levels.* Electrons can move to higher levels only if they gain energy; when they drop to lower levels, they give off energy (usually as photons); the greater the difference between levels, the greater the energy in the emitted photons. Electron transitions between energy levels produce (downward) and absorb (upward) light.

7. Spectroscopic analysis reveals the properties of stellar atmospheres, whose physical environments and chemical compositions can be inferred like those of the sun.

8. A key rule in the operation of the universe is that energy is conserved, although it can mutate into many forms. Atoms work to transform potential and kinetic energy into radiative energy and also do the reverse. Molecules can do the same.

9. Atoms are the basic units of matter. The number of positively charged protons in an atom's nucleus determines the element; the number of electrically neutral neutrons determines the isotope. The number of negatively charged electrons attached to the nucleus equals the number of the protons. The electrons and their energy levels give atoms their ability to manage light, and so reveal the universe to us.

Table 5.2 Basic Types of Spectra

Spectrum Type	Appearance	Physical Conditions
Emission	Distinct bright lines against a dark background	Hot, transparent gas
Continuous	Smooth blend of all colors	Hot, opaque gas, liquid, or solid
Absorption	Distinct dark lines against a bright background of colors	Light from a hot, opaque gas, liquid, or solid passing through a cooler, transparent material

STUDY EXERCISES

1. You throw ordinary table salt, which contains sodium, into the flame of a Bunsen burner. What kind of spectrum do you see when you look at the vaporized salt with a spectroscope? (Learning Outcome 5-1)
2. How do we know that sodium exists in the sun? (Learning Outcomes 5-1, 5-2, and 5-3)
3. The spectra of most stars look like the spectrum of the sun: the absorption type (see Fig. 5.13). How can particular elements be identified in these spectra? (Learning Outcome 5-2)
4. Suppose you had a box of positive hydrogen *ions.* What kind of *line* spectrum would they produce? (Learning Outcome 5-9)
5. The Balmer lines in the sun's spectrum are absorption lines. Do electrons jump up or fall down energy levels to produce them? (Learning Outcome 5-8)
6. Arrange the following kinds of electromagnetic radiation in order from the least to the most energetic: x-rays, radio, ultraviolet, infrared. (Learning Outcome 5-5)
7. Use Kirchhoff's rules to explain why the spectrum of the moon resembles that of the sun. (Learning Outcome 5-3)
8. What happens to a hydrogen atom when it absorbs light with enough energy to knock off its electron? (Learning Outcome 5-9)
9. Use Kirchhoff's rules to predict what would happen if the emission-line spectrum from a hot, transparent gas passed through another transparent gas that was *hotter* and had a different chemical composition. (Learning Outcome 5-3)
10. Consider a planet orbiting the sun. At what point is its kinetic energy the greatest? The least? What is the value of the total energy of the planet at these two points? (Learning Outcomes 5-10 and 5-11)
11. Does light of different wavelengths travel at the same speed or at different speeds? (Learning Outcome 5-4)
12. What region of the electromagnetic spectrum contains electromagnetic radiation of the longest wavelengths? (Learning Outcome 5-6)
13. In the energy-level model, when an atom absorbs a photon, what generally happens to one of the electrons? (Learning Outcome 5-7)
14. In the most general terms, what does a spectrum show? Sketch a plot of each major type. (Learning Outcomes 5-1, 5-3, and 5-12)

PROBLEMS AND ACTIVITIES

1. Calculate the wavelengths of electromagnetic radiation at the following frequencies: (a) 100 MHz (about the middle of the FM band), (b) 1000 kHz (AM band), and (c) 10 GHz (a common frequency for a radio telescope).
2. Calculate the ratio of the energies of each frequency in Problem 1: 100 MHz to 1000 kHz, 1000 kHz to 10 GHz, and 100 MHz to 10 GHz.
3. Compare the kinetic energies of cars of the same mass moving at 10, 30, and 60 km/h.
4. What is the energy in joules of the first line in the Balmer series of hydrogen?
5. Assume that a 100-watt lightbulb converts all its electrical energy into light at a wavelength of 5500 Å. One watt equals the output of one joule per second. How many photons does the lightbulb give off every second?

Telescopes and Our Insight into the Cosmos

LEARNING OUTCOMES

After studying this chapter, you should be able to:

6-1 Describe the impact of telescopic observations on scientific models.

6-2 Outline the main functions of a telescope (light-gathering power, resolution, and magnifying power); relate each to specific optical properties of a telescope's design and sketch those relationships in graphical form.

6-3 Compare and contrast a telescope's light-gathering power, resolution, and magnifying power, and discuss the limitations of ground-based telescopes.

6-4 Compare and contrast reflecting and refracting telescopes; include a sketch of the optical layout of each in your comparison.

6-5 Compare a radio telescope to an optical telescope in terms of functions, design, and use.

6-6 Cite a key drawback of a radio telescope compared with an optical telescope and

describe how radio astronomers cope with this problem.

6-7 Describe the usefulness of a radio interferometer compared to a single-dish radio telescope.

6-8 Describe what is meant by the term *invisible astronomy.*

6-9 Contrast an infrared telescope to an optical telescope in terms of functions, design, and use.

6-10 Discuss at least two important advantages a space telescope in earth orbit has over a ground-based telescope, and the even greater advantages of telescopes on the moon.

6-11 Describe how astronomers make images, and explain the meaning of an intensity map of an astronomical object.

CENTRAL CONCEPT

Telescopes extend our perception of the cosmos by revealing faint objects and a wide range of the electromagnetic spectrum. New observations impel the development of new models and often the demise of old ones.

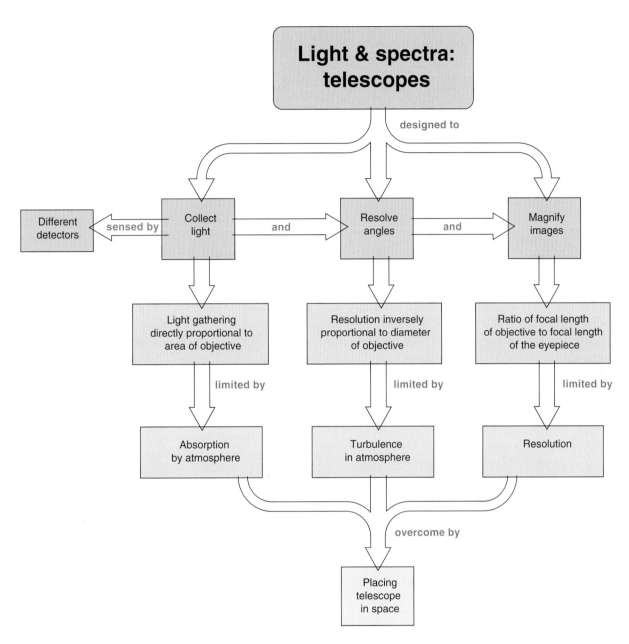

Chapter 6
Celestial Navigator™

This chapter turns from astronomical concepts to astronomical tools. Telescopes and other equipment, such as spectroscopes, spring from advances in technology. The technological development of astronomical tools affects not only what we observe but also how we observe. It expands, deepens, and sharpens our perceptions of the cosmos. Observation is a prelude to and confirmation of our models of astronomical objects.

Technical advances in this century have been so great that now astronomers rarely labor alone. Usually, they work in concert with colleagues whom they may never see during the course of a research project. The technology has grown too complex for any one observer to master, and the astronomer now often works with engineers and computer experts. This chapter looks at some of the developments that have provided new ways of observing the universe and considers how they have influenced our astronomical models.

Observations, by which the effectiveness of models is judged, accelerate the evolution of astronomical ideas. From such judgments some models are discarded, new ones are proposed, and a few are finally adopted. Not all new observations have dramatic effects. In many cases, the change of view brought on by the change of vision is slow and subtle. Slowly or swiftly, new observations compel new conceptions of the cosmos.

6.1 OBSERVATIONS AND MODELS
(Learning Outcome 6-1)

How do observations and models interact? The process usually goes like this (recall Section 2.1): models spring from observations, whether straightforward or subtle. Basic astronomical observations naturally drove people to create models of the cosmos. The Ptolemaic model marked the first detailed attempt at a scientific explanation of these observations (Section 2.3). It used geometric, physical, and aesthetic ideas to explain what was seen, and it also predicted planetary positions. To be useful, a prediction must have definite numbers attached to it. How well a model corresponds to actual observations becomes crucial to its acceptability.

The issue of "how well" revolves around the techniques of observation, as well as gut judgment of how good is good enough. For example, astronomers knew for centuries that Ptolemaic predictions often missed the actual planetary positions by several degrees. This discrepancy did not bother astronomers in those earlier times, for a few degrees of error were considered acceptable. Their observations did not aim at any greater accuracy. In particular, the Ptolemaic model provided sufficient accuracy for planetary conjunctions.

Not until Tycho Brahe (Section 3.3) did observations reach the limits imposed by the human eye. Tycho's technical achievement compelled Kepler (Section 3.5) to take a disparity of only 8 arcmin seriously. The failure of the Copernican model to fit the data with circular orbits resulted in Kepler's devising a model with elliptical orbits. The success of Kepler's modification formed a critical link in the evolutionary chain to Newton's idea of gravitation (Section 4.4).

Prior to Tycho's observations and the astronomical use of the telescope by Galileo, the Ptolemaic and the Copernican models explained the motions of the planets equally well – and both made predictions just about as badly! Because both these models were confirmed observationally, astronomers had to choose between them on the basis of aesthetic and philosophical beliefs.

In fact, the crucial observation needed to distinguish between the basic heliocentric and geocentric models – that of heliocentric parallax (Section 2.2) – was not made until the 1830s, more than two centuries after the introduction of the Copernican model! Only by then had the techniques of measuring stellar positions become accurate enough to detect heliocentric parallax, which amounts to less than 1 arcsec for even the nearest star. (Recall that 1 arcsec is only 1/3600 of a degree: that's the angular diameter of an U.S. penny placed at a distance of about 72 km.) Copernicus asserted correctly that the stars were very far from the earth compared with the earth–sun distance.

TO SUM UP. Observations form the building blocks for model making, and they can also act as the driving force that causes models to be discarded. This destructive effect emerges when accepted models fail to account plausibly for new observations.

One more point. Astronomy, in comparison to physics and chemistry, works more as an observational science than an experimental one. In a laboratory, investigators can isolate certain traits and keep some fixed while changing others to see how the variables affect the outcome. Not so in astronomy. We cannot change or control our subjects to study their physical characteristics. We can work only with what is given – for the most part, the light from celestial objects.

You should not, however, get the impression that astronomy totally lacks experimentation. Astronomers do experiment in basic ways. First, we can observe the same objects in different ways. Usually this means using different telescopes that operate at different wavelengths. That's the importance of technological innovation, for it provides new tools for new experiments. Second, laboratory experiments such as those of spectroscopy can help us to analyze observations, and of course we make use of experiments in physics and chemistry that provide basic data on properties of matter and radiation.

Figure 6.1 The refraction and reflection of light. (a) When light rays cross the boundary between two different materials, their paths are bent. The ray here bends when it enters the glass from the air and also when it exits into air. (b) The reflection of light. Note that the incident angle and the reflected angle are the same. This holds true even if the reflecting surface is curved.

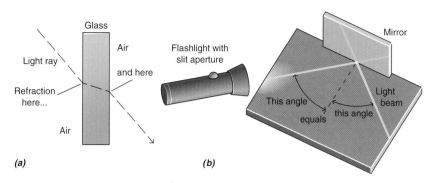

(a)   *(b)*

Third, within the solar system, we can use space probes to measure the actual conditions in interplanetary space, in the neighborhoods of planets, and on some planetary surfaces.

Finally, we can play with theoretical models backed up by whatever observations we have. Theoretical model making may also be tied to technology; for example, electronic computers can quickly manipulate very detailed models of astronomical objects. With the advent of supercomputers and computer graphics, we have begun to explore visually many realms of theoretical ideas. Such numerical simulations promote insights into the complex physics of astronomical systems – in essence, a new way to conceive them. Computer graphics also provide the power to manipulate data so that we can understand them in new ways.

6.2 VISIBLE ASTRONOMY: OPTICAL TELESCOPES

As extensions of the human eye, telescopes amplified the power of detection without, at first, extending the spectral range of our vision. Today we can sense much more than the visual part of the electromagnetic spectrum (Section 5.3). This section treats only optical telescopes, those that manipulate light detectable by the eye. Before I deal with telescopes, let me discuss a little about **optics:** how the direction of light is controlled. Knowledge of basic optics is necessary to understand how telescopes work.

The Basis of Optics

When traveling through empty space or a uniform medium, light moves in a straight line. To change its direction, we act on it in specific ways. Although light has characteristics of both waves and particles, geometric optics pictures it as particles moving along straight-line paths, which are called **light rays.** Using lenses, mirrors, and prisms, we can change the direction of light rays. We can even break white light into its component colors (light of different wavelengths). How light rays are affected by bouncing off or passing through materials is the essence of optics. Light must be manipulated to change its path from its natural way, which is a straight line.

When light crosses the boundary from one transparent material to another (from air to glass, for example), its direction changes (Fig. 6.1a). This bending of light rays is termed **refraction.** Refraction occurs because light travels at different speeds in different media. Consider what happens when a beam of light – a bunch of rays – passes through the air to hit glass at some angle (Fig. 6.1a). The first part of the beam to strike the glass enters it and slows down. The rest of the beam continues to move, but at a faster speed, and so gains on the light already in the glass. This catch-up results in the front of the beam turning toward the glass. Here's an analogy: imagine a row of musicians in a marching band turning a corner. To maintain a straight line, the people on the inside march more slowly than those on the outside. The entire row turns around some angle and remains straight.

KEY POINT. The amount of refraction depends on the wavelength of light, with blue light bent more than yellow, and yellow more than red, for most kinds of glass. The shorter the wavelength, the greater the amount of refraction.

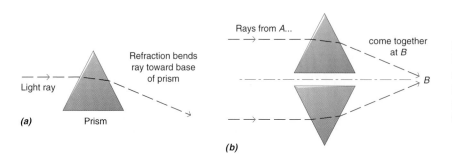

Figure 6.2 Prisms as crude lenses. (a) Because its sides are not parallel, a prism refracts a light ray toward its base. (b) Placed base-to-base, two prisms refract rays from *A* to meet at *B* in a crude focus to make an image.

You have looked in a mirror, so you know that flat surfaces return light by bouncing back the rays. This process is **reflection** (Fig. 6.1b). A light ray bounces off a polished surface the same way a ball bounces off a smooth wall: the ball rebounds at the same angle at which it hit. Similarly, when light is reflected, the incoming and outgoing angles are equal. Reflection does not depend on the wavelength of the light; red and blue light are reflected the same way at the same angle.

Note that reflection and refraction are common to all waves. You can even see these effects with water waves!

Optics and Images (Learning Outcome 6-2)

The goal of optics is to make images by refraction and reflection. An **image** forms when light rays are collected in the same relative alignment they had when they left an object. You can recognize an image of an object as a visual representation of the object itself.

How can refraction form an image? Suppose that light travels through a glass prism (Fig. 6.2a). Because the sides of a prism are not parallel, a light ray does not come out along its original path but is bent toward the prism's base. Now place two prisms base to base (Fig. 6.2b), so that two light rays from a source converge to a point. However, rays entering the two prisms at different heights and angles converge at different points. To get all rays to come to the same point requires a smoothly curved surface. Such a piece of glass is a **lens** (Fig. 6.3), which refracts light to form an image.

A lens brings rays from a very distant source to an image at its focus. From an object of finite size, the lens makes an image of the object by focusing parallel rays from each point of the source onto separate points in the image. This image is generally (but not always) smaller than the object and upside down (Fig. 6.4). For objects at large distances, the distance from the lens to the image is roughly the same for all objects. This distance, from the lens to the image, is termed the **focal length.**

How can a mirror make an image? You know that an image from a flat mirror is undistorted. An irregularly curved mirror, such as one in a fun house, creates a distorted image. A smoothly curved mirror – whose surface, for instance, follows the curve of a parabola – brings all the light to a focus (Fig. 6.5).

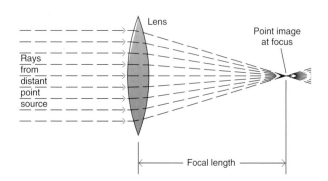

Figure 6.3 A lens forming a point image from a point source. A lens acts like base-to-base prisms (Fig. 6.2). Because its surface is smoothly curved, it forms a sharp focus at the focal point. The distance from the lens to the focal point, for a distant source, is the focal length.

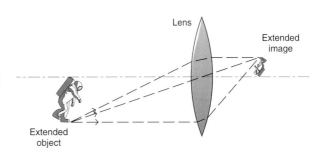

Figure 6.4 A lens forms an extended image at the focal point, upside down, of an extended object. If the object were a point source of light, such as a star, the image would also be a point.

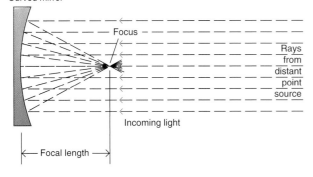

Figure 6.5 A smoothly curved mirror reflects incoming light to a focus. Astronomical mirrors have their reflecting material on their front surfaces and usually follow a parabolic curve to make a sharp focus.

This curved surface directs the light rays to a focus according to the basic law of reflection: the incoming angle equals the outgoing angle.

Telescopes
(Learning Outcomes 6-2, 6-4)

Basically, a telescope gathers up light and allows you to examine an image at a focus. To make a telescope, you need a lens or mirror, called the **objective,** to bring light to a focus. A lens called an **eyepiece** placed just beyond the focus allows visual examination of the image, or a camera can be placed at the focus to photograph the image, or the light can be diverted into a spectroscope.

There are two basic types of telescope, distin-guished by the objective: **refracting telescopes** (or refractors) use a lens (Fig. 6.6), and **reflecting telescopes** (or reflectors) use a mirror. Galileo's telescope (Section 4.1) was a refractor, Newton's (Section 4.3) a reflector. Early large telescopes of good quality tended to be refractors, but images viewed through a refractor are flawed by color haloes: that is, when one color is in focus, the other colors are not. This is because simple lenses act essentially as prisms and different colors have dif-ferent focuses. Reflected light, however, does not break up into colors, so the image formed by a reflector does not have color haloes. Later lenses have eliminated most of the refractor's color prob-lems, but Newton's solution was to design the reflector.

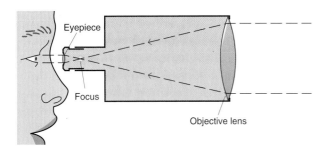

Figure 6.6 Design of a simple refracting telescope. An objective lens gathers the incoming light and brings it to a focus. An eyepiece allows viewing of a magnified image.

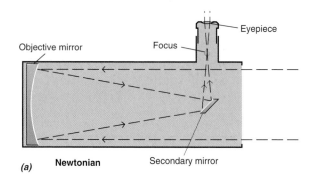

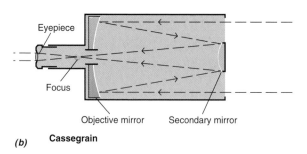

(a) Newtonian

(b) Cassegrain

Figure 6.7 Designs of reflecting telescopes. The main problem is viewing the image made by the objective mirror. Newton placed a small, flat mirror at an angle to reflect the light out to one side, where an eyepiece is positioned (a). This is called a *Newtonian reflector*. A common modern design uses a small convex mirror to reflect the light back down through a hole in the objective mirror for viewing (b). This is called a *Cassegrain reflector*.

How do we view the image formed by a reflector? Newton put a small mirror tilted at 45° to the path of light to direct the focus out the side of the telescope tube (Fig. 6.7a), where an eyepiece is placed. Such a design is called a *Newtonian reflector*. Most research telescopes today use a different optical design called a *Cassegrain reflector* (Fig. 6.7b). All large telescopes built now are reflectors. Advances in technology allow them to be lighter and cheaper to construct.

Functions of a Telescope (Learning Outcome 6-3)

1. To gather light. Whether a reflector or a refractor, a telescope has one primary function: to gather light. A telescope is basically a light funnel, collecting photons (Section 5.3) and concentrating them at a focus, so objects viewed through a telescope appear brighter than they do when seen by the eye

alone. This difference indicates a telescope's **light-gathering power.** (Enrichment Focus 6.1)

How much light a telescope can collect depends on the area of its objective, which is proportional to the square of the diameter. A mirror with twice the diameter of another can gather four times as much light. That's the reason astronomers want large telescopes – big mirrors that have plenty of light-gathering power.

Light-gathering power has no absolute standard; it's a relative measure. For example, the diameter of your eye in the dark is about 0.8 cm. A telescope with an 80-cm-diameter objective (100 times bigger) would have a light-gathering power 10,000 times (100^2) greater than your eye. Similarly, the 5-meter Hale telescope at Mt. Palomar in California outdoes the 4-meter Mayall telescope at the Kitt Peak Observatory by more than 50 percent.

2. To resolve fine detail. The next most important function of a telescope is to separate, or resolve, objects that are close together in the sky. This ability is called a telescope's angular **resolution,** usually expressed in terms of the minimum angle between two points that can be clearly separated. For example, a 10-cm telescope has a resolution of 1.4 arcsec at visual wavelengths. If the telescope is aimed at two stars that are more than 1.4 arcsec apart, you will see two separate star images. If the angular separation of the stars is less than 1.4 arcsec, you will see a single, elongated image. For the same wavelength of light, the resolution depends inversely on the objective's diameter. A mirror twice the size of another has twice the resolution; that is, it can resolve objects half as far apart in angular distance. A 20-cm telescope would have a resolution of 0.7 arcsec at visual wavelengths.

You can think of resolution in terms of size-to-distance ratios (Enrichment Focus 1.1). At best, the unaided human eye can see an angular separation of about 1 arcmin, which corresponds to a size-to-distance ratio of about 1/3600. So we can see that the moon (angular size about 30 arcmin) has a disk with features. Mars, even when closest to the earth, has a maximum angular size of 26 arcsec, so we cannot see Mars as a disk without a telescope. From the resolution just given for a 10-cm telescope, you'll note that with such an instrument, we could see Mars as a disk and some of the larger features

ENRICHMENT FOCUS 6.1
Properties of Telescopes

The main optical properties of a telescope revolve around its objective, its main lens or (more commonly) mirror. The objective will have a certain diameter, D, and a certain focal length, f, over which distance the objective brings light to a focus. The **f-ratio** is defined by the focal length divided by the objective's diameter:

$$f\text{-ratio} = \frac{f}{D}$$

For example, at the Capilla Peak Observatory of the University of New Mexico in the United States, we have a 60-cm-diameter mirror with a focal length of 900 cm; so the f-ratio is 900/60, or 15.

The smaller the f-ratios, the brighter the image at the focus, just as for a camera lens. (If you have a camera with an adjustable lens, you can see that the f-ratios are in a sequence: 2.8, 4, 5.6, 8, 11, 16, and 22, for example. Stepping up from one f-ratio to the next one, say from 22 to 16, makes the image twice as bright.) Other conditions being equal, a telescope with a smaller f-ratio produces a brighter image for viewing or for recording. However, the total light gathered still depends on the area of the objective.

The objective's diameter determines the light-gathering power (LGP) by its area (= πR^2), which is proportional to $(\pi D^2)/4$ for a circular aperture. Since LGP has no absolute standard, you need to compare the ratio of the areas (diameters squared) to find the relative performance. For example, to compare 5- and 4-meter telescopes:

$$LGP = \left(\frac{5\,m}{4\,m}\right)^2 = \frac{25}{16} = 1.6$$

The objective's diameter again plays a critical role for the theoretical resolution (TR), the smallest angle discernible by the telescope. Using units of meters for diameter and the wavelength of the light, we have:

$$TR\ (arcsec) = (2.52 \times 10^5)\frac{\lambda}{D}$$

So, for example, an optical telescope with a 0.1-m (10-cm) objective at a wavelength of about 5500 Å (5.5×10^{-7} m) has a theoretical resolution of

$$TR\ (arcsec) = (2.52 \times 10^5)\frac{5.5 \times 10^{-7}}{0.1} = 1.4\ arcsec$$

The atmosphere usually allows this resolution to be achieved in practice. At the best sites, the seeing limit of the atmosphere can be as fine as 1/3 arcsec, though not typically so.

Last comes magnifying power (MP). It depends on the ratio of the focal length of the objective (f_o) to the focal length of the eyepiece (f_e) in the same units:

$$MP = \frac{f_o}{f_e}$$

An eyepiece can have a focal length of 10 mm. If it is used with an objective that has a focal length of 300 cm, then

$$MP = \frac{(300 \times 10\ mm)}{10\ mm} = 300\ power$$

So you can change a telescope's magnifying power by changing its eyepiece. A shorter focal-length eyepiece results in a higher power.

---

on the surface. The better the resolution, the more information we can obtain about a celestial object.

Even the best optics in a telescope has limited resolution because of the wave nature of light. Water waves bend when they strike the edge of a barrier; this process is called *diffraction*. A similar bending occurs with light waves. The objective of a telescope acts as a hole in a barrier to the light waves. As the light waves travel through this opening, they bend around the edges (just as water waves do). This bending spreads out the waves and

ultimately determines the **theoretical resolution** of any telescope. The limit to a telescope's resolving power depends directly on the wavelength of light, and it varies inversely as the size of the aperture (Enrichment Focus 6.1).

The theoretical resolution of a 5-meter telescope is a fine 0.02 arcsec, but the actual resolution of a big telescope is limited not by the telescope's optics but by the earth's atmosphere. You have probably noticed that stars twinkle. The twinkling comes from turbulence in the air that makes the atmosphere act like a huge, distorted lens. The motion of blobs of air, like the shimmering above a hot road, distorts and blurs images seen through a telescope. Heat sources in or near the telescope have the same bad effect.

At a good site on a typical night, star images appear about 1 arcsec in diameter. On really good nights, stellar images fall to about 0.3 arcsec. The limit that the earth's atmosphere sets on the resolving power makes a strong case for a large telescope in space. There a telescope's resolution would be limited by the optics, not by the atmosphere.

3. To magnify the image. The *least* important of a telescope's functions is its **magnifying power,** as represented by the apparent increase in the size of an object compared to visual observation. Magnifying power depends on the focal length of the objective and the focal length of the eyepiece. Changing the eyepiece on a telescope changes its magnifying power as follows: the shorter the focal length of the eyepiece, the greater the magnifying power. For example, if you put in an eyepiece with half the focal length of an earlier one, you double the magnifying power (Enrichment Focus 6.1).

An eyepiece with a shorter focal length would give a higher magnifying power, as high as you might like, but there is no point in using a magnifying power any greater than necessary to see clearly the smallest detail in the image, which is set by the resolution. Extremely high magnification merely makes the fuzziness worse.

CAUTION. Don't develop the impression that astronomers observe the skies with their eyes. They do not. A light-sensing device, called a **detector,** is usually placed at the focus. The detector may be a photographic plate (light-sensitive materials on glass rather than film). That's how many of the photos in this book were made. Or it may be an electronic detector similar, for example, to a television camera. Astronomers rarely, if ever, use their eyes directly for observations.

TO SUM UP. The three principal functions of a telescope are to gather light, to resolve fine detail, and to magnify the image. Of these, light-gathering power ranks first. Most astronomical objects are extremely faint. Without a telescope, you can see about 6000 stars in a dark sky, but even a small 15-cm telescope allows you to see some half million stars. That is the real power of a telescope – to enable us to see objects that we otherwise would not know existed. (See the Telescopes Celestial Navigator.)

Next Generation of Telescopes (Learning Outcome 6-3)

Astronomers strive for better resolution and for the ability to detect fainter objects. Large space telescopes, such as the Hubble Space Telescope (HST), only partially alleviate the resolution problem. Although HST orbits above the blurring atmosphere, its expense and the limited number of objects it can observe each day restrict its scope. The largest ground-based telescopes can harvest more photons than HST can. Their light-gathering power from larger mirrors more than makes up for the distortions of the atmosphere.

The ultimate goal of the new generation of telescopes is to achieve high resolution with large apertures. Telescopes with awesome mirrors – up to 10 meters – are planned, under construction, or completed. In some, the mirror is made in pieces rather than as a single unit, as was done for the twin Keck telescopes. Located on Mauna Kea, Hawaii, at an elevation of 4145 meters (Fig. 6.8), the Keck has an innovative 10-meter mirror, which consists of 36 hexagonal segments with a total weight of 14 tons. The segments are individually controlled to maintain the mirror's shape under a variety of stresses. The mirror control system can move each segment as little as one-thousandth the thickness of a human hair! The dome and the telescope's structure minimize local **seeing,** or blurring from atmospheric unsteadiness, to take advantage of the site's natural potential – occasionally as good as 0.3 arcsec.

The Southern Hemisphere will also gain big eyes. The European Southern Observatory (ESO) has sited powerful optical telescopes in a desert in Chile that looks a lot like Mars – on a barren mountain reaching 2.6 km into clean, steady air. Here the Very Large Telescope (VLT) is sprouting on the mountain's top – four 8.2-m telescopes (Fig. 6.9). In contrast to the Keck's design, the mirrors are thin (18 cm thick!) but not segmented. Like the Keck, they require a computer-controlled system to sustain their shape. The goal of this massive project is to synchronize all four telescopes. When completed, the VLT is expected to see the heavens 50 times sharper than the HST.

Figure 6.8 Composite image showing meteors above the twin Keck Telescopes on Mauna Kea, Hawaii. Each telescope has a mirror that is 10 m in diameter; they can be used together as an optical interferometer to provide more detailed images.

Figure 6.9 Domes of the Very Large Telescope at Cerro Paranel, Chile. Each telescope has an 8.2-m mirror and working together, they provide the light-collecting power of a single 16-m telescope. The VLT can detect parts of the spectrum ranging from near-ultraviolet to near-infrared.

New Optical Techniques
(Learning Outcome 6-3)

One way to improve resolution is to compensate for the motion of the atmosphere during the observations. We could, in principle, deform the telescope optics to compensate for the turbulence – a tech- nique called **adaptive optics**. The engineering problems are huge. Drives at the base of the primary mirror alter its shape to cancel out the seeing as the light is collected.

How does the telescope sense the atmospheric distortion? Quickly! The information needs a response by the mirror within 1/1000 s. How?

Figure 6.10 The powerful Starfire laser beam creates an artificial guide star for adjusting a telescope with adaptive optics.

One technique is to use artificial "guide stars." Powerful lasers can project a beam 10 to 40 km up in the air, where it is reflected back (Fig. 6.10). The laser point of light serves as an artificial star, whose beam samples the same layers of turbulence as incoming starlight. This beam provides the information to compensate optically for the blurred image. Such techniques herald a renaissance for high-resolution, ground-based optical astronomy.

6.3 INVISIBLE ASTRONOMY
(Learning Outcome 6-8)

Your eye senses only a tiny sliver of the electromagnetic spectrum (Section 5.3). When you have your teeth x-rayed, the dentist's x-ray machine does not glow brightly when it's on, but the film placed in your mouth senses the x-rays and gives an internal picture of your teeth. When you stand next to an almost-dead fire, the coals look black. But your skin senses heat – infrared radiation – from the coals.

Why work on invisible astronomy? Processes different from those that generate visible light often produce electromagnetic radiation outside the visible region of the spectrum. Such processes take place under particular astrophysical environments. Invisible astronomy captures these photons and probes regions not seen in visible light, revealing different information about astronomical objects and even different astronomical objects!

Invisible astronomy has a less obvious aspect: the radiation from space may or may not get to the earth's surface. Our atmosphere effectively absorbs large blocks of the electromagnetic spectrum, especially ultraviolet, x-rays, gamma rays, some infrared, and short-wavelength (millimeter) radio waves (Fig. 6.11). What produces this **atmospheric absorption?** In the case of infrared radiation, it is primarily absorbed by the water vapor in the atmosphere, which is found concentrated in the lower portions, below 20 km. The ultraviolet and x-ray radiation is primarily absorbed in the ionosphere, at an altitude of 100 km, well above the levels that can be reached by balloons and airplanes. The absorbed radiation may have journeyed through space for millions or billions of years, only to be snuffed out in the last 0.001 second of its trip, never making it to the earth's surface!

One way to avoid atmospheric absorption is to go above it. This is **space astronomy.** (In this book, *space astronomy* refers to rockets, balloons, and airplanes, as well as satellites and spacecraft.) Invisible astronomy thus has two natural divisions: work that can be done from the ground (such as projects using radio and infrared) and that which must be accomplished in space.

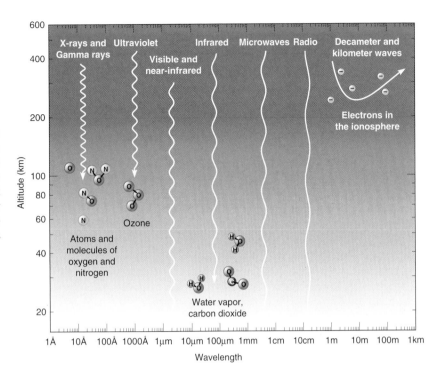

Figure 6.11 Transparency of the earth's atmosphere from short-wavelength gamma rays to long-wavelength radio waves. The scale at the left indicates the altitude down to which the radiation at various wavelengths penetrates. Water vapor does most of the absorption in the infrared. Note that the axes are powers of ten.

Ground-Based Radio (Learning Outcome 6-5)

Radio astronomy was born in 1930 when Karl Jansky (1905–1950) undertook a study for the Bell Telephone Company of the United States on sources of static affecting transoceanic radiotelephone communications (Fig. 6.12). Jansky identified one source of noise as a celestial object: the Milky Way in Sagittarius. Jansky's discovery was published in 1932 but had little impact on the astronomers of the day.

However, an American radio engineer, Grote Reber, read Jansky's work and decided to search for cosmic radio static in his spare time. By the 1940s, Reber had made detailed maps of the radio sky.

Figure 6.12 Karl Jansky of Bell Telephone Laboratories with the radio antenna in Holmdel, New Jersey, U.S.A., that he used to discover radio waves from space. Jansky's research in the 1930s led to the birth of radio astronomy.

Figure 6.13 The night sky spins above the Very Large Array (VLA). Each dish has a diameter of 25 m. The antennas move along tracks to attain different positions along the array.

Sensing a new astronomy in the making, Reber took an astrophysics course at the University of Chicago to learn more about astronomy and to discuss his discoveries with astronomers. Only a few were impressed.

World War II forced technical developments in radio and radar work. Accidentally, John S. Hey in Britain discovered that the sun emitted strong radio waves. After the war, Hey continued his astronomical pursuits at radio wavelengths. So did other groups in Britain, the Netherlands, and Australia. Radio astronomy was reborn as part of the technological fallout from research by scientists forced to deal with the practical problems of war.

A common type of **radio telescope,** a radio dish (Fig. 6.13), functions like an optical reflecting telescope. Essentially, it's a radio-wave reflector with a detector (a radio receiver) at the focus of the dish, which reflects and focuses radio waves, acting much like a mirror in a reflecting telescope. Unlike optical astronomers, radio astronomers cannot see

sources at the focus. However, the radio receiver that detects the incoming radio waves translates the signal into a voltage that can be measured and recorded. A computer then generates a map of the radio intensity over a region of the sky.

Our atmosphere allows some millimeter, centimeter, and longer wavelengths to reach the ground. Radio telescopes can observe these both day and night, and even on cloudy days at longer wavelengths. Radio telescopes are typically much larger than optical telescopes, and so they can catch more radiation. Thus, radio telescopes are usually more sensitive than optical instruments. These are a few of the advantages of radio over optical astronomy.

Resolving Power and Radio Interferometers (Learning Outcomes 6-6, 6-7)

Radio telescopes have one major drawback: poor resolution. How so? Resolution depends on both the diameter of the objective (the light-gathering surface) and the wavelength of the gathered light. Radio waves are much longer than visible light, typically 100,000 times as long, so if an optical and a radio telescope had the same diameter, the radio telescope would have 100,000 times less resolution and much less ability to show details in radio sources. For example, a radio telescope with the same resolution as the 5-meter Hale telescope would have to have a diameter 100,000 times as great, about 500 km! Clearly, radio telescopes need to be large to have decent resolving power, but it is also obvious that a single dish of this size cannot be built on earth!

There's a method for making a small radio telescope mimic a large one. Imagine two radio telescopes placed, say, 10 km apart. By synchronizing the signals received by both, the pair can be made to act like a single dish with a diameter of 10 km, but only for a thin strip across the sky. (That's because the two instruments act like two small pieces at the opposite ends of a strip of a larger radio dish.) To get good resolution in a small, more or less circular region of the sky requires an array of coordinated radio telescopes. One such telescope, called the Very Large Array (VLA), is in the state of New Mexico in the U.S.A. (Fig. 6.13). Completed late in 1980, the VLA (which can have its antennas spread out over a few tens of kilometers) has a resolution, at centimeter wavelengths, equivalent to that of a moderate-sized optical telescope. Such complex devices are called **radio interferometers.**

How much resolution does a radio interferometer have? Basically, that depends on the separation of the antennas and the wavelength observed (Enrichment Focus 6.1). If the antennas are 1 km apart and operate at a wavelength of 1 cm, the resolving power is about 2 arcsec, almost as good as optical telescopes and much better than a single dish could achieve.

How good can the resolution get? In the technique known as **very-long-baseline interferometry** (VLBI), the signals received by very distant antennas (even devices located on different continents) are recorded and combined later in a computer. The maximum baseline extends to the diameter of the earth, so a resolution of 2×10^{-4} arcsec is possible. In the future, radio telescopes in space, separated by larger baselines, will give even better resolution. Some proposals envision one element of

an interferometer on the moon – a separation some thirty times greater than is possible on the earth. Resolutions could approach 10^{-6} arcsec and we would be able to see (at radio wavelengths) the details on the disks of nearby solar-type stars!

The United States operates the Very Large Baseline Array (VLBA), centered in New Mexico. Resembling the VLA in function, the VLBA incorporates VLBI techniques in a coordinated array that spans the United States from the Virgin Islands to Hawaii and includes an antenna in Iowa. The completed array includes ten 25-meter antennas (similar to those of the VLA) with separations as great as a few thousand kilometers. The receivers are synchronized by atomic clocks, and the data are shipped to a central computer for image processing.

Optical interferometers work the same way as radio ones. Laser beams synchronize individual telescopes, such as the Kecks or the VLTs. The combined effort makes a mirror with a diameter equivalent to the separation of the telescopes. The VLT can bring 200 square meters of light-collecting area for a resolution of 0.001 arcsec at a wavelength of 1 μm. That's the ability to discern a U.S. quarter at a distance of 500,000 km! The tricky part with optical interferometers is that optical wavelengths are much shorter than radio ones, so the positions of the individual telescopes must be known much more precisely than the positions of radio telescopes.

Ground-Based Infrared (Learning Outcome 6-9)

Water and carbon dioxide in the earth's atmosphere absorb much of incoming infrared radiation. The infrared astronomer can observe at only a few restricted wavelength ranges: 2–25, 30–40, and 350–450 μm (recall that 1 μm equals 10^{-6} m). Such observations are best made from high sites in dry climates, where the atmosphere above the telescope contains little water vapor.

How does an infrared telescope differ from an optical one? Mainly in the detector at the telescope's focus. Because our eyes and photographic film sense infrared radiation poorly, special infrared detectors are required. We now have small arrays of infrared detectors that can produce infrared images.

Infrared observing has at least two distinct advantages over optical observing. First, infrared radiation is less hindered than visible light by interstellar dust. Second, cool celestial objects give off most of their radiation in the infrared. Typically, such cool objects cannot be seen in visible light but can be detected in the infrared, so infrared astronomy brings the cold universe into view.

Space Astronomy (Learning Outcome 6-10)

What about the parts of the infrared spectrum that do not penetrate to the ground? What about ultraviolet light and x-rays? These can be detected only above the earth's atmosphere, from airplanes, balloons, rockets, satellites, spacecraft, or lunar observations. To these telescopes, the sky looks dramatically different from that we perceive by eye.

For example, most of the far infrared (wavelengths longer than about 40 μm) does not make it to the ground, but at altitudes of 15 to 20 km or so, very little of the earth's atmosphere remains. Far-infrared observations can be made at these altitudes from airplanes or balloon-borne telescopes.

Optical astronomy in the future will involve large space telescopes. Their main advantage: the atmosphere does not limit a space telescope; rather, it operates at its theoretical limit of resolving power. The HST has a 2.4-meter mirror (Fig. 6.14). Its primary goal is observations of faint objects with high resolution, especially at ultraviolet wavelengths. The best resolution with HST is about one-tenth of an arcsecond. In addition, HST was intended to operate over the wavelength range from 1200 Å to 1 μm, depending on the instruments used for sensing the light.

In April 1990, the space shuttle *Discovery* deployed the HST. At a developmental cost of $1.5 billion, the HST is the most expensive astronomical project ever. (For comparison, the VLA cost a bit less than $80 million when it was completed in 1981.) Astronomers were shocked to find out in summer 1990 that the main mirror was flawed. HST was designed for a 15-year mission, however, with new and upgraded instruments ferried up by future shuttles. In December 1993, astronauts installed optics designed to compensate for the mirror's flaws. This worked superbly. The telescope now makes images very near the theoretical limit of resolution (about 0.1 arcsec). HST can now detect a firefly at a distance of about 20,000 km in a 1½-h exposure. And if two fireflies were about 3 m apart, HST could see that there were two!

The Chandra X-Ray Observatory marked a new

Figure 6.14 The Hubble Space Telescope (HST) was placed 400 km above the earth by the U.S. space shuttle *Discovery*.

addition to the space telescope team. Equipped with a 1.2-meter mirror, Chandra's forte is high-resolution x-ray images as fine as 0.5 arcsec, some hundred times better than the previous best x-ray space telescope. Chandra also does double duty: it can take spectra of x-ray sources as well as images.

Eventually, space astronomy may sport observatories on the moon. Just those aspects of the lunar environment that make it tough on people result in a great place for astronomy. First, the lack of atmosphere not only eliminates weather problems but permits access to all wavelengths at a resolution only limited by the telescope. Second, because moonquakes are weak and infrequent, the moon has a firm, seismically stable surface on which to construct telescopes. (A problem of a telescope in low earth orbit is that it is difficult to aim it and keep it on target.) Third, the background electromagnetic noise is very low (especially for radio emissions on the moon's far side, which blocks out

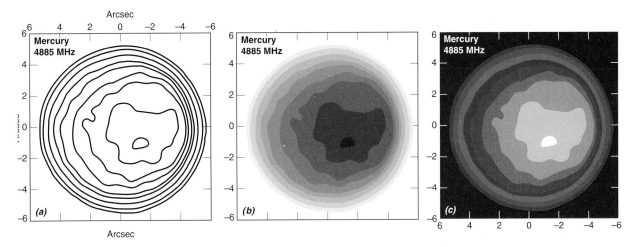

Figure 6.15 Variations on digital processing intensity maps. (a) Contour map produced by the VLA of the radio emission from the planet Mercury at a wavelength of 6 cm. Note the peak ("high"), lower right of center. (b) The same observations as in (a) displayed as a grayscale map. The darker the region, the greater the radio emission. They show the same trend. (c) The same observations as in (a) and (b) displayed as a false-color map. Here blue is the weakest emission and dark red is the strongest. Note the peak of emission to the right of center. (Courtesy of NRAO/AUI; VLA observations by J. Burns, D. Baker, J. Borovsky, G. Gisler, and M. Zeilik; image processing by J.-H. Zhao and M. Ledlow.)

terrestrial radio noise). In contrast, even low earth orbit is a dirty, noisy environment for astronomy.

What kind of telescope would take best advantage of the moon? A grand observatory would operate as an interferometer (like the VLA) at ultraviolet, optical, and infrared wavelengths. If it had telescopes spread over a diameter of about 10 km, a lunar array would achieve resolutions about 100,000 times better than ground-based telescopes. That amount of resolving power would pick out a U.S. dime on the earth's surface from the moon! Such an instrument would see earthlike planets around nearby sunlike stars.

6.4 IMAGE COLLECTION AND PROCESSING

You may have the romantic image of an astronomer working with eyes glued to the telescope to take spectacular photographs. Indeed, photography is still done, but today the astronomer usually sits in a warm room and "observes" on a TV monitor that shows an enhanced view through the telescope. Most optical telescopes do very little direct photography. Detectors of other kinds can make images. In most cases, including that of photography, a computer later manipulates the data. This procedure, called **image processing,** plays a central role in astronomy today, especially with invisible astronomy.

The computer can display an image of information otherwise not detectable by eye, and the brain can process this information rapidly by visual cues. This technique is called *computer visualization*. Computer-generated images often use false colors to show information in a visually informative fashion. These colors are false in the sense that they are not the actual colors of the object in the visual range of the spectrum. Rather, they are color codes to a specific property, such as different intensities of a signal.

Understanding Intensity Maps

Many of the figures in this book are intensity maps, pictures of how the intensity of some kind of radiation (radio, visible, infrared) varies over some region of the sky. They often may show regions of different colors as false-color maps. What do they mean?

Image processing can display astronomical data in a variety of formats, depending on what aspects the astronomer wants to emphasize. A contour map resembles a weather map that shows high- and low-pressure systems (Fig. 6.15a). A variation is a grayscale map, in which the density of filled-in areas indicates the signal's intensity (Fig. 6.15b). The most common is a false-color map, in which the computer generates *different colors for different levels of intensity*

(Fig. 6.15c). Why use false colors, which must be recognized as processing artifacts rather than as properties of the object observed? False colors are used because they convey more visual information than black-and-white or gray scales. Beware, though, that the colors rarely represent colors visible to your eyes – that is why they are "false" colors.

Photography

Photography is still the old standby for gathering a large amount of storable information in a short time. Astronomers use special light-sensitive coatings on glass plates, whose size depends on the type of telescope employed. Long time exposures add up the light, so that very faint objects appear clearly. However, even the most sensitive plates convert only a small percentage of the photons striking them into images, so photography does not make very efficient use of limited time on a telescope.

Despite low efficiency, photography still has one great advantage: the images can cover a large area of view. Hence, a single photograph's information content can be enormous, especially from a wide-angle telescope. Also, plates preserve information with little loss over a long time, providing a record that an astronomer a century later can examine – probably for a purpose completely different from the original intention! Today, specialized optical devices can convert the information on a photograph into digital form for computer enhancement.

Charge-Coupled Devices (CCDs)

The development of solid-state electronics has resulted in small computers with large memories and detectors that match the digital nature of computer processing. The natural match that would make astronomers most happy involves a two-dimensional detector that captures most of the photons striking it. Microelectronics may have produced the dream detector: **charge-coupled devices** (CCDs for short), which are used in digital and sometimes video cameras.

A CCD is a thin silicon wafer a few millimeters on a side. It contains a large number of small regions, each of which makes up a picture element, called a *pixel*. A typical CCD chip contains a million pixels arranged in a grid. Each pixel converts photons to electrons and builds up charge over a long time. A small computer then moves the charge out of the chip in a regimented way, keeping track of the number of charges accumulated in each pixel. That set of charges is then displayed (again by computer) on a video monitor, and the image is the brightest for the pixels with the greatest amount of the charge, and darkest for the pixels with the least charge. Computer processing after data acquisition brings out certain aspects in which the astronomer has the most interest.

Because they are flat, two-dimensional detectors, CCDs can gather a large amount of information during one exposure. They share this advantage with photographic plates (though they are a lot smaller than such plates right now). CCDs have two distinct advantages over plates. First, they collect electrons at a rate that is directly proportional to the number of photons. Second, they are very, very efficient – in some regions of the spectrum (notably the red), they come close to detecting 100 percent of the photons and converting them into electrons.

Along with the small computers that control their operation, CCDs are the best bet (in my opinion) for expanding the capabilities of "small" telescopes for doing frontier astronomy. One telescope peers at only a small part of the sky at a time, and the number of telescopes on the earth is extremely limited. Most of them are small, with apertures less than 100 cm. The great efficiency of CCDs enhances the power of these small telescopes and will revive the status of small observatories. For example, in a one-hour exposure, the Capilla Peak Observatory's 60-cm telescope with a CCD camera can display objects that once were visible only on a one-hour exposure with a photographic plate on a telescope 10 times larger (such as the Hale 5-meter)!

KEY CONCEPTS

1. Astronomy is primarily an observational science. Telescopes amplify astronomers' observing power by revealing faint objects, expanding the range of the electromagnetic spectrum that we can perceive, and increasing our ability to see fine detail. These instrumental abilities drive the generation and evolution of astronomical models.

2. Optical telescopes use lenses (refraction) or mirrors (reflection) to gather light and bring it to a focus; usually a photographic plate or light sensor is placed at the focus to detect the light. Telescopes that use only lenses are called *refracting telescopes,* and those that use a mirror as the objective are called *reflecting telescopes.*

3. Regardless of its type, a telescope has three basic functions: to gather light (its most important function), to resolve details, and to magnify the image. The light-gathering power of a telescope depends directly on the area of its objective; resolution depends inversely on the diameter of the objective and directly on the wavelength of light being observed (as well as on the turbulence in the earth's atmosphere); magnifying power depends on the ratio of the focal length of the objective and of the eyepiece. Large optical telescopes typically cannot reach their theoretical resolving power because of atmospheric turbulence. New techniques, however, allow ground-based observers to compensate for these image distortions.

4. Invisible astronomy uses light outside the visible region of the spectrum, such as radio, infrared, ultraviolet, x-rays, and gamma rays. Many of these wavelengths are blocked by the earth's atmosphere, so telescopes must go above it to reveal these ranges of the electromagnetic spectrum.

5. Radio telescopes are limited in resolution because radio wavelengths are very long (compared to visible light). Special electronic techniques allow widely separated radio telescopes to function as a single one; this technique has been used in the design of the VLA in New Mexico and the new VLBA, which are both radio interferometers.

6. Space telescopes have opened up our view of the infrared, ultraviolet, and x-ray heavens. They are not hindered by bad weather, the atmosphere, or the atmosphere's lack of transparency to certain kinds of electromagnetic radiation. Of locations near the earth, the moon promises to be the best all-around site for telescopes of the future.

7. New detectors (such as CCDs) and techniques for the computer processing of images have effectively transformed "small" telescopes into large ones and so extended their usefulness to probe deep into the cosmos. They have made large telescopes even more powerful in some ways that rival the current generation of space telescopes. Computers that control telescopes, collect data, and then process the data into information (such as intensity maps) now play an integral role in new telescopes and detectors of all types.

STUDY EXERCISES

1. How did Galileo's observations support the Copernican model and refute the traditional one? (Learning Outcome 6-1)
2. What telescopic observations were critical in the confirmation of the Copernican model? (Learning Outcome 6-1)
3. What is the most important function of a telescope? (Learning Outcome 6-2)
4. What are the advantages and disadvantages of radio telescopes? How do they differ from optical ones? How are they similar? (Learning Outcomes 6-4, 6-5, and 6-6)
5. Suppose you are on a television quiz show. They show you optical, infrared, and radio telescopes with the same size objective and ask you to list the devices in order of increasing resolving power. What is the correct order? (Learning Outcomes 6-3, 6-4, and 6-5)
6. Imagine that you are going before a federal government committee to justify the expense of putting a large telescope in space. What arguments would you use to persuade the skeptical committee? (Learning Outcome 6-10)
7. Imagine that you are going before another federal government committee to justify the expense of a radio interferometer. What main argument would you use to persuade the skeptical committee? (Learning Outcome 6-7)
8. Describe one advantage of an infrared telescope over an optical one. (Learning Outcome 6-9)
9. Why do x-ray telescopes have to be put above the earth's atmosphere? (Learning Outcomes 6-8 and 6-10)
10. What kinds of astronomical objects can be best studied with a space telescope? With a large (5- to 10-m) ground-based optical telescope? (Learning Outcome 6-10)
11. Imagine that you want to buy a telescope on a limited budget. What one property of the telescope should be most important in making your decision? (Learning Outcomes 6-2 and 6-3)
12. What do the different colors in a false-color intensity mean? (Learning Outcome 6-11)

PROBLEMS AND ACTIVITIES

1. Suppose you have a reflecting telescope with a 900-cm focal length and an f-ratio of 15. What is its theoretical resolving power at visual wavelengths? *Hint:* Compare it to a telescope with a 10-cm mirror. What is its light-gathering power compared to that of your eye?
2. How much greater is the light-gathering power of the 5-meter Hale telescope than that of the 2.4-meter Hubble Space Telescope? Note that the ground-based Hale telescope can never quite achieve its complete light-gathering power because the earth's atmosphere absorbs some of the light.
3. A typical telescope for amateur astronomy has a 30-cm-diameter objective and an f-ratio of 10. What is the telescope's focal length?
4. A 10-cm telescope has a theoretical resolution of 1.4 arcsec at visible wavelengths. What is the theoretical resolution of the Hubble Space Telescope at the same wavelengths? What is the size-to-distance ratio that it can resolve? How far away could it see the disk of a planet like the earth?
5. The maximum possible telescope separation of the VLA is about 40 km. What is its resolution when operating at a wavelength of 2 cm? How does this compare to a 10-cm optical telescope?
6. Imagine that you have a small telescope equipped with interchangeable eyepieces. An eyepiece with a 20-mm focal length gives you a magnifying power of 150. What focal length eyepiece do you need to have a magnifying power of 75? Of 250?
7. What is the smallest feature that you can observe on the moon with the unaided eye? With a 20-cm telescope operating at its theoretical resolution? With any telescope that is limited by the

atmospheric seeing, which under good conditions can be 1 arcsec? *Hint:* Use size-to-distance ratios; assume that the moon is about 400,000 km from the earth.

8. Imagine two lights separated by one kilometer. With your unaided eye, how far away from you could these lights be placed and still be visible to you as separate lights? If you used a 10-cm telescope at its theoretical observing power, what would be their distance from you?

9. Consider an optical interferometer placed on the moon with a maximum separation of 10 km. What would be its theoretical resolution? How far away could it see the disk of a planet like the earth?

10. Consider a radio interferometer with an earth–moon baseline. The receiver works at 5 cm. What is the resolving power of this telescope?

7

Einstein's Vision

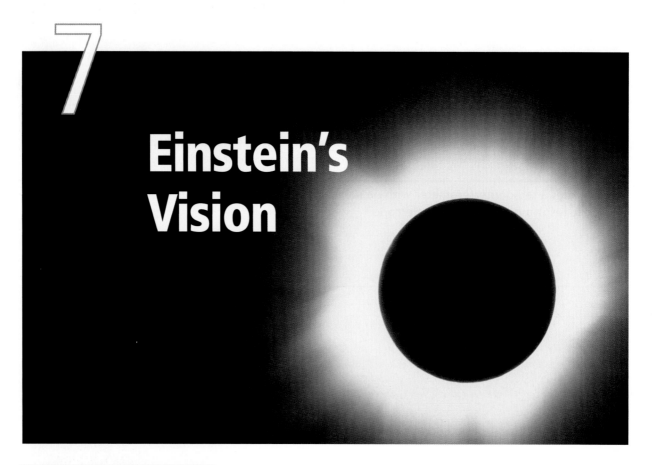

After studying this chapter, you should be able to:

7-1 State the principle of equivalence and illustrate it with a concrete example.

7-2 Show how the principle of equivalence leads to the local cancellation of gravitational forces and weightlessness.

7-3 Compare and contrast Aristotle's, Newton's, and Einstein's concepts of natural motion for bodies falling near the earth and of the motions of heavenly bodies.

7-4 Describe what is meant by the term *spacetime* and give a common example.

7-5 Argue that concepts of natural motion must be coupled to a notion of the geometry of spacetime, both locally and for the cosmos globally.

7-6 Use the concept of escape speed to relate the future of the universe to its geometry: hyperbolic, spherical (closed), or flat.

7-7 State Einstein's relation between matter and energy, apply it to astrophysical situations, and use it in algebraic form.

7-8 Describe one observational confirmation of Einstein's general theory of relativity.

CENTRAL CONCEPT

The general theory of relativity regards space and time as unified in four dimensions. The new view of gravity – radically different from that of Newton's – predicts an expanding universe that may be finite or infinite in spacetime.

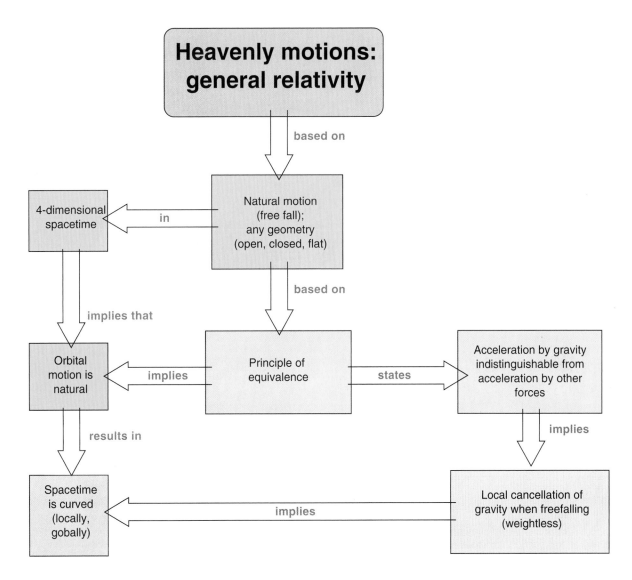

Chapter 7
Celestial Navigator™

The preceding chapters highlighted important changes in people's ideas of the universe. Two influential schemes, Aristotle's (Chapter 2) and Newton's (Chapter 4), rested on physical laws but differed in their concepts of natural motion. These distinct foundations resulted in dissimilar pictures of the nature of the universe. Newton's cosmos was infinite, yet tied together by a universal physical law. Aristotle's cosmos was finite, with different laws for the terrestrial and celestial realms. Both models shared one aspect: the universe was static, fixed in general form for eternity.

Newton's grand model gained the authority of success after success (Section 4.5). His infinite space, established firmly by his laws and intertwined with gravitation, seemed invincible. Newton had failed to discuss the still mysterious nature of gravitational force, however – though he did describe its effects. Albert Einstein (Fig. 7.1) admired Newton's synthesis but wondered about gravity. Einstein also puzzled over the nature of light. From these two seemingly separate realms – gravity and light – Einstein synthesized a new view of the universe in his theory of relativity. One startling result: in Einstein's vision, *gravity is no longer a force.*

Figure 7.1 In 1931 Albert Einstein and his wife, Elsa, visited the Hopi House at the Grand Canyon, Arizona, U.S.A., where they posed with Hopi Indians.

In Einstein's theory, space, time, mass, and natural motion take on new relationships from which a new conception of the cosmos emerges – one in which the universe itself evolves. The cosmos has two futures: the Big Bore or the Big Crunch. This chapter looks again at the concepts underlying Newtonian physics to find the themes of relativity and the results of relativistic ideas for the cosmos.

7.1 NATURAL MOTION REEXAMINED

Newton's physics transformed the Aristotelian world. Try to witness the world the way Newton did. If an object falls or moves, you assume that a force acts on it. You are then applying Newton's laws to the motion you observe. You infer that a force is acting because you see its effect, a change in motion: acceleration. You deem the force to be real because you can see the acceleration that results from it.

Newton's Assumptions (Learning Outcome 7-3)

Take gravity as an example. How do you know it's a force? Pick up a stone; you feel its weight. Drop it. As it falls, its velocity continually increases. The stone exhibits a constant acceleration. What has changed? The stone's natural motion, which according to Newton, is either at rest or moving at a constant speed along a straight line (Section 4.3). In contrast, Aristotle (Section 2.2) believed that a falling stone displayed the natural motion of earthy material, so no force was acting.

The key point: Newton's definition of natural motion led directly to his concept of force, particularly gravitational force. Newton saw natural motion as ordinary, inconsequential, and so needing no explanation. To him, accelerated motion demanded explanation. After all, without accelerated motion, apples would not fall from trees, the moon would not orbit the earth, and the earth would not revolve around the sun.

To cope with accelerations, Newton dealt with forces. These result in accelerations, changes in velocity. This emphasis on rate of *change* of velocity rather than velocity itself is critical. All velocities are relative, so you can always make any given velocity disappear. For example, when you have stopped just before entering a highway, other cars go by you at 90 km/h (55 mph). You accelerate to 90 km/h,

too. Then you have matched the velocities of the other cars, so that compared to yours, they have zero velocity. The original 90-km/h difference has vanished.

In any situation, an observer can match velocities with any other observer. Relative velocities can always be made to disappear. If not, astronauts would not be able to dock their spacecraft! Newton believed that forces were real, that they could not be made to disappear arbitrarily. By connecting forces to accelerations (and not to velocities), Newton seemed to have nailed down their absolute reality.

Motion and Geometry

However, Newton made an implicit assumption. In his definition of natural motion, he talks of velocity along a straight line. What's a *straight line?* The shortest distance between two points. You probably have an intuitive picture of a straight line drawn on some flat surface (like a blackboard). You then have in mind the same kind of straight line that Newton envisaged. That is the geometric assumption behind Newton's definition of natural motion: straight line means a line on a flat surface. Newton really did not have any choice. In his time, a flat geometry was thought to be the only possible geometry.

Can you imagine straight lines that are on a curved surface rather than a flat one? Consider the earth's surface. A straight line on the earth's surface differs from that on a flat surface. Look at a flat (Mercator) map of the world (Fig. 7.2). You might think that traveling on a straight line of the same latitude is the shortest distance between two points of the same latitude, but it's not. Airplanes, for instance, do not travel along latitude lines to go the shortest distance. Instead they travel part of a *great circle,* a circle on the earth's surface whose center is the earth's center.

You can make your own straight lines on the earth with a string. Stretch the string tightly against a globe with ends at the two places you want to connect (Fig. 7.3). Note that this path curves relative to latitude lines! On a flat map, such a line does not look straight, but it is – on a *curved* surface.

The point here is that the geometry of a curved surface differs from that of a flat surface. In his definition of natural motion, Newton assumed that our universe had the same geometric properties as a flat surface. Einstein challenged this assumption.

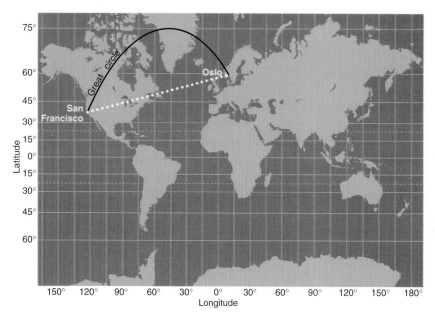

Figure 7.2 Straight lines and geometry on the earth. This Mercator map artificially makes the round earth appear flat, so straight lines on such a map are *not* the shortest distances between two points on the earth. If you use a ruler on this map to draw a straight line between San Francisco and Oslo, you have not shown the path that is the shortest distance between these cities. That path is a great circle line, which appears curved on this flat map.

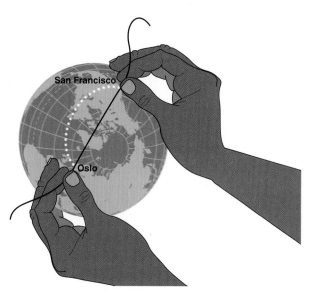

Figure 7.3 Straight lines on a sphere. You can check that great circle distances on a globe are the shortest distances (and so straight lines in a spherical geometry) by stretching a string along the globe's surface to connect two points.

7.2 THE RISE OF RELATIVITY

At the start of this century, Albert Einstein created a new vision of the cosmos. This picture emerged in two stages: the **special theory of relativity,** then the general theory. The special theory dealt with the laws of physics as seen by *unaccelerated* observers: those who experience uniform motion. The general theory went beyond the special theory to cope with the nature of gravitation and so accommodated the case of accelerated observers.

Mass and Energy, Space and Time (Learning Outcome 7-7)

Einstein proclaimed in his special theory that mass and energy are two aspects of the same essential stuff. He then derived the famous formula relating mass and energy to the speed of light:

$$E = mc^2$$

where E is the energy in joules, m is the mass in kilograms, and c is the speed of light in meters per second. This relationship means that you can envision all matter as a form of energy and all kinds of energy as possessing mass. This mass produces gravity and also shows up as inertia.

Small masses convert to large amounts of energy. For instance, the complete transformation of 1 kg of matter to energy results in 9×10^{16} J, which is enough to keep a 100-watt lightbulb lit for about 30 million years! Note that c acts like an exchange rate between matter and energy.

The special theory also resulted in the unification of the concepts of space and time. Scientists had looked at these as separate aspects of the universe (similar to the ancient astronomical separation of earth and sky). Newton, for instance, dealt with space and time as totally unrelated. In Einstein's view, all events in the universe involve space and time together: **spacetime,** which has four dimensions (three of space and one of time).

Einstein made special note of what seems so obvious: when anything happens in the world, it takes place in both space and time. He called such happenings *events*. You mark an event by noting *where* it took place (space) and *when* it took place (time). Events make up points in spacetime.

Light takes a finite time to reach us; so we do not see in real time events that are spread out in space. Instead, we see distant objects as they were at some instant in the past. Look up at the moon. You see it as it was about a second ago. When you look at the sun, your view is some 8 minutes old. The stars at night give you a deeper look into the past. Sirius, as you see it now, shines as it did some 9 years ago. As you look out into space, you look back in time. You now receive light emitted from different objects at many different times. In this sense, a telescope acts like a one-way time machine – it peers only into the past.

Einstein used these connections between mass and energy and space and time to underpin his concept of gravity, the general theory of relativity.

The General Theory of Relativity

From 1905 to 1915, Einstein tackled the case of accelerated observers in spacetime. This investigation led to the general theory of relativity, which is Einstein's theory about the nature of gravitation. (See the General Relativity Celestial Navigator™.)

Einstein rethought gravitation by questioning Newton's concepts of mass and inertia. He saw that Newton defined mass by two different operations: in the second law (Section 4.3) and in the law of gravitation (Section 4.4). Suppose you apply a known force to an object and measure its acceleration. Using the force and acceleration, Newton's second law gives you the object's mass. Mass determined this way is called **inertial mass.** Now take the same object and weigh it. Weight is a force, the amount of gravitational force with which the earth (in this case) attracts the object. Mass measured this way is called **gravitational mass.**

Newton believed (but did not make the distinction) that an object's inertial mass and gravitational mass were equal. He knew this equality from Galileo's experiments with falling bodies (Section 4.2). These results demonstrated that near the earth, all masses fall with the same acceleration. Experimentally, this equality of gravitational and inertial mass holds true to a very high degree of accuracy: no difference in the limits of sensitivity with which tests can be done has ever been detected.

The Principle of Equivalence (Learning Outcomes 7-1, 7-2)

Einstein saw the equality of gravitational and inertial mass as a fundamental fact about the universe. He gave it a special place in the general theory as the **principle of equivalence**: *you cannot distinguish accelerations due to gravitation from accelerations due to forces of other kinds.* Stated differently, the acceleration of gravity does not depend on the mass of an object.

Here is Einstein's own example of the principle of equivalence (Fig. 7.4). Imagine yourself on the earth in a spacecraft with no windows. If you were to drop objects in the spacecraft and measure their accelerations, you would find that they all fell with the same acceleration, 9.8 (m/s)/s. Now suppose that without your knowledge, you and the spacecraft were placed in space and constantly accelerated at 9.8 (m/s)/s. Repeat your experiments dropping other objects. They will accelerate at 9.8 (m/s)/s. *Where are you?* You'd probably conclude that you were still on the earth – unless you could look out of the spacecraft. You cannot, by your experiments, distinguish between the effects due to a gravitational force and those due to the force of a rocket engine.

The principle of equivalence has profound consequences: you now have a way to *cancel* out gravity locally. Put yourself in an elevator in a tall building. Let the elevator free-fall (Fig. 7.5). You find yourself weightless; gravity has vanished! Instantly transport yourself in space, far away from any large masses. Your condition is the same – you are weightless without gravity.

Now, you may think that the falling elevator is some kind of cheat: what happens when the elevator hits the ground? Abruptly everything in the elevator has weight! But imagine that the elevator were to continue to fall straight into an airless tunnel through the earth. What would happen?

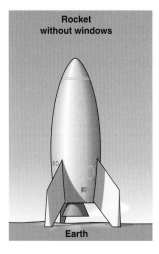
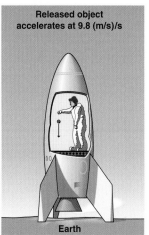

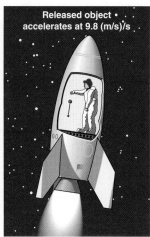

Figure 7.4 Principle of equivalence. Imagine that you are in a small spacecraft with no windows. Drop a mass and measure its acceleration. If the spacecraft is on the earth, the acceleration will be 9.8 (m/s)/s. If the ship is in space far away from the earth and its engine is accelerating it at 9.8 (m/s)/s, the same experiment will yield the same results. Under the conditions of this experiment, you could not determine whether you were on the earth or in an accelerating ship.

Figure 7.5 Local elimination of gravity using the principle of equivalence. An observer in an elevator sees no effects of gravity (a) if the elevator free-falls at the earth's surface; a dropped object has no measurable acceleration. Out in space, far away from all masses (b), the same experiment in a spaceship moving at a constant velocity produces the same result.

It would fly (with weightless conditions inside) through the earth's center and come out the other side to the same height above the ground that it had started from at the other end. It would then reverse direction and fall back, and continue to swing back and forth, taking about 84 minutes for each complete swing. In a sense, it would be orbiting through the earth. During each pass, the conditions for weightlessness would exist inside the elevator.

Weightlessness and Natural Motion (Learning Outcome 7-2)

By free-falling, an observer can make gravity disappear. How strange! Newton saw gravity as mysterious, though he was astute enough to describe its effects. To Einstein, falling objects were not mysterious at all. The objects were simply following their natural motion in spacetime. In Einstein's view, gravity was not a force. Here's a rule of thumb in relativity: when you are weightless, you are following your natural motion in spacetime. Having weight is unnatural.

CAUTION. Don't confuse mass and weight! You know that if you are far enough away from any large mass, you will be weightless. What does that mean? Simply that if you placed a scale beneath you, it would read zero. Now, when you stand on a scale on the earth, what are you measuring? In Newton's terms, you are reading the amount of gravitational force exerted on you by the earth's mass. Weight is a force.

Mass is related to the inertial properties of matter. Forces can tell you how much mass you have. For example, to move around weightlessly in a spaceship, you still have to put out an effort. Suppose you want to go from one side of the ship to the other. The easiest way is to push yourself off a wall with a small amount of force. You'll then drift to the other side, and to stop, you'll need to push against the opposite wall. If you measure the amount of force and your acceleration, you can use Newton's second law to calculate your mass – your inertial mass. No matter where you are – on the earth, the moon, or in space – your inertial mass is always the same.

You might point out the moon also free-falls, moving under the influence of gravity. If the moon's motion around the earth is its natural motion, why is the orbit curved rather than straight? To answer this question requires a look at geometry and its relationship to physics.

7.3 THE GEOMETRY OF SPACETIME

How are geometry and physics related? Newton assumed that the geometry of the universe was flat. Einstein made no such assumption. He put this question up to experimental confirmation.

Euclidean Geometry

Geometry is derived from practical surveying techniques; it was developed from experience. For years, people held that Euclidean geometry – and only Euclidean geometry – applied throughout space to physical measurements. Newton believed this assumption. He had no choice, for no other geometry had been yet devised.

Because of Euclid's *parallel-line postulate* (essentially, that two parallel lines when extended to infinity remain the same distance apart and will never meet), Euclidean geometry is flat. Thus we have the Pythagorean theorem for right triangles: that the sum of the angles of any triangle equals 180° (Fig. 7.6a). This is a key property of a **flat geometry.**

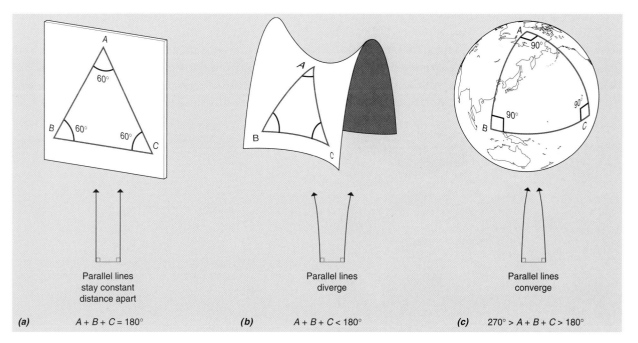

(a)   Parallel lines stay constant distance apart $A + B + C = 180°$

(b)   Parallel lines diverge $A + B + C < 180°$

(c)   Parallel lines converge $270° > A + B + C > 180°$

Figure 7.6 Properties of different geometries. (a) In a flat geometry, the sum of the angles of a triangle always equals 180°. (b) In a hyperbolic (open) geometry, the sum of the angles is always less than 180°. (c) In a spherical geometry, the sum of the angles is always greater than 180°. For example, the triangle drawn from two points on the earth's equator to the North Pole contains 270°.

Non-Euclidean Geometry

In 1829 the Russian mathematician Nikolai I. Lobachevski (1792–1856) pointed out that Euclid's parallel-line postulate was not unique. Instead, Lobachevski proposed a new postulate that allowed parallel lines to diverge. In two dimensions, his geometry has properties similar to the surface of a saddle (Fig. 7.6b), on which all parallel lines diverge when extended, affording a curved geometry sometimes termed **hyperbolic geometry.** When a triangle is drawn on a hyperbolic surface, the sum of its angles is less than 180°. Hyperbolic geometry is infinite, because extended parallel lines never meet. For this reason, both hyperbolic and flat geometries are called **open geometries.**

In 1854 the German mathematician Georg F. B. Riemann (1826–1866) devised yet another geometry with a different parallel postulate. Riemann allowed parallel lines to *converge* when extended. The surface of a sphere has this **spherical geometry** or **closed geometry** (Fig. 7.6c), in which the sum of a triangle's angles is greater than 180°. Consider, as an analogy, the earth's surface with its imaginary lines of longitude and latitude. Two lines of longitude, both perpendicular to the equator (and so parallel to each other at the equator), intersect at the poles.

Both these non-Euclidean geometries are characterized by their curvature, which does not typify a flat, Euclidean geometry. The hyperbolic geometry has a negative curvature because it bends away from itself. It extends infinitely far. Spherical geometry, in contrast, curves in on itself. Because of its positive curvature, spherical geometry is finite but unbounded: it has a definite size, but no edge.

Consider again the earth's surface. You can travel around the earth's surface as many times as you like and in any direction you want without ever discovering a boundary, yet the surface of the earth has a definite area. (Careful – I'm using two-dimensional examples here because they are easy to visualize, but the physical world exists in the four dimensions of spacetime. That's harder to picture, but the idea is the same.)

Local Geometry and Gravity (Learning Outcome 7-4)

Newton had only one geometry at his disposal. By Einstein's time, three general categories of geometry – hyperbolic, spherical, and flat – were available. Which one was the right choice to apply to the physical world? How does this choice relate to gravity? Let's look at geometry and physics in a local region and then for the entire universe.

Imagine that you live in two dimensions on the earth's surface. You have no concept of a third dimension and no experience of it. You cannot conceive of an "up" that is off the surface. You and a friend do an experiment. You both stand on the equator some distance apart (Fig. 7.7). You both walk away from the equator on paths that are at right angles to the equator. You both believe that the geometry of your world is flat.

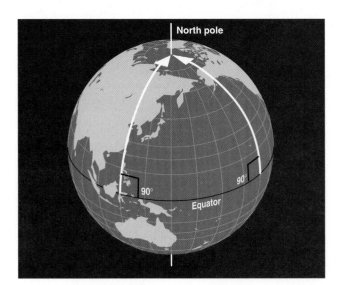

Figure 7.7 Following straight lines in two dimensions on a spherical geometry. Lines at right angles to the equator converge at the pole.

As you walk – being very careful to keep on lines at right angles to the equator – something strange happens. You are moving closer together! Yet by all the precepts of the Euclidean geometry you learned in school, you should be traveling on parallel lines and staying the same distance apart. What's happening?

You might still stick with the belief that your world is flat. Then you could explain your convergence by saying that there is a strange force of attraction bringing you and your friend together. You might even call this force "gravity" and see it as mysterious. Then you would be thinking like Newton.

Or you could say to your friend, "Perhaps our assumption about this world's being flat is wrong. Maybe it's actually curved. We are moving on straight lines at right angles to another line. These lines should be parallel, yet we move closer together. That's not what should happen on a flat world. So the experiment is telling us that the world is not flat but curved."

No need for a mysterious force! An experiment has revealed the local geometry of the world in the region in which the experiment was performed.

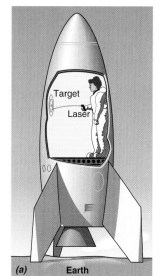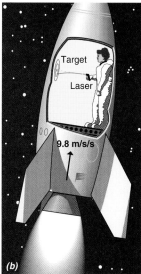

Figure 7.8 Bending of light in a gravitational field. In an accelerating spacecraft (a), a laser beam shot at a target on the opposite wall will hit below the target (b). By the principle of equivalence, the same will happen if a spacecraft is at rest on the earth.

The Curvature of Spacetime
(Learning Outcomes 7-1, 7-5)

How is spacetime curved? In Einstein's general theory, the local distribution of mass (and energy) determines the geometry of spacetime. All objects produce a curvature of nearby spacetime. That curvature shows itself by accelerated motion, which Newton would say was caused by gravitational forces.

Einstein's theory also predicts that light rays will be bent by the curvature of spacetime near massive objects. To see this effect, let's go back into our Einstein elevator (Fig. 7.8a), which is accelerating by its rocket engine at 9.8 (m/s)/s. Put a laser in the front of the elevator to fire across its width at a target level on the other side. Will the light beam hit the target? No, it will strike below it (Fig. 7.8b).

Can you see the reason? As the light travels from one side of the elevator to the other, the elevator continues to accelerate "upward." The light path appears curved to an observer inside the elevator. Now, by the principle of equivalence, if you moved the elevator to the earth's surface, you should not be able to tell the difference by an experiment

inside the elevator. The light path will also bend, and by the same amount as before, so light paths should be deflected in all regions of curved spacetime, near all massive objects.

You should notice that the amount of bending in the elevator example above would be extremely small. The reason: light travels fast and crosses the width of the elevator in an extremely short time. Suppose, for instance, that the elevator were 3 meters wide. Then the light transit time would be a mere 10^{-8} s! During this time, the elevator would have moved only 5×10^{-16} m. Testing this prediction of general relativity on the earth would be a difficult job.

Spacetime Curvature in the Solar System
(Learning Outcomes 7-4, 7-5, 7-8)

Where is the most warped region of spacetime in the solar system? Where the most mass is located: at the sun. Relativity predicts that the sun's mass deflects light rays away from Euclidean, straight-line paths. How to test this? Take a picture of the stars near the sun during a total solar eclipse. Later, when the sun is in a different part of the sky, take a picture of the stars. Compare the angular separation of two stars close to but on opposite sides of the sun.

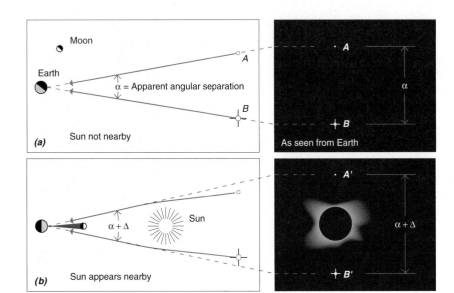

Figure 7.9 Effect of the sun on the apparent angular separation of two stars. Stars *A* and *B* have a measured angular separation (α) when the sun is not nearby. During an eclipse, when the sun lies between the two stars, they appear to be farther apart (angle α plus Δ) because of the bending of the light paths by the curvature of spacetime near the sun.

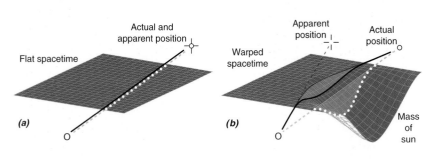

Figure 7.10 Einstein's explanation for the change in the stars' angular separation during an eclipse. (a) When the sun is not near the line of sight to the star, the region of spacetime through which the light rays travel to an observer (at *O*) is flat. (b) When the sun is near the line of sight, the region of spacetime near the sun is distorted. The light rays travel the shortest distance through this curved region. This path refracts that light so that an observer at *O* sees the star shifted to an apparent position away from the sun.

With the sun present, the angular separation of the stars is greater than when the sun is not there (Fig. 7.9). Is this shift visible? Yes! Such observations have been made over many years. They come out very close to Einstein's prediction from general relativity of 1.75 arcsec. (That's about 0.001 the angular diameter of the sun.)

It may strike you as a bit odd that the stars appear farther apart in angle when the sun lies between them. After all, aren't the light rays bent inward? Yes, after they pass the sun. On earth, we think that the light has traveled on a straight line. Our line of sight goes straight out as if the light path

had not been bent. The light has been refracted (Section 6.2). The sun acts as a *gravitational lens*. (You will come upon cosmic gravitational lenses in Chapter 19.)

What's happening? In Einstein's view, light pursues a straight line in spacetime, but the geometry in which it travels is curved by the sun's mass. You can picture the sun's mass as creating a warp in the geometry of spacetime (Fig. 7.10). Light crossing the warp appears to us to take a curved path, although it is actually traveling on the shortest distance between two points through the nonflat region of spacetime. Thus, we refer to the **curvature of spacetime.**

You can imagine the situation as analogous to that of a golf ball moving along a poor putting green filled with depressions. Between depressions, the surface is flat and the ball moves along a straight line. When it crosses a depression, the ball follows a path that bends, compared to the path on the flat surface. The amount of bending depends on the depth of the depression and the speed of the ball. In Einstein's general theory of relativity, mass and energy create local warps in spacetime. The amount and density of mass and energy determine how steep the depression is. As an object moves through a warped region of spacetime, it follows a straight-line, natural-motion path. In warped spacetime, this path appears curved to us. No force acts on the object! This is essentially how Einstein understood the nature of the gravitational force locally.

Einstein considered the orbits of the planets in the same way. All the planets move on straight-line paths in spacetime. We view these paths as elliptical orbits in three dimensions. The paths appear curved to us because spacetime is warped, not flat. The amount of warping decreases as the distance from the sun increases. That's why the earth, for example, follows an orbit more severely curved than that of Jupiter.

To sum up the difference in Newton's and Einstein's concepts of gravity: Newton concentrated on forces, Einstein on courses. Newton saw gravity as a force acting instantaneously between all matter, a force whose strength depended on the mass of the attracted object and the distance. Einstein focused on the paths of objects in spacetime; such paths, for free-falling objects, do not depend on the mass of the object.

Experimental Tests of General Relativity (Learning Outcome 7-8)

Einstein's model for gravitation would be no more than a fascinating idea if it did not make numerical predictions that could be tested experimentally, predictions of effects unknown or unexplained by Newton's gravitation. Let's look at two essential solar system tests: the deflection of light in curved spacetime and the advance of Mercury's orbit. Both these phenomena occur because, in the solar system, spacetime is strongly distorted by the sun's mass.

I have already mentioned the deflection of starlight from measurements during solar eclipses. Radio waves and light are essentially the same, so these deflections should also occur for radio signals from very distant celestial objects. The sun eclipses a few of these as it moves along the ecliptic. Radio astronomers, using interferometry (Section 6.3), can accurately measure the angular deflection of radio waves. Such experiments give an observed deflection within 1 percent of the value predicted by general relativity.

Astronomers have known for a long time that the major axis of Mercury's orbit does not remain fixed in space with respect to the stars (Fig. 7.11). The major axis rotates around in the plane of the orbit. Part of this shifting arises from the gravitational attraction of the other planets on Mercury. But when this effect and others are taken into account, there remains a residual shift of 41 arcsec per century (which means that the orbit turns through an entire 360° in about 3 million years).

Figure 7.11 Perihelion advance of Mercury's orbit. The major axis of the orbit (a) rotates in space with respect to the stars (b), covering an angle of about 16° every 10,000 years.

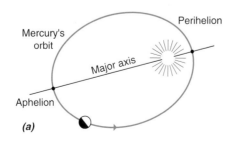

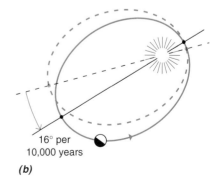

What causes the perihelion advance of Mercury's orbit? Perhaps an undiscovered planet, sometimes called Vulcan, orbiting within the orbit of Mercury. No such planet has ever been definitively observed! General relativity predicts a motion because of the strong curvature of spacetime close to the sun. (Motion of the major axis due to curvature also happens for Venus, the earth, Mars, and the other planets, but it is much smaller for these planets than for Mercury.) The predicted value for Mercury is 43 arcsec per century, so the observed and predicted results agree to within a few percent. Again observations confirm general relativity.

7.4 GEOMETRY AND THE UNIVERSE

So far I have described Einstein's picture of gravity in a small region of spacetime, such as near the sun, but his general theory applies also to the universe. With it, the geometry of the cosmos relates to the dynamics of the whole cosmos. Keep in mind that three basic geometries are possible: hyperbolic, spherical, and flat. Remember also that Einstein does not state which geometry must apply; he leaves this open to experimental confirmation. How to find out?

Cosmic Geometry (Learning Outcome 7-5)

Imagine again living in two dimensions on the surface of a sphere. Start at any point and walk in a straight line away from it. Eventually you'll return to your starting point because a sphere has so much curvature that it comes back on itself. Suppose the universe has the same geometric properties (in spacetime) as a sphere's surface. If we sent out light signals, they would eventually return to us. Why? Just as for a two-dimensional surface, our four-dimensional spacetime, if curved enough, will close back on itself. (If the universe is closed, it does not have an edge, because it has no boundary. Consider again a sphere's surface: you can go around it many times and never find an edge.) We don't know for sure that our universe is closed – that depends on whether it contains enough matter and energy to curve it sufficiently. If it does contain enough, then it will be finite and unbounded, just as the surface of a sphere is finite and unbounded.

Now turn this idea around. The density of mass (and energy) determines the curvature of space-time. If this density has a certain critical value (or greater), then the universe curves back on itself and is closed. Einstein's general theory gives this critical density: it is about one hydrogen atom for every cubic meter of space (roughly 5×10^{-27} kg/m^3).

In 1917 Einstein constructed a model of the universe with a closed geometry. How different from Newton's model, which had to be infinite (Section 4.5)! Einstein's model was also static because of special conditions he imposed on it – no expansion or contraction. General relativity actually has two natural models of the universe, one expanding and one contracting. These are *dynamic* models; in a static model, the universe does not expand or contract. No problem with general relativity, just Einstein's specific application of it. A few years after Einstein proposed his model, astronomers discovered it was wrong. They found that the universe is expanding! The discovery of this expansion caused a radical revision of Einstein's static model for the universe, which did not allow for the change in distances between galaxies at all.

Using HST and large ground-based telescopes, astronomers have put in best efforts to measure the rate of expansion. The values are finally converging, with 20 km/s/Mly a reasonable bet. That means the expansion increases by 20 km/s for every million light years in distance.

The expansion of the universe that we see now implies that it began at a finite time in the past and at the same time for all the cosmos. From the expansion rate now, we can estimate when the expansion began by imagining it running backward in time. All the stuff in the cosmos would coalesce about 15 Gy ago, if the expansion has been constant. The value 15 Gy is one estimate of the age of the universe. Astronomers call this beginning of the universe's expansion the **Big Bang.** (Note: this book uses Gy, gigayears, as the SI unit for 10^9 years.)

CAUTION. When first confronted by the expansion of the universe, people commonly have the misconception that the universe expands "into empty space." That's wrong! It is space itself that is expanding. And it does not expand into "something" or even "nothing." The expression "expand into" is *meaningless* in this context. Visible galaxies mark this expansion, as they surf the flow of spacetime.

7.5 RELATIVITY AND THE COSMOS

With the discovery of the expanding universe, cosmologists began devising models that would account for the expansion in a manner consistent with Einstein's general theory of relativity. In these models, the rate of expansion and the geometry of spacetime (whether flat, hyperbolic, or spherical) are physically linked. Let me show you how, using the concept of escape speed (Section 4.5 and Enrichment Focus 4.4).

Escape Speed and the Critical Density (Learning Outcome 7-6)

Consider throwing a ball off the earth's surface. If the ball's speed is less than the escape speed, it will slow down and fall back, regardless of how large or small its mass is. A ball thrown with a speed greater than the escape speed will slow down a bit while it is still influenced strongly by the earth's gravity and then coast outward as the earth's gravity affects it less and less. If it does not run into anything, the ball will travel out to infinity and never return to the earth.

Now consider the universe and the galaxies within it. Any one galaxy in the universe is analogous to the ball. The galaxies were once all "thrown away" from one another, which we now see as an expansion. Consider a galaxy at some distance from us. Newton showed the net effect of all matter spread within some space: it acts as if this total mass were concentrated at the center (as long as the matter is distributed uniformly). If there is enough mass within that space, the escape speed will be faster than the expansion speed. Note that "enough mass within that space" means a high enough density. (The meaning of density here is related to how closely galaxies are packed together; the closer they are, the higher the density.)

If the density is large enough, the galaxies will not have escape speed; the expansion velocities will decrease with time and eventually reverse as gravity herds all the galaxies together. If the average density is too low, the galaxies will have more than escape speed. Gravity will never bring the galaxies all together, and the expansion will continue indefinitely.

Hyperbolic, flat, and spherical geometries correspond physically to the cases of greater than, equal to, and less than escape speed. From Einstein's general relativity and the observed value of the expansion, we can calculate the density required for the cosmos to be flat, known as the **critical density.** This calculated density can then be compared to the observed cosmological density (which must include energy as well, since $E = mc^2$). If the observed cosmological density is greater than the critical density, the universe is spherical and closed. If less, the universe is hyperbolic and open. If they are exactly equal, the universe is flat.

Whether the universe is open or closed depends on the density of energy and matter within it. A large enough density curves spacetime so that it wraps around itself to close the cosmos. If we can observe the average density of matter and energy in the universe, we will have an experimental basis for finding out the appropriate geometry of spacetime! Observations of the expansion rate combined with general relativity provide a way to find out whether the universe is open or closed.

The Future of the Universe

What's the future for the universe? In each geometry, some special moment marks the start of the expansion. In a closed universe, the rate of expansion slows and eventually stops; then the universe contracts. In a hyperbolic universe, the rate of expansion does not decrease as rapidly, and it never stops. Even after infinite time, the galaxies are still moving apart at a finite velocity. In the borderline case of a flat universe, the expansion slows down just enough to permit the universe to come to a stop after an infinite time of infinite expansion (Fig. 7.12).

We seem to have two possible cosmic destinies. In one, the expansion grinds on forever, and the universe relentlessly thins out and cools down. We can expect a "Big Bore" as all the astronomical action dies. In the other fate, the expansion slows down, stops, and reverses, and the universe collapses. During the time of diminishing size, the galaxies rush together into a dense conglomeration with (theoretically) zero radius. The universe eventually would crush everything into high-energy light – and many weird forms of mass (Chapter 20).

This final collapse is sometimes called the "Big Crunch."

Results to date have been a mixed bag, some supporting an open, others a closed universe. The case is not yet settled; the issue will arise again in Chapters 18 and 20. The crucial problem is that of *dark matter,* which may fill space enough to make the actual density greater than the critical density. (Visible matter accounts for only about one hydrogen atom per *ten* cubic meters of space, far below the critical density.) We cannot see this stuff directly, though it may dictate the future of the cosmos. It is the dark side of the visible universe.

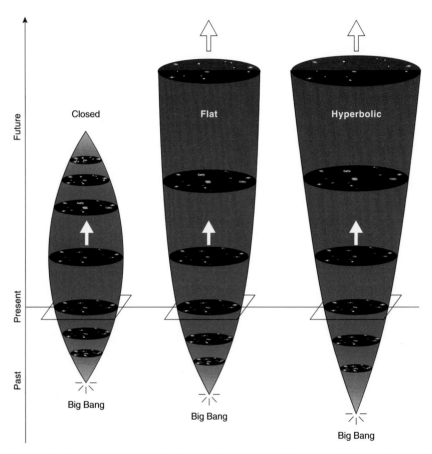

Figure 7.12 Cosmic geometry and the future of the universe. All start in a Big Bang. The closed universe collapses on itself (in a Big Crunch). Both flat and hyperbolic universes expand forever. (Adapted from a NASA diagram.)

KEY CONCEPTS

1. Newton's concept of natural motion (and so of forces) rested on the assumption that the geometry of the universe was Euclidean (flat). Einstein, in developing his theory of gravitation – the general theory of relativity – challenged this assumption by allowing any geometry to describe the cosmos.

2. Before developing the general theory, Einstein worked out the special theory of relativity, which relates to the laws of physics as seen by observers moving at constant velocities (no acceleration). Consequences of the special theory are the unification of space and time (in four-dimensional spacetime) and the intimate relationship between matter and energy and their transformation.

3. The general theory deals with gravitation; it is based on the principle of equivalence, which states that accelerations due to gravity cannot be distinguished from accelerations from other forces. The principle of equivalence implies that gravity disappears in free-fall, when one is weightless. Einstein then views free-falling as natural motion; gravity appears when the spacetime in which free-fall takes place is curved rather than flat.

4. Einstein's concept of gravity as the curvature of spacetime has been tested and found to predict accurately such effects in the solar system as the bending of starlight near the sun. Mass and energy cause the warping of spacetime.

5. The general theory allows the cosmos to have any one of three general geometries: hyperbolic, spherical (closed), or flat (Table 7.1). Which one actually applies can be determined by observations.

6. Astronomers have found that the cosmos is expanding. The rate of expansion is the same measured in different directions in space, which implies that the expansion is uniform. The beginning of this expansion was a cosmic explosion called the Big Bang. From the rate of expansion, we can estimate the time since the beginning of the expansion and so the "age" of the universe: at most 15 Gy ago.

7. If the geometry of the cosmos is hyperbolic or flat, the universe will expand forever; if closed, the expansion will stop at some time in the future, whereupon the universe will contract. To find out which is the case, compare the critical density to the actual average density of the cosmos. If the actual is greater than the critical, the universe is closed; if less than the critical, it is open.

8. The geometry of the universe determines its future; the actual density of the universe relative to the critical density indicates the overall geometry. If open or flat, the universe will expand forever (Big Bore), whereas a closed universe will eventually stop expanding and then contract (Big Crunch).

9. The concept of escape speed is one way to understand the physical meaning of the geometrical curvature of the cosmos. If the Big Bang had enough energy to accelerate matter to great enough speeds, the universe will expand forever (open). If not – that is, if the energy was too small, hence the speeds too low – matter will eventually collapse (closed). If the energy were just right, the speeds would just equal the escape speed, and the expansion could go on forever (flat).

Table 7.1 Geometry of the Universe and Observations

Aspect	Hyperbolic	Spherical	Flat
Extent in space	Infinite	Finite	Infinite
Extent in time	Infinite	Finite	Infinite
Topology	Unbounded	Bounded	Unbounded
Average density, kg/m³	Less than 5×10^{-27}	Greater than 5×10^{-27}	Equals 5×10^{-27}
Future	Expansion forever	Expansion stops, collapse	Expansion forever

STUDY EXERCISES

1. Describe standing on the earth in Newton's terms and in Einstein's. (Learning Outcomes 7-1, 7-2, and 7-3)

2. Describe orbiting the earth in Newton's terms of gravity and contrast this to Einstein's description. (Learning Outcomes 7-1, 7-2, and 7-3)

3. Describe the motions of the planets around the sun in Newton's terms and in Einstein's. (Learning Outcomes 7-1, 7-2, and 7-3)

4. In Einstein's model, how is the geometry of the universe related to its average density? (Learning Outcome 7-6)

5. Compare physically the case of the universe expanding and then contracting (a closed geometry) to the situation of firing a rocket off the earth with less than escape speed. (Learning Outcome 7-6)

6. Consider the following situations and state whether the condition of weightlessness characterizes each one: (a) a spacecraft orbiting the earth, (b) the moon orbiting the earth, (c) the earth orbiting the sun, (d) Jupiter orbiting the sun, and (e) a spacecraft orbiting the sun. For each of these cases, what can you conclude with respect to whether the object is following a straight-line path in spacetime? (Learning Outcomes 7-2, 7-3, 7-4, and 7-5)

7. In what sense is free-falling motion natural motion in Einstein's view? (Learning Outcome 7-3)

8. State verbally Einstein's relation between matter and energy. Does it violate the conservation of energy? (Learning Outcome 7-7)

9. According to Einstein's ideas, what does weightlessness indicate about your motion? (Learning Outcomes 7-2 and 7-3)

10. In what way does the observed expansion of the cosmos "confirm" general relativity? (Learning Outcome 7-8)

PROBLEMS AND ACTIVITIES

1. Imagine that the mass of your body were completely converted to energy. About how much energy would be released? Assume that this conversion takes place in one second. How many 100-watt lightbulbs could be lit up?

2. What is the ratio of the mass–energy equivalent of the earth to that of your body?

3. Consider the concept of escape speed (Enrichment Focus 4.4) applied to matter over large regions of space. Now, Newton's law of gravitation requires that if mass is spread over some spherical volume, then if you were just outside this volume, the mass would act as if it were all concentrated in the center. Assume that the average density of matter in space is 10^{-26} kg/m^3. How much matter lies in a spherical volume with a radius of 10^9 ly? What is the escape speed from this mass?

PART TWO
The Planets: Past and Present

PART OUTCOME

Outline an evolutionary scenario, with supporting evidence, for each of the major objects in the solar system and for the solar system as a whole.

INQUIRY FOCUS

How do we know:

The physical properties of the earth?

The physical properties of the other terrestrial planets?

The physical properties of the Jovian planets?

The physical properties of satellites and solar system debris?

The processes that drive the evolution of solar system objects?

A basic model for the origin and evolution of the solar system and its future and evidence to support it?

8

The Earth: An Evolving Planet

LEARNING OUTCOMES

After studying this chapter, you should be able to:

8-1 Describe and apply a method for determining the earth's mass and density; apply the concept of escape speed to the earth.

8-2 Sketch the interior structure of the earth, indicating the composition of each general region, and argue that the earth's interior structure implies that it must have been molten at one time.

8-3 Argue from at least two observations that the earth's core probably has a metallic composition.

8-4 State the estimated age of the earth, with a range of uncertainty, and explain the method by which this age is inferred using radioactive decay and half-life.

8-5 Apply the properties of magnetic fields in general to the overall structure of the earth's magnetic field, and present a possible model for the field's source.

8-6 Describe at least two ways in which the earth's atmosphere affects the earth's surface environment.

8-7 Explain how the earth's atmosphere acts like an insulating blanket that keeps the earth's surface warm.

8-8 Outline a possible model for the evolution of the earth's crust and interior, with an emphasis on heat flow.

8-9 Outline a possible model for the evolution of the earth's oceans that ties in with a broader view of the earth's history.

8-10 Outline a possible model for the evolution of the earth's atmosphere; indicate how humankind affects the atmosphere now.

8-11 Summarize the physical processes that affect the evolution of the earth's atmosphere, crust, and interior (especially the outward flow of energy) and serve as a model for planetary evolution.

CENTRAL CONCEPT

The dynamic earth is a highly evolved planet, built over thousands of millions of years by geologic processes that are driven by the slow outflow of internal energy. It serves as the model for understanding other planets.

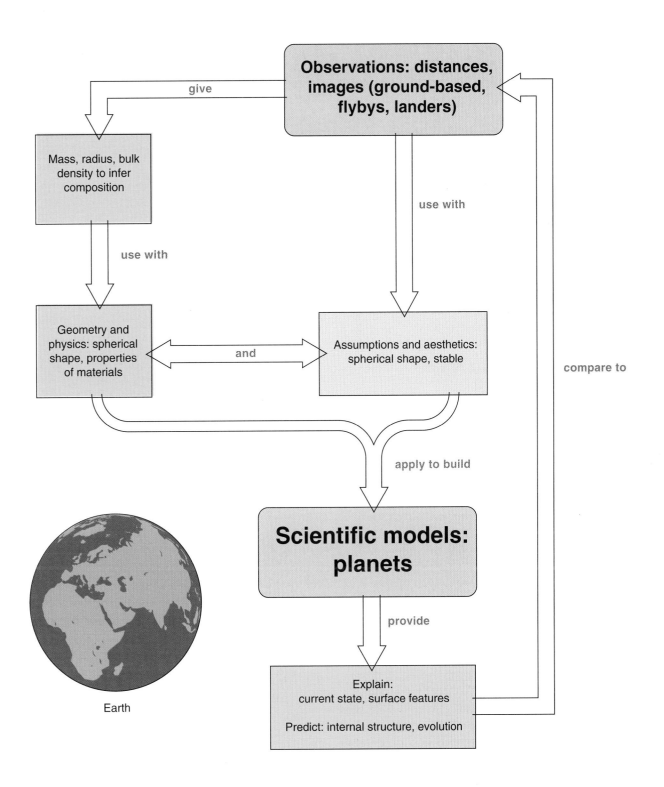

Observations: distances, images (ground-based, flybys, landers)

give

Mass, radius, bulk density to infer composition

use with

use with

Geometry and physics: spherical shape, properties of materials

and

Assumptions and aesthetics: spherical shape, stable

apply to build

compare to

Scientific models: planets

provide

Earth

Explain:
current state, surface features

Predict: internal structure, evolution

A tiny planet, the earth whirls around one ordinary star. As our home, the earth suits us well now, yet our world has modified tremendously since its birth, bustling with tireless change. A primary scientific goal of examining the earth in fine detail is to infer the history of the earth and to mull over its future and our impact on its future.

This chapter looks at the physical makeup of the earth. We live here, so we know this planet better than any other. The earth is a highly evolved planet – altered dramatically in its physical structure since its formation. How? From the flow of energy from its interior to its surface – some of that energy stored from its birth long ago. We will use our planet as the basis of comparison for understanding the makeup and evolution of the other earthlike planets in the inner part of the solar system.

8.1 THE MASS AND DENSITY OF THE SOLID EARTH
(Learning Outcome 8-1)

Have you ever hefted a ball made of unknown materials? How could you guess its composition? To measure its diameter, no problem! And its mass, too, with a scale. Then what? Well, you could drop the ball in water. If it sinks, it is denser relative to water; if it floats, it is less dense. You can then make reasonable conjectures as to the ball's composition. For planets like the earth, we estimate densities to infer compositions – a key clue to understanding planets.

The earth is basically a ball-like sphere (Fig. 8.1). Its equatorial radius, which is a standard unit of planetary size in the solar system, is about 6400 km. (Note: Most figures given in the text are rounded off to one or two significant figures; those in tables have more significant figures.)

How to find out the earth's mass? On its surface, we have a special way. Recall (Section 4.2) that Galileo found that all bodies at the earth's surface have the same acceleration due to gravity, g. Newton's law of gravitation (Section 4.4) relates this acceleration to the earth's mass and radius and to the gravitational constant, G. So if we know G (which can be measured in the lab), the earth's radius, and g (which can be measured at the earth's surface), we can figure out the earth's mass from Newton's law. It comes out to some 6×10^{24} kg – or about the mass of 10^{21} automobiles or megapetacars; see Appendix A for prefixes!

Another way to find the mass is to apply Newton's form of Kepler's third law (Enrichment Focus 4.3) to the orbital properties of an artificial satellite in earth orbit. Then we find the *sum* of the masses (satellite plus earth). But the earth's mass is much greater than any satellite's mass, so, in essence, we get the earth's mass alone.

Knowing both the earth's mass and its volume (from its radius), we divide the earth's mass by its volume to find its **average density:** 5500 kg/m^3, or 5.5 times the density of water. (Water has a density of 1000 kg/m^3 and makes a good density standard.) This average density indicates that the earth, in bulk, consists of a combination of rocky and metallic materials. (Most rocks have a density between 2 and 4; pure iron has a density of 7.8.) Different materials in general have different densities. Iron is more dense than water; wood, less so. If you place both materials in water, the wood floats but iron sinks.

Rocks near the earth's surface average 2.4, about half the average density. This difference implies that the core of the earth is denser than the surface average – the core's density may be 12 times the density of water. This high density means that the core contains dense materials, such as iron. The weight of the overlying layers compresses the core, causing the core's material to have a density higher than if it were uncompressed.

You need to be very clear about the concept of density, as you will encounter it many times in your study of astronomy. The division of the mass by the volume tells us how much of the mass is associated with *one unit* of the volume. For the unit of a cubic meter, that amount is 5500 kg – about the same mass as a few automobiles!

8.2 THE EARTH'S INTERIOR AND AGE
(Learning Outcomes 8-2, 8-3, 8-4)

From the earth's general traits (such as mass, size, and density), we develop models of the interior. It has three distinct density layers: the core, the mantle, and the crust (Fig. 8.2). The **core** contains the central zone and extends more than halfway to the surface. The core has a radius of about 3500 km and is probably a blend of iron and nickel. In its center, the temperature may exceed 6000 K.

Above the core extends the **mantle,** roughly 2900 km thick. The mantle material is rock made of iron and magnesium combined with silicon and

Figure 8.1 Earth from space. The spiral cloud patterns arise from the global circulation of the atmosphere.

oxygen (a silicate mineral called *olivine*). Rocks are generally made of *silicates*, minerals containing compounds of silicon and oxygen. The temperature within the mantle varies, from about 3800 K at the base to 1300 K at the top (Fig. 8.3). In this range of temperatures, the mantle behaves like a plastic. Under slow, steady pressure, it flows like a liquid.

Encasing the mantle is the **crust,** the solid surface layer, which varies in depth from 8 km (under oceans) to 70 km (under continents). Most of the crustal material consists of rocks that have solidified from molten lava and are called **igneous rocks.** These rocks are basalt: silicates of aluminum, magnesium, and iron. They comprise the ocean basins and the subcontinent sections of the crust. The continental masses are mostly granite.

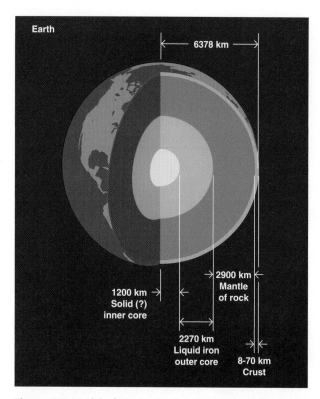

Figure 8.2 Model of the earth's interior showing the sharply defined core, mantle, and crust, each with its characteristic composition and density. The core and mantle have roughly constant thickness. The crust differs in thickness by almost a factor of 10. It is thinnest under the ocean basins (about 8 km) and thickest under the continents (up to 70 km). Note how large the radius of the core is compared to the overall radius of the earth.

Because the granite has a lower density than the basalt, the continental plates float on the basalt. And, because the mantle is denser than the basalt and granite, the entire crust floats on the mantle!

Imagine if the earth were the size of a tennis ball (diameter about 6.6 cm). Then the core would take up about the inner 3.5 cm, the mantle almost all of the rest. The crust would have a thickness of only 0.3 mm! If it had the earth's average density, its mass would be about 350 g, about 6 times the actual mass of a regulation tennis ball.

How do we know about the earth's interior, since we cannot see into it? Geologists infer the structure by studying the shock waves from earthquakes. These waves, which travel through the interior and are affected by it, convey clues to the interior's physical state. This information verifies our models. We can get a firm idea of the interiors of other earthlike planets if we also have information from shock waves going through their interiors.

TO SUM UP. The earth's interior is *differentiated* – it has distinct layers (Fig. 8.2). The least dense materials lie on the surface; the most dense at the center. Basically, the interior consists of two zones: one metal-rich (the iron core), the other silicate-rich (the mantle and crust). This separation naturally occurs in a mixture of materials of different densities that is fluid (at least in part). Shake a bottle of oil and water. What happens? The mixture separates. The less dense water floats to the surface with

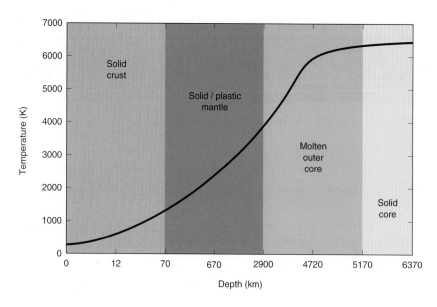

Figure 8.3 Model of the temperature profile of the earth's interior. The temperature at first rises quickly through the crust and mantle, then levels off in the core.

the oil beneath it. Gravity separates the lower-density materials from the higher-density ones.

How did the earth get this way? According to current models, when the earth first formed, its materials were well mixed up. Imagine the interior then heating up until it was mostly molten: then the denser materials (iron) would settled at the core, with the less dense materials (silicates) forming a froth on top.

What heated the interior? One source is stored internal energy from formation. Another is the energy generated by radioactive decay (Enrichment Focus 8.1). In the past, the earth had more radioactive material than now – perhaps six times as much. The heating from the decay of so much radioactive material could melt the interior enough for it to flow and separate into regions of different density.

Geologists estimate the earth's age at 4.6 giga-years (abbreviated Gy, equals 10^9 years) from **radioactive dating** (Enrichment Focus 8.1), which uses the natural decay rate of radioactive isotopes. The oldest rocks on the earth's surface are not this old – they are very close to 4.0 Gy old. These rocks are igneous rocks, which means that they solidified from once-molten material. (Small mineral grains in rocks have ages as old as 4.4 Gy.) Geologists estimate that about 0.5 Gy had to elapse for the crust to melt and then cool to solidify the first rocks. So the earth's age is that of the oldest rocks plus the time to form them – some 4.5 Gy.

This estimate falls close to that for meteorite material (4.55 Gy) and lunar material (4.6 Gy), determined by the same radioactive dating techniques. The coincidence of these ages implies that the solar system formed along with the sun about 4.6 Gy ago (Chapter 11). The exact times of these cosmic events are still debated, but most geologists agree that the earth is 4.5 to 5 Gy old, with 4.6 Gy as the best estimate (with an uncertainty of about 0.1 Gy).

8.3 THE EARTH'S MAGNETIC FIELD (Learning Outcome 8-5)

You can visualize the earth's magnetic properties by imagining a giant bar magnet located in the core. The magnetic field protrudes from the south magnetic pole in the Southern Hemisphere and returns to the north magnetic pole in the Northern Hemisphere. The magnetic axis, which connects the magnetic poles, tilts about 12° from the spin axis and does not pass through the earth's center (Fig. 8.4). The part of this magnetic field that is parallel to the earth's surface orients a compass needle so that it points to the north and south magnetic poles. The strength of the earth's magnetic field at its surface is about 4×10^{-5} T. Note the T stands for *tesla,* which is the SI unit for magnetic field strength (Appendix A). A small toy magnet is about 2×10^{-2} T – some 500 times stronger.

The magnetic fields around the earth, the sun, and some planets have two poles and so are called *dipole fields.* This attribute – dipoles – allows us to imagine these fields as arising from a giant bar magnet buried in the sun or planet (Fig. 8.5), even though they really come from circulating electric currents.

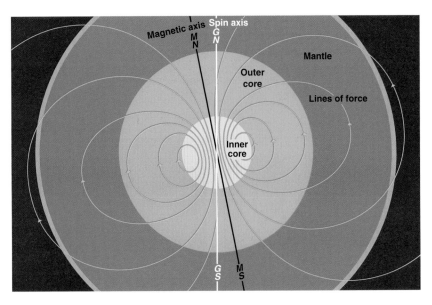

Figure 8.4 Model of the earth's magnetic field. The magnetic axis (*MN–MS*) is *not* aligned with the spin axis (*GN–GS*), but is tilted about 12°. Note the continuity of the magnetic field lines between the two magnetic poles. The convention in this book is that the direction of the field lines is out from the south magnetic pole and into the north magnetic pole.

ENRICHMENT FOCUS 8.1

Radioactivity and the Dating of Rocks

The radioactive dating technique relies on the breakdown of the nuclei of radioactive elements, such as uranium. When they decay, they break apart into simpler nuclei and release energy in the form of high-speed particles. In *alpha decay,* the isotope emits an alpha particle, which has two protons and two neutrons. (An alpha particle is the nucleus of a helium atom.) In *beta decay,* the nucleus expels a high-speed electron, which is called a *beta particle* for historic reasons. The alpha and beta particles carry kinetic energy. By collisions, they transfer this energy to electrons and atomic nuclei in the surrounding rock and heat it. The more radioactive material in a rock and the faster it decays, the more a rock will be heated.

How fast is that decay? You cannot estimate when any given atom will decay because the process is random, but for a large number of atoms, you can determine a gross rate of disintegration. (It's like popping popcorn. You cannot predict which kernel will pop next, but you can estimate when the entire batch will be finished.)

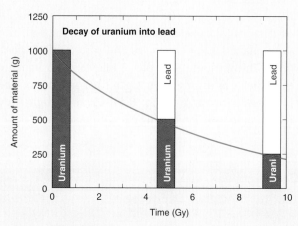

Figure F.8 Radioactive decay of uranium into lead, which illustrates the concept of half-life.

Half a piece of uranium-238 (^{238}U) decays to lead in 4.5 Gy, half again in the next 4.5 Gy, and so on. The length of time required for half the material to disintegrate is called the **half-life** of the element (Fig. F.8), so you can calculate the amount of original uranium left at any time, even though the decay time for any one uranium atom cannot be specified.

By reversing this idea, you can estimate the age of a rock. Given a rock sample containing ^{238}U and lead and knowing the half-life of the uranium, we can calculate the age of the sample by noting how long it would have taken to form the observed amount of lead by radioactive decay. (To use this technique, you must be able to estimate the amounts of various isotopes of lead originally in the rock at formation. It also involves a sequence of complicated and precise chemical measurements.)

Uranium is not the only element that can be used in radioactive dating. Others that also can serve as radioactive "clocks" include rubidium (^{87}Rb), which decays to strontium (^{87}Sr), with a half-life of 47 Gy, and potassium (^{40}K), which decays to the inert gas argon (^{40}Ar), with a half-life of 1.3 Gy. Other isotopes are measured as well in order to find out the relative composition of the original rocks. Whatever elements are used, the derived age is the time elapsed since the rocks last solidified.

For example, suppose a rock sample now contains equal numbers of potassium-40 and argon-40 atoms. If there were no argon atoms in the rock originally, they must all have come from decay of potassium-40. Exactly half have decayed (and half remain), so the rock must be one half-life old, 1.3 Gy. How old is a rock that contains seven times as many argon-40 atoms as potassium-40 atoms? If all the argon came from decay of potassium, the remaining potassium-40 is one-eighth of the original. So three half-lives must have elapsed ($\frac{1}{8} = \frac{1}{2} \times \frac{1}{2} \times \frac{1}{2}$), and the rock must be $3 \times 1.3 = 3.9$ Gy.

Note that the age inferred in this way is the time since the rock last solidified. If a rock has melted and resolidified, the age given by radioactive dating will err on the low side, since it will not have "counted" the portion of the argon gas that escaped during melting.

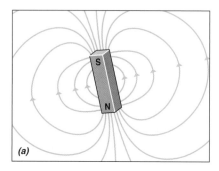

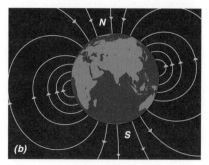

Figure 8.5 Magnetic field configurations for dipoles. The spacing of the field lines indicates the strength of the field: the closer the lines, the stronger the field. (a) The magnetic field of a bar magnet is a dipole with a magnetic north (*N*) and south (*S*) pole. (b) The earth's magnetic field has a dipole-like form with north and south poles.

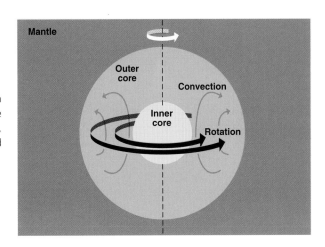

Figure 8.6 Fluid motions in the earth's core. The fluid currents in the core are driven by the overall nonuniform rotation, where the inner parts rotate at a higher angular speed than the outer parts, combined with the convection in small regions. These organized currents, in turn, are believed to generate the earth's magnetic field.

Origin

We can apply a **dynamo model** to explain the source of the earth's magnetic field. Why? Because its metallic core is both liquid (in part) and a good electrical conductor. It acts like a giant electrical generator – a dynamo – and a large electromagnet. The liquid core generates electricity and creates a magnetic field. (In an electrical dynamo in a generating plant, the reverse process occurs – a spinning magnet moves electrons to make an electrical current.)

Basically, organized fluid motions in a conductor will generate a magnetic field. The hot, liquid part of the earth's core contains convective flows, where hotter materials rise and cooler materials sink. Because it is liquid, the outer core also rotates nonuniformly – the inner regions are carried around faster than the outer one. This kind of motion causes the convective regions to curl around rather than flowing up and down (Fig. 8.6). These motions – mere millimeters per second! – generate magnetic fields. The earth's rotation stirs the core to generate electrical currents.

Magnetosphere

Earth-orbiting satellites have surveyed two doughnut-shaped belts of protons and electrons trapped by the terrestrial magnetic field. These are called the *Van Allen radiation belts* (Fig. 8.7), after their discoverer, James A. Van Allen. The sun supplies the charged particles that are trapped in the Van Allen belts. (The flow is called the *solar wind,* Section 12.5.)

The earth's magnetic field alters the flow of charged particles for many hundreds of earth radii out into space. The region so affected is called the earth's **magnetosphere** (Fig. 8.7). The magnetos-

phere acts like a buffer between the earth and the solar wind, which flows over and around it. In turn, the solar wind compresses the earth's field on the day side and stretches it on the night side in a long magnetic tail. We expect a magnetosphere to engulf any planet with an ample magnetic field.

8.4 THE BLANKET OF THE ATMOSPHERE (Learning Outcomes 8-6 and 8-7)

Our atmosphere is a blessing. It provides oxygen for breathing, shields us from cancer-causing ultraviolet radiation of the sun, furnishes a thermal blanket to keep the surface warm, and spreads the heat around the earth. Understanding the earth's atmosphere provides invaluable information for the study of other planetary atmospheres.

Relative to the total number of atoms and molecules available, our atmosphere contains approximately 78 percent molecular nitrogen (N_2), 21 percent molecular oxygen (O_2), 1 percent argon (Ar), 0.03 percent carbon dioxide (CO_2), and traces of other elements. Water vapor (H_2O) occurs in variable amounts, as much as 4 percent near the surface.

The weight of the upper atmospheric layers makes the lower portion denser than the upper, just as a sandwich at the bottom of a pile is squashed by the weight of the sandwiches above it. The gas pressure at sea level on the earth's surface, where the entire atmosphere is piled above it, is called *one atmosphere* (1 atm) of pressure. (The *pressure* of a gas is the force it exerts on an area of surface.) The atmospheric pressure and density decrease with height, rapidly at first, then more slowly. High in the atmosphere (about 1000 km up), molecules and atoms can escape into space.

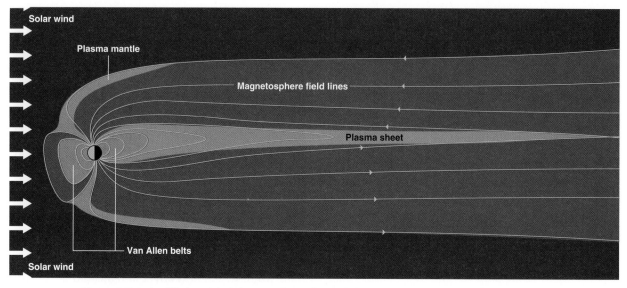

Figure 8.7 A simplified model of the Van Allen radiation belts and magnetosphere, which encircle the earth. The earth's magnetosphere is created by the interaction of the earth's magnetic field with the flow of charged particles in the solar wind. A long magnetic tail forms downstream, with a thin plasma sheet in its core and a plasma mantle around the earth, both trapped by the magnetic fields. The impact of the solar wind compresses the Van Allen belts on the sunward side.

At sea level, the pressure is about 10^5 newtons per square meter (or about 15 pounds per square inch) – a car tire is inflated to about twice this pressure. The SI unit for pressure is the *pascal,* which equals one newton per square meter, so the atmospheric pressure at the surface is about 10^5 Pa (Appendix A). The tires on my car state that their maximum rated pressure is 240 kPa.

The Greenhouse Effect

Sunlight heats the ground, and the air in contact with it also becomes warmer. If sunlight were the only source of heat, the temperature at the ground would reach a frigid 255 K (or –18° C). Water would always be frozen, and life wouldn't exist! The average temperature at the earth's surface is actually higher, about 288 K (+15° C).

What happens? Visible light from the sun gets through to the surface and heats the earth. The earth emits in turn infrared radiation. If the infrared radiation simply escaped into space, the earth would be too cold for life, but this radiation does not escape completely. Instead, the infrared is absorbed by the earth's atmosphere (mostly by water vapor and carbon dioxide). The atmosphere heats up by absorbing this radiation. Some energy goes off into

space; the rest radiates back to the ground and adds to the heating. Both direct sunlight and infrared radiation from the atmosphere heat the earth's surface. Water vapor plays the major role in the earth's atmosphere, with carbon dioxide a minor role, in blocking the escape of infrared emitted by the ground.

Clouds in the atmosphere reflect to space about 30 percent of the energy in incoming sunlight (Fig. 8.8). The air absorbs about another 30 percent, and the remaining 40 percent or so reaches the earth's surface. The earth's surface reflects about 15 percent of the solar energy that reaches it back toward space. The remaining sunlight heats the lands and seas. These then send most of the energy back into the atmosphere, chiefly as infrared. Some escapes to space (about 10 percent). The rest stays trapped in the atmosphere and heightens the surface temperature.

This warming of the surface by the atmospheric trapping of infrared radiation is often called the **greenhouse effect** by analogy to one process that keeps a greenhouse warm: glass is transparent to visible light but opaque to infrared. Sunlight enters the greenhouse and warms the interior, which in turn emits infrared. This heat cannot radiate through the glass, so it stays to help warm up the interior.

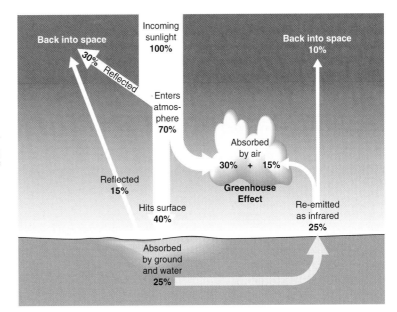

Figure 8.8 Energy budget for the earth's greenhouse effect. About 45 percent of the incoming energy from the sun heats the air. The balance of gain and loss sets the surface temperature.

KEY POINT. Any planetary atmosphere that is more or less transparent to visible sunlight but opaque to infrared will act to keep the planet's surface warmer than it would be if the planet had no atmosphere.

CAUTION. The greenhouse effect is misnamed! The absorption of infrared radiation is *not* the main process that keeps the greenhouse warm. The air inside, heated by contact with the hot inside of the greenhouse, cannot escape. This inhibition of cooling occurs with any roof. However, the expression "greenhouse effect" is so ingrained in the astronomical vocabulary that this book will use it.

Atmospheric Circulation

Sunlight heats the ground and drives the winds; global wind and cloud patterns result from the earth's rotation. These processes work to propel the atmospheric circulation on any planet.

Differences in heating of air result in **convection,** which carries heat from hotter to colder regions (Fig. 8.9). The process of convection carries energy in fluids (gases and liquids). Sunlight heats the ground and the air in contact with it. The heated air expands and its density decreases, so it rises. Denser, cooler air descends to replace it, and a convective develops (Fig. 8.9a). Local convective flows produce cumulus clouds. You have felt the convective gusts if you have

flown on a plane that has pierced through a cumulus cloud layer – the ride gets bumpy!

Such convective currents work globally as well as locally. The air over the poles is cold and dense; air over the equator is hot and less dense. Why? Because the tropics receive more solar energy than the poles. The temperature differences result in a north–south circulation. Convective cells result in these regions (Fig. 8.9b). The pattern of earth's atmospheric convective cells is sculpted by land and ocean masses.

The earth's rotation then produces spiral motions of these flows: counterclockwise in the Northern Hemisphere and clockwise in the Southern Hemisphere. In general, the faster a planet rotates, the stronger the spiral flows. The earth rotates rapidly; this, combined with convective flows, fixes the atmospheric circulation. Similar processes dictate the atmospheric flow of other planets.

8.5 THE EVOLUTION OF THE CRUST
(Learning Outcome 8-8)

The earth's surface divides roughly into two levels: the continents and the ocean basins, with an average height difference of about 5 km. Erosion and water transport of materials in a few million years should erase the difference in height between the oceanic and continental plains. Given the great age of the earth (Section 8.2), the fact that these levels remain distinct implies that the mountain heights and ocean depths are regularly replenished. How?

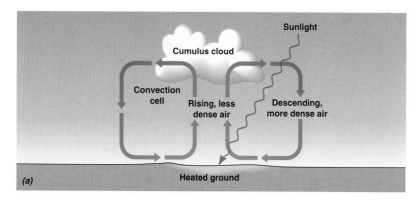

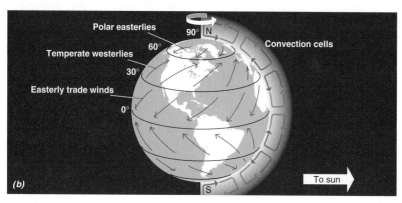

Figure 8.9 Atmospheric circulation. (a) Convection currents cause the development of cumulus clouds in the earth's atmosphere as heated (and so less dense) air rises and displaces cooler air, which has a higher density and so sinks. (b) Solar heating and the earth's rotation result in a global air-circulation pattern. The large convection cells have their air-flow twisted by the earth's rotation to create an overall flow of winds near the earth's surface. Note the symmetry about the equator.

Planetary Evolution and Energy

Energy flow from the earth to space drives planetary evolution. Each major part of the earth – interior, crust, atmosphere, and oceans – has a different temperature. Energy flows between them (from hotter to cooler) and activates their changes. All the internal energy gradually flows to the surface. There, some energy is added by sunlight to the air, ground, and oceans. Finally, this energy radiates into space, which is much colder than the earth.

Today, the total outward energy flow averages a mere 0.06 watt per square meter at the surface. If you could capture all of it over a square meter, it would take about two weeks to raise a cup of water to boiling temperature, 100° C! Yet, because the earth has a large surface area, the amount flowing out per second is large (about 10^{13} joules per second). The earth stores a vast amount of thermal energy to keep evolving.

Volcanism and Plate Tectonics

Earthquakes and volcanic activity conjoin with mountain and island building. This activity shows up in the ocean basins. A midoceanic ridge winds almost continuously through the ocean basins; it indicates that crustal evolution takes place there.

The idea of *continental drift* relates to crustal evolution. The present continents were at one time a larger landmass that fragmented into pieces, which drifted apart. The face of the earth looked quite different only 120 million years ago (Fig. 8.10), when the present continental masses were crowded together.

Evidence from the magnetic characteristics of the ocean floors near ridges supports the continental drift model. If the continents do move apart, then the seafloor spreads. When lava solidifies to form igneous rock, the iron minerals in the rock align with the earth's magnetic field. The directions and reversals of the direction of the earth's magnetic field in the past are preserved in the rock. So the seafloor rocks act as a magnetic tape (a very slow-moving one!) that preserves the record of the past changes in the earth's magnetic field.

This record indicates that new material emerges from a rift in the center of the ridge at a rate about 3 cm per year at its fastest speed across the mid-

240 million years ago   120 million years ago   Present

Figure 8.10 Continental drift. These computer-generated models show the trend of continental drift from 240 million years ago through 120 million years ago to now. The position of the continents is based on magnetic evidence. The present configuration of the continents will change just as dramatically during the next few hundred million years. (Based on maps prepared by A. M. Ziegler and C. S. Scotese of the University of Chicago's Paleographic Atlas project.)

Atlantic ridge. (That is about the amount your fingernails grow in a year!) This rate amounted to enough movement to push apart the Old World and the New World in a few hundred million years.

New material oozing out from the earth's interior along oceanic ridges accounts for the renewal of the ocean plains. This process marks one example of **volcanism,** by which molten material (produced by internal heating) rises through a planet's mantle and crust to the surface. We tend to think of volcanism as violent and sporadic, but it also acts slowly as a key process to modify the earth.

The continental plates float like large rafts on the basaltic basin material. Where one continental plate crashes into another, the impact raises mountains (Fig. 8.11). This process builds up mountains; erosion wears them down. In some regions, one plate may force another to fold under and descend into the mantle in seafloor trenches. The plate's descent eliminates surface material essentially at the same rate it is created. The plates' creation and destruction zones make natural fault areas. The earth's ocean basins, because they have been recently formed, are the youngest parts of the earth's crust, while the centers of continents are usually the oldest.

This model of the earth's crustal activity and evolution – seafloor spreading and the creation and destruction of the crust – is called **plate tectonics.** Much of the geologic action takes place where plates collide, such as the arc where the Pacific plate encounters the North American one. Earthquakes occur along fault lines (such as the San Andreas fault) and new, volcanic mountains thrust above the ground. The stretching, thinning, and splitting of the earth's crust produce rifts in the continents.

What moves these plates? Energy flow! One model (Fig. 8.12) pictures the upper part of the mantle as divided into large convective whirls of flowing rock in the upper 700 km. The mantle's plasticity allows a slow flow upward, horizontally, and downward. At the region of horizontal flow, friction between the plate and the mantle drags the plate along with the mantle's flow. The upwelling fluid rock supplies new materials to the plate. The energy source for such convection comes from heating by radioactive decay or from internal currents left from the time when the earth's core formed. The flows here are very slow, averaging only about a few centimeters per year. All the major tectonic features are signs of this flow.

KEY POINT. The earth's crust has evolved since the planet's formation and is changing right now. The two main processes are *volcanism* and *tectonics*. When investigating other earthlike planets, you should look for evidence of their crustal evolution.

8.6 EVOLUTION OF THE ATMOSPHERE AND OCEANS
(Learning Outcomes 8-9 and 8-10)

The earth's favorable environment for life results from the transfer of solar heat around the globe by the atmosphere and the oceans. This fine thermostat has worked for many years. However, the atmosphere and the oceans have changed over time, influenced by and influencing the earth's biological evolution. The future of life on the earth depends critically on the fate of its fluid systems.

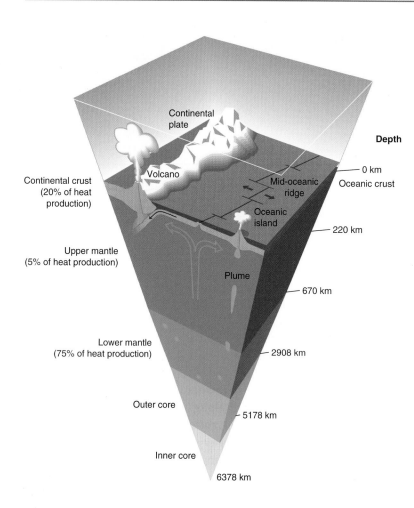

Figure 8.11 Interactions of oceanic and continental plates. The oceanic plates gain material from the outflow at oceanic ridges. As these plates expand, they crash into continental plates. Here mountains ascend and earthquakes occur. Most of the heat is generated in the lower mantle, which also releases plume material that flows upward into the upper mantle and crust.

Figure 8.12 Simple model of convection in the mantle. Hotter, less dense material rises from the lower mantle, flows horizontally, and then descends to create large convection cells in the mantle. The upflow adds material to the crust at the ocean ridges and pulls down material at the seafloor trenches. The horizontal flow moves the crustal plates.

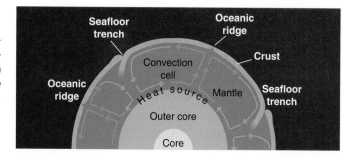

Origin and Development of the Oceans

By the current model of the formation of the planets (Chapter 11), the earth had no oceans when it formed 4.6 Gy ago. It was hot! The primeval surface may have been about a few thousand kelvins. When the surface had cooled to 373 K (100° C), so that water could condense, only one continent existed on the surface, and the rest was the initial ocean basin. It contained very little water then, only a few percent of the current volume.

The rest of the oceanic water probably came from the earth's interior. Volcanism conveys to the surface a variety of gases, such as carbon dioxide and water vapor. The steam issues from water that was trapped in the solid earth when the planet formed. The present rate of water gassing out from the interior accounts for the amount of water in the oceans today, if it has been constant for 4.5 Gy. It may have been faster in the past, when the earth was hotter. The earth's oceans now cover 70 percent of its surface with an average depth of 4 km, but the array of the oceans was very different in the past because of plate tectonics.

Evolution of the Atmosphere

Outgassing from the earth's interior also drove the development of the earth's atmosphere. The earth's atmosphere evolved from the chemical interplay of the solid and fluid earth.

Here is one reasonable model of atmospheric evolution. Our first substantial atmosphere arose mostly from volcanism, which spews out carbon dioxide, sulfur dioxide, hydrogen, nitrogen, water, methane (CH_4), and ammonia (NH_3). (In addition, some inert gases, such as helium and argon, come from the decay of radioactive materials.)

The earth's atmosphere may well have started out with a large amount of carbon dioxide. In about 1 Gy, the carbon dioxide ended up in the oceans and rocks, so 3 Gy ago the atmosphere consisted mostly of methane and other hydrogen–carbon compounds.

Then the atmosphere contained little free oxygen. Ultraviolet light readily penetrated and broke up methane, ammonia, and water. The hydrogen from these molecules fled into space. Some of the oxygen freed from the water combined with some of the carbon from the methane and gradually eliminated the elemental carbon. Some of the oxygen atoms eventually formed an ozone layer. Nitrogen from the ammonia became the main constituent of the atmosphere.

Biological activity produced (and now maintains) the high abundance of atmospheric oxygen. Geologic evidence indicates that a transformation to an oxygen-rich atmosphere began roughly 2 Gy ago, when plant activity and photosynthesis bloomed. About 600 million years ago, the oxygen content suddenly increased to present levels, with a proliferation of life.

Evolution of the Earth's Surface Temperature

How hot it gets at the earth's surface depends on how much energy is received from the sun and how effectively the greenhouse effect operates. Less solar energy results in lower temperatures. A larger greenhouse effect (more carbon dioxide and water vapor in the atmosphere) delivers higher temperatures. The balance of the input and outgo of energy fixes the temperature.

People have disturbed the natural carbon dioxide balance by extracting fossil fuels from the earth and

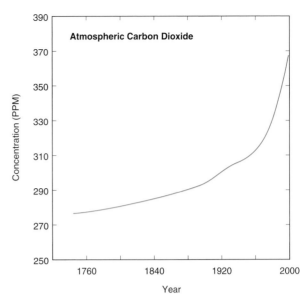

Figure 8.13 Growth of atmospheric carbon dioxide with time since the Industrial Revolution. This graph shows the yearly average *without* the seasonal variations. The carbon dioxide concentration is measured in parts per million (ppm).

burning them for energy, which adds to the carbon dioxide in the atmosphere. Our destruction of forests has eliminated a substantial portion of green plants, which take in atmospheric carbon dioxide, and has added some of their carbon to the atmosphere. The ocean can absorb only a part of the excess.

Our activities have a net result of increasing the percentage of carbon dioxide in the earth's atmosphere (Fig. 8.13). Observations indicate an increase of 5 percent over the past 30 years. Although we are not sure what the impact of this increase will be, it may result in a global temperature increase of about 2° C by A.D. 2020 if the overall water vapor and cloudiness do not change. This heating could have serious effects on climate over the long term.

8.7 THE EARTH'S EVOLUTION: THE BIG PICTURE (Learning Outcome 8-11)

The earth has evolved more than any of the earth-like planets. We live on an active world now (powered by internal heat), and it shows no signs of quitting. Geologic evidence implies that it's been lively since its hot birth.

Let's step back to scan the earth's overall evolution (Fig. 8.14). It falls into four large-scale stages (Table 8.1).

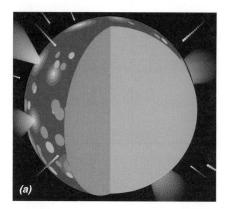

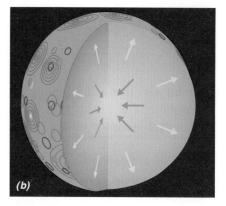

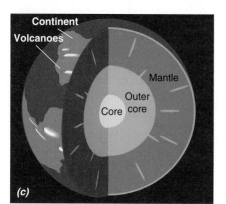

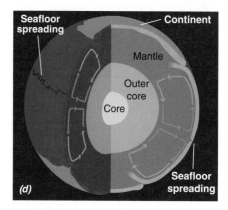

Figure 8.14 One model for the overall evolution of the earth. (a) The earth forms by the accretion of planetesimals; late-arriving objects crater the surface. (b) The interior, heated by radioactive decay, differentiates. (c) Water condenses, the first continents form, and intense volcanic activity occurs. (d) The crust thickens and crustal plates roam across the surface.

Table 8.1 A Possible Sequence for the Earth's Evolution

Stage	Events
I. 4.6 Gy ago	1. Formation of earth by accretion from the solar nebula 2. Rapid internal heating from accretion and radioactive decay 3. Blowing off of primeval atmosphere
II. 4.5 Gy ago	1. First differentiation of interior; formation of crust and core 2. Outgassing of carbon dioxide, water, methane, ammonia, etc., to make a second atmosphere 3. Large objects fracture crust in places 4. Ocean basins form; begin filling with water
III. 3.7 Gy ago	1. First tectonic movements, seafloor spreading, formation of volcanoes and mountains 2. Cooling and thickening of crust 3. Ocean basins mostly filled
IV. 600 million years ago	1. Continuation of processes of stage III at slower rate 2. Enlargement of ocean basins 3. Formation and subsequent breakup of original continental mass

One, the accretion to the earth of smaller bodies in the cloud of gas and dust that eventually formed the sun and planets (Fig. 8.14a). The buildup probably took place quickly, in only a few million years. An intense bombardment of large extraterrestrial objects played a main role then, leaving a pockmarked planet of more or less uniform composition (since each body was made of about the same combination of materials). The barrage halted about 4 Gy ago.

Two, the combination of radioactive heating and the heat from formation melted the interior, which differentiated (Fig. 8.14b). Denser materials sank to the core, lighter ones rose to the crust. The earth's surface cooled enough for rain to fall and the oceans to form.

Three, about 1 Gy later, the first continents appeared (Fig. 8.14c). Plate tectonics began; mountains grew, only to yield to the weathering of wind and rain. The atmosphere evolved.

Four, roughly 2 Gy ago, crustal cooling thickened the crust enough to allow plate activity as we see it today. At the end of this stage, which began 600 million years ago (Fig. 8.14d), the earth looked much as it does today. During this time, plate tectonics reordered the continents and ocean basins.

TO SUM UP. The earth's atmosphere, crust, and interior have changed greatly since the formation of the planet. The interior's evolution is driven by internal energy: some from radioactive decay, some stored from its formation by accretion. The heat caused internal layering. As energy flows outward (and finally out to space), it drives tectonics and volcanism. In the distant past, the crust was also shaped by impacts of solid bodies from space. Few such impacts occur now, but if the incoming object is large enough, it would forge a catastrophic explosion!

8.8 MODELS OF PLANETS
(Learning Outcome 8-11)

Why build models of planets? So that we can understand and predict their physical properties and their evolution over billions of years. That is how we know more about planets than we can observe directly. (See the Planetary Models Celestial Navigator™.)

How to build models of planets? First, we observe them. Next, we find key physical properties (mass and radius) from which we determine the bulk density. From this, we infer the general composition (ices, rocks, metals in some combination).

What next? We assume that planets have spherical shapes. That follows from geometry and aesthetics – a sphere is the simplest and most symmetrical solid (as you saw in Part 1). Newton's law of gravitation requires that a body with enough mass becomes spherical from its own gravity.

Now comes the hardest and most uncertain step. We must have an understanding of the properties of the specific materials that comprise a planet. That is, we need to know how various solids (ices, rocks, metal), liquids, and gases behave under different temperatures, densities, and pressures. Gases are rel-

atively simple; so are liquids. Solids are another matter. We deal with them under conditions that we cannot duplicate in a laboratory. In most cases, we model their behavior based on limited information.

Our planetary models aim at explaining the current state and surface features (if the planet has a surface; otherwise, the properties of the top of the atmosphere). Also, models try to predict the internal structure. To verify interior models is tough. We really have done so only for the earth and moon. Most details of interior models will change as we acquire more information.

Generally, models show that planets have distinct layers of materials of different densities. The highest-density material lies at the center; the least dense at the surface. As you plunge into a planet, densities, temperatures, and pressures increase. The state and kinds of materials also change.

What drives the evolution of these models? Energy, in the form of heat! Because thermal energy must move from hotter to cooler regions, energy flows from the core outward. Eventually, it radiates into space from the surface or atmosphere – gone forever. When a planet loses enough internal energy so that its internal temperature becomes the same throughout, its evolution stops. The planet dies.

The internal thermal energy arises from two sources. One is the radioactive decay of isotopes in the planet when it formed. Two is the formation process, which was likely to be **gravitational accretion.** Imagine a collection of solid debris in space, with a range in sizes. Consider when gravity brings all the stuff together. Their potential energy transforms to kinetic energy. They collide and stick together – a process called *accretion.* The resulting mass gains internal energy, and so its temperature increases. Any planet formed by accretion was originally hot; we think they all were.

How fast a planet loses its internal heat sets its **evolutionary lifetime.** The lifetime depends on the store of energy and its rate of loss to space. For a simple model, the lifetime rests primarily on one physical property: size. *The greater the size, the longer and more vigorous the evolutionary lifetime.* For terrestrial planets (or any bodies with solid surfaces), we find evidence of the extent of internal evolution by surface modification, especially by volcanism. Impacts by solid objects can also shape the surface, though you will see that most of these took place about 4 Gy ago.

CAUTION. Do not confuse evolutionary *lifetime* with *age*. Bodies that formed with our solar system all have the same age – 4.6 Gy. But some, such as the moon, have already expired. Their evolutionary lifetimes ended, while the earth still evolves.

KEY CONCEPTS

1. We use the physical properties of the earth, such as mass and density, to develop a model of its interior, which can be probed by analyzing earthquakes. The earth is divided into zones that are denser the deeper they lie below the surface – the interior is differentiated. The core is the densest part and probably consists of metals such as nickel and iron. The mantle and the crust are made of rocky materials.

2. From radioactive dating of the oldest rocks using isotopes of known half-lives, we estimate that the earth's age is 4.6 Gy, give or take about 0.5 Gy. The earth formed with the rest of the solar system at about this time.

3. The earth's atmosphere consists mainly of nitrogen molecules (78 percent) and oxygen molecules (21 percent). The pressure at the surface at sea level (1 atmosphere) comes from the weight of all the air above it. In its outermost region, the atmosphere slowly leaks into space where particles move faster than the earth's escape speed and the density is low so that collisions are infrequent.

4. Carbon dioxide and water vapor trap the infrared radiation emitted by the ground and so keep the earth warmer than it would be if the infrared escaped directly into space. The water vapor has a greater effect than the carbon dioxide. This warming is called the *greenhouse effect* and makes our planet habitable. Carbon dioxide added to the atmosphere by people may result in global warming.

5. The earth's crust evolves by seafloor spreading and the motions of crustal plates, driven by convective flows below the surface that result from the transport of internal thermal energy.

The ocean basins are the youngest parts of the earth's crust; the continents are older in comparison. Tectonics and volcanism, the main processes driving the evolution of the crust, are powered by the heat flow from the interior.

6. The water in the oceans and the gases in the atmosphere probably came from the release of materials from the earth's crust and interior. Such water and gases interact and have also changed because of the development of life on our planet.

7. Outgassing from the crust by volcanism, escape of gases into space (if the particles of the gas move faster than escape speed), biological activity, and sunlight influence the evolution of the atmosphere. The atmosphere also interacts with the oceans.

8. The interior and crust of the earth have been greatly modified by the generation and outflow of heat since the planet's formation some 4.6 Gy ago by the gravitational accretion of small masses. The heating from the transformation of gravitation energy and radioactive decay caused the interior to differentiate. Some of that stored heat now flows out to the surface from the hotter core.

9. A dynamo set up by regular fluid motions in its metallic core generates the earth's magnetic field. The overall structure is that of a dipole, similar to a bar magnet. The field interacts with the interplanetary solar wind to create a magnetosphere. The field has varied dramatically over tens of millions of years by switching polarities.

10. The earth is the detailed model for the other planets like it in the solar system; these comprise the terrestrial planets, which we believe were formed by accretion. The energy released in gravitational accretion, plus that from radioactive decay, drives volcanism as it flows to the surface. Impacts by interplanetary debris and erosion, along with volcanism, have modified the surfaces. Atmospheres are renewed by outgassing from the interior; some gases escape into space. The most important factor in the vigor and length of a planet's evolution is its mass.

STUDY EXERCISES

1. Explain how you could determine the earth's mass by jumping off a building. (Explain it; don't do it!) (Learning Outcome 8-1)
2. Contrast the composition of the earth's core to that of its crust and mantle. (Learning Outcomes 8-2 and 8-3)
3. Explain how the earth's interior became differentiated. (Learning Outcome 8-2)
4. Make a simple argument to establish that the earth's core must be denser than its crust. (Learning Outcome 8-3)
5. Describe two effects of the earth's atmosphere on sunlight passing through it. (Learning Outcomes 8-6 and 8-7)
6. Suppose the amount of water vapor in the atmosphere suddenly increased by a large amount. What would happen to the earth's surface temperature? (Learning Outcome 8-7)
7. How can volcanoes affect the evolution of the earth's oceans and atmosphere? (Learning Outcomes 8-8, 8-10, and 8-11)
8. What was the composition of the earth's first atmosphere? What happened to it? What was the composition of the earth's second atmosphere? Where did it come from? What happened to it? (Learning Outcome 8-10)
9. Give two ways in which the oceans affect the atmosphere. (Learning Outcome 8-9)
10. Where did the oceans' water come from? (Learning Outcome 8-9)
11. What part of the earth's crust is the youngest? The oldest? (Learning Outcome 8-8)
12. How do we know the
    a. earth's mass? (Learning Outcome 8-1)
    b. earth's age? (Learning Outcome 8-4)
    c. composition of the core? (Learning Outcomes 8-2 and 8-3)
13. What model describes pretty well the properties of the earth's magnetic field? (Learning Outcome 8-5)

PROBLEMS AND ACTIVITIES

1. Consider uranium-238, which has a half-life of 4.5 Gy. If this isotope of uranium arrived on the earth's surface when the earth formed, how much of the original amount remains today? How much will remain 4.5 Gy from now?
2. The acceleration due to gravity at the earth's surface is 9.8 (m/s)/s. The earth's radius is about 6400 km. Use Newton's second law and law of gravitation to calculate the earth's mass. *Hint:* $F = ma = mg$, where g is the acceleration due to gravity at the earth's surface.
3. Imagine that the earth's rotation is very gradually slowing, so that gigayears from now, the earth will rotate at only one-thirtieth (1/30) its present rate. If the dynamo model is correct, what will happen to the earth's magnetic field when this substantially slower rotation has been reached?
4. Imagine that the mass of the earth were increased four times but the radius remained the same. What would happen to the earth's density? Its escape speed?
5. The earth's mass is about 6×10^{24} kg. If its average density is 5500 kg/m^3, what is the earth's total volume?
6. Find the radius of the earth's core relative to the total radius if the core density is 10,000 kg/m^3, the mantle density is 2500 kg/m^3, and the average density is 5500 kg/m^3.

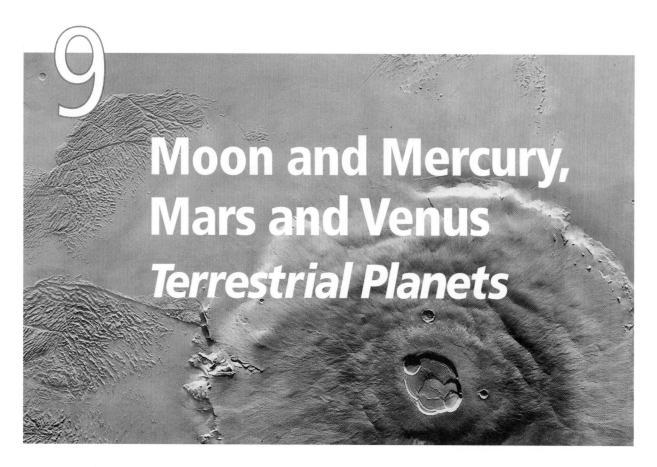

9

Moon and Mercury, Mars and Venus
Terrestrial Planets

LEARNING OUTCOMES

After studying this chapter, you should be able to:

9-1 Compare the moon, Mercury, Mars, Venus, and the earth in terms of their general surface and physical properties (such as mass and density), with a special focus on how we know this information.

9-2 Describe each planet's major surface features and indicate a possible formation process for these features.

9-3 Compare and contrast the surface environments (such as temperature, atmosphere, surface features, escape speed) of the terrestrial planets.

9-4 Sketch a model for the structure of the interior of each terrestrial planet and compare them.

9-5 Compare and contrast the magnetic fields (or lack thereof) of the terrestrial planets.

9-6 Compare and contrast the evolution of the terrestrial planets, applying the general processes for evolution, such as cratering, volcanism, and tectonics.

9-7 Compare and contrast the evolutionary lifetimes of the terrestrial planets; argue that the moon is the least evolved, the earth the most.

9-8 Describe the process of cratering of planetary surfaces and tell how craters can be used to infer the relative ages of surfaces.

9-9 Use Newton's law of gravitation to explain the nature of tidal forces, and apply tidal forces to astrophysical situations.

CENTRAL CONCEPT

The evolutions of the moon, Mercury, Mars, and Venus have been driven by processes similar to those that have created the earth but have not operated as long or as vigorously.

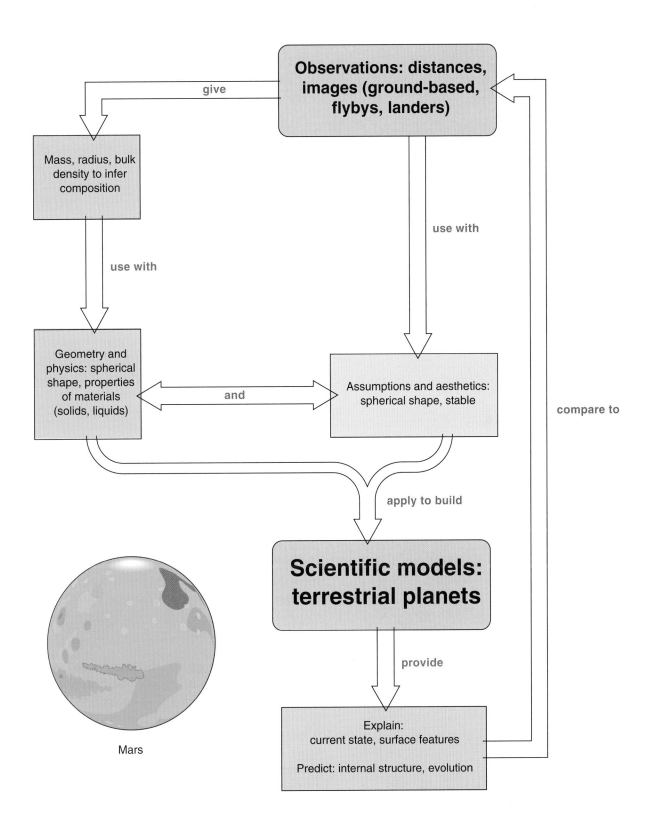

Observations: distances, images (ground-based, flybys, landers)

give

Mass, radius, bulk density to infer composition

use with

use with

Geometry and physics: spherical shape, properties of materials (solids, liquids)

and

Assumptions and aesthetics: spherical shape, stable

apply to build

Scientific models: terrestrial planets

compare to

provide

Explain: current state, surface features

Predict: internal structure, evolution

Mars

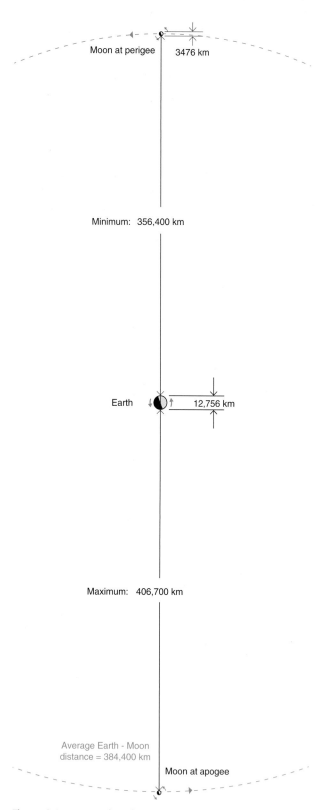

Figure 9.1 Moon's orbit relative to the earth: looking down on the orbital plane. The minimum perigee and maximum apogee distances are shown for this fairly eccentric orbit. (The actual values of apogee and perigee vary monthly.) The earth and moon are drawn to the same scale as the distances.

Through a small telescope, the moon strikes you as a stark world, tantalizingly close, its surface ravaged by explosive impacts. The solar system has another body with a similar face: Mercury. Like the moon, Mercury is a small, airless world pockmarked with many craters. Both the moon and Mercury are now dead worlds. Their interiors are now cooler than the earth; no heat drives surface activity. Without atmospheres, these planets have no wind or water to erode their landscapes. Their heyday of evolution has passed.

Not so for Mars and Venus. These planets resemble the earth more than any others do; yet the differences go much deeper. Volcanism abounds on Venus, reshaping the surface. And it's blisteringly hot –even the rocks are plastic! Mars shows a cold desert where the water ceased to flow tens of millions of years ago, and volcanism ceased even earlier.

How did these planets evolve? What forces shaped their evolution? This chapter deals with the comparative evolution of these four earthlike worlds. They have a family resemblance to the earth but each also has a unique character.

9.1 GENERAL ORBITAL AND PHYSICAL CHARACTERISTICS (Learning Outcomes 9-1 and 9-4)

We need to know the overall physical properties of the terrestrial planets in order to compare them and so infer their evolutionary status. Let's look at each individually first, and then contrast them. (See the Terrestrial Planets Celestial Navigator™.)

Moon

The moon revolves around the earth in an elliptical orbit once a month at an average distance of 30 earth diameters (Fig. 9.1), or about 384,400 km. The moon shows us the same face (Fig. 9.2) because it rotates on its axis with a period of 27.3 days, the same as its period of revolution about the earth with respect to the stars. This matching of rotation and revolution rates is called **synchronous rotation.** It results from **tidal forces** between the earth and the moon (Enrichment Focus 9.1).

You can prove to yourself how the moon must rotate once each revolution to keep the same face to the earth. Place a globe on a table. Start to circle around the table, imagining your head to be the

ENRICHMENT FOCUS 9.1

Tides and Tidal Friction

You may be familiar with the periodic rise and fall of ocean tides, which are controlled by the moon, so that typically two high tides and two low tides occur each day. I'll give you a general explanation of tides using Newton's law of gravitation (Section 4.4).

Imagine that the earth's surface is level and covered completely with a layer of water. Consider for a moment the moon's gravitational attraction at three points lined up with the moon (Fig. F.9a). Recall that the force of gravity decreases as the inverse square of the distance between masses: the moon's gravitational force must be greater at A than at B and greater at B than at C. The greater the force acting on the same mass, the greater its acceleration, so a mass at A has a greater acceleration than a mass at B, and a mass at B has a greater acceleration than a mass at C.

The difference between these accelerations is crucial. Because of the difference, the water at A bulges ahead of the earth (point B), and the water at C lags behind the earth and forms a bulge on the side of the earth opposite the moon. Thus, we have two high tides: one on the side of the earth toward the moon, and one on the opposite side. At any point in the earth's oceans, these two high tides take place about a half day apart.

Tides result in **tidal friction.** As water moves about in the ocean basins, both high-tide peaks almost line up with the moon (one on the side of the earth under the moon, and one on the side opposite). However, the tidal bulges are not right on a line to the moon; the closer bulge (A in Fig. F.9b) lies ahead of the moon, the farther bulge (B in Fig. F.9b) lags behind. This displacement occurs because of the earth's rotation and the friction of the water flowing in the ocean basins. Because the earth rotates faster than the moon revolves around it, the friction pulls the line between the tidal bulges ahead of the line between the center of the earth and moon. In return, that friction slows down the rotation rate of the earth. Also, the gravitational attraction of bulge A on the moon causes the moon to accelerate in its orbit. This acceleration results in the moon moving away from the earth.

Note that it is the *difference* in gravitational forces that accounts for the tides; this difference is what is referred to as a **tidal force.** It pulls bodies apart. The curious aspect of tidal forces is that they vary as the *inverse cube of the distances between the masses;* so tidal forces grow *very* strong as two bodies come closer together. You will come across tidal forces in many other situations in the astronomical universe.

Note that tidal forces are interactions, so the earth also exerts a tidal force on the body of the moon. This interaction keeps the same side of the moon facing the earth.

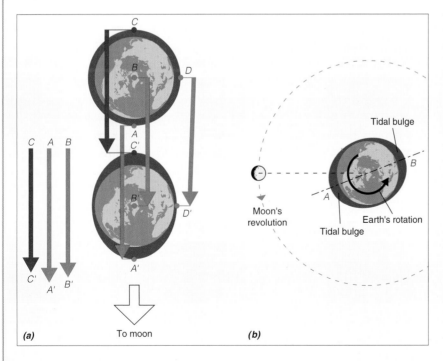

Figure F.9 Tides and the moon. (a) Imagine a smooth earth covered with water. Consider the water at A, C, and D pulled by the moon's gravity. How far the water falls in a given time depends on its distance from the moon; water at A falls farther than that at C. Water at D and water at the earth's center (B) fall the same distance because the acceleration on them is the same. As the earth falls toward the moon (from B to B'), water on one side falls a bit more (from A to A') and on the other side a bit less (from C to C'). Water flows from D to supply the bulges; the depth at D' is less than at A' or C'. (b) The tidal bulges on the earth and the moon's motion. The bulges spin ahead of the moon because the earth rotates faster than the moon revolves and friction carries the bulges with the earth's surface. This friction slows down the earth's rotation. Meanwhile, the bulge at A acts gravitationally to accelerate the moon in its orbit.

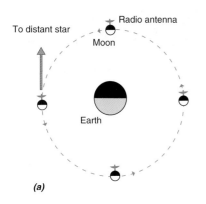
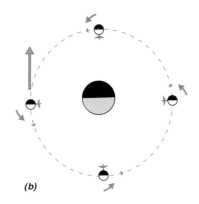

Figure 9.2 The moon's rotation. The moon keeps the same face toward the earth because its rotation period is the same as its period of revolution with respect to the stars. Imagine a radio antenna on the moon. (a) If the moon did not rotate with respect to the stars, the antenna would not always point at the earth. (b) Because it does rotate at a rate equal to its revolution, our imaginary antenna always points to the earth. The moon exhibits *synchronous rotation*.

moon and keeping your eyes on the globe. You will soon see that your head (and body) need to turn around once for one revolution around the table. Almost all the moons in the solar system feature synchronous rotation.

Knowing the distance from the earth to the moon, you can find its physical diameter from its angular size. Recall that the angular diameter of the moon is about ½ degree, which means its actual size-to-distance ratio is about 1/110 (Enrichment Focus 1.2). The moon lies about 110 of its diameters from the earth. So its diameter is 3476 km, about one-fourth the earth's diameter (Table 9.1). If the earth were the size of your head, the moon would be about the size of a tennis ball. On the same scale, the diameter of its orbit would be about 12 meters.

The moon's mass is small, only 1/81 times the mass of the earth, or 7.4×10^{22} kg. We calculate from the mass and radius that the bulk density is a little more than 3.3, about the same density as the rocks in the earth's mantle. This low density implies that the moon's interior contains only a small percentage of metallic materials (Fig. 9.3). The moon is mostly rocky, similar to those of the earth's mantle.

Its core is enriched with iron and is probably cool and solid, not liquid like the earth's.

Mercury

Mercury speeds around the sun once every 88 days in a very eccentric orbit. That means its distance from the earth can vary a lot. For instance, in 1997 Mercury ranged from 0.55 to 1.42 AU from the earth. With the actual distance known (Enrichment Focus 9.2), we can calculate a planet's actual diameter from its angular diameter, which gives the diameter-to-distance ratio (Enrichment Focus 1.2) to find its physical diameter. Mercury turns out to be a tiny world, only 4900 km in diameter, or about 40 percent larger than our moon.

In 1965 radar astronomers discovered that Mercury rotates in about 59 days, so Mercury's rotation period is two-thirds its orbital period. This ratio results from tidal interactions between the sun and Mercury (much as the tidal interactions of the earth and moon have resulted in the moon keeping one side toward the earth). How long, then, is the solar day on Mercury – the time from noon to noon?

Table 9.1 Comparison of the Terrestrial Planets

Planet	Diameter (earth = 1)	Mass (earth = 1)	Density (water = 1)	Surface Pressure (earth = 1)
Mercury	0.38	0.055	5.4	10^{-15}
Venus	0.95	0.82	5.2	100
Earth	1.0	1.0	5.5	1.0
Mars	0.53	0.11	3.9	0.01
Moon	0.27	0.012	3.3	10^{-15}

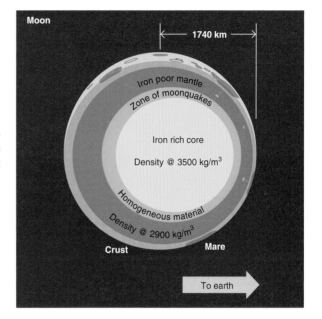

Figure 9.3 Interior structure of the moon inferred from Apollo space-craft's seismic and heat-flow measurements. Note the relatively large core of iron-rich silicates, the rocky mantle, and the relatively thick crust. Metals make up a small percentage of the total composition.

ENRICHMENT FOCUS 9.2
Distances in the Solar System

The distance between a planet and the earth can be calculated by triangulation from various points in the earth's orbit (Section 3.2). The distances of the planets from the sun can be computed from observations of their synodic periods, transformed to sidereal periods (Section 3.2) and slipped into Kepler's third law (Section 3.5). Both of these methods give distances in Astronomical Units (AU), that is, relative to the earth–sun distance. You can draw up a scale map of the solar system with the planets' positions all neatly laid out in AUs.

To work out the distance scale in kilometers requires that only one piece, already known in AU, be measured in kilometers. An analogy: suppose you are given a map of your region of the country with no distance scale in kilometers but with all locations laid out correctly relative to one another. Hop in a car and drive between two points on the map while keeping close track of the distance in kilometers between these two points, which are separated on the map by so many centimeters. You have found the distance scale for the map.

To do the same for the solar system, we use the known speed of light. Radar signals, which travel at light speed, are bounced off Venus, and the time between transmission and reception is accurately measured and converted into kilometers. (This is analogous to driving the car between two locations on a map.) Kepler's laws give the earth–Venus distance in AU at the time of the radar measurement.

(This is analogous to having a distance scale on a map in centimeters.) The radar measurement is usually made when the earth and Venus are close together on the same side of the sun. The distance at that time is some fraction of an AU (about 0.3 AU), so we know a fraction of an AU in kilometers. One AU in kilometers: 1.496×10^8 km. (Remember it by rounding off to 150 million km.)

Example: From the measured return time and the accurately known speed of light, we can readily calculate the distance to Venus, d, in kilometers; it is the return time, t, divided by two, times the speed of light, c:

$$d = \frac{t}{2} \times c$$

Then we know the earth–Venus distance (a part of an AU) in kilometers. If we then calculate the number of kilometers per AU, we have established the distance scale for the entire solar system map.

The best time to make this observation occurs when the earth and Venus lie closest together on the same side of the sun. Their separation is then about 0.28 AU. A radar signal takes roughly 280 seconds for the round trip. The distance is then

$$d = \frac{(280 \text{ s})}{2} \times (3.00 \times 10^5 \text{ km/s})$$
$$= 4.20 \times 10^7 \text{ km}$$

This equals 0.28 AU, so

$$1 \text{ AU} = \frac{4.20 \times 10^7}{0.28} = 1.5 \times 10^8 \text{ km}$$

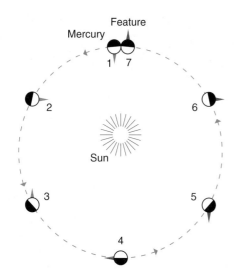

Figure 9.4 Mercury's rotation and solar day. Radar measurements in 1965 showed that the rotation period is 59 days with respect to the stars, so relative to the sun, a feature on Mercury at noon (1) ends up at midnight (7) after one revolution. After the next revolution, it is at noon again (1). The solar day on Mercury is thus twice the orbital period, or 176 days.

It's about 176 terrestrial days, just twice the length of Mercury's year (Fig. 9.4).

Mercury has no natural satellite, so it is hard to find its mass. Mariner 10 sped past Mercury three times in 1974 and 1975. The spacecraft's acceleration from Mercury's gravity was accurately measured, and we could calculate Mercury's mass from Newton's second law: about 0.06 that of the earth, or 3.3×10^{23} kg.

For such a small planet, Mercury has a fairly large mass. That means its bulk density is high: 5.4, essentially the same as the earth's. Comparing Mercury's interior with that of the earth, we expect a large metallic core (Fig. 9.5a) and a rocky mantle. Although we have not yet measured seismic waves to confirm this interior model, it's probably correct by analogy to the earth.

Mars

Though looming large in myth, Mars is actually a small planet. It has a diameter of some 6800 km, only 53 percent the earth's size. If the earth were the size of a tennis ball, Mars would be about the size of a Ping-Pong ball.

Surface markings (Fig. 9.6) visible through telescopes tag the Martian rotation rate: 24 hours, 37 minutes. A day on Mars lasts only a bit longer than on the earth. The rotation axis of Mars tilts at almost the same angle as the earth's, with the result that the seasonal variations are similar (but Mars is much colder than the earth overall).

Mars has two moons (called Deimos and Phobos), so we can use Newton's form of Kepler's third law (Enrichment Focus 4.3) to find its mass from their orbits: it is only 6.4×10^{23} kg, only 11 percent of the earth's mass.

Mars' bulk density measures 3.9, only a bit higher than the moon's (3.3) and much less than the earth's (5.5). This comparatively low density implies that Mars' interior (Fig. 9.5d) differs from the earth's. Its core is smaller and probably consists of a mixture of iron and iron sulfide, having a lower density than the materials in the earth's core. The Martian mantle probably has the same density as the earth's; it could be made of olivine (an iron–magnesium silicate), iron oxide, and some water.

These physical properties show that Mars falls between the larger (earth and Venus) and smaller (moon and Mercury) terrestrial planets. We expect Mars to provide key clues about a midstage of evolution, and it does.

Venus

Viewed from the earth, Venus outshines every celestial body except the sun and the moon. Venus' brilliance comes from its unbroken swirl of clouds (Fig. 9.7). This cloud cover completely frustrates any attempt to view the planet's surface features through optical telescopes, yet we have pierced Venus' cloudy veil to uncover the surface. Radar beams paint its terrain. Spacecraft have sped by the planet, orbited it, and plunged onto its forbidding surface.

Second planet from the sun at an average distance of 0.72 AU, Venus completes one orbit in 225 days. The orbit of Venus conveys it closer to the earth than any other planet. For instance, in 1998, Venus moved to within 0.28 AU of the earth. Its physical diameter almost matches the earth: 12,100 km, only 5 percent smaller.

Thwarted by the lack of a surface view, astrono-

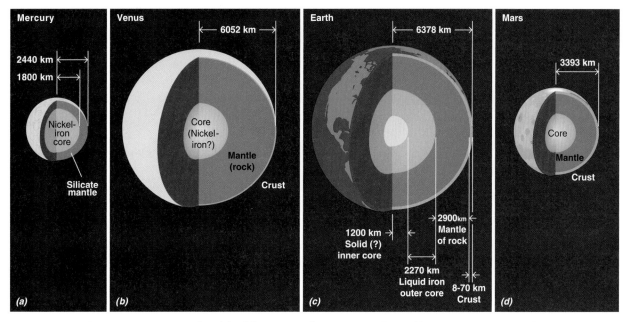

Figure 9.5 Comparison of models of the interiors of the terrestrial planets. (a) Mercury's interior. Note the large metallic core (relative to the size of the planet) and the small, rocky mantle. (b) Venus: note how closely it resembles that of the earth with its large, rocky mantle and metallic core. (c) Earth. (d) The Martian interior: note the small size of the metallic core (iron and iron sulfide) relative to the rocky (olivine) mantle.

mers before 1961 had no real idea of Venus' rotation rate. Radar astronomers found then that Venus rotates once every 243 days, retrograde. **Retrograde rotation** means that the planet spins from east to west, rather than west to east, as the earth does; so on Venus, the sun (though invisible through the clouds) rises in the west and sets in the east. This long retrograde rotation results in a solar day on Venus of 117 terrestrial days.

Like Mercury, Venus has no known natural moon, so we can accurately find the mass only when a spacecraft passes or orbits it. For orbiters, we apply Newton's version of Kepler's third law (Section 3.5). During a flyby, we measure the acceleration of the spacecraft and use Newton's law of gravitation. Venus' mass turns out to be about 0.82 times that of the earth, or 4.9×10^{24} kg.

With the mass and diameter known, we can calculate Venus' bulk density: 5.2, almost the same as that of the earth. We guess that the interior of Venus closely resembles the earth's interior (Fig. 9.5c): a rocky crust, a large rocky mantle, and a metallic core. Because Venus has a lower bulk density than the earth, we infer that it has a somewhat smaller core.

Figure 9.6 HST true-color image of Mars at opposition in 1995, when Mars was about 100 million km from the earth.

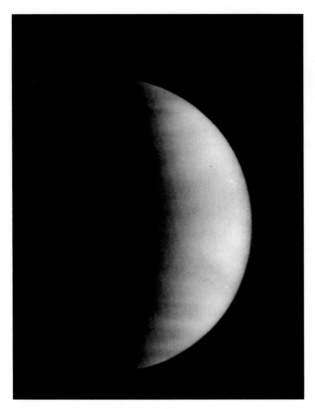

Figure 9.7 HST false-color image of Venus taken in ultraviolet light, when Venus was about 114 million km from the earth.

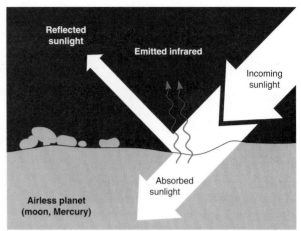

Figure 9.8 Energy budget for an airless planet. The ground reflects and absorbs incoming sunlight. The absorbed light heats the ground, which emits mostly infrared to space. The balance of loss and gain fixes the local surface temperature.

9.2 SURFACE ENVIRONMENTS
(Learning Outcomes 9-2, 9-3, 9-6, and 9-8)

Could you survive unprotected on the surfaces of the terrestrial planets (aside from the earth)? No! All the environments are hostile, and some more so than others.

Moon

The moon's **surface gravity** rates less than that of the earth's. A planet's surface gravity is simply the acceleration of any object at its surface from the planet's gravitational pull. The moon has a mass 1/81 that of the earth and a radius 0.27 that of the earth, so the moon's surface gravity amount to one-sixth that of the earth. Objects weigh one-sixth as much on the moon as they do on the earth (a fast way to lose weight – a trip to the moon).

The moon has only a very tenuous atmosphere. Why? Atmospheric escape! Gravity grips any atmosphere of a celestial body. An atmosphere consists of a gas of molecules and atoms, moving at various speeds. The temperature of the gas measures the average speed of the particles in it. Typically, gas particles collide frequently, but in the thin upper layers of an atmosphere, far fewer collisions occur. If a gas particle has escape speed here, it speeds off into outer space. This is how **atmospheric escape** occurs.

For the moon, the escape speed attains only 2.4 km/s. At typical lunar temperatures, even fairly massive gas particles at the surface attain escape speed. The main natural constituent of the lunar atmosphere is neon, which has a high mass. All the gases, added up, give a surface density that is a mere 10^{-15} that of the earth's. A basketball stadium on the earth contains as much total mass of air as does the entire lunar atmosphere!

Most gases have escaped from the moon's gravitational grasp since its formation. The exhaust from one Apollo landing dumped more gases into the atmosphere than had formally existed there! These gases will not stay around very long: for example, oxygen dumped at the moon's surface escapes in about 100 years.

We can estimate an airless planet's surface temperature from its energy budget. Suppose that only sunlight heats the planet's surface (Fig. 9.8). The surface absorbs some of the incoming sunlight and reflects some back into space. The moon reflects 7 percent of the light that hits it and absorbs the remaining 93 percent. (Even though a full moon seems to shine brightly in the sky, the moon's surface is really quite dark.) The absorbed sunlight

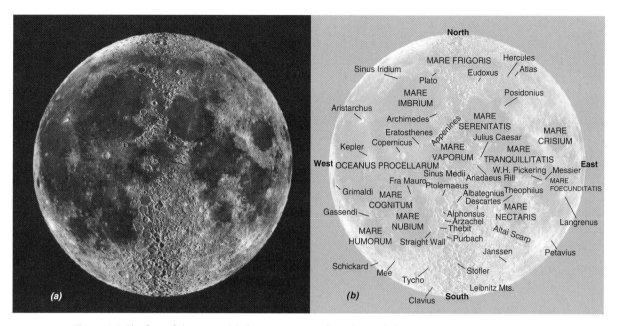

Figure 9.9 The face of the moon. (a) The moon as seen from the earth through a telescope at full phase. The darker regions are lowlands; the lighter areas, highlands. Note that most of the craters appear in the highlands. Some of the larger craters have rays emanating from them. This composite photograph shows relief that is invisible under the flat lighting at full moon (when it is noon on the moon). (b) Map and names of the major features on the moon's near side at full phase.

heats the surface; it radiates mostly infrared. The balance between incoming sunlight and outgoing infrared sets the surface temperature of the sunlit side, which hits 375 K (about 100° C).

At night, solar heating stops, and the infrared – because no atmosphere traps it – radiates away into space. Since the lunar night is so long (about 15 earth days), the surface has time to cool, and the temperature plummets to 100 K (about −175° C). The large noon-to-midnight temperature difference occurs because the moon rotates slowly and has no atmosphere. We expect the same kind of large temperature variation from any airless, slowly rotating planet.

The moon has a grim surface. On it, the lunar **maria** ("seas") appear dark compared with the rest of the surface (Fig. 9.9). Most maria look circular, and some are interconnected. They have smooth surfaces (Fig. 9.10) compared with the brighter, cratered regions. Also, the maria have lower elevations, by about 3 km, than the rest of the surface. Hence, the maria are called the lunar **lowlands;** the other areas, the **highlands.** The lowlands cover some 20 percent of the moon's total surface; the highlands, more than 70 percent.

The vast flat extents of the maria are solidified lava flows. The lunar maria fill up large, shallow **basins** on

the moon's surface, which range in size from 700 to 3000 km. The moon's most striking basin is Mare Orientale (Fig. 9.11) with its concentric rings of mountains. The outer rim of mountains has a diameter of about 970 km and rises to 7 km. Surrounding this basin for about 1000 km outward lies a blanket of lighter material covering the older lunar surface.

Craters (from the Greek word meaning "cup" or "bowl") are round depressions that litter the moon almost everywhere. Their interiors range widely in size: many are smaller than a pinhead; five have diameters greater than 200 km (about the north–south width of Cuba). The heights of the rims of lunar craters are small compared with the diameters, and the floors are depressed compared with the surrounding landscape.

Impacts forged most lunar craters. Solid objects from space slammed into the moon's surface and punched out the craters (Enrichment Focus 9.3). An impact blasts a crater with wavy slopes, usually covered by debris, with a rippled terrain around it. If the debris and crater wall were put back into an impact crater, they would fill it up. Material blasted out by an impact falls in streaks, leaving a raylike pattern. Large chunks of ejected material create small secondary craters around the main impact crater.

Figure 9.10 Close-up of the surface of Mare Imbrium. Note the smoother, darker, and lower terrain here compared to the highlands.

Figure 9.11 The moon's far side, showing the Orientale Basin (center). The Cordillera Mountains ring the basin like a bull's eye, an indication of the impact process of the basin's formation.

ENRICHMENT FOCUS 9.3

Impact Cratering

Craters scar every terrestrial planet and nearly every satellite in the solar system. That cratering has happened throughout the solar system provides clues to the processes that formed the terrestrial planets. Cratering also serves as a tool by which to infer the relative ages of planetary surfaces.

Massive objects crashing onto a planet's surface create impact craters. The great abundance of craters on the planets and satellites implies that a storm of ancient impacts blasted them long ago. Since that time, erosion and crustal evolution have modified the original pattern and changed the planets' surfaces. If we can estimate the rate at which cratering occurred, the number of craters will give a guide to the age of the surface on which they are found.

On any planet's surface, small craters greatly outnumber the larger ones. The size of a crater is related to the kinetic energy of the projectile that formed it. The largest of these must have been rocks greater than 100 km in diameter. These objects strike planetary surfaces with speeds of 10 km/s or more. They bang into the surface with enormous amounts of energy: some 10^{23} J

for the mass that formed the Copernicus crater on the moon. For comparison, the eruption of Mount Saint Helens in May 1980 blew with an energy of 10^{17} J, an explosive yield equivalent to 35 megatons of TNT. Large craters involve explosions a million times greater!

Upon impact, the projectile's kinetic energy converts into an explosive force below the ground at a depth only a few times the diameter of the infalling mass (Fig. F.10). A shock wave from the impact spreads through the rock, deforming it and throwing it outward. The volume blown out is much larger than that of the projectile, so small objects result in big holes. Because the explosion occurs below the ground, even projectiles hitting at oblique angles leave round craters. (Examine the photos and notice how lunar craters are round.) Even on an airless world, craters, once formed, can be wiped out by a number of processes: later impacts, materials thrown up from younger, nearby craters, and lava flows. Large craters are less likely to be obliterated by these processes than small ones, so generally the largest craters on a surface are the oldest. Overall, the oldest part of the surface will have the highest density of craters on the surface. We can rank the planets and moons according to the relative amounts of modifications of their surfaces.

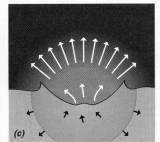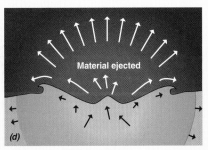

Figure F.10 Formation of a crater by impact of a solid body. (a) A meteoroid strikes the surface at a speed of some tens of kilometers per second. (b) The initial explosion ejects surface material at a high speed. (c) Shockwaves through the rock compress and fracture it. The rebound of the rock throws material out of the crater. (d) The rim of the crater is folded back by the rebound. Ejected material can form rays and smaller craters surrounding the main impact.

What projectiles made these impacts? Myriad small, solid bodies orbit the sun; they are called **meteorites** when they strike the earth. If they pass close enough to the moon, they will collide with it. We have seen no new, large craters form recently. (One of the youngest is some 2 million years old.) So the era of heavy-impact cratering finished long ago.

Apollo samples reveal the physical and chemical nature of the lunar surface. The very top layer (1 to 20 m deep) extends as a porous, somewhat adhesive layer of debris. It consists of fine particles, called **lunar soil,** and larger rock fragments. The soil samples contain a large amount of mostly round pieces of glass, which makes the surface slippery, as the Apollo astronauts found by slip-sliding around!

The moon rocks mostly fall into three categories: (1) dark, fine-grained rocks (Fig. 9.12a) similar to terrestrial basalts (magnesium–iron silicates), called **mare basalts;** (2) light-colored igneous rocks (Fig. 9.12b) with visible grains, which are called **anorthosites** (aluminum–calcium silicates) – by far the most common rock on the surface; and (3) **breccias,** rock and mineral fragments bound together (Fig. 9.12c).

What do these characteristics imply? Moon rocks are igneous rocks, so they formed from the hardening of lava. The rate at which lava cools determines the grain sizes of igneous rocks: fast cooling results in small grains, slow cooling in large ones. The dark rocks (found in the lowlands) therefore cooled faster than the light-colored ones (found in the highlands). In addition, the light-colored rocks are less dense than the dark ones. These two density classes arise from different compositions. The anorthosites formed from partial melting and slow cooling of low-density silicates at a relatively low temperature (about 1300 K) inside the moon. In contrast, the mare basalts were hotter, they cooled more rapidly, and they were formed from more dense materials.

How did the breccias form? Imagine new igneous rocks on the moon's surface. Bodies from space pound into these rocks, fragmenting and heating them. The heat binds some fragments together to make breccias. The loose material left over makes up the lunar soil.

In a few key ways, moon rocks differ from the earth's igneous rocks. For example, they contain more titanium, uranium, iron, and magnesium. Compared with earth rocks or meteorites, moreover, they are depleted of elements that would condense at relatively low temperatures (less than 1300 K). These elements, called **volatiles,** include sodium, potassium, argon, and chlorine. Moreover, the moon rocks turned out to be bone dry, whereas earth's surface rocks always have some water locked up in their minerals.

Moon rocks are dated by the same radioactive decay techniques used to date earth rocks (refer to Enrichment Focus 8.1). Only a few rocks and some fragments from the lunar soil have ages as great as 4.6 Gy. The anorthosites from the highlands generally are the oldest: 4 Gy. The mare basalts are the youngest: some only 3.2 Gy old and a few as old as 3.8 Gy. These ages imply that the moon formed a

(a)

(b)

(c)

Figure 9.12 Main types of lunar rock. (a) Typical mare basalt. The holes were air bubbles in the lava. (b) Rock containing anorthosite (white area). This sample from Apollo 15 is known as the "Genesis Rock." (c) Lunar breccia. The dark areas are melted rock. Large pieces of other rock are visible embedded in it.

little more than 4.6 y ago. After formation, the present highlands solidified about 4 Gy ago. Then the lava flows that made the maria took place, some 3 Gy ago.

Mercury

Suppose that you stood on Mercury's equator at perihelion at noon. You would need to be made of sturdy stuff, because the surface temperature is about 700 K! The surface temperature drops to 425 K at sunset and reaches about 100 K at midnight. This range of temperatures is the widest known in the solar system, as you'd expect from a slowly rotating, airless world. (On earth, the day/night variation rarely exceeds 20 K because of the greenhouse effect.)

The long, hot solar day and low escape speed (4.3 km/s) make it unlikely that Mercury has any atmosphere. Gas molecules, even the most massive ones, would easily reach escape speed, so no atmosphere could be expected to last long. Overall, sodium vapor makes up the major part of Mercury's very thin atmosphere, along with potassium. The sodium and potassium are probably released from the rocks, which absorb ultraviolet light from the sun. The surface atmospheric pressure is very small, a little less than 10^{-15} that of the earth.

Mercury's surface overall takes after that of our moon (Fig. 9.13). Its highlands are riddled with craters like the moon's bleak highlands. Light-colored rays spring from some of the craters, an indication that these formations resulted from violent impacts during Mercury's stormy past. Some craters are larger than 200 km in diameter, comparable to the biggest lunar craters.

And what of large basins? The Caloris Basin ranks as the largest (Fig. 9.14). It probably has an overall diameter of some 1300 km. Rings of mountains about 2 km high rim the basin. The Caloris Basin has a crinkled floor, perhaps representing fractures from rapid cooling of lava. Also visible are older craters flooded by the lava outpouring from the Caloris impact. In size and structure, the Caloris Basin resembles the moon's Mare Orientale. Mercury's surface has more than 20 large, multiringed basins – many very old and covered with later impact craters.

These similarities do *not* mean that the moon and Mercury are identical. Their surfaces differ in at least three ways: (1) Mercury's surface has *scarps* (scalloped cliffs) hundreds of kilometers long; (2) even the most heavily cratered regions are not completely saturated with craters; and (3) Mercury has fewer small craters than would be expected if the craters had formed at the same rate as on the moon.

Figure 9.13 Cratered surface of Mercury photographed by Mariner 10. Note the heavy cratering, which resembles that of the light-colored regions on the moon.

From the last two points, we infer that the cratering objects had different ranges of sizes for Mercury and for the moon. That is likely because the gravitational effects of the sun on nearby Mercury would influence the distribution in the sizes of orbiting bodies.

Mercury's scarps vary in length from 20 km to more than 500 km and have heights from a few hundred meters to 1 km. Individual scarps often travel over different types of terrain. These characteristics of the scarps imply that Mercury's radius has shrunk some 1 to 3 km (at which time the scarps formed). How? Probably from cooling of the planet's core, its crust, or both – much as the skin of a baked apple wrinkles as it cools.

Mars

The two Martian hemispheres have different topological characteristics: the southern hemisphere is relatively flat, older, and heavily cratered; the northern hemisphere is younger, with extensive lava flows, collapsed depressions, and huge volcanoes. Near the equator, separating the two hemispheres, lies a huge canyon, called Valles Marineris (Fig. 9.15). This chasm stretches 5000 km in length (about the east–west length of Canada) and some 500 km wide in places. It is likely a rift valley – the only one on Mars.

Close up, the Martian surface looks bleak and dry (Fig. 9.16). Large rock boulders are strewn about amid gravel, sand, and silt. The boulders are basaltic. Some contain small holes from which gas has escaped – the holes make the rock look spongy. On earth, such basalts originate in frothy, gas-filled lava; the Martian rocks probably had a similar origin.

A number of sinuous outflow channels were cut in the surface by running water (Fig. 9.17). The largest ones have lengths up to 1500 km and widths as great as 100 km. These channels resemble the arroyos commonly found in the southwestern United States. An **arroyo** is a channel in which water flows only occasionally. They provide evidence of water erosion.

How so? The evidence is by analogy to the landforms cut by water on earth: the flow direction is downhill; the flow patterns meander; tributary structures branch off; and sandbars are cut

Figure 9.14 Mercury's Caloris Basin. Only a part shows in the left half of this photo; the rest is in shadow. The basin is about 1300 km across and rimmed by mountains 2 km high.

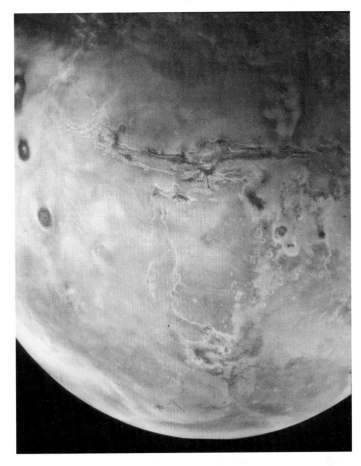

Figure 9.15 Valles Marineris (across the middle and center) and the Tharsis ridge, with its huge shield volcanoes (at the left).

by smaller flow channels, as is commonly found in arroyos on the earth. Most such channels originate within the cratered highlands, just north of Valles Marineris. Some sections of the ancient terrain in the southern hemisphere show networks of small valleys, which resemble terrestrial runoff channels.

The formation of the Martian arroyos required extensive running water for at least a short period of time. Different conditions occurred in the past: a warmer climate and a denser atmosphere. Brief water and mud flows during this period could have cut out the arroyos and valleys we see today.

Where is the water now? The Martian polar caps do contain water ice, but not much. If all the water in the polar caps could cover the surface as a liquid, it would form a layer only about 10 m deep. More water exists, much in a frozen water layer just below the surface. The total would cover the planet with a layer only a few hundred meters deep.

Figure 9.16 Panorama view of the Martian surface from Sagan Memorial Station, May 14, 2000. Many of the rocks have small holes, which indicates a volcanic origin. The rocks are 10 to 25 cm in size. The sandy soil, blown by the wind, piles up between the rocks.

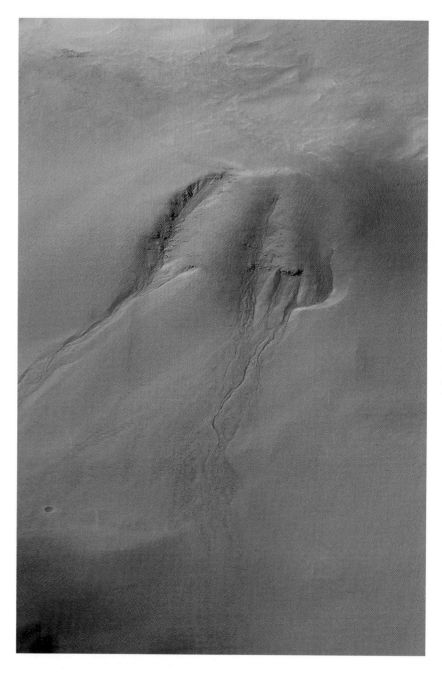

Figure 9.17 Gullies on Mars. Only recently observed, these gullies are rare and may result from groundwater seepage and runoff. Here they cut into the wall of a crater, showing that they appeared later; further evidence of their recent development is the fact that no surface features overlay them. Are the flows that formed them still active?

The Mars Global Survey, orbiting the planet, has sent high-resolution images, some of which have prompted speculations of liquid water flowing on Mars now, at depths of a few hundred meters (Fig. 9.17). A few craters may have been recent lakes; one appears to have dark deposits implying that underground liquid water flowed out of the crater wall and ran down into the floor.

By far the most awesome Martian features are the **shield volcanoes** clustered on and near the Tharsis ridge – clear evidence of Martian volcanism.

The largest is Olympus Mons, some 600 km across at its base (Fig. 9.18). It would span the islands of Hawaii, which formed from several volcanoes. If put down in California, Olympus Mons would stretch from San Francisco to Los Angeles! It soars 21 km above "sea level." Indirect evidence from the nearby lava flows implies that the volcanoes formed about a few hundreds of millions of years ago. Most of the lava plains appear about 1 to 3 Gy old. The youngest flows may be a mere 20 million years old!

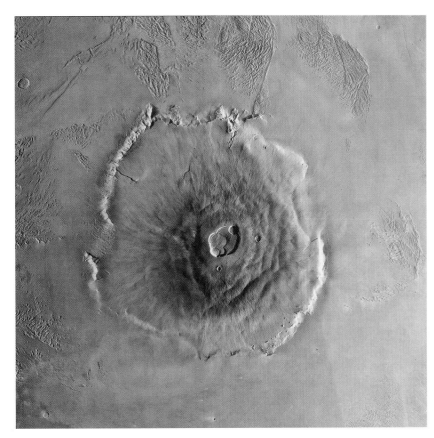

Figure 9.18 Olympus Mons. This image shows the volcano rising above the surrounding terrain; the base spans some 600 km, the ridge around the base is 6 km high, and the summit lies 21 km high. The central depression at the summit is about 3 km deep and 25 km across. Old lava flows drape over the sides of the volcano.

The huge mass of Olympus Mons requires that the Martian crust beneath it be thicker than the crust beneath smaller such volcanoes on the earth. Geologists estimate the thickness of the Martian crust to be some 120 km, about twice that of the earth's. This thick crust may well explain why Mars never developed a global plate tectonic system.

Olympus Mons crowns a string of volcanoes situated on the Tharsis ridge. Occasionally, thin ice clouds decorate the tops of the volcanoes there. These clouds might result from erratic spurts of outgassing. On earth, volcanic activity spews forth gases (including water vapor) from the earth's mantle. Such outgassing by the giant Martian volcanoes in the past may have added significantly to the Martian atmosphere.

The southern hemisphere of Mars sports a cratered terrain that resembles the ancient highlands of the moon or the plains of Mercury. The landscape contains impact craters that range in size from huge, lava-filled basins down to some only a few meters across. The Martian craters come in the same varieties as lunar and Mercurian ones, some with central peaks that mark their impact origin. In general, the Martian craters are shallower than the ones on the moon and Mercury. Windblown dust fills many. Wind scours the craters and piles the dust in dunes within the craters' bowls, so we know that wind erosion has worked for a long time on Mars.

Venus

The atmosphere of Venus differs remarkably from ours. Since 1932 we have known that Venus' atmosphere contains carbon dioxide, but we did not know how much until fairly recently. Interplanetary probes indicate that the atmosphere contains about 96 percent carbon dioxide, 3 percent nitrogen, some argon, and traces of water vapor (varying from 0.1 to 0.4 percent), oxygen, hydrogen chloride, hydrogen fluoride, hydrogen sulfide, sulfur dioxide, and carbon monoxide.

The yellowish white clouds in the air conceal the surface (Fig. 9.19). The cloud tops reach about 65 km above the surface. (The highest clouds above the earth go up to only about 16 km.) The cloud tops flow with the upper atmosphere, in patterns similar to jet streams of the earth. Ringing the equator, the clouds whiz around at roughly 100 m/s, fast enough to orbit the planet in only 4 days. Winds also blow from the equator to the poles in large cyclones 100 to 500 km in diameter. They culminate in two giant vortices that cap the polar regions. What drives such fierce winds in the upper atmosphere of Venus? Solar heating does the trick, as on the earth.

What are the clouds made of? The best idea sees the upper-level clouds as concentrated solutions of sulfuric acid. (Sulfuric acid, H_2SO_4, is used in car batteries.) Models of the clouds imply that they contain a solution of 90 percent sulfuric acid mixed with water.

Although the atmosphere and clouds of Venus do contain some water vapor, it doesn't amount to very much compared with the total amount of water on the earth. If all the earth's water (in both the atmosphere and oceans) were spread in a uniform layer over the earth's surface, that sheet would be 3 km thick. All the water in the atmosphere of Venus (none exists on the surface now because it's so hot) would amount to a layer only 30 cm thick. Our earth's surface and atmosphere have about 10,000 times as much water as Venus does.

Do not dream of visiting the surface of Venus! There the pressure is 90 atm, and the surface temperature about 740 K (about 470° C)! That's about the temperature of a home electric oven during its hot self-cleaning cycle. This high temperature results from the very effective trapping of surface heat (Section 8.4), because carbon dioxide and water vapor absorb infrared radiation well. An extreme greenhouse effect keeps Venus very hot because only about 3 percent of the sunlight hitting the upper atmosphere of Venus hits the surface. The atmosphere must be a very good insulator to result in the high temperatures.

Atmospheric winds on Venus blow from the day side to the night side and from equatorial to polar regions. The wind flow carries heat. Along with the very effective atmospheric insulation, this flow helps to keep the temperatures fairly constant over Venus' surface. They vary about 10 K or less from day to night side, so it doesn't cool off much at night.

The Active Surface of Venus

We've investigated Venus' surface by examining material brought back by landing probes and by analyzing the reflections obtained by bouncing radar (which penetrates the clouds) off the surface. The former USSR landed spacecraft that sent back close-up photos. The pictures showed slabby rocks about tens of centimeters in size (Fig. 9.20). A few rocks have small holes that were once filled with gas, implying a volcanic origin. The rocks rest on loose, coarse-grained dirt. Direct measurements showed that most of the rocks are volcanic basalt, like those lining the earth's ocean basins or lying on the earth's ocean floor near midoceanic ridges. These observations provided evidence of local volcanism.

What about the general lay of the land? Radar mapping reveals a varied terrain: mountains, plains, high plateaus, canyons, volcanoes, ridges, and impact craters. Overall, Venus looks fairly flat. Elevation differences vary little, only 2 to 3 km, except for a few highland regions. (The continents there reach up to only some 10 km, compared with a 25-km difference on Mars and a 20-km one on the earth.) Only some 10 percent of the surface extends above 10 km. In contrast, about 30 percent of the earth's surface reaches above 10 km (from the bottom of the ocean basins).

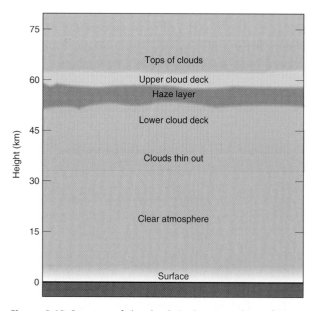

Figure 9.19 Structure of the clouds in the atmosphere of Venus. The clouds, consisting of sulfuric acid and water, make up two decks (extending from 45 to 65 km) with a layer of haze between them. The air is clear from the surface up to the lower cloud deck.

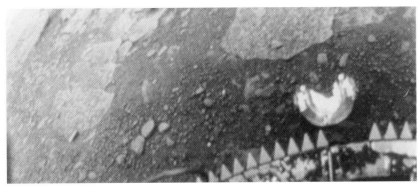

Figure 9.20 Close-up of the surface of Venus, taken by the Russian spacecraft Venera 13, which touched down in the lowlands southeast of Beta Regio in 1982. The color here is real; the atmosphere removes blue and yellow from sunlight, leaving red and orange. Note the slabs of rock, patches of dark soil, and pebbles close to the edge of the lander (lower right).

The southern and northern halves of the face of Venus differ remarkably. The northern region is mountainous, with uncratered upland plateaus; these resemble continents on earth. The southern part has relatively flat rolling terrain, which appears to consist of vast lava plains.

The great northern plateau of Venus, called Ishtar Terra, measures some 1000 km by 1500 km (Fig. 9.21). That's larger than the biggest highland plateau on the earth, the Himalayan plateau. This great plateau may have been built up from thin lava flows over an uplifted section of older crust. Mountain ranges border Ishtar on the east, northwest, and north. The eastern range, called Maxwell Montes, contains the highest elevations on Venus: up to 11 km. A volcanic cone lies near Maxwell's center – evidence of volcanism here.

Most of the surface consists of flat, volcanic plains that are imprinted by tens of thousands of volcanic domes and shields. The rolling lava plains are punctuated by craters (Fig. 9.22), both large (over 100 km in diameter) and smaller ones (less than 10 km). About a thousand have been identified. They are impact craters created by solid bodies from space crashing into the surface (Enrichment Focus 9.2). Small craters – less than 3 km – do not form on Venus because the dense atmosphere completely vaporizes small pieces of incoming debris before they can reach the ground. Amid these spread stunning impact craters. They generally show central peaks, terraced walls, shocked surfaces, and flooded floors; because of the dense atmosphere, they lack extensive ray systems of exploded debris.

The volcanoes of Venus often develop complexes amid fracture and rift zones. These volcanoes have gentle slopes (Fig. 9.23); they are shield volcanoes. Instead of exhibiting a sharply uplifted cone, shield volcanoes are relatively flat, like an armor shield. Shield volcanoes form when lava erupts out of the sides rather than at the summit. Whether any volcanoes are now active is unclear. Volcanoes discharge gases from inside a planet. On earth, this **outgassing** contains mostly water vapor, carbon dioxide, nitrogen, and sulfur gases – all found in the atmosphere of Venus. The Magellan orbiter (see below) did *not*, however, find firm evidence of current activity.

Plate tectonics sustain the large difference on the earth between ocean basins and continental masses. The lack of such overall differences on Venus suggests an absence of planetwide crustal plates. Tectonic work on Venus seems localized to zones spread around the planet, especially near highland regions. That's a surprise, because Venus and earth should have lost heat at the same rates. It's the heat flow from the mantle that pushes the plates in the crust.

Overall, Venus displays a largely young surface, no more than a few hundred million years. (That's about twice the average age of the earth's surface.) The degree of volcanism seems more widespread than on the earth. The entire surface may have been reshaped over the past several hundred million years by runny lava flows. Rift valleys, which form at the separations of crustal plates, show extensive surface fractures. Also visible are folded and faulted regions that resemble some mountain-building regions on the earth. Such features indicate local tectonic activity. The high degree of activity that seems to typify Venus now should continue for eons to come.

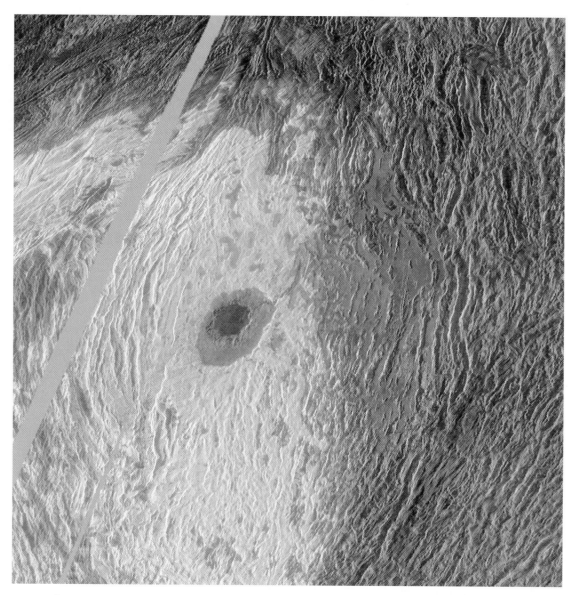

Figure 9.21 Radar image of Maxwell Montes in the highest volcanic mountain range on the planet (11 km above the mean surface). In its center, Cleopatra crater (diameter, 105 km) can be seen.

But where are the crustal plates? Perhaps it is a "one-plate planet," with a crust that never became rigid. Like dough, Venus' crust is not stiff enough to slide around in pieces. Instead, it wrinkles and puckers locally from heat and lava upflow.

9.3 MAGNETIC FIELDS
(Learning Outcome 9-5)

The earth's magnetic field tells us about its interior: it has a liquid, conducting core. The dynamo model also implies that the earth rotates rapidly to gener-ate the currents that make the field. What about the other terrestrial planets?

Moon

The moon's magnetic field is very weak, only 10^{-4} times as strong as that of the earth. If the dynamo model for a planet's magnetic field is correct (Section 8.3), the moon's core cannot be completely molten or composed mainly of metals. (Also, the moon's low density rules out a substantial metallic core.) Yet, some surface rock samples are magne-

Figure 9.22 Three large impact craters visible in the Lavinia region of Venus; they range in diameter from 30 to 50 km. Note the bright material, ejected upon impact, surrounding each crater and the central peak in each. Howe crater appears centered in the lower part of the image. It is 37.3 km in diameter. Danilova, an even larger crater, is at the upper left; Aglaonice, with a 62.7-km diameter, is to the right of Danilova. The color is simulated.

Figure 9.23 Sif Mons volcano on Venus. It is massive, with a diameter of 300 km and a height of 2 km. Lava flows extend for hundreds of kilometers across the fractured plains in the foreground.

tized much more than you would expect from such a weak magnetic field. Recall (Section 8.3) that iron minerals in an igneous rock preserve the magnetic field present at the time of solidification. This implies the moon's magnetic field was stronger in the past than now.

Mercury

Mariner 10 detected a weak planetary magnetic field, about 0.01 times the strength of the earth's magnetic field. Small as this sounds, it's sufficient to carve out a magnetosphere in the solar wind (Fig. 9.24).

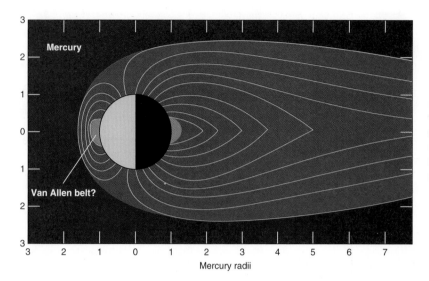

Figure 9.24 Inferred from Mariner 10 observations, Mercury's magnetosphere may trap solar wind particles to form Van Allen-type radiation belts. (Adapted from a diagram by D. Baker, J. Borovsky, G. Gisler, J. Burns, and M. Zeilik.)

Mercury's field is a dipole, more or less aligned with its spin axis. In general, Mercury's magnetic field is similar to the earth's, only weaker. And that's the problem! Mercury, like the earth, has a metallic core. But it's presumed to be relatively cold and solid now because a small planet loses heat quickly. Recall (Section 8.3) that the earth's magnetic field arises from swirling motions in its hot, liquid metallic core. The churning is driven by the earth's spinning. Because Mercury rotates much more slowly than the earth and is thought to have a cool core, no one really expected it to have a planetwide magnetic field.

What's the explanation? No one knows for sure. Perhaps part of the core is fluid. Or the dynamo model may be incorrect! After all, Venus, which has a much larger and presumably hotter core than Mercury, has no detectable magnetic field. (But Venus rotates even more slowly.) Or, perhaps the field remains from an older time. Maybe it originated from the planet's interaction with the solar wind. Or perhaps small planets with cool metallic cores produce magnetic fields by a mechanism not yet imagined.

Mars

Mars has an extremely weak planetwide magnetic field, less than 0.002 times the strength of the earth's. A value that small presents a puzzle if the dynamo model correctly describes planetary magnetic fields. Mars rotates as fast as the earth. Though the Martian core is smaller, it should contain an ample amount of metals. We have no direct proof that the core is liquid, but the evidence for past volcanic activity implies a hot mantle; hence, probably, a hot, liquid core. Mars should have a moderately strong magnetic field, but it does not.

Perhaps, as for Venus, Mars falls in the middle of a magnetic field reversal, but it is unlikely that we would catch both planets in the process of changing polarity. Again, the dynamo model is called into question by the weak field of another earthlike planet.

Venus

A metallic core, liquid in part, implies that Venus should have a planetary magnetic field. Because Venus rotates 243 times more slowly than the earth, we expect its internal dynamo to be weaker; hence the magnetic field should be less intense than the earth's but still there. No probe to date has detected any magnetic field. If one exists, it must be at least 10,000 times weaker than the earth's magnetic field! That's much weaker than we would expect from a simple dynamo model.

What a magnetic mess – tiny, high-density Mercury has a planetary field, and Venus does not! Perhaps a simple dynamo model does not apply to all planets.

A very weak magnetic field means that the interaction of Venus with the solar wind differs from that of the earth. Without the buffer of a magnetic field, the solar wind runs right into the upper atmosphere of Venus. As the wind flows around and past the planet, it carries off some of the atmosphere.

9.4 HISTORY AND EVOLUTION
(Learning Outcomes 9-6 and 9-7)

You have seen that the terrestrial planets share some characteristics with the earth and differ in others. How did they achieve these stages of evolution?

Keep in mind a general concept: *the larger a planet, the longer its evolutionary lifetime.* An analogy: consider two potatoes, one large and small. Stick a thermometer into each and put them into a hot oven. Let each heat up to the same temperature; then take them out. Which will cool off to room temperature the fastest? The smaller one! This applies to the cooling of planets, too, and so their evolutionary lifetimes. How so? A planet's total internal energy basically depends on its volume. The rate at which a planet radiates energy into space depends on its surface area. If you divide amount stored (volume) by the loss rate (surface area), you get the evolutionary lifetime (which depends directly, then, on the radius).

Moon

From the Apollo results, we can concoct a scenario of the moon's history. The inferred sequence of events (Fig. 9.25) relies heavily on the dating of the lunar rocks.

About 4.6 Gy ago, the moon formed by the accretion of chunks of material. These pieces continued to plunge into the moon after most of its mass had gathered. During the first 200 million years after formation, these projectiles bombarded the surface and heated it enough to melt it. Less dense materials floated to the surface of the melted shell; volatile materials were lost to space. The crust began to solidify from this melted shell about 4.4 Gy ago. From 4.1 to 4.4 Gy ago, the crust slowly cooled as the bombardment from space tapered off. The debris from this later bombardment made many of the craters now found in the highland areas.

Below the surface, the moon's material remained molten. About 4 Gy ago, a few huge chunks smashed the crust to produce basins that later became maria. (Most of these large impacts happened on one side of the moon, which now faces us today because of tidal forces.) For example, the Mare Orientale Basin

formed some 4 Gy ago when an object about 25 km across smashed into the moon. Only later did the basins fill in. As the crust lost its original heat, short-lived (rapidly decaying) radioactive elements reheated sections of it. From 3.0 to 3.9 Gy ago, lava from the radioactive reheating punctured the thin crust beneath the basins, flowing into them to make the maria.

For the past 3 Gy, the lunar crust has been inactive. However, small particles from space have incessantly plowed into the surface since its solidification. These sand-sized grains scoured the surface, smoothed it down, and pulverized it. Continued bombardment by larger bodies churned the fragmented surface. Impacts melted the soil, which swiftly cooled to form breccias and glass spheres. The moon's surface today resembles a heavily bombarded battlefield – constantly blasted, fragmented, stirred up, and melted.

The moon is a low-density, rocky world that cooled quickly. Why? Because it has a large heat-radiating surface compared to its heat-producing (or storing) volume and mass. The basic rule: the smaller the planet, the faster it loses thermal energy. Compared to the earth, the moon had only a brief episode of volcanism, and it never developed tectonic activity.

Mercury

What forces have shaped the surface of Mercury? Using the moon as the basis for comparison, the most important was impact cratering. Both tiny worlds have cool interiors compared to the earth's interior. Neither has much (if any) volcanic activity now, and neither has undergone the kind of continual surface evolution that the earth experiences from tectonics (Section 8.5).

The lack of an atmosphere and the short period of crustal evolution are related to their small masses. Their surface gravities fall so low that most gases have escape speeds, and any atmosphere slips away. The small masses also imply that internal heating from radioactive decay are less than that for the earth, and the flow of energy outward would be so fast that both bodies would cool off quickly. The earth's hot interior promotes the outward flow of heat to set up currents in the plastic mantle. Both Mercury and the moon lack this combination to drive the evolution of their crusts.

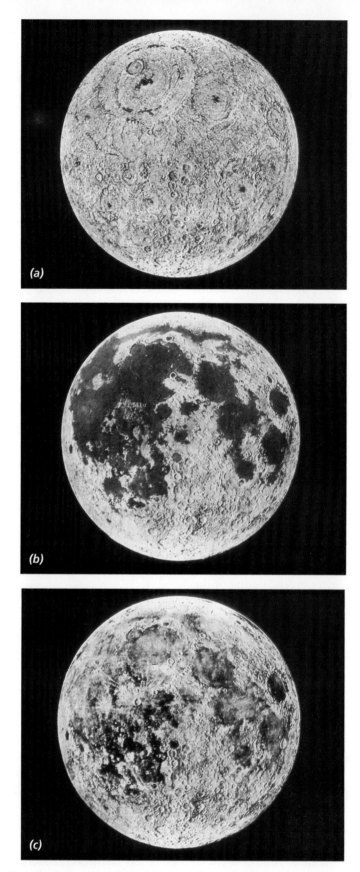

Figure 9.25 Model for the moon's evolution. (a) 3.9 to 4 Gy ago: the surface has solidified and is cratered by the infall of planetesimals; most of the surface is saturated with craters. (b) 3.0 to 3.2 Gy ago: fractures in the surface allow magma from the interior to flow out and fill the lowland basins with material darker than the highlands.(c) Now. Many of the large rayed craters blossomed on the surface after the formation of the maria. Parts of the maria have grown lighter in color from the material blown out by crater formation.

Without surface erosion and crustal evolution, the moon and Mercury retain evidence of their early years. The similarities of surface features suggest similar histories. For example, the Caloris Basin on Mercury resembles maria on the moon. The lunar maria were formed by the impacts of large bodies, which created large basins. The inflow of lava from the interior filled them to make the maria. Such processes probably built the Caloris Basin.

Using lunar analogies, we can set up a working model for the evolution of Mercury. Once formed by accretion, Mercury probably went through the following general stages: (1) heating of the surface (by impacts and/or radioactive decay) and formation of a solid crust, (2) heavy cratering, (3) formation of impact basins, (4) filling in of basins by volcanism, and (5) low-intensity cratering. We cannot date this sequence as we can for the moon because we do not have rock samples from Mercury's surface. But a lunar comparison suggests that the intense sculpting of Mercury's surface took place about 4 Gy ago, not long after the planet formed.

The moon and Mercury are dead worlds; their dramatic evolution ended early. Compared with the earth, they are fossil planets rather than living ones (Table 9.2).

Mars

The Tharsis ridge labels Mars' northern hemisphere, which differs so dramatically from the southern one. The ridge rises about 10 km above the average surface height for the planet and contains numerous volcanic structures. Over time, impact craters have been wiped out by volcanic resurfacing, and very few are visible now. In contrast, the southern hemisphere appears as basically a desert, pockmarked by old, eroded craters. The geologic inference from this difference is that about 3 Gy ago in the northern hemisphere, a huge mass of lava oozed out from under the surface, creating the volcanic plains and the volcanoes over a long period of time. This initial flow destroyed the older deserts and craters in this region. Other flows followed later, perhaps as recently as 20 million years ago.

Mars has ancient impact craters, somewhat eroded, in the southern hemisphere. The northern shows evidence of extensive volcanism. Between

Table 9.2 Main Stages in the Evolution of Terrestrial Planets

Stage	Processes
I	Formation by accretion, heating of crust and interior, crust formation
II	Crust solidification, intense impacts, cratering of surface
III	Basin formation and flooding, lowlands formation
IV	Low-intensity impacts, atmosphere formation by outgassing
V	Volcanoes, crustal movement, continent formation

them lies the giant rift valley of Valles Marineris – the only suggestion of local tectonics. Mars lacks mountain ranges, and so any evidence for plate tectonics. Because the Martian volume-to-surface ratio lies between those of the earth and the moon, we expect that the planet lost internal energy at an intermediate rate.

A Martian planetary history may have followed a sequence like the early part of the earth's. First, after the formation of Mars by accretion, impact craters covered the surface. Shortly afterward, the planet differentiated (as did the earth; Section 8.5) to form a crust, a mantle, and a core. Regions of thicker crust rose to higher elevations. In the second phase, thin regions of the crust fractured, and the Tharsis ridge uplifted, cracking the surface around it.

During this time, a primitive atmosphere, denser and warmer than now, held large amounts of water vapor from the volcanic outgassing. Rainfall may have eroded the surface in furrows and then percolated to a depth of a few kilometers. Decreasing temperatures formed ice at shallow depths. When heated (perhaps by volcanic activity), this ice melted, leading to the formation of collapse and flow features. Planetwide water erosion carved the surface.

In the next phase, extensive volcanic activity occurred, especially in the northern hemisphere. The Tharsis region continued to uplift, generating more faults. Valles Marineris formed from expansion of the crust. Finally, volcanism – most of it concentrated on the Tharsis ridge – broke the surface and spewed out great flows of lava. Since that last time of great eruptions, wind erosion has mainly sculpted the Martian surface.

This geologic activity occurred abundantly during the first two-thirds of Mars' history. We have new indications that some activity has continued at a scaled down rate for the last few percent of Martian time – maybe even today!

We have scant indication for plate tectonics on Mars. On earth, volcanoes tend to form in chains, where one crustal plate encounters another, on Mars, volcanoes are pretty much clustered in one highland region. On earth, plates also show up as continental masses with basins in between. Mars does not have such crustal configurations. The implication is that because of its small mass and size, Mars cooled quickly and developed a thick, inactive crust and mantle. So it never experienced the large convective motions in the mantle that drive the plate tectonics on the earth. Because Mars cooled off faster, it did not evolve as much as our planet. It just reached stage V (Table 9.2) and then stopped.

Venus

We expect the earth and Venus to be very similar, yet in some vital respects, they are not. How did the differences develop, given that the two planets have about the same size, density, mass, interior composition, and structure? Probably from Venus' proximity to the sun and consequent lack of liquid water on the surface. On earth, the oceans and carbonate rock from the remains of ocean life contain a large amount of the carbon dioxide. On Venus, the carbon dioxide cannot be trapped in this way. It stays in the atmosphere, keeping up the severe greenhouse effect.

Although Venus has a high degree of overall volcanism and local outbreaks, it does not show the global tectonic pattern we see on the earth. The crust of Venus has experienced some localized crustal plate movement, but most of the crust is intact. That's a real puzzle.

We can speculate about the history of Venus using the earth as a guide. We infer that with the other terrestrial planets, Venus formed by accretion about 4.6 Gy ago (Section 8.5). Venus' interior differentiated as a result of internal radioactive and impact heating. During the first 500 million years, a crust formed and solidified. About 3 to 4 Gy ago, large masses bombarded the surface (as they did the earth) and fractured the crust. Volcanoes erupted. A heavy bombardment by smaller bodies from space cratered the surface; this activity ended about 3 GY ago. That ancient surface has been wiped out.

Since the initial bombardment, huge volcanoes vented through cracks in the surface; their cones form the shield volcanoes visible today. Lava flows have crossed hundreds of kilometers of terrain. Volcanism has reworked the surface almost completely in the last 400 million years.

Venus may well resemble the earth at an early age, from 4.5 to 2.5 Gy ago. The Venus of today appears like a view of the earth's distant past. Our current notion is that Venus seems to have evolved in a sequence similar to that of the earth, but more slowly for unknown reasons.

Earthlike worlds have five general stages of evolution (Table 9.2). The moon and Mercury have evolved the least, up to stage IV. Their evolutionary lifetimes were only about 3 Gy. Mars is the intermediate case, through stage IV; its evolutionary lifetime lasted about 4 Gy. Venus has changed somewhat more, into the beginning of stage V. Only the earth has made it through stage V, so its evolutionary lifetime is at least 4.6 Gy and grows as long as our planet is active.

KEY CONCEPTS

1. The moon and Mercury are the two smallest terrestrial planets. Their bulk densities indicate that they are made of mostly rocks (moon) and metals (Mercury). Their small sizes mean that they cooled off quickly.

2. The most prominent surface features of the moon are large basins and craters, most formed by the impacts of solid bodies from space billions of years ago. The surface is made of highlands (covered with low-density, light-colored rocks) and lowlands (covered with darker, higher-density rocks). The Apollo samples, dated by using radioactive decay, show that the highlands are older than the lowlands. The maria are lava flows that filled in huge impact basins.

3. The moon may or may not have a well-defined core. If such a core exists, it consists simply of dense rocks with few metals. The core is hot enough to produce a fair heat outflow at the moon's surface. The moon's mantle is now cool and rigid.

4. Mercury has a bulk density almost the same as that of the earth, so it probably has a large metallic core, which is cool. The mantle is likely made of rocky materials. The differentiation of Mercury's interior indicates that in the past it was hot enough to be liquid. The most prominent features on Mercury's surface are craters (formed by impacts), basins (formed by larger impacts), and scarps (formed by the shrinkage of the planet as it cooled).

5. The surface of Mercury resembles that of the moon, so the processes that shaped it and the timing of the sequence of important events probably follow those of the moon. Impact cratering played the major role. Volcanism lasted for a brief period as these masses cooled rapidly. No plate tectonics have occurred.

6. Venus and Mars, the terrestrial planets most like the earth (Table 9.1), have bulk density close to that of earth and so have somewhat similar interior structures (Venus more so than Mars).

7. The atmospheres of Venus and Mars have essentially the same composition: almost all carbon dioxide. Both atmospheres are also very dry. The Martian atmosphere is very thin compared to that of Venus. Even so, wind erosion takes place on Mars. Venus has a high surface temperature because of an extreme greenhouse effect from its thick carbon dioxide atmosphere.

8. The surface of Venus has mountains, volcanoes, upland plateaus, rolling plains, large valleys, and impact craters. Most of the surface is very flat and lacks large impact basins. Almost the entire surface has been remolded by volcanism, so it is relatively young (though older than the earth's youngest surface regions).

9. Venus, because it has almost the same size as the earth, went through a similar evolutionary sequence; Mars, with smaller size, did not evolve nearly as much. Neither planet developed global tectonics, though both show evidence of volcanism, Venus more so than Mars.

10. The surface of Mars has giant volcanoes, arroyos cut by water, a huge rift valley system, and impact craters. The northern hemisphere is volcanic and younger than the southern. The arroyos and eroded craters indicate that in the past Mars had an atmosphere extensive enough to permit abundant liquid water to exist on the surface (now both the surface pressure and the temperature are too low). Underground rivers may flow on Mars now.

11. Terrestrial planets go through similar evolutionary sequences, with radius mostly determining how far a planet will evolve: the smaller the radius, the less the mass, and the faster a planet loses internal heat and its atmosphere, and the less it evolves. Volcanism and tectonics play the major role in the evolution of the surface. Wind and water erosion are secondary processes.

STUDY EXERCISES

1. Suppose you stepped out of Transplanetary Spacelines Flight 101 onto the moon's surface. You look slowly around, at both the ground and the sky. How does what you see differ most from what you would see on the earth? (Learning Outcomes 9-2 and 9-4)

2. Suppose you stepped out of the same carrier's Flight 102 onto the surface of Mercury. How would the scene differ most from that on the earth? (Learning Outcomes 9-3 and 9-4)

3. Suppose you stepped out of the same carrier's Flight 103 onto the surface of Venus. How would the scene differ most from that on the earth? (Learning Outcomes 9-3 and 9-4)

4. Suppose you stepped out of the same carrier's Flight 104 onto the surface of Mars. How would the scene differ most from that on the earth? (Learning Outcomes 9-3 and 9-4)

5. Argue from a comparison of average densities that the moon cannot have a metallic core like the earth's; that Mercury should have a metallic core; that the core of Venus should be most like the earth's. (Learning Outcomes 9-1 and 9-6)

6. How were most of the craters on the moon formed? On Mercury? On Venus? On Mars? Back up your statement with specific evidence. (Learning Outcomes 9-2, 9-3, and 9-8)

7. What specific evidence do we have that the moon's lowland regions (maria) formed after the highlands? What about on Mercury? (Learning Outcome 9-8)

8. You are writing a grant proposal to NASA to do research on the origin of the moon. Describe the theory you plan to support in the best light possible. (Learning Outcome 9-10)

9. Compare the characteristics of the Orientale Basin on the moon to the Caloris Basin on Mercury. (Learning Outcomes 9-2, 9-3, and 9-4)

10. In one sentence, describe how the surfaces of the moon and Mercury differ. (Learning Outcomes 9-2, 9-3, and 9-4)

11. Neither the moon nor Mercury has a substantial atmosphere. Why not? (Learning Outcome 9-4)

12. In one sentence, describe the difference between the interiors of the moon and Mercury. (Learning Outcomes 9-5, 9-6, and 9-7)

13. In one sentence, compare the evolution of the surface of the moon and the surface of Mercury. (Learning Outcome 9-9)

14. Mariner 10 photographed only about half of Mercury's surface. Imagine that in the future a new flyby mission photographs the rest. It finds a region almost devoid of craters. What statements could you make about the age and history of this region? (Learning Outcome 9-8)

15. How does it happen that the earth has two tidal bulges, one on the side facing the moon and another on the opposite side of the earth? (Learning Outcome 9-9)

16. Imagine you are in space near the earth. You have with you two spheres of equal mass. You place them in a line, one closer to the earth than the other. How would the tidal force of the earth on these spheres affect the distance between them? (Learning Outcome 9-9)

17. How do we know the masses Mercury and Venus? (Learning Outcome 9-1)

PROBLEMS AND ACTIVITIES

1. Calculate the surface gravity of the moon. Then calculate the surface gravity of Mercury compared to that of the moon.

2. Calculate the escape speed from the moon. Then compare the escape speed from the moon to that of Mercury.

3. When a mass falls into a place from a great distance, it hits the surface with the escape speed from that planet. Imagine a 100-kg mass falling onto the moon and Mercury. Describe the kinetic energy at impact on each planet. Which body do you expect to have larger craters?

4. At Mercury's evening (eastern) elongation on March 13, 1992, the planet had an angular diameter of about 8 arcsec. What was its size-to-distance ratio then? Use Mercury's physical diameter given in this chapter to calculate Mercury's distance on this date.

5. Imagine that the moon were twice its current distance from the earth. By how much would the moon's tidal force on the earth change? What implication does this have for tides on the earth as the moon moves away from the earth?

6. Compare (using ratios) the escape speed of Mars to that of the earth and Venus.

7. Use the properties of ellipses and the orbital characteristics of Mars and the earth to show that the closest the two planets can come at opposition is about 56 million km.

8. What is the angular diameter of Mars when viewed at closest opposition, as given in Problem 7?

9. Compare (using ratios) the surface gravity of Venus to that of the earth and Mars.

10. In the 1993 opposition, Mars reached a maximum angular diameter of 15 arcsec. What was its size-to-distance ratio? How far was it from earth at this opposition? How does this value compare to the closest opposition possible (Problem 7)?

10

The Jovian Planets: Primitive Worlds

LEARNING OUTCOMES

After studying this chapter, you should be able to:

10-1 Compare and contrast the Jovian planets as a group to the terrestrial planets, emphasizing the greatest differences.

10-2 Contrast the Jovian planets to one another in terms of their relative sizes, relative masses, bulk densities, atmospheric compositions, internal structures, and unique features.

10-3 Compare and contrast the interior and composition of Jupiter to those of the earth.

10-4 Compare the rings of Saturn with those of Uranus, Neptune, and Jupiter in terms of size, shape, and possible composition.

10-5 Present the unique characteristics of Pluto that make it neither a Jovian nor a terrestrial planet.

10-6 Describe the general properties of the Pluto–Charon double-planet system.

10-7 Compare and contrast the general characteristics, surface features, and evolu-

tion of the Galilean satellites of Jupiter: Io, Europa, Ganymede, and Callisto.

10-8 Compare and contrast the Galilean satellites to the earth's moon and to Pluto and Charon, especially in terms of evolutionary processes.

10-9 Compare the magnetic fields and magnetospheres of Jupiter, Uranus, Neptune, and Saturn to those of the earth, and apply a dynamo model to them.

10-10 Contrast Saturn's largest moon, Titan, the Galilean moons of Jupiter, the largest moons of Uranus, and the earth's moon.

10-11 Argue that the Jovian planets have evolved since their formation by heat flow from their interiors, but in contrast with the terrestrial planets, they leave behind no record of their changes.

CENTRAL CONCEPT

The Jovian planets, compared to the terrestrial ones, have greater masses and sizes but lower densities. Today they pretty much resemble their early states because they preserve no records of the history of their evolution.

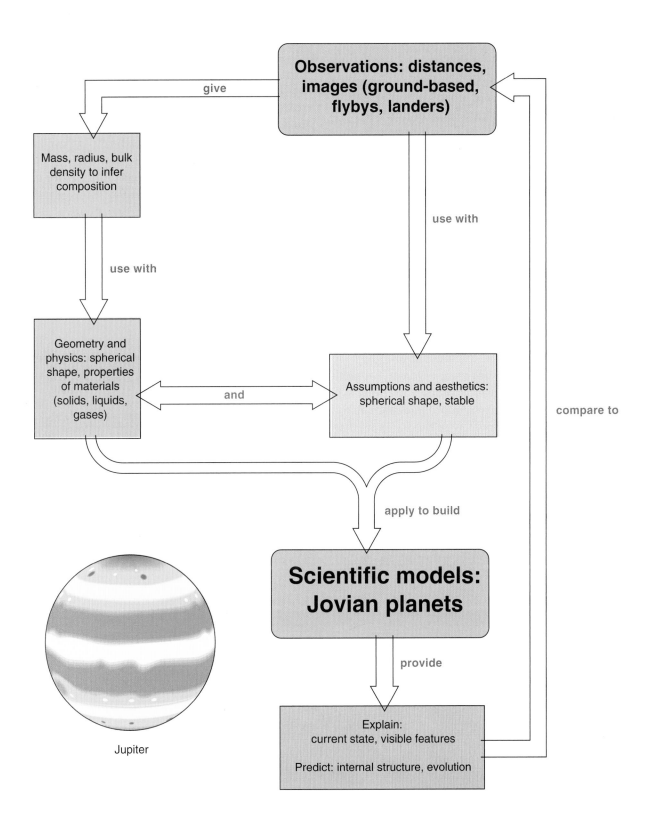

Observations: distances, images (ground-based, flybys, landers)

give

Mass, radius, bulk density to infer composition

use with

Geometry and physics: spherical shape, properties of materials (solids, liquids, gases)

and

use with

Assumptions and aesthetics: spherical shape, stable

apply to build

compare to

Scientific models: Jovian planets

provide

Explain: current state, visible features

Predict: internal structure, evolution

Jupiter

Chapter 10
Celestial Navigator™

Jupiter and Saturn are impressive, awesome planets – two giant worlds languidly circling the sun. Banded Jupiter drags along its coterie of satellites. Saturn is set in its rings like a prize gem of space. Uranus and Neptune seem like twins at a distance but differ greatly when viewed close up. You can barely grasp the weird environments of these giant worlds: planets many times the diameter of the earth, mostly gases and liquids, and without breathable atmospheres.

The realm of the Jovian planets – Jupiter, Saturn, Uranus, and Neptune – marks a region completely different from that of the terrestrial planets. The difference includes not only the physical properties of the planets but also their multiple moons and graceful rings. The satellites here are worlds in their own right, some as large as the smaller terrestrial planets. These moons have surface environments that differ remarkably from those in the inner solar system, so their evolutionary histories were different – a point that illuminates another facet of the early solar system. The double-planet system of Pluto and Charon may have more in common with these moons than with the Jovian planets.

This chapter investigates the features of the Jovian planets that set them apart from the terrestrial ones. The major differences between the Jovian and terrestrial planets furnish additional clues about the evolution of the planets and the formation of the solar system.

10.1 JUPITER: LORD OF THE HEAVENS (Learning Outcomes 10-1, 10-2, and 10-11)

The **Jovian planets** (Table 10.1 and Fig. 10.1) are those whose physical properties resemble those of their prototype – Jupiter. It ranks as the largest and most massive body in the solar system, after the sun. Jupiter's total mass is about 2.5 times that of all other planets put together (and more than 300 times the earth's). Eleven earths placed edge to edge would stretch across Jupiter's visible disk, and more than a thousand earths would be needed to fill its volume – truly an immense planet! (See the Jovian Planets Celestial Navigator™.)

Table 10.1 Comparison of the Jovian Planets

Planet	Diameter (earth = 1)	Mass (earth = 1)	Bulk Density (water = 1)
Jupiter	11.0	318	1.3
Saturn	9.5	95	0.7
Uranus	4.1	15	1.2
Neptune	3.9	17	1.7
Pluto	0.17	0.002	2.1

EARTH

Figure 10.1 Giant planets. The Jovian planets, shown to scale relative to the earth. Note the banded atmospheres of Jupiter (far left) and Saturn. Uranus, to the right of Saturn, and Neptune both have methane gas in the atmosphere that absorbs red light and so gives an overall bluish color. The rings of Jupiter, Uranus, and Neptune are not visible.

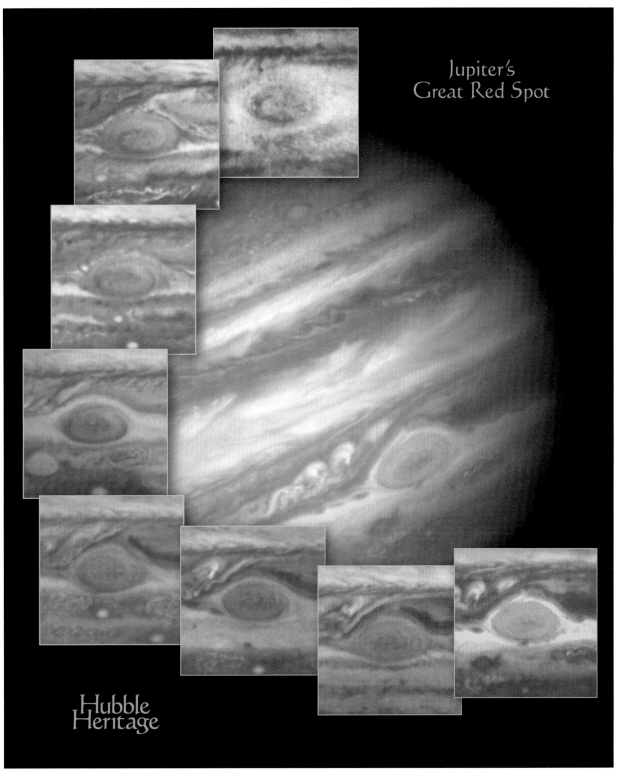

Figure 10.2 Jupiter and its Great Red Spot imaged by HST from 1992 to 1999. The small images show how it changed over this period. The Great Red Spot is a giant storm that has whipped around the atmosphere for hundreds of years.

Physical Characteristics
(Learning Outcomes 10-1, 10-2, and 10-11)

Just as Jupiter's mass is the largest among the planets, so is its diameter – 143,000 km. For all this size, however, Jupiter's material is much less concentrated than the earth's, for Jupiter's density is only 1.3 – not much more than water!

All the Jovian planets have low densities compared with the terrestrial planets – one key difference between the two classes. These low densities imply that the Jovian planets are made of quite different stuff. The terrestrial planets are basically rocks and metals, made of elements such as iron, aluminum, oxygen, and silicon. Jupiter, in contrast, is made mostly of hydrogen and helium in a liquid and gaseous state.

This hydrogen and helium came together when Jupiter formed, and the giant planet has lost little, if any, since then. Why not? Jupiter's huge mass means that its escape speed (discussed in Section 4.5) is large, about 60 km/s. Remote from the sun, Jupiter's upper atmosphere is cold, only about 125 K, and hydrogen molecules there move at about 1 km/s, so even hydrogen molecules do not have escape speed. If the hydrogen cannot escape from the upper atmosphere, more massive atoms and molecules also cannot get away. Jupiter has retained its atmosphere for eons and will hold it for eons to come. What you see now is basically the atmosphere and mass with which Jupiter was born. (The earth and other terrestrial planets lost atmospheric materials early in their histories.)

Atmospheric Features and Composition
(Learning Outcome 10-2)

The visible disk of Jupiter (Fig. 10.2) is *not* the planet's surface but its upper atmosphere, which shows alternating strips of light and dark regions that run parallel to the equator. The light regions, called **zones,** have lower temperatures than the dark regions, called **belts.** The lower temperatures imply that the zones are higher than the belts because, in general, the temperature in a planet's atmosphere decreases with altitude. These differences in temperature suggest that the zones flag the tops of rising regions of high pressure, and the belts mark the descending areas of low pressure (Fig. 10.3). This convective atmos-

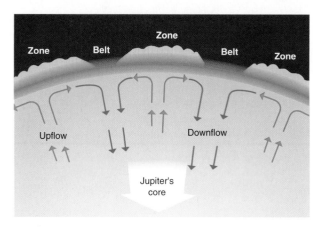

Figure 10.3 Circulation in Jupiter's upper atmosphere. Rising air creates high-pressure regions (called zones) and the downflow makes low-pressure areas (called belts). The zones generally appear lighter in color than the belts because they are higher up in the atmosphere. The convection is driven by heat flowing out of the planet's interior.

pheric flow transports energy out to space from the planet's interior and so implies that its interior is hot. These convective flows generate colossal thunderstorms, many times mightier than the earth's.

Markings in the clouds allow measurement of Jupiter's rate of rotation. (It can also be measured from the **Doppler shift** of the light emitted by the approaching and receding edges of Jupiter; see Enrichment Focus 10.1.) The rate varies with latitude: Jupiter spins faster at its equator than at its poles. Such variation of rotation rate is called **differential rotation** – it indicates that the body is fluid. (A solid body like the earth rotates so that each point in its surface has the same rotational period.) This rapid rotation and Jupiter's large radius produce an equatorial speed in excess of 43,000 km/h and makes the planet fairly oblate.

Such an enormous rotation speed drives the circulation in Jupiter's atmosphere. It causes the permanent high-pressure zones and low-pressure belts to stretch out completely around the planet. Jet streams zip along at the boundaries between the belts and zones, creating atmospheric disturbances. Typical wind speeds are 100 m/s – about three times faster than the earth's jet streams. These complex streams and swirls of Jupiter's upper cloud layer indicate a turbulent atmospheric flow.

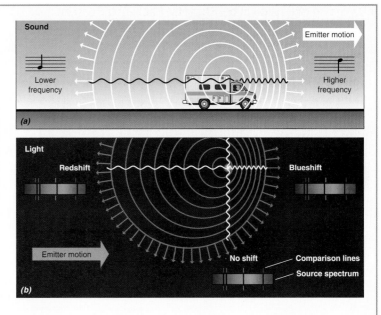

Figure F.11 The Doppler shift for waves of different types. (a) Sound waves from a moving truck's horn undergo a similar effect, increasing and decreasing in pitch as the truck approaches and then passes. (b) For light waves, the Doppler shift causes a change in the position of lines in the spectrum: a blueshift for approach and a redshift for recession.

ENRICHMENT FOCUS 10.1
The Doppler Shift

Essentially, the Doppler shift provides us with a cosmic speedometer that allows us to measure the speeds of objects moving toward or away from us. Let's see how it works.

The Doppler shift occurs with light and sound waves. Here's the essence of the Doppler shift: when you are moving *toward* a wave source, the waves appear to be more frequent and shorter in wavelength; in contrast, when you move *away* from a wave source, the waves appear to be less frequent and the wavelength longer. It's only the *relative velocity along the line of sight* – called the **radial velocity** – that causes the Doppler shift (for speeds much slower than that of light). Because velocities are *relative*, it makes no difference whether you're moving, the source is moving, or both. The Doppler shift is an apparent shift in the received wavelength when the source and receiver have a relative radial velocity. The Doppler shift does *not* depend on the distance between the observer and the source, only on their relative radial velocities.

You have probably heard the Doppler shift with sound waves (Fig. F.11). As an ambulance approaches you with its siren blasting, its pitch is higher than when it is at rest. Just as the ambulance passes you, its pitch is unshifted. Then as the ambulance moves away, its siren seems to put out a lower pitch. The same effect happens with light waves (Fig. F.11b), where we can measure the Doppler shift by the *change in the wavelength* of lines in a spectrum. The emission or absorption lines are shifted toward the blue end of the spectrum for an object approaching you and toward the red for one that is receding. These shifts are called a **blueshift** and a **redshift.**

Now apply this idea to radar observations of a planet. We observe the visible spectrum of Jupiter, for example. If we look at the lines in its spectrum, we see that those from the *receding* edge of the planet are redshifted (Fig. F.12). Those from the *approaching* edge are blueshifted. We measure the size of these shifts to find the speed of rotation. We apply the same concept to other planets and to their rings.

What about Venus, whose surface we cannot see? We then send radar signals of an exactly known frequency from the earth. They strike Venus (or any rotating planet) over its surface. Some waves bounce off and return to the earth for reception. Imagine viewing the planet along its equator. Then some of the reflected waves come from the edge of the planet that is approaching us. These are blueshifted. Waves reflected by the receding edge are redshifted. The *difference* in frequency between the original signals and the received ones tells us how fast the planet is rotating.

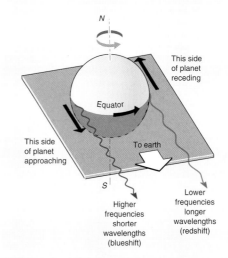

Figure F.12 Doppler shift from a rotating planet.

Figure 10.4 Close-up of the Great Red Spot and the turbulent region near it, taken by the Galileo spacecraft in 1995. The color of a cloud in this false-color image indicates its height above the planet: blue or black are deep clouds, pink are high, thin clouds, and white are high, thick clouds.

The most permanent and famous atmospheric disturbance is the Great Red Spot (Fig. 10.4), first observed in 1630. The Red Spot changes in size, averaging some 14,000 km wide and up to 40,000 km long – it could easily swallow the earth! The Red Spot is a few degrees cooler than the surrounding zone and pokes about a few kilometers above it, so it is a rising region of high pressure. The Red Spot rotates counterclockwise, just as expected from a high-pressure zone in Jupiter's southern hemisphere. It turns once in 7 days, a huge vortex pushed by the surrounding atmospheric flow. The Red Spot deflects nearby clouds and forces them around it like leaves in a whirlpool. Behind the Red Spot runs a region of turbulent flow from the atmosphere flowing past it. The Red Spot stands out as a long-lived (hundreds of years) atmospheric eddy.

Why is the Red Spot red? (Many other spots on Jupiter are white.) It is likely that colored compounds are made by chemical reactions, driven by the outflowing heat. In general, the colors seen in Jupiter's upper atmosphere (blue, red, and yellow) probably result from various chemical compounds formed there.

Jupiter's upper atmosphere contains by mass about 82 percent hydrogen, 18 percent helium, and traces of all other elements – essentially the same composition as the sun. Most of this material is in the form of molecules. Examples are methane, ammonia, molecular hydrogen, and water. The visible clouds at the tops of the zones are most likely ammonia ice crystals. Below them may float liquid ammonia and water ice clouds. The entire atmosphere may be 1000 km thick. In fact, no distinct boundary separates the atmosphere and interior. The atmosphere gets denser and hotter farther in, gradually merging into the liquid interior.

A Model of the Interior (Learning Outcomes 10-2, 10-3, and 10-11)

We can infer Jupiter's internal structure from physical models that include two key pieces of information: (1) Jupiter's low density and its atmospheric composition imply a solar mix of material throughout, and (2) Jupiter radiates into space more energy than it receives from the sun (about twice as much), so it must be hot inside. The internal heat is probably left over from Jupiter's formation. Large, massive planets lose heat very slowly.

Models of Jupiter, assuming a solar composition and internal heat, come up with a differentiated interior (Fig. 10.5a). The atmosphere covers the planet like a thin skin and consists mostly of molecular hydrogen. As one dives into the planet, the density, temperature, and pressure increase, so the hydrogen exists in a liquid state. At a pressure of about 2 million atmospheres, the hydrogen is squeezed so tightly that the molecules are separated into protons and electrons that move around freely and can conduct electricity. (This state, called *metallic hydrogen*, has been observed in a laboratory on the earth.) It continues to within about 14,000 km of the planet's center. Here, perhaps, if Jupiter does have a solar composition, lies a core of heavy elements (perhaps mostly rocky materials).

Most of Jupiter is hydrogen, and most of that

hydrogen is liquid – quite a contrast to the earth's interior (Section 8.2) and that of the other terrestrial planets. The core temperature may be about ten times hotter than the earth's core. The flow of the heat outward from the core, along with the rapid rotation, drives the circulation to produce the beautiful banded atmosphere.

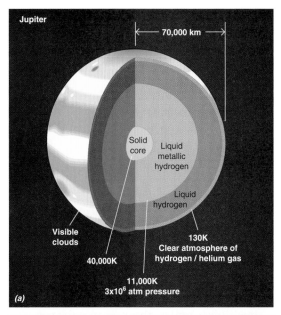

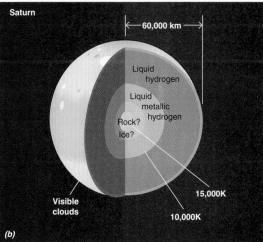

Figure 10.5 Jovian interiors. (a) Jupiter. Below the atmosphere is a thick layer of liquid molecular hydrogen; below that, a region of metallic hydrogen because of the high pressure (above 3×10^6 atm). The core is a dense material, perhaps silicates, that may be molten because it is quite hot, perhaps 40,000 K. The core may have a mass about 10 to 20 times that of the earth. (b) Saturn. Note that the metallic hydrogen zone is smaller compared to that of Jupiter. The inner, rocky core may have a mass of about 20 times that of the earth.

Magnetic Field
(Learning Outcomes 10-2 and 10-9)

Jupiter has a magnetic field some ten times as strong as the earth's. At the cloud tops, the field strength is about 4×10^{-4} T. Recall that the earth's magnetic field at the surface is about 5×10^{-5} T. (The **tesla,** T, is the SI unit of magnetic field strength: see Appendix A.) The magnetic field axis is tilted about 10° with respect to the rotation axis. The magnetic field rotates with Jupiter. Its period, 9 hours, 55 minutes, 30 seconds, is taken as the rotation of Jupiter as a body. This technique is used to infer the actual rotation periods of the Jovian planets.

Jupiter's strong magnetic field produces a magnetosphere much larger than that of the earth, though the formative processes are basically the same. As the solar wind plows into this intense field, it creates an enormous shock wave, called a *bow shock* that enshrouds the planet (Fig. 10.6). The magnetic field traps plasma particles from the sun in belts, similar to the Van Allen belts around the earth, close to Jupiter. As an "object" in the solar system, Jupiter's magnetic field is larger than the sun!

The sunward side field acts as a buffer to deflect the solar wind around Jupiter. On the night side, a magnetic tail stretches out to a length of a few astronomical units and may reach as far as Saturn! Electrons moving close to the speed of light in the inner regions of the magnetosphere generate radio emission. This radio source "outshines" all others in the sky.

How does Jupiter generate such an intense magnetic field? Recall that a dynamo model (Section 8.3) pictures currents in the liquid metal core generating the magnetic field like an electromagnet. For a large part of Jupiter's interior, liquid metallic hydrogen can conduct electric current. The conditions for a planetary dynamo to operate are here: a fluid able to conduct electricity, filled with convective currents driven by heat and rapid rotation. A dynamo in the metallic hydrogen zone could produce Jupiter's magnetic field – the model seems to work well here.

10.2 THE MANY MOONS AND RINGS OF JUPITER

Jupiter possesses an entourage of at least 28 moons (Appendix B, Table B.7). The brightest and largest, whose orbits lie within 3° of Jupiter's equatorial plane, were observed and reported by Galileo. These

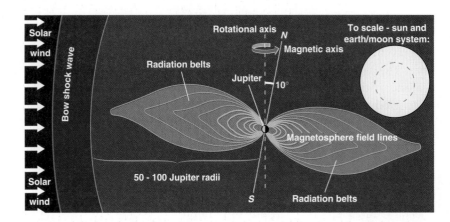

Figure 10.6 Model for Jupiter's magnetosphere based on spacecraft data. The size and shape vary according to the strength of the solar wind. The solar wind forms a bow shockwave where it interacts with the magnetic field; behind it lies the magnetosheath, a region of relative calm. Note that the tilt of the magnetic axis to the rotation axis (10°) is about the same as that for the earth. The sizes of the sun, and within it, the earth and moon system, are shown on the same scale at the upper right.

huge **Galilean moons** orbit within 2 million km of Jupiter. From the closest outward, they are Io, Europa, Ganymede, and Callisto. Like our moon, each is locked in synchronous rotation and so keeps one face toward Jupiter. These moons are among the largest in the solar system, along with Saturn's Titan and Neptune's Triton (Fig. 10.7).

Each Galilean moon is a world of its own; together, they exhibit dramatic differences in surface features and internal structures. Key differences are hinted at by the bulk densities of the moons: Io, 3.5; Europa, 3.0; Ganymede, 1.9; and Callisto, 1.8. Note the pattern: the density of the moons decreases with increasing distance from Jupiter. Such density differences show that the compositions of Io and Europa resemble that of our moon – mostly rock, with perhaps a little icy material. In contrast, Ganymede and Callisto contain substantial amounts of water ice or other low-density icy materials, and much less rock.

Io (Learning Outcomes 10-7 and 10-8)

Io has a thin atmosphere, composed mainly of sulfur dioxide. (Only two other satellites, Saturn's Titan and Neptune's Triton, are known to have an atmosphere.) At the surface, the atmospheric pressure is about 10^{-10} atm. Io's atmosphere gives off a bright, continuous glow of emission from sodium atoms. This sodium glow surrounds Io like a yellow halo out to a distance of about 30,000 km.

What produces Io's sodium cloud? Volcanic eruptions, at least in part. Io is the most active body in the solar system (Fig. 10.8); fuming lava lakes cover its surface. This activity implies that the interior is hot; sulfur and sulfur compounds melt and are vented in volcanic outflows. The colors of the surface indicate that the outflows contain liquid sulfur. Io's escape velocity (2.5 km/s) is greater than the speed with which the volcanic dust and gases erupt, so little direct loss of the material to space occurs.

Io's volcanoes have a different shape from those commonly found on the earth, Venus, and Mars. Few appear as cones or shields; rather, they resemble collapsed volcanic craters. Lava simply pours out of a crater vent and spreads outward for hundreds of kilometers. Dark-colored lava lakes, including one about the size of the island of Hawaii, surround many of Io's volcanoes. The temperature of these lava lakes is about 400 K. (The melting point of sulfur is 385 K.)

Figure 10.7 Family portrait of the Galilean moons of Jupiter to scale.

Figure 10.8 Volcanic plumes on Io, imaged by the Galileo spacecraft. One is on the bright edge, erupting from the Pillan Patera caldera, and one is towards the center (near the boundary between day and night), erupting from the Prometheus volcano. The Pillan Patera, a recent eruption, reached a height of about 140 km.

Why is Io's interior so hot? The gravitational effects of the other Galilean moons force Io into a slight perturbation of its circular orbit. The tidal forces of Jupiter (which are large because Io is close) cause Io to flex. These tidal stresses continually act on Io generating tides some 100 meters high in the solid surface! Its interior heats up from the recurring push and pull of the tidal forces, just as the continuous squeezing of a rubber ball will heat it. This heating drives the volcanism. The surface of Io must be very young, because volcanic activity continually alters it. No impact craters appear on Io; volcanic flows have submerged them. Io's surface seems to be the youngest in the solar system, probably less than 1 million years old.

Europa (Learning Outcomes 10-7 and 10-8)

Europa's surface shows bright areas of water ice among darker, orange-brown areas. It is crisscrossed by stripes and bands that are filled fractures in the moon's icy crust (Fig. 10.9). The dark markings crisscross its face, making it look like a cracked eggshell.

Some of these shallow cracks extend for thousands of kilometers, split to widths up to 200 km, but they reach depths of only 100 meters or so. Europa's surface is really incredibly smooth. In relation to its size, its dark markings are no deeper than the thickness of ink drawn on a Ping-Pong ball. This moon has the smoothest surface we have seen so far in the solar system.

Europa's surface is almost devoid of impact craters, so Europa's surface cannot be a primitive one; it has been resurfaced into a tortured landscape. Fault-crazed rafts of ice some 10 km across jumble across the terrain. Features here may be only tens of millions of years old. Beneath the icy veneer now lies a layer of liquid water – perhaps an ocean 10 km below the surface! Europa may possess more water in total than the earth.

Ganymede (Learning Outcomes 10-7 and 10-8)

Largest moon of Jupiter and in the solar system, Ganymede dons two basic types of terrain: cratered and grooved. Craters up to 150 km in size densely

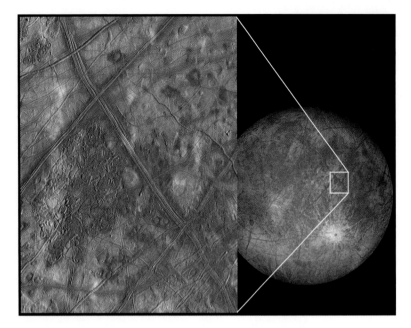

Figure 10.9 False-color image of Europa's crust shows blocks that have broken apart. Geologists ascribe this movement to subsurface oceans, which may still exist on Europa today. Icy plains appear blue; non–ice material is reddish brown. Long, dark lines are ridges and fractures in the crust.

mark the surface of the cratered terrain. Their abundance indicates that the cratered terrain is old, some 4 Gy. Compared with those on the moon and Mercury, the craters are shallow for their diameter – an indication of an icy rather than rocky surface. Very bright rays extend from many of the craters (Fig. 10.10). These features attest to impacts on an icy surface at the time of formation.

The grooved terrain consists of a mosaic of light ridges and darker grooves where the ground has slid, sheared, and torn apart. Long cracks, where the ground has moved sideways for hundreds of kilometers, also cover the surface.

No large mountainous regions or large basins relieve the terrain on Ganymede; no relief is greater than about 1 km. This suggests that the crust of Ganymede is somewhat plastic, probably because of the large amount of water ice (perhaps 90 percent) in the crust. Ganymede's bulk density is only about 1.9, so it contains about half water and half rock overall. Occasional stresses on the water-rock crust have created the fracture patterns. Some ridges and grooves overlie others, an indication that there may have been many episodes of crustal deformation in the past. Basically, Ganymede has been geologically active for less of its life than Io and Europa, but more than Callisto.

Callisto
(Learning Outcomes 10-7 and 10-8)

Farthest out of the Galilean moons, Callisto has a surface riddled with craters in a wide range of sizes. Some have bright ice rays; others are filled with ice. Callisto's craters are shallow – several hundred meters deep or less. Why? Because the surface is a mixture of ice and rock, the surface slowly flows, flattening out the land. Overall, Callisto has not had the spirited internal activity of the other Galilean moons.

Callisto sports a huge, multiringed basin called Valhalla (Fig. 10.11a). Its central floor is 600 km in diameter (about the east–west width of Germany). Twenty to thirty mountainous rings that have diameters of up to 3000 km surround it like a bull's eye. The rings look like a series of frozen waves. They were formed in a stupendous collision that melted subsurface ice, causing the water to spread in waves that quickly froze in the 95 K surface temperature. The ripples are preserved as rings. The central floor has fewer craters than the rest of the terrain, an indication that the impact occurred after much of the initial cratering had been accomplished (Fig 10.11b).

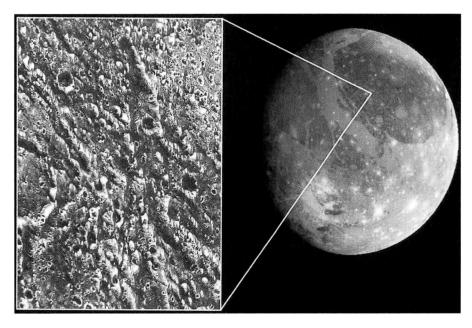

Figure 10.10 Modification of impact craters on Ganymede, imaged by the Galileo spacecraft, indicates the vast age of the moon – several billion years. The surface is dirty water ice. Dark furrows gouged into the crust may be fragments of material collected from meteorite impacts; light furrows are more recent.

The Rings of Jupiter
(Learning Outcome 10-4)

Jupiter has millions of moons – tiny ones that make up its **ring system.** The rings are so thin (less than 30 km thick) that they are essentially transparent. The ring particles scatter light well when backlit; this implies that the particles are small (about 10 μm). The best view of such low-density rings of small particles occurs when we view them edge-on and backlit.

Dramatic pictures of the back-lit rings (Fig. 10.12) show the rings' structure. The main ring lies about 125,000 km from Jupiter's center and is 7000 km wide. Within that ring lies a faint sheet of material that extends down to the cloud tops. A faint, outer ring surrounds the whole system and reaches out to 225,000 km – more than three times the radius of Jupiter.

10.3 SATURN: JEWEL OF THE SOLAR SYSTEM

Saturn bears a marked affinity to Jupiter, but its fantastic ring system outranks in splendor that of the larger planet (Fig. 10.13). Saturn is only slightly smaller, but it has only about a third the mass of Jupiter. Why? Because Saturn has the lowest density of any of the planets – about 0.7, which is much less than that of water. Saturn could float! It must have an interior rich in light elements and lacking in rocky materials.

Atmosphere and Interior
(Learning Outcomes 10-1, 10-2, and 10-11)

The atmospheric structure of Saturn resembles that of Jupiter: belts running parallel to the equator, ≠driven by its rapid rotation. Saturn's rotational period is 10 hours and 14 minutes at the equator and varies with latitude; it, too, shows differential rotation. Disturbances in the belts rarely occur compared with the frequency of such events on Jupiter. Mighty storms do strike at intervals of about 30 years, roughly at midsummer in the northern hemisphere.

The atmosphere of Saturn has roughly the same composition as that of Jupiter: mostly hydrogen and helium. Spacecraft observations indicate that about half the percentage of helium is found in Jupiter's upper atmosphere. Methane, water vapor, and ammonia make up a minority of the gases.

Saturn's clouds appear far less colorful than those of Jupiter – mostly a faint yellow and orange.

Figure 10.11 Callisto. (a) Giant ringed basin on Callisto. Centered on the smooth, bright area near the center of the image, this impact basin is surrounded by concentric rings up to 1700 km across. Note the lack of ridges or mountains surrounding this impact basin, which is called *Asgard*.

Because of the lower temperatures on Saturn compared to Jupiter, the clouds lie lower in the atmosphere, and a high-altitude haze subdues our view. However, jet streams produce complex cloud patterns (Fig. 10.14), with wind speeds up to 500 m/s near the equator. Weather patterns change weekly.

Saturn's interior (go back to Fig. 10.5b) probably reflects Jupiter's composition – roughly the same as that of the sun. Saturn may have a small, rocky core some 20,000 km in diameter and a mass of about 20 earth masses (or it may have no such core at all!). It probably has a large zone of liquid hydrogen and a smaller one of metallic liquid hydrogen. So, like Jupiter, much of Saturn's interior is probably in a liquid state.

Similarities to Jupiter
(Learning Outcomes 10-2 and 10-9)

Saturn resembles Jupiter in two other important respects. First, infrared observations show that Saturn emits more energy, as infrared radiation, than it receives from the sun – about twice as much. As with Jupiter, this excess heat may be left over from the planet's formation.

Second, radio and spacecraft observations of Saturn show that it, too, has a strong magnetic field and a large magnetosphere. The magnetic axis aligns within one degree of Saturn's rotation axis. The magnetic field is probably produced by a dynamo effect in the liquid metallic hydrogen zone

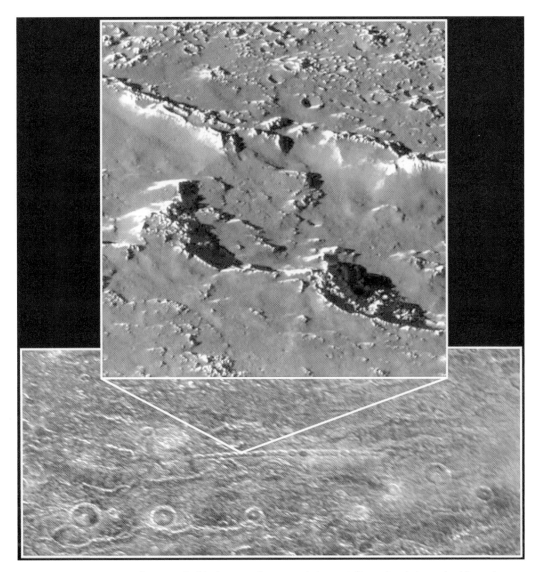

Figure 10.11 Callisto *(continued)*. (b) Close-up of a crater chain on Callisto. The chain resulted from the impact of a split object. The view is 13 km across; the smallest visible crater is about 130 m in diameter.

of Saturn, in the same way it is presumably produced in Jupiter. The magnetic field creates Van Allen-like belts around Saturn, which trap charged particles from the sun.

10.4 THE MOONS AND RINGS OF SATURN
(Learning Outcomes 10-8 and 10-10)

Saturn's retinue of moons totals at least 30 (Appendix B, Table B.8). With two exceptions (Phoebe and Iapetus), all the moons stick close to Saturn's equatorial plane. Masses for some smaller moons were determined from their gravitational attraction on spacecraft. The densities range from 1.2 for Tethys and a few others to 1.9 for Titan, similar to the densities of the outer, low-density Galilean moons of Jupiter.

The moons of Saturn fall into three groups: big Titan by itself; the six large icy moons (Mimas, Enceladus, Tethys, Dione, Rhea, and Iapetus, in their order outward from Saturn); and the small moons (Phoebe, Hyperion, and the rest).

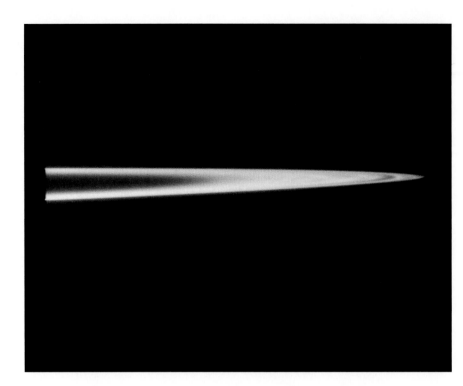

Figure 10.12 Main ring of Jupiter, imaged by the Galileo spacecraft in 1996. Note the bright, sharp edge and diffuse inner region. The small ice particles in the ring reflect light to make it visible.

Figure 10.13 Saturn viewed by the HST in 1998. This false-color image was taken in infrared light. The bands in the atmosphere are much less prominent than those of Jupiter.

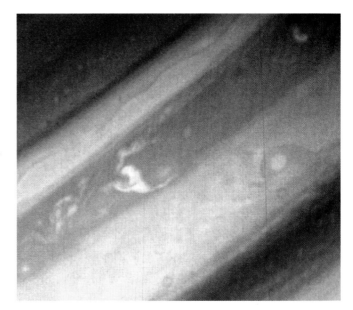

Figure 10.14 Image of the northern hemisphere of Saturn, showing jet streams and turbulent wind flows; winds here can reach speeds of 100 m/s.

Overall, their densities are less than 2.0, which implies that the moons are mostly ice (60 to 70 percent) with some rock (30 to 40 percent). In contrast to the Galilean moons, the densities do *not* decrease with distance from Saturn. Like Jupiter's moons, all except one (Phoebe) keep the same face toward Saturn.

Most of the moons are cratered. Some cratered terrain has been modified on the larger moons, which implies that internal heating has melted parts of the icy surfaces. In contrast, the smaller moons, which are also densely cratered, show no changes – they still have their original surfaces.

With these basics in mind, let's look at selected moons in some detail. This inquiry will reveal a key aspect of the outer solar system: many bodies contain ices (frozen volatiles of various kinds) and a dark, carbon-rich rocky material. These light and dark substances dominate the cold regions of the solar system. They are very likely primitive materials from the era of planetary formation.

Titan
(Learning Outcomes 10-8 and 10-10)

Titan, Saturn's largest moon, has a mass about twice that of our moon and a diameter of 5150 km, about 50 percent larger than our moon. Titan's density is about 1.9, which implies a 50-50 composition of ice and rock.

Titan was the first moon found to have an atmosphere. It consists mostly of molecular nitrogen (some 80 percent), with about 1 percent methane and perhaps a trace of argon. Several hydrocarbons other than methane have also been detected, including ethane, acetylene, and ethylene. The surface temperature is 94 K; the atmosphere's surface pressure, about 1.5 atm. That's an incredibly thick atmosphere for its size!

Spacecraft images (Fig. 10.15) showed a layer of orange smog as well as a blue color along Titan's edge. This hue indicates that the atmosphere varies in composition. The surface was completely obscured, but the pressure and temperature data, along with the spectroscopic detections of nitrogen and hydrocarbons, have led to models of a surface covered with a frigid ocean of ethane, methane, and nitrogen up to a depth of 1 km, beneath which may reside a layer of acetylene.

Other Moons

Saturn's four largest moons, after Titan, are Iapetus, Rhea, Dione, and Tethys, with diameters ranging from some 1100 km to 1500 km. (Mimas and Enceladus are much smaller than these four.) Their surfaces are heavily cratered. Except for Iapetus, they have bright, icy surfaces.

The rest of the moons are small, heavily cratered bodies, 300 km or less in diameter. We presume that all these bodies are basically ice, as are the larger moons (except for Titan).

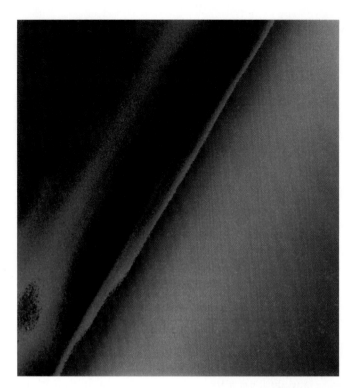

Figure 10.15 Haze layers in the upper atmosphere of Titan appear in this false-color image. The upper level appears orange and lies about 500 km above the clouds.

Figure 10.16 Schematic diagram of Saturn's ring system and the orbits of some of its moons as seen from above Saturn's north pole. The satellites Janus and Epimetheus essentially share the same orbit. Prometheus and Pandora act as shepherd satellites for the F ring. Distances are given in units of the planet's radius (R_s).

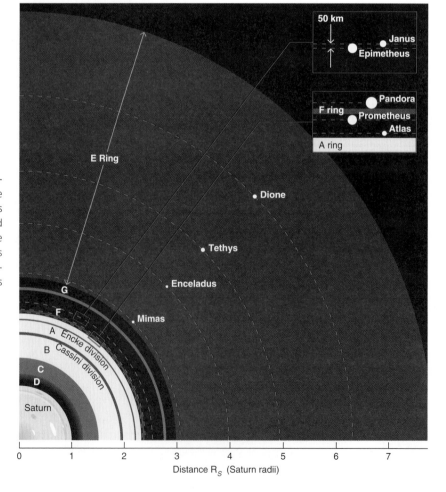

Figure 10.17 False-color image of Saturn's B and C rings, taken by Voyager 2 from a distance of 2.7 million km. The image processing emphasizes the difference in color between the B ring (yellow) and the C ring (blue). These are not the actual colors of the particles in the rings; rather, the differences indicate that the surface compositions of the particles in these two rings must be different.

Ring System (Learning Outcome 10-4)

Saturn's rings lie tipped about 26° to the orbital plane. Because of their tilt, they change their appearance when viewed from the earth during the course of Saturn's revolution about the sun. The near disappearance of the edge-on rings indicates that they are very thin, no more than a few kilometers thick. At their edges, the rings are only a few hundred meters in thickness. Although thin, the rings are wide; the three main rings visible from the earth (called A, B, and C) reach from 71,000 to 140,000 km from Saturn's center (Fig. 10.16).

Voyager images revealed spectacular detail in the ring system. Although the A ring is relatively smooth, the B and C rings break up into numerous small **ringlets** (Fig. 10.17). Many hundreds, perhaps a thousand, of light and dark ringlets surround the planet, with widths as small as 2 km (and possibly smaller). Even the Cassini division, apparently empty as seen from earth, contained at least 20 ringlets.

The rotation rate of the rings varies in a regular way, as measured by the Doppler shift (Enrichment Focus 10.1). The speeds range from 16 km/s at the outer boundary of the A ring to 20 km/s at the inner boundary of the B ring. These speeds agree with those expected from Kepler's third law for individual masses placed at the ring distances from Saturn. This conformity indicates that separate particles comprise the rings and orbit Saturn just as the planets orbit the sun. The same is true for all the rings of the Jovian planets. (If the rings were solid like a record, the rotational speed of the outer edge would be higher than that of the inner edge.)

Infrared observations show that Saturn's rings are made of particles of water ice or rocky particles coated with water ice. The ice does not evaporate because the surface temperature of the particles is only about 70 K. The particles average about one meter in diameter, but they range in size from centimeters to tens of meters – ice lumps the size of golf balls to boulders as big as a house.

How do the rings remain stable when they consist of myriad particles following their own paths? The gravitational effects of satellites play a key role in the dynamics of the rings. Consider the two satellites that straddle the F ring (Prometheus and Pandora). These F-ring moons, in particular, are called the **shepherd satellites** because they keep the ring particles herded together in a narrow range of orbits. The inner moon accelerates the inner ring particles as it passes them (as expected from Kepler's third law, it has a shorter orbital period). These particles spiral outward to larger orbits just as the tidal force of the earth on the moon forces the moon into a larger orbit.

In a physically similar way, the more slowly moving outer moon decelerates outer-ring particles as they pass by, with the result that they spiral inward. The balance of these interactions constrains the particles' motions and preserves the narrowness of the F ring. Because tidal forces tend to spread rings out, shepherd satellites work to preserve the sharp edges. The "herding" of ring particles seems common for the rings of the Jovian planets.

Although deceptively solid in appearance and covering a large area of space, the rings have a total mass estimated to be just 10^{16} kg, only about 10^{-6} the mass of our moon, and a mere 10^{-10} the mass of Saturn.

10.5 URANUS: THE FIRST NEW WORLD

On March 13, 1781, the then-unknown English astronomer William Herschel perceived a star "visibly larger than the rest" in the constellation of Gemini and "suspected it to be a comet." Observations later in March and in April proved that the object's orbit was not like that of a comet. Herschel concluded that he had found a new planet – the seventh in the solar system and the first to be discovered with a telescope. He named it Uranus after the mythological father of the Greek god Saturn.

Atmospheric and Physical Features (Learning Outcomes 10-1, 10-2, and 10-11)

At an average distance of 19.2 AU, it takes Uranus 84 terrestrial years to journey around the sun. Far from the sun, the upper atmosphere is very cold – 58 K. Like Jupiter and Saturn, the atmosphere contains mostly molecular hydrogen and helium. Uranus has a distinctive bluish-green color, which comes from sunlight that penetrates deep into the planet's atmosphere; some red light is absorbed in the atmosphere, and much of the blue and green is reflected back into space. This color is expected from an atmosphere that contains methane gas.

Spacecraft images (Fig. 10.18) showed ammonia clouds lying below a thick layer of haze. The clouds have a delicately banded structure. Winds blow the clouds in the same direction as the planet rotates. Occasional plumes of clouds appeared in the upper atmosphere, perhaps produced by violent upflows. The rotation at the cloud tops varies from 17 hours at the equator to 15 hours near the poles. Uranus' rotation axis lies almost in the plane of its orbit – it spins on its side! Journeying around the sun in this lopsided manner (Fig. 10.19), Uranus exposes each pole to sunlight for 42 years at a time; night at the opposite pole lasts equally long.

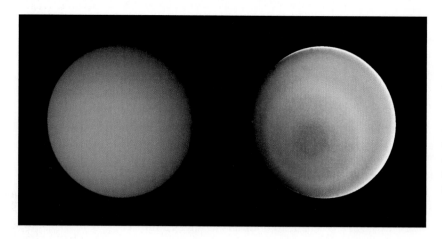

Figure 10.18 Uranus. The image at left is processed to show the planet as it would appear to the eye; the bluish green color results from the absorption of red light by methane in the atmosphere. The right-hand image uses false colors to bring out details in the structure of the upper atmosphere. Note the dark hood over the south pole with a series of concentric rings.

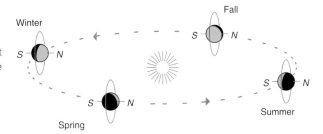

Figure 10.19 Orientation of the spin axis of Uranus as the planet orbits the sun, and the seasons in the southern hemisphere. Note how the axis of rotation lies in the plane of the orbit.

The low bulk density of Uranus, 1.3, implies that it contains mostly lightweight elements (Fig. 10.20). Uranus consists of roughly 15 percent hydrogen and helium, 60 percent icy materials (water, methane, and ammonia), and 25 percent earthy materials (silicates and iron). Its internal structure differs greatly from that of Jupiter or Saturn. It may have a very small icy-rock core, encased in an icy or watery mantle (perhaps with some rocky material).

Moons and Rings
(Learning Outcomes 10-4, 10-8, and 10-10)

Five moons were known for Uranus before spacecraft missions: Miranda, Ariel, Umbriel, Titania, and Oberon – all named for characters in Shakespeare's plays. They all move in the planet's equatorial plane and revolve in the direction of the planet's rotation.

Because the moons lie in the same plane as Uranus' equator, their orbits as seen from the earth are alternately edge-on, then fully open, every 21 years; in 1966 they appeared edge-on, but in 1987 they appeared face-on.

Miranda is the smallest moon and the closest to Uranus. The others range in diameter from some 1100 km to 1600 km. Their surfaces appear to be made of a dirty ice, very much like that of Saturn's Hyperion. The bulk densities range from about 1.4 to 1.7, which implies these are bodies made of rock and ice.

Miranda has the most complex surface, with many different types of terrain jumbled together, including grooved regions, faults, and 5-km-high cliffs – certainly the strangest surface seen to date in the solar system (Fig. 10.21). How could such a small moon be so deformed? At some time, it must have had a hot interior, but the source of that heating has not yet been figured out. Tidal stress is a possible culprit.

Figure 10.20 Model for the interior of Uranus. It consists of three main regions: a very small rocky core (density about 8000 kg/m³); an icy (or watery) region with perhaps some rocky material (density from 5000 to 1000 kg/m³); and a gas layer, composed mostly of hydrogen and helium, with enhanced concentrations of methane, ammonia, and water (density about 300 kg/m³). The rocky core, if it exists, may have a mass about equal to that of the earth.

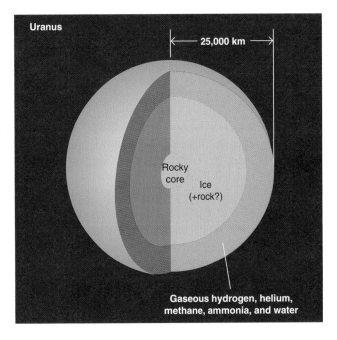

Figure 10.21 Surface of Miranda, imaged by Voyager 2 in 1986. Two distinct types of terrain are visible: grooved, low-elevation area (left) and older, high-elevation, heavily cratered area (right). The largest impact crater (bottom) is about 25 km in diameter. The area of the image is about 250 km across.

The other large moons appear similar in surfaces and evolution. Oberon has a dense cover of impact craters (an ancient surface) and at least one mountain (possibly volcanic) about 5 km high. Scarps and faults also spit the surface. Titania also has a surface plastered with many impact craters, strewn with valleys 100 km wide and hundreds of kilometers long. Also visible are multiringed impact basins. Ariel has impact craters, large fractures, and valleys. Some regions appear to be resurfaced by ice floes. Finally, Umbriel has the least dramatic surface, with overlapping impact craters. The heavy cratering is more evidence of an era of torrential impacts early in the solar system's history.

Ten other smaller moons have been named after other characters in Shakespeare. The largest (called Puck) is only 170 km in diameter and orbits inside the already-known moons. Six moons (Appendix B, Table B.9) orbit between Uranus and Puck. These inner moons have diameters between 40 and 80 km. The other moons are near the edge of the ring system and have dark surfaces with low albe-

dos. These serve as shepherd satellites to keep the rings stable. At least eleven rings are now known. The smallest rings have widths of only a few kilometers; the largest is about 100 km wide (the epsilon ring, Fig. 10.22).

The material that makes up Uranus' rings is extremely dark – the albedo is 5 percent. In contrast, because they are covered with water ice, the particles in the rings of Saturn have an albedo of more than 80 percent. The particles in the rings of Uranus, therefore, are probably bare of ice and made of dark carbon materials. Most of the particles in the rings are a few meters across, but many smaller particles are distributed throughout the Uranian ring system.

Magnetic Field (Learning Outcome 10-9)

Uranus' magnetic field is tilted 59° with respect to the rotational axis, with the north magnetic pole closest to the south geographic one. The magnetic field turns once in about 17 hours, 20 minutes, and

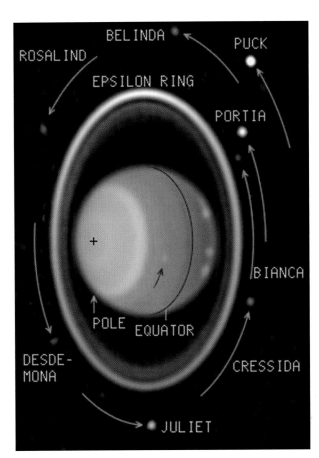

BELINDA

ROSALIND

PUCK

EPSILON RING

PORTIA

BIANCA

POLE   EQUATOR

DESDE-
MONA

CRESSIDA

JULIET

Figure 10.22 False-color HST images of Uranus with rings and satellites, taken in 1997. The rotation of the planet and revolution of eight satellites in a 90-minute period is visible.

that has been taken as the planet's rotation period. (Why the magnetic field has such a large tilt is a puzzle. Moreover, the field does not center on the core of Uranus!) The magnetic field of Uranus is strong enough to develop a stable magnetosphere.

10.6 NEPTUNE: GUARDIAN OF THE DEEP

The eighth planet from the sun, Neptune was discovered in 1846 because of its gravitational effects on the orbit of Uranus (Section 4.5). Calculations of Neptune's orbit show that it should have been very close to Jupiter in the sky in January 1613. Galileo's journals have entries showing that he observed an object in the vicinity of Jupiter near Neptune's predicted position on December 27, 1612, and again on January 28, 1613, when he detected a small motion of Neptune with respect to a nearby star. Inexplica-

bly, Galileo never followed up on this discovery and so failed to recognize the object as a new planet.

Physical Properties
(Learning Outcomes 10-1, 10-2, 10-9, and 10-11)

Far from the sun, Neptune revolves once in 165 years. Like Uranus, Neptune shows a light bluish color (from methane in the atmosphere). The main constituents are molecular hydrogen and helium; methane makes up a minor amount. The upper atmosphere displays distinct cloud bands (Fig. 10.23).

Infrared observations show that Neptune's temperature is about 60 K. If Neptune were heated by the sun alone, we would expect a value of 44 K, so Neptune, unlike Uranus, has internal heat. It gives off three times as much energy as it receives from the sun; the heat flow is about 0.3 W/m^2. The thermal energy emitted most likely was left over from the planet's formation.

Moons and Rings
(Learning Outcomes 10-8 and 10-10)

The largest moons of Neptune are Triton and Nereid. Triton has a diameter of about 2700 km. Nereid has a size of only some 340 km. Triton is one of the few moons with an atmosphere – a very thin one that contains mostly nitrogen with traces of methane.

Triton (Fig. 10.24) displays a fascinating pink and blue face. The cratering here is not too heavy, which means that the surface must be relatively young and recently modified – subject to meltings and refreezings. The overall temperature is low, a mere 37 K. On parts of the surface lie frozen ice lakes, shaped like lunar maria. Some are stepped, which suggests a series of meltings and refreezings, but in general the surface relief is quite low – less than 200 meters. Triton's atmosphere is about 800 km thick; it contains mostly nitrogen with a trace of methane.

Near Triton's south pole (which is now in a summer season), the surface ice, consisting of methane and nitrogen, appears to have evaporated in spots. In other regions, small flows have filled valleys and fissures – slow-moving glaciers of methane and nitrogen. In other sections, the icy surface appears to have melted and collapsed.

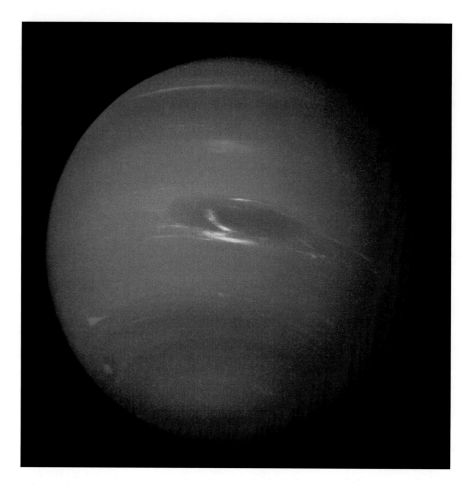

Figure 10.23 Neptune. The planet's bluish color is caused by the presence of methane in its atmosphere. Note the lack of a general banded structure in the atmosphere. The Great Dark Spot is near the center.

Figure 10.24 High-resolution mosaic of the portion of Triton's hemisphere that faces Neptune. The southern pole cap (at bottom center), containing methane ice, is slightly pink. The bluish green area consists of nitrogen frost; Triton is so cold that most of its nitrogen is condensed as frost.

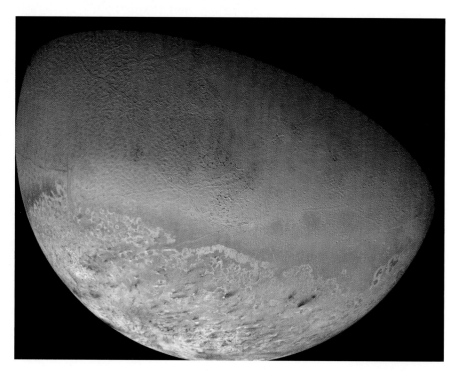

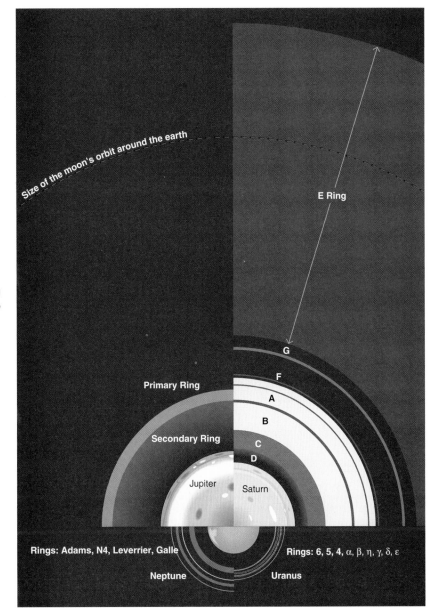

Figure 10.25 Comparison of the ring systems of the Jovian planets, drawn to relative scale.

Dark elongated streaks tens of kilometers across seem to be the trails of ice volcanoes. Just 30 meters below the surface, the pressure is high enough to liquefy nitrogen ice. When the surface cracks, the liquid can burst out and turn to gas, which shoots up several kilometers above the surface. The vapor condenses and falls back down, creating a thin, dark layer on the surface.

Triton's bulk density turns out to be 2.1 (about the same as Pluto). That implies a rocky core surrounded by a mantle of methane and water ice. The icy surface, which is primarily molecular nitrogen (N_2) ice, results in an average albedo of roughly 70 percent.

Neptune's orbit results in a seasonal cycle of 165 years. So far from the sun, the surface remains so cold in summer that regions of nitrogen ice are still visible.

Neptune's ring system contains four individual rings. The two brightest, outer rings have radii of 53,000 and 62,000 km with a tenuous ring spread between them. Closer in lies a wide ring some 2000 km across. The outer ring is clumpy, with three brighter segments appearing along a fainter but complete ring, like sausages on a string. We now know that Saturn's ring system is the only exception among the Jovian planets in terms of its broad extent (Fig. 10.25).

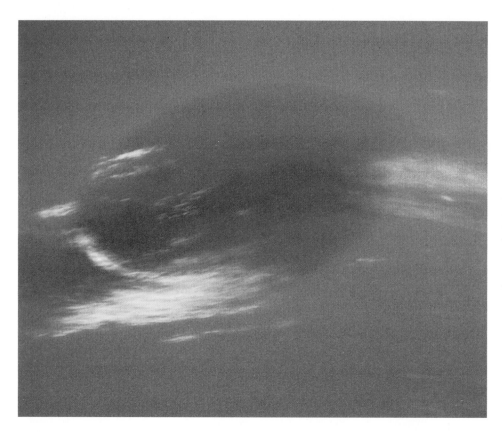

Figure 10.26 Neptune's Great Dark Spot, accompanied by white, high-altitude clouds. These clouds change their appearances during periods as short as two hours. Relative to the planet's core, the Dark Spot moves westward; it rotates counterclockwise.

One other point about the ring systems of the Jovian planets: all of them lie within the tidal limit for the parent planets. So it is unlikely that any of the rings' pieces will accrete to make a larger mass. We then have limited scenarios for the origin of the rings. One, they could be material left over from the planet's formation. Since they lie within the tidal limit, they would not accrete to make satellites. Two, they could result from a body or bodies that passed within the tidal limit and were tidally ripped into pieces. Of these two notions, I feel that the first is the most workable.

Atmospheric Features
(Learning Outcome 10-2)

The most conspicuous marking in the upper atmosphere of Neptune is the Great Dark Spot (Fig. 10.26), which is a storm some 30,000 km across, rotating counterclockwise in a few days. A region of high pressure, the Dark Spot lacks the typical atmospheric methane; here, we are looking deep into Neptune's atmosphere.

Bright, cirruslike clouds accompany the Dark Spot and also appear in some other latitude bands. Most of these clouds change size or shape from one rotation to the next. Believed to be condensed methane, the clouds lie about 50 km above the general cloud layer, which consists of hydrogen sulfide. Compared to that on bland Uranus, the atmospheric activity on Neptune came as a surprise. It is likely driven by the outflow of Neptune's internal heat. A few other dark and bright spots are present, but the complex swirls and banded structure seen on both Jupiter and Saturn seem to be absent on Neptune.

Magnetic Field (Learning Outcome 10-9)

Surprise! The magnetic axis is tilted about 47° from Neptune's axis of rotation, almost as much as the tilt of Uranus' magnetic axis. The reason for

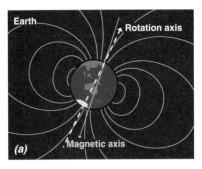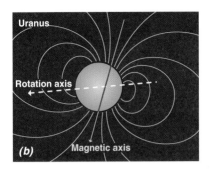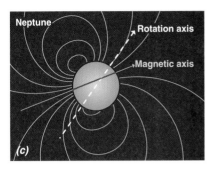

Figure 10.27 Comparison of the magnetic fields of the earth, Uranus, and Neptune. The fields are modeled as if a bar magnet were set in the planet's interior, with the angle and position inferred from the magnetosphere. Note how the fields of Uranus and Neptune are offset from the centers of the planets and grossly misaligned with the rotation axes.

these large tilts is not yet known. The magnetic field strength at the equator is about the same as that of the earth's. The dipole is strangely offset about four-tenths of a Neptune radius toward the south pole from the planet's center (Fig. 10.27). The magnetosphere has a very low density of trapped charged particles. Tracking radio signals from Neptune's magnetosphere gives an accurate measure of the rotation period: 16 hours, 3 minutes.

10.7 PLUTO AND CHARON: GUARDIANS OF THE DARK

Early in the twentieth century, Percival Lowell became fascinated with the problem of a planet beyond Neptune (called Planet X) and initiated a search program at Lowell Observatory in Arizona. When Lowell died in 1916, the search for Planet X was terminated pending the completion of a new telescope, which occurred in 1929.

Clyde W. Tombaugh (1906–1997) worked at the new search, which started on April 1, 1929. Bearing in mind the way Neptune was discovered (Section 4.5), most astronomers assumed that Planet X would resemble Neptune because of irregularities in Neptune's orbit, so they searched for a visible disk of a new planet. Instead of concentrating on a disk, however, Tombaugh looked for Planet X's motion relative to the background stars.

The photographic search was tedious, but on February 18, 1930, Tombaugh noted two images on different photographs, in the area near a star in Gemini, that had shifted slightly. The shift was such that the object had to be a body orbiting the sun. The discovery was announced within a month, on March 13 – Lowell's birth date. Lowell Observatory officially accepted the name Pluto.

Orbital and Physical Properties (Learning Outcomes 10-2, 10-5, and 10-6)

Pluto's average distance from the sun is 39.44 AU in a highly eccentric orbit. It is never closer to the earth than 28.7 AU at closest approach (opposition). In January 1979, Pluto edged closer to the sun than Neptune. It orbited closer to the sun than Neptune until March 1999. However, because of the high inclination of its orbit, Pluto is actually well above Neptune's orbital plane. So there is no danger of a collision!

Methane ice coats Pluto's surface, which means that the surface is bitter cold, no more than 60 K even during the daytime. Recent observations show nitrogen and carbon monoxide ices as well as methane. The overall albedo is about 50 percent. Observations of Pluto's brightness have uncovered a cycle (because of patches of the ice) about every 6.4 days. This variation is generally accepted as Pluto's rotation period.

Pluto has an atmosphere, which stretches over 600 km from the planet's surface. This atmosphere probably consists of nitrogen, carbon monoxide, and methane gas (with a surface pressure of a mere 10^{-8} atm or so) that has been released from the ice on the surface as the planet is heated by its closest approach to the sun in 248 years.

Charon: Pluto's Companion Planet (Learning Outcome 10-6)

In June 1978, James Christy of the U.S. Naval Observatory in Flagstaff, Arizona, noticed what appeared to be a bump on Pluto's image in a photo. Checking older photos, Christy found seven showing the same bump, always oriented approximately north–south. He proposed that the bump was the faint image of Pluto's moon partially merged with the image of the planet. Christy named this moon Charon, after the mythological Greek boatman who ferried the souls of the dead across the river Styx to Hades, where Pluto sat in judgment. HST took an image that clearly showed Pluto and Charon (Fig. 10.28) but could not reveal any details on their surfaces.

A few years after discovery, Pluto and Charon eclipsed each other as seen from earth. (Such alignments occur only twice each 248 years.) These eclipses indicate that Pluto has a diameter of some 2300 km; Charon's diameter is roughly 1200 km (Fig. 10.29). The observations of Charon show a revolution period of 6.4 days (the same as Pluto's rotation period, so it is in synchronous rotation) at a distance of 19,100 km from

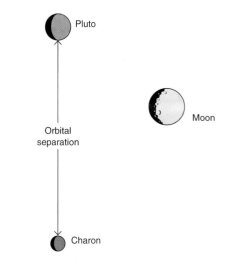

Figure 10.29 Pluto–Charon system drawn to scale, with the earth's moon shown on the same scale.

Pluto. That the revolution period of Charon is the same as Pluto's rotation indicates that the two bodies are tidally locked. We can use Charon to find Pluto's mass by Kepler's third law. The result: Pluto has a mass about 0.002 that of the earth or one-fifth that of our moon. Charon's mass is about one-fifth that of Pluto. This mass ratio of 5 to 1 for Pluto and Charon means that the center of mass of their binary system lies outside the body of Pluto, so they really do make up a double-planet system! (In contrast, the center of mass of the earth–moon system lies within the earth.)

Together, Pluto and Charon have an average density of some 2.1 – much higher than earlier estimates. Pluto may contain as much as 75 percent rocky material and may have a density of 2.0. Contrast the Jovian moons, which have a larger percentage of water ice. Charon's density might fall as low as 1.2 (or even less); recent infrared observations revealed water frost on its surface, so Charon-like moons of Saturn – may well consist mostly of icy materials. Charon has a dark gray surface in contrast to Pluto's reddish one. In addition, the equatorial region of Pluto is darker than its polar caps.

The idea that Pluto is a small, rocky-ice planet fits with a speculation that Pluto suffered a giant impact early in its history. A collision with a large

Figure 10.28 Pluto and Charon. At the time of this HST observation done with the Faint Object Camera (FOC) in 1994, Pluto (lower left) was 4.4 Gkm from earth. Charon (upper right) is bluer than Pluto.

body stripped Pluto of some of its icy materials and changed the rock-to-ice ratio from its original lower-density (like an icy) Jovian moon. The analogy here is to a giant impact stripping the rocky mantle of Mercury, leaving behind a denser object. So Pluto's density is higher than expected.

The search at Lowell Observatory for planets beyond Neptune ended in 1945 with no results. Tombaugh's hunt, which continued for 13 years, would have uncovered a planet like Neptune as far as 100 AU from the sun. Later searches also failed to find a body as large as Pluto.

But Pluto and Charon are not lonely hearts at the ends of the solar system. In recent years, about a dozen small bodies have been discovered near or beyond Neptune. They seem to resemble asteroids. A few have orbits that cross those of Neptune and Pluto, in fact, with semimajor axes of about 39 AUs and periods about 246 years! Though their motions have a lot in common with Pluto, they are much smaller in size and mass.

KEY CONCEPTS

1. As a group, the Jovian planets differ most from the terrestrial ones on the basis of their lower densities, greater diameters, and greater masses. Also, in chemical composition they are more like the sun than the earth. Jupiter and Saturn are mostly fluids in their differentiated interiors.

2. The Jovian planets are large masses that consist mainly of hydrogen and helium, with common molecules of methane, hydrogen, ammonia, and water. Their visible disks show the tops of their atmospheres; they have no solid surfaces. The large escape speeds and low temperatures of the Jovian planets imply that their atmospheres have changed little since their time of formation.

3. Jupiter, Saturn, and Neptune emit into space more heat than the energy they receive from incoming sunlight, so their interiors must be hotter than would be the case if the sun were the sole source of their heating. Their internal heat was probably generated during their formation. The outflow of this heat drives the evolution of these planets. However, because they do not have solid surfaces, evidence of an evolutionary history has not been preserved.

4. Jupiter, Saturn, Uranus, and Neptune have strong planetary magnetic fields, which implies that they have liquid, conducting interiors – but their composition rules out a liquid metal core like the earth's. For Jupiter and Saturn, dynamos likely arise from a zone of liquid metallic hydrogen, with organized flows set up by these planets' rapid rotations.

5. Jupiter's largest moons decrease in density outward from the planet, an indication that at some time in the past Jupiter was hot enough to affect them. These Galilean moons also have undergone different amounts of crustal evolution, as indicated by the extent of their unmodified impact cratering – Io the most and Callisto the least. Io, the most volcanically active body in the solar system, has a hot interior. Its surface is very new, modified by volcanic eruptions.

6. The rings of Jupiter are thin and consist of small particles; those of Saturn are wide, consist of larger, icy particles, and contain many small ringlets; those of Uranus and Neptune are thin and narrow and consist of very dark, small particles. The particles in these rings orbit the planet following Kepler's laws. Small moons in orbits near the rings may help to keep the rings stable.

7. Except for Titan, Saturn's largest moons are basically ices. They have evolved somewhat since the time of their formation, as indicated by their valleys and modified craters. The smallest moons, also of dirty ice, are heavily cratered and unevolved. Titan is the only moon in the solar system with a thick atmosphere.

8. The atmospheres of Uranus and Neptune are essentially the same (hydrogen and helium); but their interior structures are different, as implied by the difference in their densities. Neptune has more violent weather, driven by its heat outflow. Uranus has no or little outflow of heat.

9. Except for Miranda and Triton, the moons of Uranus and Neptune are dark, icy bodies with little evidence of crustal evolution. Their surfaces are heavily cratered. Miranda has a weirdly modified surface, and Triton shows signs of volcanism.

10. Pluto is a small icy world, with a frosty surface and a very thin atmosphere. Its moon, Charon, is about half Pluto's size and one-fifth its mass. Pluto and Charon orbit a common center of mass as a double-planet system. Akin to the nuclei of comets in their physical characteristics, Pluto and Charon are nevertheless much larger in size.

11. The solid materials in the outer solar system are largely ices and carbon-rich rocks. For the most part, the solid bodies were modified 4 Gy ago by an intense bombardment of solid debris. Minor heating and surface alterations have taken place in a few instances. Volcanism appears on Io and Triton. None of these solid bodies shows evidence of plate tectonics.

STUDY EXERCISES

1. In what significant respect is Jupiter most different from the other Jovian planets? (Learning Outcome 10-1)
2. Suppose you flew very close by Jupiter. What outstanding features would you see in the atmosphere? (Learning Outcomes 10-2, 10-4, and 10-7)
3. How do we know the bulk density of Pluto? (Learning Outcomes 10-2 and 10-5)
4. How do we know that the rings of Saturn are thin? (Learning Outcome 10-2)
5. What fact makes it relatively easy to find the masses of the Jovian planets? (Learning Outcome 10-2)
6. In one word, state the greatest difference between the Jovian and terrestrial planets. (Learning Outcome 10-1)
7. In two sentences, compare the rings of Saturn to those of Jupiter and to those of Uranus. (Learning Outcome 10-4)
8. How is Jupiter's magnetic field similar to the earth's? How is it different? Answer the same questions for Saturn. (Learning Outcome 10-9)
9. What features does Pluto have in common with the Galilean moons? (Learning Outcome 10-8)
10. In one short sentence, describe the interior compositions of the moons of the Jovian planets. (Learning Outcomes 10-7, 10-8, and 10-10)
11. What, in general, does a heavily cratered surface indicate about a body's evolution and age? (Learning Outcome 10-11)
12. In what way is the magnetic field of Uranus different from that of Jupiter or Saturn? (Learning Outcome 10-9)
13. In what way is the magnetic field of Neptune very different from that of Jupiter or Saturn? (Learning Outcome 10-9)
14. What is the main difference in the composition of the interior of the earth compared to the interior of Jupiter? What is the major difference in the physical state of the materials? (Learning Outcome 10-3)
15. In what sense can Pluto and Charon be considered a double-planet system? (Learning Outcome 10-6)

PROBLEMS AND ACTIVITIES

1. Although Jupiter does not have a solid surface, you can calculate its "surface" gravity at the cloud tops. Compare it to the surface gravity of the earth. *Hint:* For any spherical mass, its surface gravity, g, is $g = GM/R^2$, where R is the radius.
2. Apply Newton's version of Kepler's third law to Pluto and Charon and find the sum of the masses of these two bodies. What additional information do you need to find the individual masses?
3. Use the orbital properties of any of the Galilean moons to calculate the mass of Jupiter, using Newton's form of Kepler's third law.
4. At closest approach to the earth, Jupiter has an angular diameter of about 50 arcsec. What is its distance? Pluto has an angular diameter of a mere 3 arcsec at closest approach to the earth. What is its distance then? In both cases, what is the alignment of these planets with the earth?
5. Compare the escape speeds of Jupiter and Pluto to that of the earth.
6. What is the total heat output, in watts, from Neptune? *Hint:* What is the total surface area of Neptune?

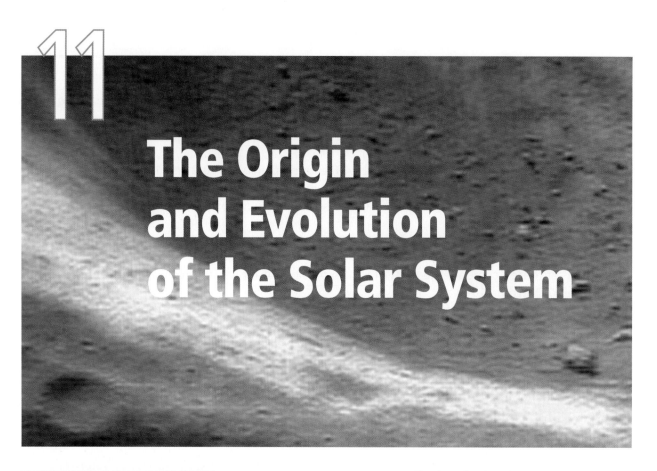

11

The Origin and Evolution of the Solar System

CENTRAL CONCEPT

The planets formed from an interstellar cloud of gas and dust as a natural outgrowth to the formation of the sun. They then evolved by common processes into the planets of today.

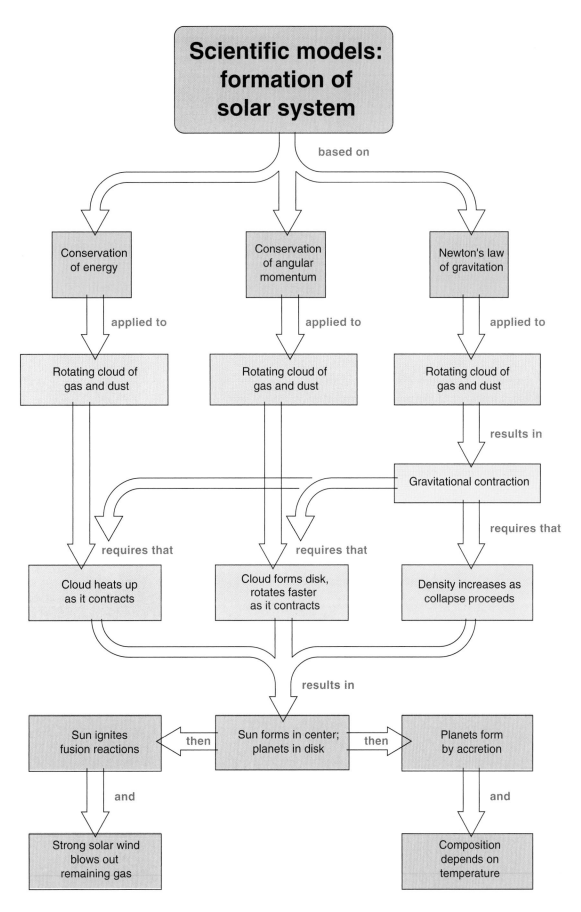

How did the solar system originate? This puzzle has challenged astronomers for centuries! Many models have been proposed. None has been completely successful. One scientific justification of the space program rested on finding information to support or refute theoretical ideas of the solar system's formation, and data from spacecraft have provided some key clues.

To unravel the puzzle of the solar system's genesis requires more than astronomy. It demands the interplay of astronomy with physics, chemistry, and geology. The question refuses to be answered simply. This chapter investigates in some detail the current approach, in which the formation of the solar system is seen as a natural result of the formation of the sun from an interstellar cloud of gas and dust.

The general outlines of this process reasonably explain the basic features of the solar system. Many details remain to be resolved; no one model yet fits them all together. We have painted the broad strokes of the big picture, and it has a profound implication: that other solar systems should be fairly common.

11.1 DEBRIS BETWEEN THE PLANETS: ASTEROIDS

The space between the planets is cluttered with comets, meteoroids, and asteroids. These bodies, along with gas and dust, make up the interplanetary debris. Scooped up, the total mass would amount to less than the mass of the earth; yet, as archaeologists have discovered, a trash heap holds valuable clues to a city's history, even though it contains much less mass than the city itself. Likewise, the solar system's debris is a fruitful hunting ground for clues to its past.

This section deals with debris in the form of asteroids; the next two sections cover comets and meteoroids. You will see again the trend that cropped up with the satellites of the Jovian planets: they're made of light-colored ices and dark-colored silicates.

Asteroids: Minor Planets
(Learning Outcome 11-2)

An **asteroid** is an irregular, rocky hunk, small both in size and in mass compared to a planet. Ceres, the largest known asteroid, has a diameter of

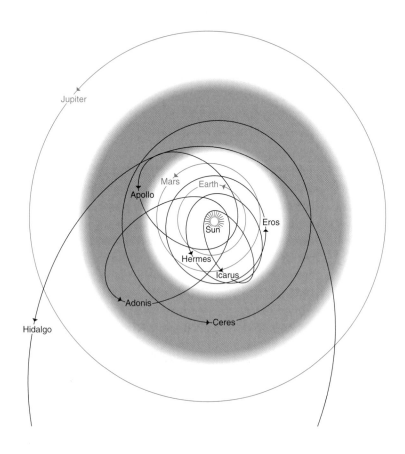

Figure 11.1 Orbits of asteroids. Most orbits are located in the zone called the asteroid belt between Mars and Jupiter. Some are found outside the belt, such as Icarus, which comes very close to the sun (and the earth). Note how eccentric the asteroidal orbits are compared to those of the planets. Some, such as Icarus and the Apollo asteroids, actually cross the orbit of the earth.

only about 900 km and a mass of perhaps 10^{21} kg. (That's about one-third the size of the moon and 1/100 its mass.) Its density is about 3.0. All asteroids combined may only amount to a mass that's a few percent of our moon. Most asteroids orbit the sun in a belt between Mars and Jupiter, at an average distance of 2.8 AU (Fig. 11.1); these bodies make up the **asteroid belt.** Some 5500 asteroids have been discovered so far, and many await future sightings.

Most asteroids fluctuate in brightness from their rotation – proof that they have irregular surfaces or shapes (or both). For example, Gaspra (Fig. 11.2a) has a lumpy shape, 19 km by 12 km by 11 km, and a cratered surface. The smallest craters are 300 meters wide. Gaspra rotates once every 7 hours. Large chunks of the asteroid have been struck off in past collisions – probably typical events in the life of an asteroid.

Deimos and Phobos, the moons of Mars, resemble Gaspra. Deimos and Phobos have lumpy shapes. Phobos, the larger, is about 27, 21, and 19 km long; Deimos is only 15, 12, and 11 km. They have cratered surfaces, too. The sizes and numbers of these craters suggest that the surfaces of these satellites are at least 2 Gy. The satellite surfaces are very dark, ranking among the blackest objects in the solar system. They reflect only 2 percent of the light that strikes them. We infer that Deimos and Phobos are unevolved, primitive bodies made early in the solar system's history – it is likely that they are captured asteroids. We expect other asteroids to look like these bodies, showing rough, pitted surfaces (Fig. 11.2b).

Do binary asteroids exist, loosely held together by gravity against tidal forces of the sun and planets? In accordance with Kepler's laws, each asteroid in a pair would revolve around a common center of mass. For instance, Ida (about 56 km long; Fig. 11.2) is orbited by Dactyl – a mere 1.4 km in diameter. Such observations support the notion of binary asteroid systems. Frequent collisions and fragmentation may create such systems, which have short lifetimes (maybe 10^5 years).

Close-up images of asteroids such as Eros (Fig. 11.2c) have transformed asteroids from points of light to geologic objects. Plus a new insight: some (most?) asteroids over a kilometer across may be rubble conglomerates of several major pieces. As smaller asteroids bang into large solid ones, large debris remains; gravity gently hauls these together. Over time, collisions turn the solid rock into a rubble pile. A few asteroids have densities between 1 and 2, implying that voids thread beneath the trashed surface.

Composition (Learning Outcome 11-3)

On the basis of their albedos, asteroids fall mainly into two classes. Some are relatively bright (reflect about 15 percent), and others are much darker (reflect only 2 to 5 percent), an indication that they contain a substantial percentage of carbon compounds. Those in the lighter class are dubbed **S-type** (stony) asteroids; the darker ones have been christened **C-type** (carbon). Ceres is a C-type asteroid, and Ida is an S-type. The S-type, in addition to having higher albedos, show spectral features indicative of silicate materials. Applying Newton's version of Kepler's third law to the orbit of Dactyl, we find that Ida has an average density of 2.5. That makes it about the same density as Deimos and Phobos. A third class, called **M-type** (metallic), has attributes that suggest the presence of metallic substances. They reflect about 10 percent. Only 5 percent of the total number of asteroids belong to this class.

The compositions of asteroids vary with their distance from the sun. Near the orbit of Mars, almost all asteroids have S-type characteristics. Farther out, we find fewer high-albedo asteroids and more that are dark. At the outer edge of the belt, 3 AU from the sun, some 80 percent of the asteroids are C-types. In fact, the C-type may well be the most abundant kind of asteroid overall.

11.2 COMETS: SNOWBALLS IN SPACE

In July 1995, Alan Hale in New Mexico and Thomas Bopp in Arizona both spotted a new comet, which was named after them. Luckily, they discovered Hale-Bopp while it was still 7 AU from the sun (beyond the orbit of Jupiter!), yet it was bright enough to be seen in modest telescopes. Comets are fickle by nature. Many appear bright far from earth but then fizzle by the time they near the sun. (Remember Comet Kohoutek in 1973?) Hale-Bopp, though, came through as a spectacular sight, reaching a peak in April 1997, and will not return again for 3000 years (Fig. 11.3).

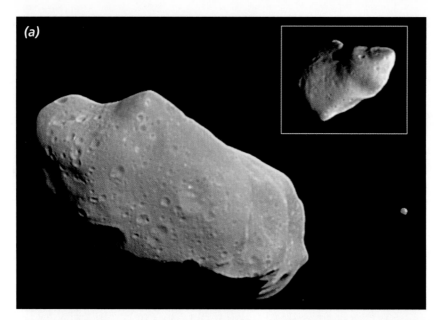

Figure 11.2 Images of asteroids. *(a)* The asteroid Ida and its satellite Dactyl, imaged from a distance of 10,500 km by the Galileo spacecraft in 1993. *(b)* Northern hemisphere of Eros, taken by the NEAR Shoemaker spacecraft in 2000. Note the saddle-shaped depression at the bottom and the crater at the top.

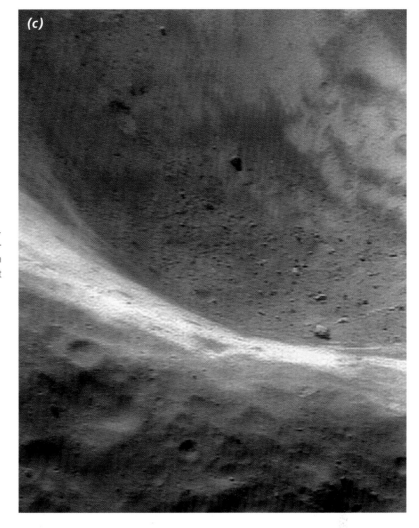

Figure 11.2 Images of asteroids *(continued)*. *(c)* Close-up of the crater: reddish hues represent older rock and regolith that have been chemically altered by weathering, while bright patches represent newer rock and regolith.

Figure 11.3 Comet Hale-Bopp. Note the separation of the tail into a gas (plasma) plume and a dusty part, which spreads in a wide fan. The gas tail appears blue from its emission; the dust tail reflects sunlight and so appears yellow. The plasma tail is blown back by the solar wind; the dust particles fall behind in their own orbits around the sun.

You probably associate comets with long, graceful tails. Not all comets exhibit tails, even at perihelion; and, when far from the sun, comets do not have visible tails. When first sighted telescopically, a comet typically appears as a small, hazy dot. This bright head of the comet is called the **coma;** the tail develops from the coma. The coma contains a small, starlike point called the **nucleus.** We think that cometary nuclei are smaller than 10 km across; none was ever viewed as more than a point of light until the spacecraft missions to Halley's Comet. Those flybys confirmed a size of roughly 10 km.

During a comet's charge toward perihelion, it grows brighter and sprouts a tail as its material is heated by the sun and vaporizes. A comet's tail may stretch for millions of kilometers, and the gaseous part of it points away from the sun because it is blown by the solar wind.

Composition
(Learning Outcomes 11-2 and 11-3)

Comets have two types of tail: gas and dust (Fig. 11.3). The physical differences between the two show up in their spectra. The spectrum of the gas or **ion tail** has emission lines, indicating that it is ionized, so the ion tail is a plasma. The **dust tail** shows a spectrum of sunlight, reflected from dust expelled out of the coma. The pressure from sunlight detaches dust from the coma to form a tail. (Recall that light can be thought of as particles – photons. Photons reflected off a dust particle apply a pressure to the surface and push a particle if it has a low enough mass.)

The ion tail's spectrum shows gases such as carbon monoxide, carbon dioxide, and molecular nitrogen. These tails consist of ions carrying magnetic fields (discussed in Section 8.3) at high speeds through interplanetary space. The magnetic fields interact with the plasma of the gas tail and impart to a comet its distinctive cast.

At great distances from the sun, the coma also shows a reflected solar spectrum. The coma may reach a diameter of a million kilometers. Infrared observations confirm that comets contain considerable amounts of silicate dust.

For all their stunning length in the sky, comets have very small masses. Halley's Comet, one of the largest, has an estimated mass of only about 10^{14} kg (roughly 10^{-8} that of our moon) and releases about 10^{11} kg during each perihelion passage, only about

0.001 of its total mass. With so little mass, a comet achieves its spectacular display only by spreading itself very thin.

The mass expelled from a comet, mostly in the form of gas, flies off into space. The nucleus supplies this material, but what is the nature of the nucleus? In 1950 American astronomer Fred L. Whipple developed the **dirty snowball comet model** (Fig. 11.4). It pictures cometary nuclei as compact, solid bodies made of frozen ices of water, ammonia, and methane, embedded within rocky material. When a comet nears the sun, the icy material vaporizes, forming the coma – a cloud of mostly gas and some dust. Continued vaporizing enlarges the coma and creates the tail. As the ice evaporates, a thin coating of rocky material remains to form a solid, but fragile, crust on the nucleus. Jets burst through this crust as ices vaporize and expand. From the missions to Halley's Comet, we now know that this idea is basically correct. Most comets are snow (water ice) and dust.

As a comet rounds the sun, the nucleus can break up from tidal forces. Comet Ikeya-Seki, the great sun-grazing comet of 1965, passed within 470,000 km of the sun and survived, but some comets skirting the sun are not so lucky. Comet West (1976) split into at least four pieces after its perihelion passage. Such a breakup leaves debris with a cometary composition.

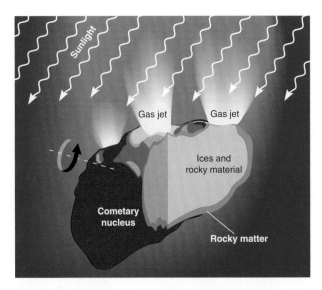

Figure 11.4 Simplified "dirty snowball" comet model. The nucleus is a pudding of ices and rocky material covered by a thin, rocky shell. Sunlight heats the nucleus and vaporizes the ices. Gas streams out to make the coma and tail. The nucleus has an irregular shape and rotates about its center of mass.

Orbits (Learning Outcome 11-3)

Comets that make regular returns to the sun have elliptical orbits and are called **periodic comets.** (Halley's Comet is the most famous periodic comet.) The periodic comets fall into two groups: long-period and short-period, with the dividing line at orbital periods of about 200 years. The short-period comets tend to orbit in the same direction as the planets, to stay in the plane of the solar system, and to have smaller orbital eccentricities. In contrast, long-period comets swing around the sun in highly eccentric orbits, cutting into the plane of the solar system from all angles.

A periodic comet eventually (in a million years or less) suffers one of three fates: it dissipates or breaks up; it collides with a planet; or it is ejected from its elliptical orbit and out of the solar system by a close encounter with a planet. The periodic comets we see must therefore be very recent additions to the solar system ("recent" compared to 4.6 Gy!). Hence, some reservoir supplies them, as explained below.

Not all comets have elliptical orbits. Some pass the sun once and never return – unless the planets jolt their open orbits into elliptical ones. These comets are seen on their first trip to the inner solar system, and then never again. They are one-shot celestial visitors.

The Comet Cloud (Learning Outcome 11-3)

Many comets have been observed in the past, but where are they all now? What is the origin of these flimsy bodies? Their orbits suggest that comets are attached gravitationally to the sun, except for those thrown into escape paths by encounters with the planets. From the observed orbits of long-period comets, we find that the average semimajor axis is about 50,000 AU; the corresponding orbital period is about 10 million years. The eccentricity of these orbits is close to 1, and the aphelion distances are about 100,000 AU. These comets, in accord with Kepler's second law, travel very slowly, only a few thousand kilometers per day, at aphelion. They spend most of their time coasting through the depths of space, 100,000 AU from the sun.

If you could step back, far from the sun, and take a wide-angle photograph at any given time, you would find a large number of comets hovering in a cloud far beyond Pluto. This cometary cloud, proposed by the Dutch astronomer Jan H. Oort (1900–1992) in 1950, and sometimes known as the

Oort Cloud, makes up the solar system's cometary reservoir.

According to Oort's model, most comets never come near the sun, and we never see them directly. Occasionally, the gravitational action of passing stars or giant interstellar clouds of gas and dust pushes a comet into an orbit that moves it closer to the sun. In passing by Jupiter or Saturn, an infalling comet may be captured into a periodic orbit. All comets are eventually lost because of vaporization of the nucleus. The comet supply must be replenished. To ensure sufficient input to make up the losses, Oort's picture requires at least 10^{11} – perhaps as many as 10^{12} – comets to be clustered in the cloud.

Where did the Oort Cloud come from? The comets in it formed along with the rest of the solar system. It's unlikely that they formed at the present distance of the Oort Cloud, because the density of material there at the time of the solar system's formation would have been very low. Some astronomers suggest a possible origin in the asteroid belt between Mars and Jupiter; others think they formed just beyond the orbit of Pluto, about 100 AU from the sun.

Both these suggestions run into the same difficulty: how did the comets get from their place of origin to their present location in the cloud? According to one idea, the gravitational effects of Jupiter kick comets out of an asteroid belt into long-period orbits. Another idea: a compact cloud of comets lies just beyond the rim of the orbit of Pluto. This inner cloud would survive encounters that prune the Oort Cloud – cometary nuclei stored here might drift into the outer cloud.

Interactions with the Solar Wind (Learning Outcomes 11-2 and 11-3)

In 1951 the German astronomer Ludwig Biermann (1907–1986) suggested that the solar wind interacts with the coma to generate a comet's ion tail. This idea has been confirmed by spacecraft observations that the magnetic field carried by the wind can indeed drag ions from the coma. (A plasma such as the solar wind can carry along a magnetic field.)

Comets are now viewed as solar wind telltales, with magnetic fields playing a major role in their structure. The two plasmas, that of the solar wind and that of the tail, interact by magnetic fields to create a comet's shape. Without such magnetic fields and plasma interactions, we wouldn't have comets!

Figure 11.5 Halley's Comet as seen over Europe in 1066, from a section of the Bayeux tapestry.

Halley's Comet
(Learning Outcomes 11-2 and 11-3)

When Edward the Confessor died in January 1066, he left no direct heir to the English throne, and the nobles chose Harold as their king. In the same year, a bright comet traveled across the sky – the commonly accepted sign of a ruler's death and misfortunes to follow. Meanwhile, across the Channel, William of Normandy cleverly interpreted the comet as a sign portending his victory (Fig. 11.5). With this psychological edge for his men, he sailed to the British Isles and conquered Harold's armies near Hastings. By the end of the year, William the Conqueror was crowned king. The comet of 1066 turned out to be one that returned on a regular basis.

Cometary orbits remained unknown until Edmond Halley (1656–1742), England's Astronomer Royal, calculated the orbits of the comets of 1531, 1607, and 1682 by a method devised by his friend Isaac Newton. Halley found, to his amazement, that the orbits were almost identical. Noting that the comets appeared at intervals of approximately 75 or 76 years, Halley concluded that these several comets were in fact one and predicted that it would return

around 1758. He was right. The comet that is named in his honor was sighted on Christmas night in 1758, after Halley had died. Halley's Comet was the first comet to be recognized as a permanent member of the solar system, with an elliptical, periodic orbit (Fig. 11.6). The spectacular comet of 1066 was none other than Halley's Comet.

Halley's Comet is the granddaddy of all known periodic comets. Its passages have been recorded at least 30 times, as far back as 239 B.C. Halley's Comet moves in an extremely elongated ellipse. Having passed aphelion beyond Neptune's orbit in 1948, it returned to the earth's neighborhood in 1985–1986, when it reached perihelion on February 9, 1986, at a distance of 0.59 AU. After rounding the sun, the comet passed by closest to the earth (0.42 AU) on April 11, 1986.

Unfortunately, this return of Halley's Comet presented the worst viewing in the past 2000 years! Even from the Southern Hemisphere, the sight was not very spectacular – except on the night of April 24, 1986, when a total eclipse of the moon took place. I viewed the comet from a high mountain in New Zealand. With the moon's light dimmed in the eclipse, the southern Milky Way stood out as a broad flow of stars. Between the Milky Way and the

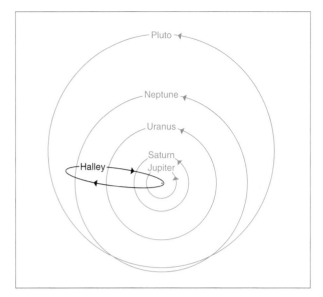

Figure 11.6 Orbit of Halley's Comet relative to orbits of the outer planets. Note how eccentric the orbit is, even compared to that of Pluto.

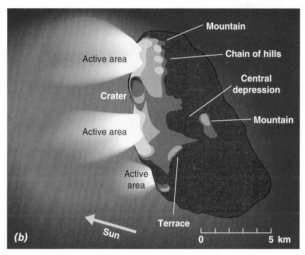

Figure 11.7 Halley's Comet. (a) Composite, detailed image of the nucleus from the Halley Multicolour Camera exposures on March 13, 1986. (b) Schematic drawing of the major surface features of the nucleus visible in the image.

moon, Halley's Comet hung fastened to the sky, its tail spread in a ghostly fan.

The most spectacular results of the spacecraft missions came from the European Space Agency's Giotto, which swooped within 600 km of the core of Halley's Comet on March 14, 1986. Well before that closest approach, Giotto sampled the ionized zone. Most of the ions are related to water; in fact, most of the comet's gas is water.

Giotto's last image, from a distance of about 1700 km, revealed a dusty realm (Fig. 11.7a). The nucleus itself was peanut-shaped, about 8 km wide and 16 km long. Its surface was dark – black as velvet – with an albedo of only a few percent. The surface also appeared rough, with at least one hill a few hundred meters high and about a kilometer wide at its base. Craterlike structures about a kilometer in size were noted in some regions (Fig. 11.7b). These structures showed the nucleus rotating once every two days.

Most dramatic were the dust jets. The sunlit side of the comet ejects heated materials – ice that vaporizes to gas, and the dust with it. These gases and particles blow off in jets only a few kilometers wide. The surface sources of the jets were less than a kilometer in diameter, and the outspray contained about 80 percent water vapor and 20 percent dust. That dust was almost all carbon (like the lead in a pencil), with a small amount of sandlike material mixed in. No one had predicted before the mission that the dust would contain so much carbon – and it is the black carbon that makes the surface of the nucleus so dark, like that of a C-type asteroid.

As expected from the dirty snowball model and confirmed by observations, Halley's Comet sheds a layer a few meters deep from its nucleus on each passage. The comet is not immortal, but it will survive for a good number of future orbits.

Figure 11.8 HST image of Comet Shoemaker-Levy 9 fragments before their collision with Jupiter.

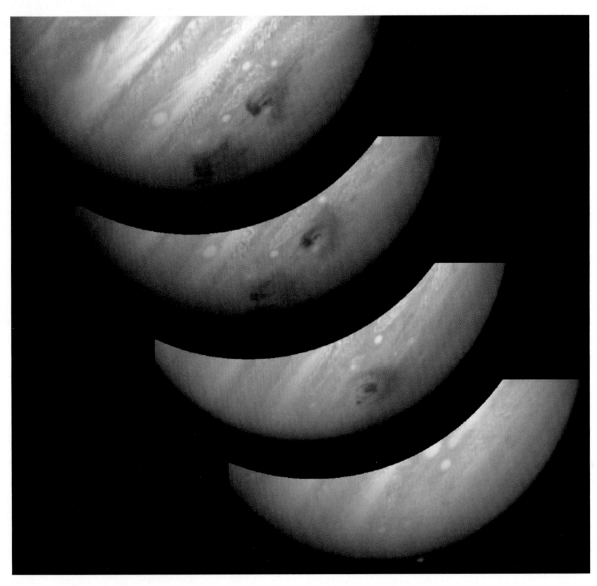

Figure 11.9 Impact! Time-lapse sequence of two fragments of SL9 striking Jupiter: from bottom left to upper right, as the cometary fragments plunged into the planet's surface, dark spots were created, which then faded.

Figure 11.10 A bright fireball of the Leonid meteor shower in November 1999, taken from Dagali, Norway.

Comet Shoemaker-Levy 9 and the Great Comet Crash on Jupiter (Learning Outcome 11-3)

U.S. astronomers Carolyn Shoemaker, Gene Shoemaker, and David Levy discovered Comet Shoemaker-Levy 9 (SL9) in March 1993 near Jupiter. HST observations (Fig. 11.8) revealed that the nucleus had broken up into many fragments. How? Probably by tidal forces (Enrichment Focus 8.1) during a close approach to Jupiter. SL9 turned out to be orbiting Jupiter as a satellite of the giant planet. Jupiter's gravitational reach has snatched some 150 comets into short-period orbits of about 8 years. SL9 was a doomed member of Jupiter's cometary family. It showered its parent in July 1994 in the solar system spectacle of the century.

On July 18, fragment G (one of the largest) struck. A fireball appeared as a giant bubble of superhot gas that rose some 3000 km above the cloud tops. Its bruise looked like a black eye about twice the diameter of the earth (Fig. 11.9). Subsequent impacts left similar features. By late August, the stratospheric winds had blown away the stains.

In the aftermath of the impacts, astronomers concluded that the largest fragments of SL9 were no more than 2 km in diameter. Still, pounding Jupiter at 60 km/s, a typical impact conveyed an energy equivalent to 25,000 megatons of TNT. (Such a bash to earth could wipe out civilization with its global destruction.) These blasts occurred in Jupiter's upper atmosphere, so they were air blasts like those that made the craters on Venus. Temperatures reached at least 10,000 K. As plumes splashed back down to the atmosphere, they heated large regions to perhaps 2000 K and prompted chemical reactions to make the dark stains.

Comets like SL9 strike Jupiter once per century. Luckily, on average, they batter the earth only every few hundred thousand years or so. Perhaps such shocks altered the course of life on earth.

11.3 METEORS AND METEORITES

A **meteor** is the flash of light from the vaporization of a solid particle in the earth's atmosphere (Fig. 11.10). As it plunges through the air, the particle is burnt up by air friction, leaving behind it a bright trail, a glowing column of light. Before the particle enters the atmosphere, it is called a **meteoroid:** a solid object traveling through interplanetary space – destined eventually to strike another object. Of course, other objects (comets, asteroids, and planets) also travel through the interplanetary void; a meteoroid differs from those chiefly in its small size, no more than a few meters in diameter, usually much less.

If a meteoroid survives its plunge through the atmosphere and strikes the earth's surface, the body is then called a **meteorite.** Most meteoroids are fragile, delicate particles (Fig. 11.11) that crumble quickly in their contact with the air. A meteoroid is typically a flimsy dust speck whose demise is its only remarkable aspect. Few hit the earth; most fall into the sun.

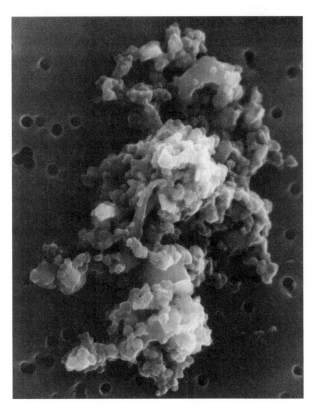

Figure 11.11 A particle of interplanetary dust about 10 μm in size. Note its open, fragile structure. This dust is material from a comet.

The dirty snowball comet model explains the intimate connection between comets and meteors. During a comet's passage by the sun, solar heating causes a continual loss of icy material from the cometary nucleus. The dust and solid particles interspersed in the ices flake off and scatter in an untidy array around the comet. This solid debris has a low density. The older the comet and the greater its number of passages by the sun, the greater the loss of icy material and release of meteoroid material. Most meteors are of cometary origin; few of these leave meteoritic debris. (Those that do probably are associated with asteroids.)

The cometary debris tends to follow the orbit of the original comet; at certain times of the year, the earth crosses the path of the debris and a meteor shower takes place. During a shower, many meteors are visible over a limited period of time. The most reliable of these is the Perseid shower, visible in August. The name of a shower identifies the constellation from which the meteors appear to radiate. The best time to observe most showers is in the early morning, although not all the showers peak then.

Types of Meteorites (Learning Outcome 11-2)

In terms of physical and chemical composition, meteorites fall into three broad classifications: irons, stones, and stony-irons. The **irons,** which are generally about 90 percent iron and 9 percent nickel, with a trace of other elements, are the most commonly found. They are easy to identify because of their high density and melted appearance. The **stones** are composed of low-density silicate materials similar to the earth's crustal rocks. Many stones contain silicate spheres, called **chondrules,** embedded in a smooth matrix. These stones are known as **chondrites** (Fig. 11.12a). Finally, **stony-iron** meteorites represent a cross-breed between the irons and the stones and commonly exhibit small stone pieces set in iron.

A most curious kind of chondrite is the **carbonaceous chondrite** (Fig. 11.12b). The chondrules in such meteorites are embedded in material that contains a large fraction of carbon compared with other stony chondrites – typically about 2 percent carbon compared to the total mass. Their carbon content gives these meteorites a dark appearance.

Carbonaceous chondrites also contain significant fractions of water (about 10 percent) and volatile materials. In addition, if some gas were extracted from the sun and cooled, the condensed elements would be chemically very similar to carbonaceous chondrites. This similarity suggests that carbonaceous chondrites formed out of the same primordial material as the sun and have suffered no major heating or changes since that time. The pristine nature of chondrites makes them a major focus in attempts to study the origin of the solar system.

Origin of Meteorites (Learning Outcome 11-3)

Most meteorites are too dense to have derived from comet-related meteoroids; that's why they survived the plunge to the ground. They resemble the inferred physical characteristics of asteroids. Orbits of meteorites prove to be like those of asteroids rather than of comets. Collisions between asteroids fragment them, and these small pieces provide a source of meteorites.

What about iron meteorites? An important clue appears when the polished surface of an iron meteorite is etched with acid. Large crystalline patterns, called **Widmanstätten figures,** become visible

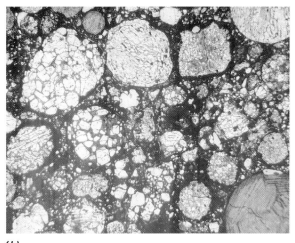

(a)

(b)

Figure 11.12 Chondritic meteorites. (a) A slice of Leoville carbonaceous chondrite. Note the dark overall appearance resulting from the high percentage of carbon. The round chondrules contrast with irregularly shaped inclusions rich in calcium and aluminum. (b) A thin-section photograph of Ragland ordinary chondrite. Some chondrules are only a few millimeters in diameter, embedded in a black matrix composed of grains of iron and iron sulfide.

(Fig. 11.13) upon such treatment. Terrestrial iron does not show these patterns when etched.

A nickel–iron mixture, when cooled slowly under low pressures from a melting temperature of about 1600 K, forms large crystals. The key point is that the cooling must be very gradual (about 1 K every million years). But metals conduct heat well, and in the cold of space, a molten mass of nickel and iron would cool rapidly without forming large crystals. So how could nickel–iron meteorites grow Widmanstätten figures? They need protection from the cold. It's likely that nickel–iron meteorite material solidified inside small bodies, termed **parent bodies.** To allow a cooling of only 1 K every million years, these objects must have been at least 100 km in diameter.

Most parent bodies were only a few hundred kilometers across. Once formed, they could be heated by the radioactive decay of short-lived isotopes, which would cause them to differentiate – the densest material sinking to the center, and the least dense frothing to the surface. Such an object would end up with a core of metals and a cover of rocky material, which would cool to form a crust. Insulated by the crust, the molten metals would be able to cool slowly and form Widmanstätten figures. If, later, the parent bodies were to collide and fragment, pieces from the outer crust would make stony meteorites, pieces from farther down would become

Figure 11.13 The polished surface of a 1-cm cube of the Toluca iron meteorite shows the Widmanstätten figures (crystal patterns).

stony-iron meteorites, and the core would produce the iron meteorites. (Based on the same reasoning, we expect the earth's core to be nickel–iron.)

Are any parent bodies around now? Yes – as asteroids! Recall the three main types of asteroid: the dark C-type, containing much carbon; the lighter S-type, composed of silicate materials; and the intermediate M-type, with metallic characteristics. These types probably are related to the carbonaceous chondrites, the stony meteorites, and the irons, respectively.

Table 11.1 Interplanetary Debris

Material	Location	Composition	Origin
Asteroids	Generally found between Mars and Jupiter in the ecliptic plane	Rocky material (silicates), iron, nickel	Uncollected debris from formation of the solar system
Comets	Short-period: elliptical orbits in ecliptic plane with aphelions near Jupiter and Saturn Long-period: orbits not confined to the ecliptic plane	Ices of water, methane, ammonia, carbon dioxide, and rocky material (silicates)	Uncoalesced building blocks of Jovian planets and fragments from collisions among them
Meteors and meteoroids	Swarms or streams in elliptical orbits generally in ecliptic plane	Flaky silicate materials, possibly some ices	Remains of exhausted comets (if dirty snowball model is correct)
Meteorites	On earth's surface	Silicates, iron, nickel	Parent meteor bodies, asteroids
Interplanetary gas and dust	Throughout the solar system, mostly in ecliptic plane	Gas: mostly hydrogen and helium Dust: silicates, graphite	Solar wind, ashes of comets, fragmentation of asteroids and meteoroids

The view of meteorites outlined here implies that their parent bodies were among the first solid objects to form in the origin of the solar system, so the ages of meteorites provide a direct indication of the age of the solar system. Using radioactive decay techniques (Enrichment Focus 8.1) can date meteorites. Such methods give ages very close to 4.6 Gy. Meteorites provide astronomers with a direct, reliable estimate of *when* the solar system formed. The real issue is *how*.

Table 11.1 summarizes our understanding of interstellar debris. Now that we have investigated the remains of the solar system's formation, let's turn to the puzzle of how it came about.

11.4 PIECES AND PUZZLES OF THE SOLAR SYSTEM

The dynamic and chemical properties of the solar system impose crucial limitations on any model of its formation. These features serve as broad templates for shaping more specific questions.

Chemistry (Learning Outcome 11-1)

Chemically, the solar system falls into three broad categories of material: gaseous, rocky, and icy (Table 11.2). Each group is distinguished by its melting point. The gaseous and icy materials together, sometimes called **volatiles,** are generally gaseous under the conditions expected during the solar system's formation. The bodies of the solar system are com-

Table 11.2 General Classes of Solar System Materials

Class	Examples of Materials	Solidification Temperature (K)
Rocky	Iron, iron sulfide, silicates	500–2000
Icy	Water, methane, ammonia	100–200
Gaseous	Hydrogen, helium, neon, argon	0–50

posed of various combinations of the three groups. Though one class of materials may dominate, any solar system body contains some of each group. For example, the sun contains mainly gaseous materials and also icy/rocky materials, but as gases (because the sun is hotter than 2000 K), not solids. The terrestrial planets and the asteroids are mainly rocky/metallic; Jupiter and Saturn mostly gaseous; Uranus, Neptune, Pluto and Charon, and comets mostly icy.

Dynamics

Dynamically, the solar system displays a regular structure in terms of its motions. Viewed from above the sun's north pole, the solar system shows the following regularities:

1. The planets revolve counterclockwise around the sun; the sun rotates in the same direction.
2. The major planets, except Mercury and

Pluto, have orbital planes that are only slightly inclined with the plane of the ecliptic (the plane of the earth's orbit); that is, the orbits are *coplanar*.

3. Except for Mercury and Pluto, the planets move in orbits that are very nearly circular.

4. Except for Venus, Uranus, and Pluto, the planets rotate counterclockwise, in the same direction as their orbital motion.

5. The planets' orbital distances from the sun follow a regular spacing; roughly, each planet lies twice as far out as the preceding one.

6. Most satellites revolve in the same direction as their parent planets rotate and lie close to their planets' equatorial planes.

7. Some satellites' orbital distances follow a regular spacing rule.

8. The planets together contain much more angular momentum than the sun.

9. Long-period comets have orbits that come in from all directions and angles, in contrast to the coplanar orbits (in the same plane) of the planets, satellites, asteroids, and short-period comets.

10. All the Jovian planets are known to have rings.

Models of Origin
(Learning Outcomes 11-6, 11-7, and 11-9)

A successful model must explain as many of the foregoing dynamic and chemical properties as possible. It must account for the greatest number of the listed characteristics, and explain them in some internally consistent, simple fashion (Section 2.1). That is what is meant by the requirement that a model be "aesthetically pleasing."

The sun contains most of the mass of the solar system (99.9 percent). The rest lies close to the plane of the solar system. In terms of the layout of mass, the solar system is quite thin. If it were the size of an average pancake, the solar system would be only a centimeter thick, with the orbits of the planets contained inside it.

Finally, a successful model must deal not only with the dynamic and chemical regularities of the system but also with the interplanetary debris: comets, asteroids, and meteoroids. Contemporary models consider these bodies to be important relics of the solar system's early history.

Most models today are variations of **nebular models,** in which the sun condenses from an interstellar cloud of gas and dust that also forms a disk, a **solar nebula,** out of which the planets condense. The nebular models view the solar system as a natural outcome of the sun's formation and, perhaps, of any star's formation. If nebular models are correct, planetary systems are very common in our Galaxy and in other galaxies. (See the Formation of Solar System Celestial Navigator™.)

The sun represents nearly all the mass of the solar system. Does it also contain the lion's share of angular momentum?

11.5 BASICS OF NEBULAR MODELS

The essential feature of nebular models is that the sun and then the planets form from a cloud of interstellar material. The sun's formation takes place in the center of a flattened cloud. The planets grow from the disk of the cloud. That's the basic picture and the nub of the problem, as developed in this section.

We know that the solar system is now basically flat, with the sun in the center. The rings of the Jovian planets are flat and consist of small particles orbiting their parent planet. You can imagine that if somehow the small particles could be stuck together, Jupiter would end up with another moon. So the problem has two parts: how to make a flat solar system, and how to get the planets to grow out of the cloud.

Angular Momentum
(Learning Outcomes 11-1, 11-4, and 11-5)

To tackle the problem of making a flat solar system, we need to consider a basic physical idea: the conservation of angular momentum (Enrichment Focus 11.1). The basic point is this: once a spinning body has started spinning, will keep on spinning as long as no outside influence affects it. The amount of angular momentum depends on how much mass the body has and how much it is spread out. If, by itself, the body changes size – for instance, if it contracts gravitationally – it will naturally spin faster to keep its angular momentum the same. (A mass contracts gravitationally when it pulls itself together by the gravitational forces between the particles that make it up.)

ENRICHMENT FOCUS 11.1

Momentum and Angular Momentum

The concept of momentum rests on that of the inertia of matter. Consider a bicycle and a truck coming at you at the same speed. Which would be easier for you to stop? The bicycle, because it has less mass. Now imagine two bicycles coming at you, one with twice the speed of the other. Which is easier to stop? The one moving more slowly. These examples show you that momentum depends on both the mass and the velocity of the object involved. In fact, **momentum** is defined as the product of an object's mass and velocity:

momentum = mass × velocity

If we let p be the momentum, m the mass, and V the velocity, then

$$p = m \times V$$

Note that velocity has a direction, so momentum does too, the same as that of the velocity. You can think about momentum like this: once you get a mass moving, you have to put out an effort to stop it.

Consider a spinning object, such as the earth rotating about its axis. What keeps it spinning? Its inertia about its spin axis. The faster it spins and the more mass is spinning, the harder the object is to stop. This spinning momentum is called **angular momentum.**

You can think of angular momentum as the tendency for bodies, because of inertia, to keep spinning (rotating) or orbiting (revolving). Angular momentum of a body is determined by its mass (but now the distribution of the mass, around the center of motion, complicates the picture), the velocity (around the center of

motion), and the radius (the distance of the mass from the center of motion). For a single particle moving in a circle, we have

angular momentum = mass × circular velocity × radius

Let L be the angular momentum, m the mass, V the circular velocity, and r the radius. Then

$$L = m \times V \times r$$

For a body such as the earth, which is made up of many particles moving at different velocities at different distances from the axis of rotation, we must add up the angular momenta of all the particles.

The key aspect of angular momentum is that it is *conserved.* If no twisting forces, called *torques,* act on an object, its angular momentum remains the same. (A torque is a force applied through a lever, such as a torque wrench used to tighten spark plugs in an engine. The owner's manual for my truck specifies a torque of 20 newton-meters.)

You can test the **conservation of angular momentum** by performing a simple experiment (Fig. F.13). Tie a ball to the end of a string and whirl it around your head at a constant speed. Now, with your free hand, grab the loose end of the string. Very slowly pull the end through your hand, shortening the string until it is at half the distance you started with. The ball will move with double its circular speed. Note that as the distance from the center of spin decreases, the rate of spin increases, but no torques have been applied, so the angular momentum is the same with the string at the different lengths. This is an example of the conservation of angular momentum.

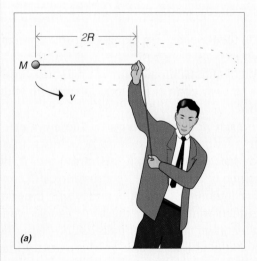

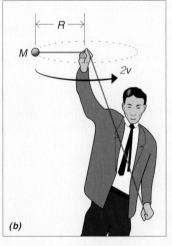

Figure F.13 Example of the conservation of angular momentum. (a) A mass at the end of a string is made to swing in a circle. (b) When the length of the string is decreased, the circular speed of the mass increases. At half the radius, the mass moves at twice its original circular speed.

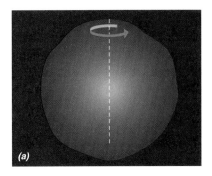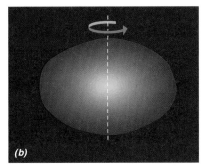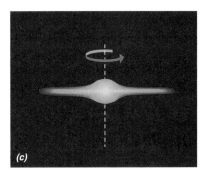

Figure 11.14 Gravitational contraction of a spinning cloud. (a) The process begins with the cloud slowly spinning. As it contracts (b), material falls in along the spin axis to form a disk with a central bulge (c). The cloud ends up spinning faster than at the start of the contraction because of the isolated cloud's conservation of angular momentum.

A familiar example of the conservation of angular momentum is the spin-up of an ice-skater. The skater goes into the spin with arms outstretched. As the skater pulls his or her arms in, the rate of spin increases. Why? Because to conserve angular momentum, the skater must rotate faster around the body's spin axis as mass is brought closer to it.

Consider now a large spherical cloud of gas and dust particles, slowly spinning. Imagine that it pulls itself inward by its gravity (Fig. 11.14). What happens? It will spin faster and collapse down along the spin axis to make a flat disk with a fat center. That's just what we need to form the solar system. As a natural result, we get the planets' orbits aligned in a thin disk, and the sun rotating in the same direction as the revolution of the planets.

With this neat solution comes a serious problem: the present distribution of angular momentum. Although the sun holds 99 percent of the system's mass, it contains less than 1 percent of the angular momentum. The outer planets have the most, 99 percent of the total. If all the planets with their present angular momenta were dumped into the sun, it would spin once every few hours rather than once a month. The sun forms from the central part of the cloud, so it should be spinning very rapidly. Actually, the sun spins 400 times more slowly than this rate. The angular momentum is there, but not in the right place! To adopt a nebular model, we must account for the present distribution of angular momentum, a problem still being worked out.

One solution involves the interaction of magnetic fields and charged particles to rearrange the distribution of angular momentum. Basically, the spin of the central part of the nebula must be decreased and transferred to the outer regions. Charged particles and magnetic fields interact so that the particles spiral along the magnetic lines of force. Such interactions provide a way to transfer spin from the young sun to the outer parts of the nebula.

Here's an analogy. Imagine standing in a swimming pool, up to your neck in water, with your arms extended. Spin around as fast as you can. Your arms will drive the water around and force it to swirl. In response, you'll feel a drag of the water on your arms. If you didn't keep yourself spinning, you'd rapidly slow down as angular momentum was transferred from you to the water.

As the sun formed, it heated the interior regions of the nebula. Here the gas was ionized, and magnetic field lines trapped these particles. As the sun rotated, it carried its magnetic field lines with it; these dragged along the charged particles, which in turn united with and dragged along the rest of the gas and dust. The magnetic field spun around the material in the nebula near the sun. At the same time, the mass of the nebula resisted the rotation. This drag on the magnetic field lines stretched them into a spiral shape (Fig. 11.15). The magnetic field linked the material in the nebula to the sun's rotation. So the nebular material gained rotation (and angular momentum) and in the process causes a drag on the sun's rotation, which slowed it down.

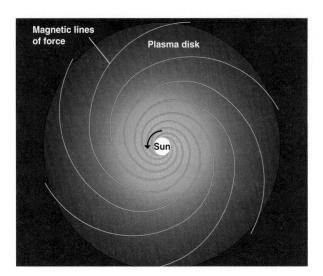

Figure 11.15 Top view of a possible primeval solar magnetic field. One end of the field lines rotates with the sun; the ionized material (plasma) in the disk drags the other end, so the spin of the central regions is transferred to the outer parts.

Whatever process transferred the angular momentum, the transfer must have taken place before large solid objects formed in the nebula. The transfer mechanism just described works effectively only on plasmas.

Heating of the Nebula
(Learning Outcomes 11-1 and 11-6)

Finally, one other fundamental process occurs: that of **gravitational contraction.** Whenever a mass pulls itself together by its own gravity, it gets hotter, as a result of the transformation of energy from one kind (gravitational potential energy) into other kinds (heat and light). The basics are just like those for accretion, but a larger number of smaller particles is involved.

Here's how. Consider a ball held above the earth's surface. Its speed is zero and it has no kinetic energy, but it does have potential energy, as you can tell when you drop it. The ball falls, and as it does, its speed increases – it accelerates. Its kinetic energy increases the more it falls.

Instead of a ball, consider a cloud of gas containing a large number of particles. Think of each particle as a small ball. Imagine that the cloud contracts gravitationally from the combined attraction of all

the particles. As it contracts, the particles gain speed. Though at the start the velocities are directed inward, collisions will soon distribute them in random directions, with only a slow net motion inward. (The collisions slow down the gravitational contraction.) So the average kinetic energy of the particles increases. Temperature measures the average kinetic energy of the particles involved, so the temperature of the gas increases.

Meanwhile, the density also increases, so the particles collide more often from the combined effects of the increases in velocity and density. The collisions excite some atoms; these emit photons. The net result of the gravitational contraction is that some of the initial gravitational energy is converted to heat (a higher temperature means greater average speeds of the particles of the gas) and some ends up as photons. In fact, half goes into raising the temperature and half into light. The key point is this: *as the cloud contracts, it gets hotter.*

With angular momentum and gravitational contraction in mind, let's turn to the next basic problem: how to form the planets.

11.6 THE FORMATION OF THE PLANETS

A successful nebular model must account in some detail for four important stages in the solar system's evolution: the formation of the nebula out of which the planets and sun originate, the formation of the original planetary bodies, the subsequent evolution of the planets, and the dissipation of leftover gas and dust. Modern nebular models give tentative explanations for these stages, but many details are lacking. No one model today is entirely satisfactory.

Making Planets
(Learning Outcomes 11-6, 11-7, and 11-8)

How can planets grow? There are three main methods: gravitational contraction, accretion, and condensation. *Gravitational contraction* works if regions in the nebula have enough mass to be able to contract by their own gravity to form a planet. *Accretion* occurs when small particles collide and stick together to form larger masses that eventually grow into planets. (An example: as snowflakes fall

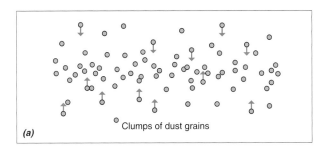
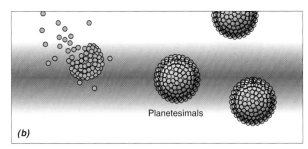
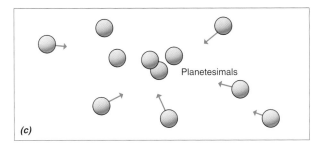
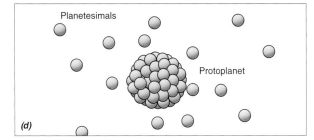

Figure 11.16 Possible scenario for the formation of the planets. (a) Dust grains collide and stick to form small objects that fall into a thin disk. (b) Gravity pulls these together to make asteroid-sized bodies, the planetesimals. (c) The planetesimals collect in clusters to form the cores of the protoplanets (d), which evolve into the planets.

through the air, they may collide and stick to form clusters of snowflakes.) *Condensation* involves the growth of small particles by the sticking together of atoms and molecules. (An example: water molecules combine in clouds to form raindrops.)

The formation of the planets was a multistep process: first small bodies formed, followed by larger ones, which eventually evolved into the planets. Accretion was probably the main process in building most of the planets and their satellites. The first fairly large bodies, from a few kilometers to a few hundred kilometers in size, are called **planetesimals.** Then, the planetesimals collided and accreted to make planet-sized masses (Fig. 11.16), called **protoplanets.** These objects formed into the planets of today.

How do small dust grains beget large protoplanets? Grains collide and accrete to form larger, pebble-sized objects, which quickly fall into the plane of the nebula. The pebbles then accumulate into planetesimals by gravitational attraction. Whatever materials happen to be available at a certain distance from the center of the nebula make up a planetesimal, so the planetesimals reflect the local compositions of material.

Once the first planetesimals have formed, they may gather into larger bodies, perhaps almost as large as the moon. These objects end up in a few protoplanets. Here gravity helps. By the time the planetesimals have gathered, in a few tens of thousands of years into several somewhat larger bodies (500 km in size), they will have enough mass to help pull in other smaller masses from a distance. A growing planet will sweep clear a zone of the nebula to feed its mass. For the terrestrial planets to grow to their present sizes, calculations indicate a time of roughly 100 million years for the protoplanets to form by the accumulation of smaller masses. This clearing out of a region around each protoplanet has led to the spacings of the planets that we see now.

Chemistry and Origin
(Learning Outcome 11-6)

How did the protoplanets acquire differences in chemical composition? The basic concept is that of a **condensation sequence.** The nebula's center was hot, a few thousand kelvins. Here solid grains, even iron compounds and silicates, could not condense. Elsewhere, the identity of the materials that would

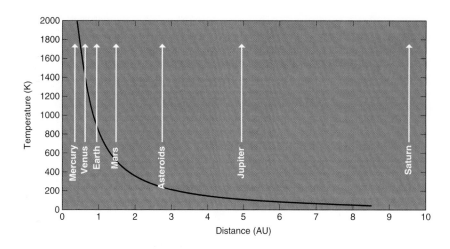

Figure 11.17 How the temperature in the solar nebula might have decreased with increasing distance from the sun: a cross section of the solar nebula and a possible distribution of temperatures within it at the beginning of planetary formation.

condense as new grains depended on the temperature (Fig. 11.17). Just below 2000 K, grains made of terrestrial materials would condense (Table 11.2); below 273 K, grains could form of both terrestrial and icy materials.

The exact sequence in which compounds could condense from a heated nebula is crucial to this idea. At different temperatures, the gases available and the solids present react chemically to produce a variety of compounds. The key result of the condensation sequence: the densities and compositions of the planets can be well explained with the condensation sequence *if* the temperature in the nebula drops rapidly from the center outward. Under this assumption, then, it is feasible to suggest that at different distances from the sun, different temperatures allowed different chemical compounds to condense and form grains that eventually made up the protoplanets. If a material could not condense because the temperature was too high, it would not end up in the protoplanet.

Unlike the Jovian planets, which are much like the original nebula in composition, the terrestrial planets contain little in the way of icy and gaseous materials. At the earth's distance, the temperature was about 600 K. Here, silicates of both iron and magnesium could condense, along with iron oxide and iron sulfide. So the model predicts a planet with a smaller nickel–iron core than Mercury's, and a larger mantle, rich in silicates and oxides of iron and magnesium. The earth has such an interior. What about Mars? Here the temperature was lower still. The resulting core would consist mostly of iron sulfide, and the mantle would be olivine, rich in iron oxide and water. We don't know the interior of Mars very well, but the prediction from the condensation sequence is compatible with our general model about the Martian interior.

CAUTION. It is how *low* the temperature falls that determines what chemical reactions operate. In general, the condensation sequence requires that a certain minimum temperature be reached to account for the known chemical composition of the planets. Roughly, these temperatures 245are 1400 K for Mercury, 900 K for Venus, 600 K for the earth, 400 K for Mars, and 200 K for Jupiter.

When the central regions of the nebula finally formed the sun, solar energy evaporated any icy material in the nebula and around the inner protoplanets or planetesimals. Intense solar radiation pressure and a strong solar wind (a solar gale!) pushed leftover gases out of the solar system.

Leftover planetesimals bombarded the planetary surfaces, leaving remnant craters on the terrestrial planets. This bombardment liquefied the surfaces of the planets. Radioactive decay heated the interiors and melted them. Dense elements, such as iron and nickel, sank to form a core. Less dense materials, such as silicates, floated to a sur-

Figure 11.18 Artist's conception of the giant impact model of the moon's origin. A Mars-sized body hits the protoearth at grazing angle. Some material from both bodies remains in orbit in a disk around the earth and accretes to form the moon.

face froth, which cooled to become the crust. The planets became differentiated – the first step in their evolution.

Not all planetesimals end up in larger bodies. Some rocky/metallic ones remain; they become asteroids. The icy ones hang around as the nuclei of comets. Most are tossed into the outer solar system by the gravitational influences of large masses, such as Jupiter. The Oort Cloud now contains these primordial icy bodies.

The condensation sequence implies that myriad small bodies rich in volatiles and carbon formed in the cold reaches beyond Mars – the moons of the Jovian planets and even Pluto and Charon are samples. The icy-carbon objects were the original supplies of the solar system. Those that ran into each other accreted or fragmented. Those that plunged toward the sun heated and in part vaporized, yet they have evolved less than the planets. The survivors today make up the interplanetary debris.

Jupiter and Saturn: A Different Story? (Learning Outcome 11-8)

Jupiter and Saturn may have followed a different path. They may have condensed gravitationally from single large blobs of material in the nebula. Their internal heat comes from the conversion of gravitational potential energy into heat during gravitational contraction.

Theoretical models of the evolution of Jupiter start with a large mass and a surface temperature of about 1000 K, as well as a total energy output per second of almost 0.01 the sun's. The early high power may explain, using the condensation sequence again, why the Galilean satellites decrease in density going outward from Jupiter (Section 11.2). If Jupiter were hot at the time of the satellites' formation, the inner ones would not have accreted as much icy materials as the outer ones, so the Galilean moons may mimic the condensation and accretion of the terrestrial planets.

Saturn may have progressed through a similar sequence.

In fact, the early evolution of Jupiter and Saturn may follow that of low-mass stars generally.

Our Moon: A Different Story! (Learning Outcome 11-8)

Finally, we come to the **giant impact model,** a hybrid that relies heavily on binary accretion ideas. This model envisions a glancing impact of a Mars-sized body into the young earth after the core had formed (Fig. 11.18). The tremendous impact vaporized the earth's mantle and spewed some of the earth's mass (and a large fraction of the impacting body's mass) into orbit around the earth. The metal-rich core of the impacting body remained intact and joined up with the earth; what was its rocky mantle remained in orbit. The orbiting material condensed, creating a ring of debris. The moon accreted from this material. The essential events were over in about a day.

Note that the chemical differences and similarities of the earth and moon are cleverly taken care of: the similarities arise from the ejected terrestrial material, the differences from those of the impacting body. The vaporization releases volatiles (such as water) from the material that forms the moon, so explaining their absence. The material ejected from the earth's mantle ensures that the oxygen isotopic abundances match. The angular momentum works out if it is assumed that the body struck the earth a glancing blow at a speed of about 10 km/s.

Note that the formation of the moon must take place far enough from the earth so that the earth's tidal forces do not inhibit the accretion of masses. Recall (Enrichment Focus 9.1) that tidal forces pull objects apart, so tidal forces will separate two small masses if they are too close to the earth. In fact, if a single mass were placed too close to the earth, tidal forces could rip it apart.

Evaluation of the Nebular Model (Learning Outcomes 11-1 and 11-10)

By and large, the nebular model makes a firm basis for exploring the origin of the solar system. Here's how it deals with the ten points listed in Section 11.4.

1. The planets' revolution and sun's rotation: well explained from the original rotation of the nebula before collapse.
2. Coplanar orbits: well explained by the conservation of angular momentum applied to the nebula's rotation during the collapse.
3. Circular orbits: explained fairly well by the interactions of planetesimals.
4. Planets' rotation: expected from the spin of the nebula; exceptions may be the result of violent collisions.
5. Planets' spacings: fairly well explained by the sweeping out of zones of planetesimals by the gravitational attraction between the protoplanets.
6. Satellite systems: nebular processes on a smaller scale; giant impacts may cause exceptions.
7. Satellite spacings: sweeping up of material during early accretion of protosatellites.
8. Angular momentum distribution: involves magnetic fields and plasmas early on, to shift distribution from center to disk of nebula.
9. Comets: explained fairly well as icy planetesimals thrown out of orbits near Jupiter, Saturn, Uranus, and Neptune.
10. Planetary ring systems: explained vaguely as unaccumulated debris, perhaps as the result of impacts among satellites.

KEY CONCEPTS

1. Any model for the origin of the solar system must account for its general chemical and dynamic properties in a unified way.

2. The key dynamic properties include the following: the system is flat; most of the mass is in the center (the sun); most of the angular momentum is in the planets; the planets' orbital directions are all the same and also most of their rotational directions; and the planets are regularly spaced.

3. The key chemical properties include the general division of the terrestrial and Jovian planets, the differences among planets in each group, and the compositions of asteroids, meteoroids, and comets.

4. Interplanetary debris provides important clues about the origin of the solar system, especially from the physical properties of comets (dirty snowballs), asteroids (rocks and metals), and meteorites (rocks and metals).

5. The crystalline patterns found in iron meteorites indicate that their material cooled slowly from a molten state, which implies that they formed in larger parent bodies.

6. A spinning cloud of interstellar material will flatten out as a result of the conservation of angular momentum and also heat up as it contracts gravitationally.

7. The condensation sequence works if the temperature in the solar nebula decreases rapidly from center to edge. Which materials vaporize and condense depends on how high and then how low the temperature gets at a certain distance from the sun.

8. Magnetic fields can transfer the spin of the sun to the rest of the solar nebula (before the planets form) to account for the distribution of angular momentum now.

9. The general process of planet formation may have been condensation of grains, followed by accretion of grains into planetesimals, and clumping of planetesimals into protoplanets, culminating in the evolution of the protoplanets into the bodies we see today.

10. Jupiter and Saturn may have been formed by the gravitational contraction of large blobs of material rather than by planetesimal accretion.

11. Comets and asteroids are very likely left-over planetesimals formed in different regions of the solar system.

12. Comets interact with the solar wind to generate long, magnetized plasma tails; their icy nuclei are irregular in shape and very dusty.

STUDY EXERCISES

1. Suppose you hopped in your spaceship on Saturday night and flew to an asteroid. What would you see? (Learning Outcomes 11-2 and 11-3)

2. Then you speed off to a comet. What would you see? How would the comet's appearance differ from the asteroid you just visited? (Learning Outcomes 11-2 and 11-3)

3. After many perihelion passages of the sun, what happens to a comet in the dirty snowball comet model? (Learning Outcome 11-2)

4. How can you tell an iron meteorite from a piece of terrestrial iron and nickel? (Learning Outcome 11-2)

5. What is the major weakness of the nebular model with respect to the dynamic properties of the solar system? (Learning Outcomes 11-4 and 11-5)

6. What one planet fits least well with the general chemical and dynamic properties of the solar system? (Learning Outcomes 11-1 and 11-10)

7. Use the chemical condensation sequence to explain the general chemical differences between the earth and Jupiter. (Learning Outcome 11-6)

8. How is it that the solar system is flat rather than spherical? (Learning Outcomes 11-4 and 11-10)

9. In what way did space probes confirm the dirty snowball model of cometary nuclei? (Learning Outcomes 11-2 and 11-3)

10. Suppose you found a dense, irregular fragment on the ground. How could you tell that it was or was not a meteorite? (Learning Outcome 11-2)

11. What is the possible source of material for the intense bombardment of planetary surfaces some 4 Gy ago? (Learning Outcome 11-7)

12. How did the protoplanets acquire their original internal thermal energy? (Learning Outcome 11-7)

13. In what key way may the formation of Jupiter have differed from that of the earth? (Learning Outcome 11-8)

14. By what process does a contracting cloud of gas and dust heat up? (Learning Outcome 11-9)

15. If an initially spherical, spinning cloud of gas and dust contracts gravitationally, into what shape will it finally end up? (Learning Outcome 11-9)

PROBLEMS AND ACTIVITIES

1. Suppose that the asteroid Ceres is made entirely of rocky material and has a spherical shape. What would its mass be? Its escape speed?

2. Suppose that the nucleus of a comet was composed entirely of icy materials and had a radius of 10 km. What would its mass be? Its escape speed? Compare these values to those for Ceres.

3. Imagine a double-asteroid system with an orbital period of 4 hours and a separation of 100 meters. What would be the sum of the masses of the two asteroids?

4. If Jupiter perturbed icy planetesimals out to the Oort Cloud (with a typical aphelion of 100,000 AU), what would be their orbital periods? Assume that they now orbit the sun.

5. The asteroid Toutatis was observed at a distance of 2.5 million km from the earth. What was the resolving power needed to show that it was two separate objects?

6. When Ceres and the earth are closest together (Ceres in opposition to the earth), what is Ceres' angular diameter?

PART THREE
The Universe of Stars

PART OUTCOME

Argue that stars of different types are born out of interstellar matter and evolve as they shine; their evolution and deaths depend on their mass.

INQUIRY FOCUS

How do we know:

The physical properties of the sun and its source of energy?

The distances to stars and their physical properties?

How to organize the physical properties of stars in a meaningful way?

The processes of starbirth?

The evolutionary cycles of stars?

The ends of stars as stellar corpses?

12

Our Sun:
Local Star

CENTRAL CONCEPT

The sun produces its life-giving energy by nuclear fusion reactions transforming hydrogen to helium in its hot core. The outward flow of this energy determines the sun's physical structure.

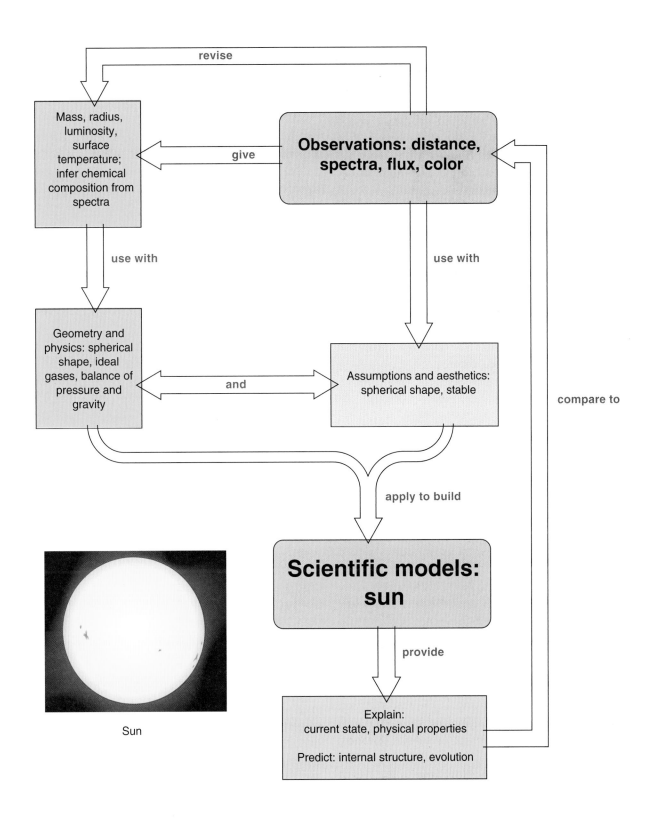

Mass, radius, luminosity, surface temperature; infer chemical composition from spectra

revise

Observations: distance, spectra, flux, color

give

use with

use with

Geometry and physics: spherical shape, ideal gases, balance of pressure and gravity

and

Assumptions and aesthetics: spherical shape, stable

compare to

apply to build

Scientific models: sun

provide

Explain:
current state, physical properties

Predict: internal structure, evolution

Sun

What is the sun? Basically, a hot, huge globe of gas with fiery fusion reactions in its core. There, deep in the heart of the sun, the energy bound up in matter is unleashed. Slowly, over about two hundred thousand years, that energy (as light) flows to the sun's surface. Free of the sun's material, it flies out to space. Only 8.3 minutes later, a very small part of that light strikes the earth, and miracles occur.

The sun serves as our local link to the stars. Our sun has the same basic structure as the other stars in the sky, which we can see directly only as pinpoints of light. How to find out the physical characteristics of these distant lights? By using our sun as a guide. The sun serves astronomers as the local laboratory for testing ideas about stars in general; our understanding of stars hinges on what we understand about our sun. That analogy marks the essence of the solar–stellar connection.

12.1 A SOLAR PHYSICAL CHECKUP

To infer the physical properties of nearby stars, we learn about the nature of the sun (Fig. 12.1). The crucial properties are mass, energy output, radius, surface temperature, and chemical composition. To tackle some of these, we need an essential fact: the distance of the earth from the sun. (You will find throughout this book that a critical question – perhaps *the* critical question in all astronomy – is how to find the distances to celestial objects.) An analysis of sunlight provides other facts without requiring knowledge of the distance. Let's see what we know more or less directly about the sun. (See the Sun Celestial Navigator™.)

How Far? (Learning Outcome 12-1)

The earth's orbit about the sun is elliptical (Section 3.5 and Enrichment Focus 3.2). The semimajor axis of its orbit, which is the average earth–sun distance, has a length called an *astronomical unit* (AU). The earth is 1 AU from the sun. How large is the AU in basic physical units, such as kilometers?

To find out, we bounce radar signals off Venus when Venus lies a known distance in AUs from the earth (Enrichment Focus 9.1). The radar measurement is made when the earth and Venus are close together on the same side of the sun. The distance at that time is some fraction of an AU (about 0.3 AU). So we know a fraction of an AU in kilometers, hence one AU in kilometers: 1.496×10^8 km. Remember it by rounding off to 150 million (150×10^6) km.

Why bother bouncing radar off a planet? Why not bounce it off the sun and measure the AU directly? Mainly because the sun, which is a hot gas and thus lacks a solid surface, does not reflect radar signals very well.

How Big? (Learning Outcome 12-2)

Viewed from the earth, the average angular diameter of the sun is 32 arcmin. (That's about 0.5°, or half the tip of your finger at arm's length.) Because we know the AU in kilometers, we get the sun's actual size from its angular size (which provides the size-to-distance ratio). For 0.5°, that ratio is about 1/110, so the sun's diameter is roughly $1/110 \times (150 \times 10^6) \approx 1.4$ million kilometers (Fig. 12.2). Imagine the earth to be the size of a U.S. dime (18-mm diameter). The sun would be about 2 meters in size and 200 meters from the coin-sized earth.

How Massive? (Learning Outcome 12-2)

To find the sun's mass, you need the earth–sun distance, the earth's average orbital velocity, Newton's second law of motion, and Newton's law of gravitation (Section 4.4). Or you can use Newton's form of Kepler's third law (Enrichment Focus 4.3), knowing the earth's distance and orbital period. The sun's mass comes out to be about 2×10^{30} kg, or more than 300,000 times the earth's mass.

How Dense? (Learning Outcome 12-2)

The sun is a big, massive object. Yet its density (the mass divided by the volume) is low: 1.4, only 40 percent denser than water! This low average density, along with its hotness, implies that the sun is a gas. In fact, the sun is a gas throughout, even in its dense core.

CAUTION. Don't confuse density with mass or size! For example, the sun is much more massive than the earth, and larger, yet it has a much lower average density. Why? The sun is a gas, while the earth is solid rock and metals.

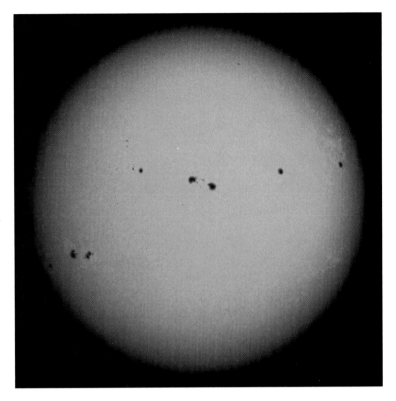

Figure 12.1 The sun's photosphere in visible light. Note that the photosphere has a granular look and fades out at the edge. The dark regions are sunspots.

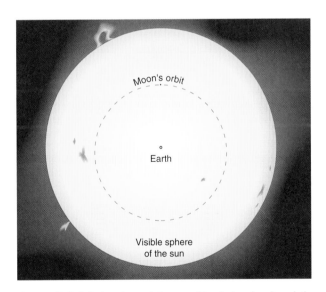

Figure 12.2 Relative sizes of the sun (its photosphere) and the moon's orbit, drawn to scale.

12.2 ORDINARY GASES
(Learning Outcome 12-12)

Because stars are gaseous, to know what happens inside the sun (and other stars), you must have some idea of how gases behave. Let me describe a simple model for ordinary gases that you can apply in most situations. (You will come across extraordinary gases in Chapter 16.)

Temperature

A gas consists of particles – atoms, molecules, or ions, or perhaps all three. Simplify the situation in a real gas by imagining all the particles to be the same, and like hard spheres. Picture these spheres trapped in a small box. Once set into motion, the spheres keep moving, bounding off the walls of the box and colliding with one another (Fig. 12.3).

These collisions push any one sphere at times faster or slower, but over time a sphere has a definite average speed. Also, collisions over time result in all spheres in the box having the same average speed. We use this average speed of particles in a gas as a measure of the average kinetic energy of its particles – a measure of its temperature. If all the particles were motionless, a gas's temperature would be zero – absolute zero on the kelvin temperature scale (Appendix A).

In the kelvin temperature scale, one kelvin is the same amount of a temperature difference as one degree in the centigrade or Celsius scale.

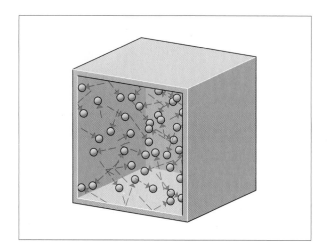

Figure 12.3 Hard-sphere model of a gas in a box. The temperature of the gas is a measure of the average kinetic energy of the particles. The internal pressure is created by many collisions among the particles.

The zero points, however, differ. The kelvin scale uses absolute zero; the Celsius scale uses the freezing point of water. On the kelvin temperature scale, the freezing point of water is 273.15 K. At room temperature, about 300 K, the average speed in a gas of hydrogen is about 3 km/s, or about 11,000 km/h! For more massive air molecules, it is about 450 m/s. (A fast baseball travels at about 45 m/s.) Higher temperatures mean higher average speeds of the particles.

Pressure

Consider putting a partition in the gas container. The spheres ram and bounce off both sides of this partition; each collision exerts a small force on it. The combined force of all collisions is the pressure of the gas. This book uses the pressure of the atmosphere at sea level as the basic unit of pressure – it's called one **atmosphere** (atm). The SI unit for pressure is the **pascal,** abbreviated *Pa* (see Appendix A). One atmosphere amounts to about 100 kPa. For comparison, the pressure of the tires on my truck is about 200 kPa. The compression pressure in a gasoline engine is roughly 1000 kPa, or about 10 atm.

Imagine that you increase the temperature of the gas. The average speed of the particles increases, so they collide with one another and strike the partition more frequently and with greater force. The pressure increases. How, in relation to the increase in temperature? For ordinary gases, the increase is a

direct one, so if you *double* the temperature (in kelvins), the pressure *doubles.*

What happens to the pressure if you increase the number of particles in the box? (You have done this when pumping up a bicycle tube.) Adding more particles to the same space increases the density of particles – on the average, each cubic meter contains more particles. For a gas, the number of particles in a unit volume is called the **number density.**

Suppose you increase the number density four times without changing the temperature. Each cubic meter now contains four times as many particles. So four times as many collisions occur, on the average, against the partition, and each collision has the same average force as before. The pressure increases; it is four times greater. Increasing the number density directly increases the pressure of an ordinary gas.

This hard-sphere model shows that gas pressure depends *directly* on both the number density and the temperature. With these basics about simple gases in mind, let's turn to that huge hot ball of gas that is the sun and infer some of its important properties.

12.3 THE SUN'S CONTINUOUS SPECTRUM

Most of what we know about the sun and other stars comes by way of light. Decoding the message of sunlight and starlight takes up much of the time, energy, and ingenuity of astronomers. Let's look at how the message is read.

Luminosity (Learning Outcome 12-2)

The sun's **luminosity** (or *power*) is its total output in radiative energy each second. By the time the light reaches the earth, it has spread out over a large region of space – in fact, over a sphere whose radius is the earth–sun distance. You cannot directly measure this radiative energy, the sun's luminosity, all at once.

How to find the sun's luminosity? To avoid the earth's atmosphere absorbing some of the sunlight, place a special detector in a satellite orbiting the earth. Point the detector directly at the sun. Measure the radiant energy absorbed each second by the detector: it amounts to 1370 watts for each square meter of the detector's area (Fig. 12.4). On a surface 1 AU in radius (an imaginary sphere surrounding the sun), every square meter catches this amount of

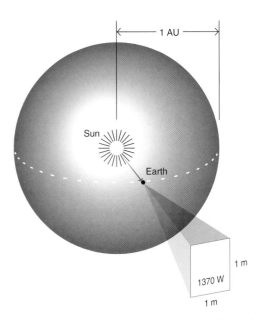

Figure 12.4 Measuring the sun's luminosity from the earth from the sun's flux at the earth and the earth–sun distance. An imaginary sphere with a radius of 1 AU would capture all the sun's radiative energy output. At the earth's distance, then, 1 square meter captures only a small part of this total energy output, or about 1370 watts.

power, which could run an electric hair dryer if converted to electricity. Totaled up over the entire surface, the emitted energy per second amounts to roughly 4×10^{26} W (Enrichment Focus 12.1).

Astronomers use a special name for the amount of energy passing through one given unit of area (such as a square meter) each second: **flux.** The sun's flux at the earth, 1370 W/m², serves as the standard of comparison for the flux from other celestial objects. When you go out on a sunny day, your skin is sensing the sun's flux.

Surface Temperature and Blackbody Radiators (Learning Outcomes 12-2, 12-3)

The sun's color (yellowish white) provides a clue to its **surface temperature.** We can assign a temperature to the sun's surface by examining its continuous spectrum, ignoring the lines for the moment. (Recall, from Section 5.1, that the sun has an absorption-line spectrum.)

To see how, consider heating a piece of metal in a very hot flame. At first the metal gives off infrared radiation before it emits any visible light. Eventually, like the coils of an electric stove at its highest setting, it emits a dull red light. As it gets hotter, it glows more brightly reddish, then orange, yellowish, yellowish green, white, and, finally, bluish white. The overall color change in the visible region of the spectrum corresponds to the temperature of the metal. Its continuous spectrum changes with temperature in a special way.

How are color and continuous spectra related? Let's look at the metal's emission in some detail (Fig. 12.5). Measure the flux at a range of wavelengths (from ultraviolet to infrared) for the metal at different temperatures. Then plot the metal's spectrum of emitted flux versus wavelength. Note three features in these spectra: the emission *peaks* at some wavelength, the peak shifts to *shorter* wavelengths as the metal gets hotter, and the metal emits more intensely at *all* wavelengths at higher temperatures.

Your eye, because it responds only to the visible region of the spectrum (violet to red), sees only a small portion of the complete spectrum. (For example, your eye doesn't sense any infrared. When an electric stove is on low, you don't see the coils glow, yet your hand can detect the infrared emitted by them.)

ENRICHMENT FOCUS 12.1

The Sun's Luminosity

The sun's luminosity, or power, is its total output of radiative energy each second. How can we determine it? By measuring the solar flux at the earth. The flux, f, equals about 1370 W/m². This amount of energy crosses each square meter on an imaginary sphere of radius d surrounding the sun, where d is the earth–sun distance in meters.

We must know this distance to calculate the luminosity. Since the area of this sphere is $4\pi d^2$, the luminosity L of the sun is given by

$$L = 4\pi d^2 f$$

Putting in the numbers, we write

$$L = 4\pi\,(1.50 \times 10^{11} \text{ m})^2\,(1.370 \times 10^3 \text{ W/m}^2)$$
$$= 3.9 \times 10^{26} \text{ W}$$

The earth intercepts only about 4.5×10^{-10} of this total. That's still a large amount, about 2×10^{17} W!

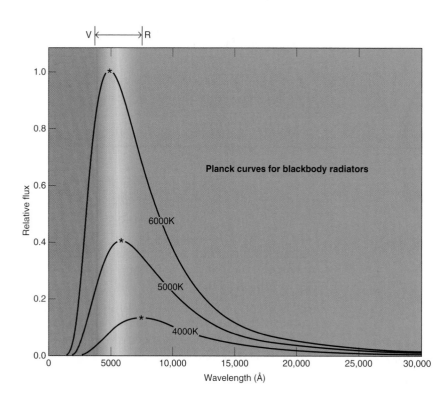

Figure 12.5 Spectrum of a hot metal (or opaque solid, liquid, or gas) as it varies with temperature (3000 to 5000 K). The visible region (from violet, *V,* to red, R) is indicated; the long-wavelength end of the curve goes out into the infrared. The asterisks mark the peak of the emission for each temperature; note that peaks occur at *shorter* wavelengths for higher temperatures. The shapes of these curves, called *Planck curves,* are typical of the continuous emission from blackbody radiators.

For the temperatures considered here, however, your eye does reasonably discern the shift in the balance of colors at the peak emission.

An object whose spectrum has this characteristic shape (Fig. 12.5) and the variation with temperature just described is called a **blackbody radiator,** or simply a **blackbody** (Enrichment Focus 12.2). Its spectrum is sometimes called a **Planck curve,** in honor of Max Planck, a pioneer in quantum physics. The spectrum of a blackbody depends only on its temperature and not on other properties (such as composition).

Perfect blackbodies do not actually exist, but the sun and similar stars emit radiation enough like an ideal blackbody to make the model worth using. (This is especially true for stars with surface temperatures about the same as the sun's.) The wavelength at which a star's continuous spectrum peaks relates directly to its temperature: *the higher the temperature, the shorter the peak wavelength.*

A common question about blackbodies: why are they called black when they give off light? Blackbodies are named for their light-absorbing abilities. They absorb light at any wavelength completely and reflect none. When a blackbody absorbs radiative energy, it heats up and emits at all wavelengths,

even though the peak of emission may not be visible to our eyes. Physically, a good absorber, when heated, is a good radiator. To conserve energy, it must emit as much energy per second as is put in. When that balance is achieved, the blackbody will have a certain temperature. Note that any *opaque* and *hot* solid, liquid, or gas will produce a blackbody spectrum, as you would expect from Kirchhoff's rules (Section 5.2). The key word here is *opaque.*

Let's examine the sun's continuous spectrum as measured at the earth's surface (Fig. 12.6). Note that when the spectrum is measured at sea level, the earth's atmosphere absorbs parts of it – especially the ultraviolet (absorbed by ozone) and the infrared (absorbed by water vapor and carbon dioxide). This absorption in the infrared causes the greenhouse effect (Section 8.4) and also forces astronomers to place infrared telescopes above the earth's atmosphere (Section 6.3). Very little of the visible light is absorbed, however. You see that the spectrum peaks at about 5000 Å. Without the atmospheric absorption (Fig. 12.6), the sun's spectrum follows a more or less continuous curve with this one peak in the yellow-green part of the spectrum. To produce this peak requires a surface temperature of about 5800 K.

ENRICHMENT FOCUS 12.2

Emission from Blackbodies

A blackbody is an object that completely absorbs any radiation that falls on it. The radiation from a blackbody has a characteristic shape (spectrum) that depends only on the temperature of the body and not on any other property, such as composition.

A blackbody has a number of special characteristics. First, a blackbody with a temperature greater than absolute zero emits some light at all wavelengths. Second, a hotter blackbody emits more light at all wavelengths than does a cooler one. Third, a hotter blackbody emits a greater proportion of its radiation at shorter wavelengths than a cooler one. Finally, the amount of radiation emitted by the surface of a blackbody depends on the fourth power of its temperature (in kelvins). If one blackbody is twice as hot as another of the same size, it emits $2^4 = 2 \times 2 \times 2 \times 2$, or 16 times as much power in total. The hotter of them also emits 16 times as much power as the cooler one from each square meter of surface.

Two of the foregoing properties of blackbodies are described by simple equations. First, the amount of power emitted for every square meter of a blackbody at temperature T kelvins is called the flux,

$$F = 5.67 \times 10^{-8}\, T^4 \text{ W/m}^2$$

This relation is called the **Stefan–Boltzmann law.** For example, every square meter of the sun's photosphere

(5800 K) emits about 6.3×10^7 W. Note that F is the flux emitted at the surface of the blackbody. Explicitly,

$$
\begin{aligned}
F &= 5.67 \times 10^{-8}\,(5780)^4 \\
&= 5.67 \times 10^{-8}\,(1.12 \times 10^{15}) \\
&= 6.33 \times 10^7 \text{ W/m}^2
\end{aligned}
$$

Second, the wavelength at which the energy output from a blackbody peaks is

$$\lambda_{max} = \frac{2.90 \times 10^{-3}}{T}$$

where T is the temperature in kelvins and $\langle wav \rangle_{max}$ is the wavelength, in meters, at which the peak output occurs. This relation is called **Wien's law.** For the sun, the peak for a temperature of 5800 K is about 4.9×10^{-7} m (0.49 μm, or 4900 Å).

Note that both the Stefan–Boltzmann and the Wien equations can be used to infer the sun's surface temperature. To use the first, you need to find out how much power each square meter of the surface puts out. For the second, you have to determine the wavelength at which the sun's spectrum peaks. These methods of determining temperature can be applied to other stars, if they also radiate roughly like blackbodies.

Note that if a star is spherical and assumed to radiate like a blackbody, we can find its luminosity from the radius and the surface temperature. The luminosity is the total surface area ($4\pi R^2$ for a sphere) times the flux, or

$$L = 4\pi R^2 F = 4\pi R^2\,(5.67 \times 10^{-8}\, T^4)$$

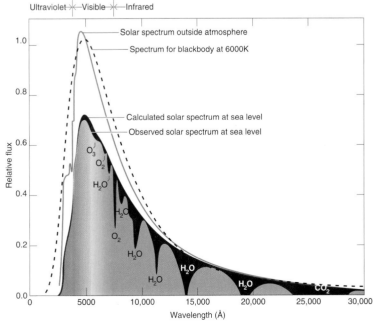

Figure 12.6 Absorption of the sun's radiation by the earth's atmosphere. The solid line indicates the sun's radiation measured above the earth's atmosphere. (Note how close this curve comes to a blackbody curve at a temperature of 6000 K, indicated by the dashed line.) The actual measured spectrum at sea level is labeled according to whether the carbon dioxide (CO_2), water vapor (H_2O), oxygen (O_2), or ozone (O_3) is responsible for the absorption. Water vapor is responsible for most of the absorption in the infrared – the major cause of the greenhouse effect.

Opacity (Learning Outcome 12-4)

If the sun's continuous spectrum has a blackbody shape, the emitting region must also be a good absorber of light. Recall (Section 5.4) that atoms can absorb light by electron transitions to higher energy levels. What specific transitions make the sun a good absorber? To see this point, you first need to understand the concept of **opacity.** On a clear day, you can see through the air to far distances (many kilometers). The air is transparent to light; its opacity is low. On a very foggy day, you cannot see far at all (only a few meters in San Francisco!). The air is opaque to light.

Interactions of light with atoms and electrons make a gas opaque. When a photon is absorbed, it no longer exists and so can't carry energy any farther. The photon's energy is not destroyed; it is transferred to an electron. When the electron loses the energy, it emits a photon. But – and this is the key point! – that photon can be emitted in *any* direction, including back in the direction from which it originated. It heads off and moves only a short distance before it is reabsorbed. When another photon is reemitted, it probably zips off in a different direction. Photons in an opaque gas travel very short distances before they are absorbed. In a gas of lower opacity, they travel greater distances. When a gas is opaque, photons bounce around, hence travel slowly through it; when it is transparent, the photons zip straight through it.

If we define the visible surface of the sun as the layer at which the gases become visibly transparent, that layer is only a few hundred kilometers thick and has an average temperature of 5800 K.

This region defines the sun's **photosphere** – the place at which photons can escape from the sun's atmosphere. Photons bouncing out from the center reach this region and then fly straight into space.

12.4 THE SOLAR ABSORPTION-LINE SPECTRUM

Recall (Section 5.2) that a spectroscope shows the solar spectrum as a dark-line spectrum. How is this absorption-line spectrum produced in the sun's atmosphere?

The Origin of Absorption Lines (Learning Outcome 12-5)

Use the concept of opacity again – for an absorption line. Consider the 4383-Å line of iron (Fig. 12.7). When we measure the absorption or emission of this line, we find that it is somewhat spread out. Its width, in terms of wavelength, is centered on 4383 Å, where iron atoms absorb the best. At somewhat shorter or longer wavelengths, say 4379 Å or 4387 Å, the iron atom can still absorb light, but not as well. In other words, a gas containing iron has a much higher opacity for 4383-Å photons (at the line's center) than for those with wavelengths shorter or longer than the central one.

CAUTION. Absorption lines are *not* perfectly black; they do contain some energy. They are dark only relative to continuous emission at neighboring wavelengths.

In the photosphere, the temperature rises quickly as you go down into it: a few thousand kelvins in a few hundred kilometers. This sharp temperature change results in the sun's dark-line spectrum.

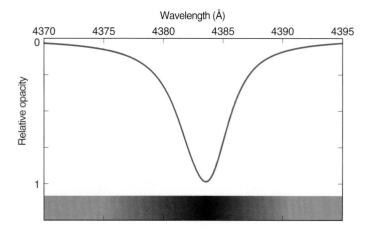

Figure 12.7 Absorption profile of the 4383-Å line of iron in the photosphere of the sun. This profile results from the ability of the iron atom to absorb photons at wavelengths near 4383 Å: that is, the relative opacity. The opacity is the greatest at the center of the line. Note how rapidly it decreases away from the center of the line.

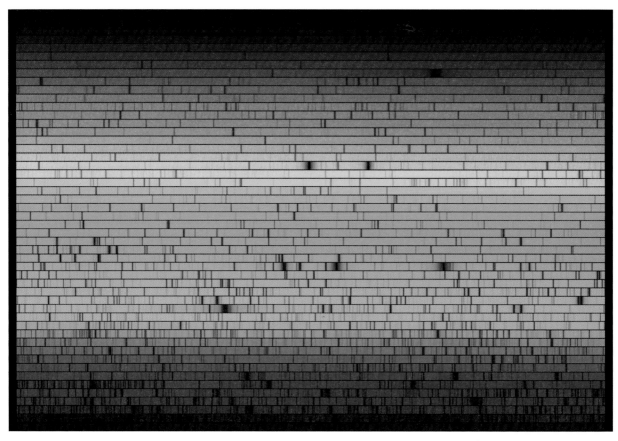

Figure 12.8 The sun's spectrum. This high-resolution image shows wavelengths increasing from left to right along each strip and from bottom to top. Each of the 50 slices covers 60 Å for a complete spectrum across the visual range, 4000–7000 Å.

You can regard the atmosphere as overlying a hot surface, just the situation needed, according to Kirchhoff's third rule (Section 5.2), to produce an absorption-line spectrum (Fig. 12.8).

The Chemical Composition of the Photosphere (Learning Outcome 12-6)

Astronomers have analyzed tens of thousands of absorption lines in the solar spectrum to find the chemical composition of the atmosphere. The intensity of the dark lines from a particular chemical element depends on the amount of that element found in the atmosphere. The line's intensity also depends on the temperature and density in the photosphere. Iron produces most of the absorption lines; other strong lines come from hydrogen, calcium, and sodium (Section 5.4). Identifying particular elements requires matching their "bar code" patterns. Like products in a store, each has a unique bar code to identify it.

It is a harder task to find out an element's actual *abundance*. To determine the abundance, you need to know how atoms of a particular element absorb and emit light. Then you must make a model of the sun's photosphere. You can construct this model from theoretical calculations and basic physical concepts. It consists of a list of temperatures, densities, and pressures at different depths in the photosphere. Astronomers try to match the calculated intensity to the observed intensity by playing with different values of the abundance. A match between observed and calculated intensities gives the correct abundance.

Most of the mass is hydrogen (74 percent) and helium (25 percent). All other elements (loosely called metals or heavy elements by astronomers) make up a mere 1 percent. These abundances are relative to the total mass in the photosphere.

CAUTION. This procedure gives only the abundance in the *photosphere*. It does not tell you the abundance in the *interior*. (Why not?)

If an element's absorption lines do not show up in the visible spectrum, does it mean that this element does not exist in the sun? Not necessarily! Perhaps so little of the element exists that it does not produce detectable absorption lines. Or the strongest absorption lines may be in a region of the sun's spectrum that we cannot observe through the earth's atmosphere. Or the temperature, pressure, and density of the sun's atmosphere may inhibit the formation of the element's spectral lines. If the conditions are too hot or too cool, the lines do not form.

Take helium, for example. The sun's visible spectrum does not show any absorption lines of helium. Why not? In the photosphere, the temperature is too low to excite helium to levels that can absorb photons of visible wavelength. Above the photosphere, the temperature rises to 40,000 K, which is high enough to excite helium atoms, and during an eclipse you can find bright lines from helium. In contrast to the continuous spectrum, however, too few excited helium atoms exist in this hot region to produce emission lines strong enough to be seen directly.

12.5 ENERGY FLOW IN THE SUN

The energy flow controls the internal structure and environment of the sun. The sun that we see directly embodies the sun's outer layers, together known as the *atmosphere*. Because the sun's atmosphere is a gas, it does not have sharp boundaries. But we have been able to discover that the atmosphere has three different zones: the photosphere, the chromosphere, and the corona. These regions are the visible expressions of the sun's energy outflow. How is energy transported from place to place?

Conduction, Convection, Radiation (Learning Outcome 12-10)

In general, conduction, convection, or radiation carries energy. If you have ever sat in front of a roaring fire, you have experienced these processes.

You are directly warmed by the fire's heat, infrared radiation that travels directly from the fire to be absorbed by your skin. That's energy transported by radiation.

Much of the energy from the fire, however, ascends into the air. The fire heats the air just around it. Air is gas, which expands and becomes less dense when heated. Cooler, denser air flows in and pushes the hotter air up and out of the house, where it cools off. This transport of energy by mass motions of a gas (or liquid) is convection.

Finally, you may have left the poker in the fire. If you grab the handle without thinking, you will find it very hot. This is a result of conduction: the poker's electrons that were actually in the fire were heated and so moved around at high speeds. They banged into their neighbors, agitating them. These particles collided with their neighbors, and so on throughout the poker, from one end to the other. The kinetic energy was transferred by direct collisions along the poker.

In ordinary stars like the sun, radiation or convection can transport energy. Conduction does not play an important role because the sun's material is a gas. However, conduction does become important in extremely dense stars, where matter is in a state different from the ordinary.

Which process is at work? That depends on the local conditions in the gas. The general rule is this: the process that works most efficiently is the one that takes over. This efficiency depends on how well radiation transports energy through a material, and that rests on opacity. Basically, opacity gauges how *efficiently* some material absorbs photons.

In the sun's interior, photons travel about a millimeter between absorptions. The gas here is fairly opaque, but radiation is still more efficient than convection for transporting the energy. As photons journey toward the surface, they run into a region with a temperature low enough to make the gas extremely opaque. The photons are bouncing around so much that they make little progress outward, and so they do not transport energy very well. Convection by flow of the gas transports energy more efficiently. A **convection zone** forms. This zone makes up the final 0.3 of the sun's radius.

Photosphere (Learning Outcome 12-10)

The photosphere has a bubbly look, like the surface of a pot of boiling oatmeal (Fig. 12.9). Each bubble has an irregular shape about 1000 km across (about

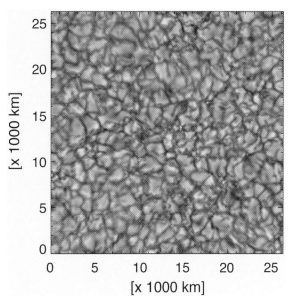

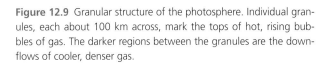

Figure 12.9 Granular structure of the photosphere. Individual granules, each about 100 km across, mark the tops of hot, rising bubbles of gas. The darker regions between the granules are the downflows of cooler, denser gas.

the width of Egypt!) and lasts for about 10 minutes. You can visualize the gas here as percolating over the sun's surface – millions of bubbles at a time.

A convection zone below the photosphere creates the bubbles. The situation resembles that of the formation of cumulus clouds in the earth's atmosphere. Here on earth, hot gases rise, carrying moisture. As they rise, they cool, and eventually the water condenses to form clouds. The photosphere marks the top of the sun's convective zone.

Chromosphere (Learning Outcome 12-10)

The chromosphere, the layer just above the photosphere, is a transparent zone (otherwise we could not see the photosphere!) marked by emission lines. The temperature, density, and pressure in the chromosphere determine the intensities of its emission lines. These line intensities provide clues to the physical conditions there.

The chromosphere begins a few hundred kilometers above the top of the photosphere and extends only about 2000 km higher, where it merges into the corona. The chromosphere has a density about a thousand times less than the photosphere but, surprisingly, it gets much hotter. The clearest way to distinguish these different regions of the atmosphere is by their temperatures (Fig. 12.10). The temperature rises from 4300 K to above 400,000 K in the first 2500 km of the chromosphere above the photosphere. This rise to high temperatures produces the emission lines from this region.

Why is the chromosphere generally hotter than the photosphere? The process probably works from the energy bound up in the magnetic fields that project up from the photosphere. You know that magnetic fields store energy if you have ever pulled two magnets apart: you have to put out an effort – work – to separate them. The trick, then, is to learn how the magnetic energy moves upward from the photosphere. Magnetic fields can transport energy by moving material or by generating electric currents that move through conducting materials. Currents flowing upward from the photosphere to the corona can leave some energy in the chromosphere.

Corona (Learning Outcome 12-10)

You can see the sun's splendid corona directly during a total solar eclipse (Fig. 12.11). Although as bright as the full moon, the corona is normally obscured by the sunlight scattered in the earth's atmosphere. During a total eclipse, when the moon blocks the photosphere out, the sky becomes dark enough for the corona to be visible.

Spectroscopy reveals that the corona emits bright lines from highly ionized atoms of iron, nickel, neon, and calcium. (Recall that when an atom loses an electron, its energy levels change.) Because it takes large amounts of energy to rip many electrons from an atom, the corona must be very hot.

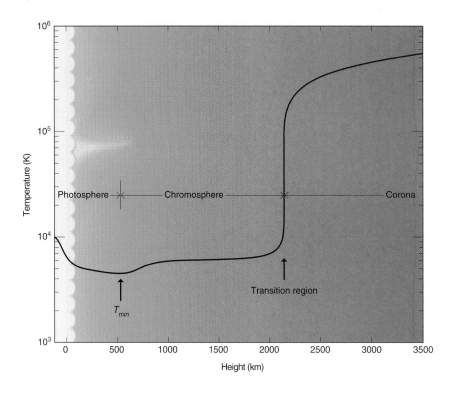

Figure 12.10 Temperature profile of the sun's atmosphere. The height is taken as the distance from the base of the photosphere. Note how the temperature decreases in the photosphere and drops to a minimum value. It then rises fairly gradually in the chromosphere. A sharp rise marks the transition between the chromosphere and the corona. Note the powers-of-ten scale for the temperature axis. (Adapted from a diagram by Eugene H. Avrett.)

For example, to strip iron of 16 of its normal 26 electrons requires a temperature greater than 2 million kelvins, and temperatures this high occur in parts of the corona.

What makes the corona so hot? At the base of the corona, the temperature rises rapidly, roughly 500,000 K in a few hundred kilometers, in a thin transition region between the chromosphere and corona (go back to Fig. 12.10). This sharp rise may result from the same physical process as in the chromosphere: transport of energy by magnetic fields.

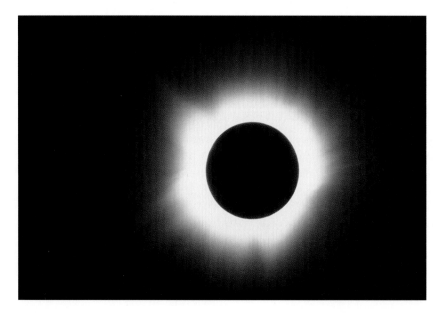

Figure 12.11 The sun's corona, seen during a total eclipse. Note the structures in the outer regions.

Solar Wind (Learning Outcome 12-10)

The solar wind pervades the solar system. It springs from the sun's corona, which extends into space to distances greater than those indicated by its visual appearance. The outflow of material occurs because the high temperature of the coronal gas causes it to exert a large outward pressure – a greater force than the inward pull of gravity. As a result, gas from the corona streams away from the sun with greater than escape speed. The wind is a high-speed plasma flow, carrying trapped magnetic fields along with it.

At the earth's orbit, the solar wind typically whips by at 500 km/s. The speed varies as the solar wind blows in gusts. The particles in the solar wind (mostly protons and electrons) travel the distance from the sun to the earth in about 5 days. The earth swims through the solar spray and catches some of the solar wind particles in its magnetic field to create its magnetosphere (Section 8.3).

The plasma of the solar wind carries along remnants of the sun's magnetic field. This transported field makes up the magnetic field between the planets. The field interacts with planets (to make their magnetospheres) and with comets (to make their tails). The wind travels as far as 100 AU, well beyond the orbit of Pluto.

12.6 THE SOLAR INTERIOR

The interior of the sun contains the bulk of the sun's mass and the furnace that generates its energy. All the features of the sun, from photosphere to solar wind, result from the energy flow out from the interior. How does the sun generate its energy?

Energy Sources (Learning Outcome 12-7)

During the nineteenth century, when the earth was found to be very old, attempts to explain the source of the sun's power became embarrassingly difficult. The crux of the problem is not the rate of energy production but its longevity. The sun has been at roughly the same luminosity for at least 3.5 Gy, according to present geologic evidence. Ordinary chemical reactions could *not* provide the amount of energy necessary for so long a history. If the sun were composed entirely of oxygen and coal, it would have burned to a dark cinder in only a few thousand years.

In the middle of the nineteenth century, both Hermann von Helmholtz (1821–1894) and William Thomson, Lord Kelvin (1824–1907), proposed that the sun shone because it was releasing gravitational energy by shrinking. That is, gravitational contraction converted gravitational potential energy to radiative energy (Section 11.5); half goes into heat energy and the other half comes out as light.

Because of the sun's substantial mass, a contraction rate of only tens of meters a year would liberate the required energy. Gravitational energy stored in the sun would last for about 50 million years, far longer than the earth's age as believed in the nineteenth century. When later geologic investigations expanded the earth's age to gigayears, however, the sun's resources of gravitational energy no longer sufficed to account for the sun's shining. What does?

At the beginning of the twentieth century, Albert Einstein provided the key idea to explain the sun's energy. In the special theory of relativity, mass and energy are related by

$$E = mc^2$$

where E is the energy (in joules) released in the conversion of mass m (in kilograms), and c is the speed of light (in m/s). Because c^2 is a large number (about 9×10^{16}), a minute mass stores enormous energy. For example, the conversion of 1000 kg of matter into energy unleashes about 9×10^{19} J, roughly equal to the total energy consumption of the United States in one year!

Nuclear Transformations (Learning Outcome 12-7)

How is matter changed into energy? Two operations in nature unleash the energy frozen in the nucleus of an atom: **nuclear fission** and **nuclear fusion.** In the process of nuclear fission, the nucleus of an atom of a heavy element (such as uranium or plutonium) splits into two lighter nuclei. The mass of the remnants adds up to less than that of the original nucleus. The deficit in mass is released as energy. In nuclear fusion, the nuclei of atoms of lighter elements are fused together to create a heavier nucleus. However, the product has less mass than the original particles. The missing mass has been converted to energy.

The sun produces energy by *fusion*. Hydrogen is its most abundant element and also has the smallest nuclear charge (one proton). The fusion of hydrogen nuclei results in the production of helium. To make a helium nucleus from hydrogen nuclei releases 4.2×10^{-12} J of energy. That's a minuscule amount – lighting a match produces 10^3 J! Many, many hydrogen nuclei must react each second to supply the sun's power.

Fusion Reactions
(Learning Outcome 12-8)

Two sets of fusion reactions are possible to transform hydrogen to helium: the **proton–proton chain** (*PP chain* for short) and the **carbon–nitrogen–oxygen cycle** (*CNO cycle*). The CNO cycle contributes a minor amount to the energy of the sun, but acts as a key source in more massive stars.

Let's look at the steps in the PP chain. A collision between two protons starts it off (Fig. 12.12). If these nuclei collide with enough energy (a temperature of at least 8×10^6 K), the protons stick together. A heavy hydrogen nucleus (^{2}H) forms, consisting of a proton and a neutron, n. The other positive charge breaks away as a positron. (A **positron** is the antiparticle of the electron and carries a positive instead of a negative charge.) Almost lost in the shuffle is a neutral, supposedly massless particle called a **neutrino, ν,** which zips away at light speed. The dense solar interior offers no barrier to the neutrino's travel. In about 2 seconds, the neutrino escapes into space, carrying away some energy.

In the solar interior, the positron quickly collides with an electron, and the two antiparticles annihilate to form two gamma rays. (When matter and antimatter collide, they are transformed from matter to light.) Meanwhile, the heavy hydrogen crashes into another proton and forms light helium (^{3}He) and a gamma ray. This sequence is repeated. Another light helium nucleus is created; it collides with the one just formed. In this final reaction of the PP chain, the usual result is ordinary helium (^{4}He) plus two protons and another gamma ray.

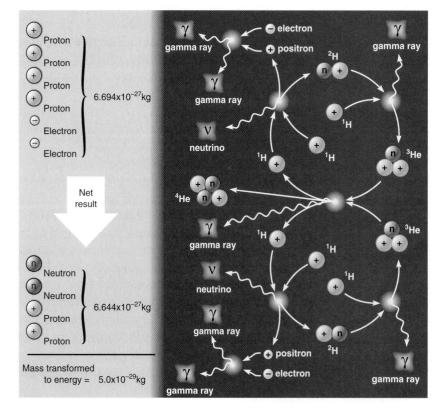

Figure 12.12 Primary proton–proton reaction that occurs in the core of the sun. Net result is the formation of one helium nucleus that has a mass of about 5.0×10^{-29} kg less than the unbound particles that go into the reaction. The missing mass has been released as energy in these fusion reactions.

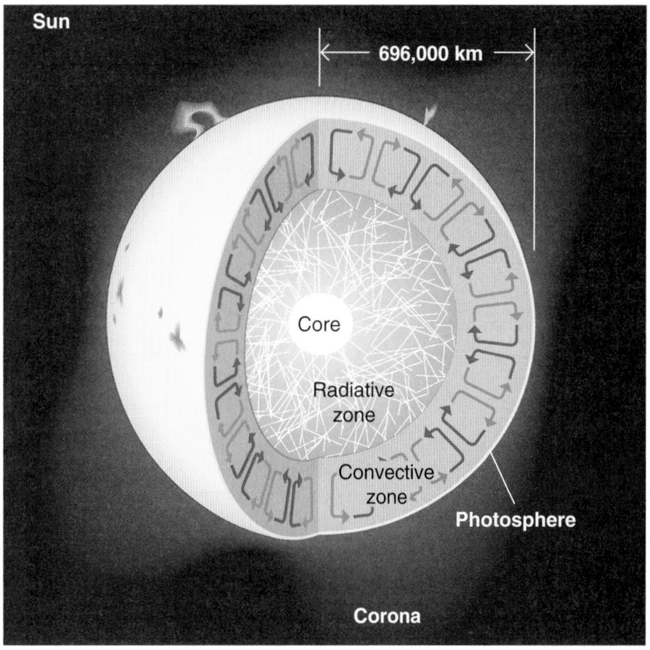

Figure 12.13 Model of the general interior structure of the sun. The fusion reactions take place in the core; the energy released here flows as radiation through the radiative zone. When the opacity becomes high enough, energy transport occurs by convection. Above the convective zone lies the photosphere. The corona marks the outer limit of the sun's atmosphere.

Keep in mind that very high temperatures (at least 8×10^6 K) and high densities are needed for protons to collide, not only with enough energy to fuse, but frequently enough to generate the sun's energy. This requirement restricts the energy production to the sun's core, about the inner 25 percent of its radius (Fig. 12.13). Only here does the temperature hit some 8 million K; it reaches almost 16 million K at the very center of the core.

Each completed PP chain unlocks about 4.2×10^{-12} J. To account for the sun's luminosity, 1.4×10^{17} kg of matter must be converted to energy each year. That amounts to 4.5×10^6 metric tons of matter transformed each second, or about the mass of a million automobiles!

TO SUM UP. The input for the PP chain is six protons and two electrons, and the usual output is one helium nucleus, two protons, two neutrinos, and miscellaneous gamma rays. Two protons come out, so only four are involved in the net result. Note that the net result of this nuclear cooking is the creation of helium and energy.

The neutrinos fly right out into space with about 2 percent of the energy, but the gamma rays, which carry off the bulk of the energy, find the sun's inte-

rior opaque. Photons bounce along random paths as they are absorbed and reemitted. Slowly, the photons creep out toward the sun's surface, from regions of higher to lower temperature (Fig. 12.14). Radiation carries the energy through a large fraction of the interior – called the **radiative zone.** Because of the temperature decline, an average photon's energy declines. The original gamma rays are degraded, by interaction with the sun's material, into many lower-energy photons. In about two hundred thousand years, the photons reach the convective zone and then finally break out of the photosphere in the form of less energetic, visible radiation.

The sun is a fusion furnace, forging helium from hydrogen in its hot core. Slowly, the core's helium abundance increases, and its hydrogen abundance decreases. The energy emitted now was produced in the core an average of some 200,000 years ago.

How long can the sun survive at this rate? The sun has enormous hydrogen supplies. However, only the hydrogen in and close to the core can burn. This amounts to about 10 percent of the total. Also, the PP chain transforms only 0.7 percent of the mass into radiant energy. Even with these restrictions, the sun's fusion energy can last about 10 Gy; it is now about 5 Gy old. Fusion solves the energy problem for the sun.

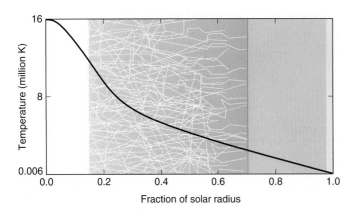

Figure 12.14 Temperature profile of the sun's interior. Note the sharp drop from the core to the photosphere. (Based on a theoretical model by John Bahcall and R. Ulrich.)

Solar Neutrino Problem
(Learning Outcome 12-9)

How to visualize the sun's interior when we cannot see below its surface? We have some idea of the interior conditions because astrophysics can make theoretical models of the sun. We do this by applying simple physical laws to a ball of hot gas, mostly hydrogen and helium, at whose center proton–proton reactions take place. High-speed computers calculate the models, which match the observed characteristics of the sun. These models show, for example, how the sun's temperature increases from edge to center.

These models of the sun's interior have been constructed as carefully as possible. However, ongoing experiments cast some doubt on them. Remember the neutrinos produced in the PP chain? These particles fly directly out of the sun's core. If we could detect these neutrinos, we could see into the sun's interior. (During the time you took to read this sentence, trillions of neutrinos have passed through your body!)

The experimental results? Detected solar neutrinos amount, on average, to only about 50 percent the number predicted from standard models of the sun's interior and our current knowledge of the nuclear reactions involved! To put the result another way, the experimental result implies that PP reactions take place now in the core at one-half the rate needed to account for the sun's luminosity.

These results suggested that standard solar models were incorrect or that something unknown happened to neutrinos in transit from the sun. We were stumped. Now the problem appears solved! The Sudbury Neutrino Observatory in Ontario, Canada, has been detecting one kind of solar neutrino since

1999 – at the rate of about ten a day. The telescope's detector is heavy water, in which one hydrogen atom has been replaced by deuterium. Recall (Section 5.1) that deuterium contains a proton plus a neutron. Very rarely the neutron converts to a proton when heavy water absorbs a neutrino.

The measured count confirms predictions of standard solar models, for neutrinos that are associated with electrons. The inference: it's the neutrinos, not the sun! These electron neutrinos transform into other types during their sun–earth trip. That conversion also implies that neutrinos have mass, albeit a very small one.

Solar Vibrations and Interior

A way to decipher the nature of the sun's interior comes from a surprising source: sound waves. Wave motions (somewhat like the seismic waves in the earth; Section 8.2) resonate in the sun's interior and appear on the surface. The sun rings like a bell at different frequencies, just as the earth's interior rings from earthquakes. The vibrations cause the gases of the photosphere to move up and down (Fig. 12.15) – motions visible as Doppler shifts of the photospheric gas.

The vibration first observed by solar astronomers involves 5-minute oscillations of the photosphere. They grow and decay over about a week. Such vibrations are excited by the turbulence in the convective zone. They involve millions of tones turned on by the noise of convection and allow us to probe the physical properties of the convection zone. The results indicate that the convective zone goes somewhat deeper than we had first thought – the outer 30 percent of the radius.

With suitable computer models of the vibra-

Figure 12.15 Computer model of some of the oscillations of sound waves in the solar interior. Blue represents approaching (expanding) regions; red, receding (contracting) ones. The patterned region just below the surface represents the sun's convection zone.

tions, we can infer some of the sun's interior properties. For example, the helium abundance of the interior affects the tuning of the sun's vibrations. The best match to the observations requires about 29 percent helium – higher than that inferred from the atmosphere. But this high-helium model requires an even higher production rate of solar neutrinos, aggravating that problem. The results pretty much rule out a rapidly rotating solar core or extremely strong internal magnetic fields – two options that could decrease the neutrinos.

12.7 THE ACTIVE SUN (Learning Outcome 12-11)

Events of the active sun are localized, short-lived phenomena on or near the solar surface. The most important of these – **sunspots** and **flares** – are associated with the locally intense solar magnetic field. In general, these areas of solar activity are called **active regions.**

One other solar property contributes to active-sun phenomena: the sun's nonuniform rotation. Because the sun is a gas, it spins faster at the equator (one rotation every 25 days) than at the poles (one rotation every 31 days) and thus shows differential rotation. As you travel from a pole toward the equator on the sun, the photosphere's rotation rate increases. Differential rotation can distort solar magnetic field lines and drive the development of active regions.

Sunspots

Sunspots appear as dark blotches on the solar disk (Fig. 12.16). With a temperature of about 4200 K, a sunspot is relatively cooler than the photosphere and so appears dark in contrast. In fact, a sunspot is almost four times fainter than the photosphere (it emits one-fourth the flux).

Sunspots tend to form in groups. They are born in an active region where photospheric granules separate, and a tiny spot appears between them as a dark pore. Such pores have magnetic field strengths averaging 0.25 T – thousands of times stronger than the sun's overall magnetic field. Usually more pores soon become visible and coalesce over a period of several hours to form a sunspot. A large, single sunspot group may contain a hundred individual spots, persisting for two or more solar rotations.

Sunspot Cycle

The sunspot numbers vary periodically. On the average, about 11 years pass between sunspot maxima (times of greatest numbers of sunspots) and sunspot minima (times of least numbers of sunspots). As the sunspot numbers vary during a cycle, the sunspots' positions also change. Sunspots usually appear only in the zone between the solar equator and 35° north or south latitude. At the start

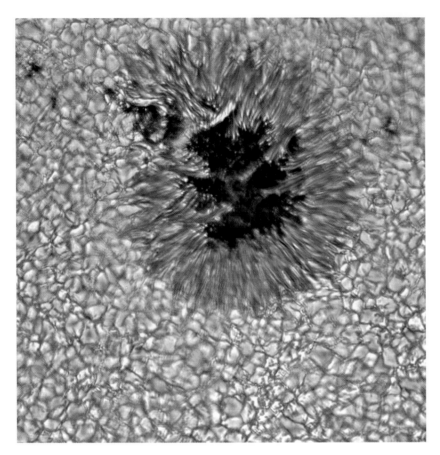

Figure 12.16 False-color image of a large sunspot. Note the dark central region; here the magnetic fields are most intense. Smallest features visible are about 100 km across.

of a sunspot cycle, a few spots emerge at high latitudes. As the cycle progresses, the sunspot latitude zone migrates toward the equator; new spots appear in this band. As the survivors of one cycle expire near the equator, new spots from the next cycle form at the higher latitudes.

In 1908 George Ellery Hale (1868–1938) detected intense magnetic fields associated with sunspots. The strongest magnetic sunspots have field strengths exceeding 0.4 T. (That's about 8000 times stronger than the earth's field at its surface.) Hale also noticed that sunspot groups contain spots of opposite magnetic polarity. For example, the east and west spots of a twin group might have north and south polarity. If this arrangement appears in the northern hemisphere of the sun, the situation at the same time in the southern hemisphere is reversed (Fig. 12.17).

Later, Hale discovered that the magnetic polarities for the west and east spots have a 22-year average cycle, just double the 11-year average cycle of sunspot numbers. That is, if a spot group has the west spot with a south polarity, in the next sunspot cycle the west spot will have a north polarity (Fig. 12.17), and in the following cycle, south polarity again.

Current investigations indicate that there is little historical evidence of an 11-year cycle in sunspot activity before 1700. Hardly any sunspots were seen in the 70-year period from 1645 to 1715 (Fig. 12.18). This period corresponded to an unusual cold spell sometimes called the Little Ice Age (the average temperature of the earth dipped about 0.5 K) that extended from the sixteenth to the eighteenth century.

The relative consistency of the cycle in modern times may be a brief phase that recurs over longer times. Evidence in the layering of Australian rocks some 700 million years old hints at an activity cycle very much like that recorded since A.D. 1700. Overall, solar activity behavior may be much more complex than we have inferred from the limited time spans investigated so far. It may well affect the earth's climate, albeit in complex ways.

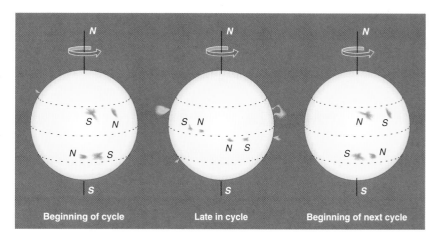

Figure 12.17 Magnetic cycle for sunspots. At the beginning of a sunspot cycle, if spots with north polarity are leading groups in the sun's northern hemisphere, then south polarity leads groups in the southern hemisphere. At the beginning of the next cycle, the polarity of the leading spots is reversed for each hemisphere.

Physical Nature of Sunspots

Enormous electric currents generate sunspot magnetic fields, much as the field of an electromagnet is produced. Exactly how this formation happens is unclear. One possibility is that the sunspots' magnetic fields suppress the hot gas rising from the convective zone. The hot gas runs into a magnetic thicket and has trouble breaking through the surface in the sunspot's center (Fig. 12.19). A convective downflow may draw away some of the hot gas, removing heat from the region. As a result, the sunspot is cooler and darker than the surrounding photosphere.

Flares

Sunspots are floating islands of electromagnetic storms. Associated with them in active regions are short-lived, violent discharges of energy called *solar flares*. These energetic bursts appear with sunspots and sometimes bridge the gap between two close spots. Near large sunspots, about a hundred small flares occur each day. The elapsed time between the birth of a flare and its rise to peak intensity is only a few minutes. Emitting myriad forms of energy – x-rays, ultraviolet and visible radiation, high-speed protons, and electrons – a large flare blows off about 10^{25} J (Fig. 12.20), the equivalent of the energy released by a bomb of 2 gigamegatons! (A *megaton* is the energy equivalent to that of an explosion of a million tons of TNT.)

Because large active regions are the most frequent locations for large solar flares, astronomers think that the concentration of energy arises from local twists and kinks in the magnetic field loops. The energy accumulates until its release is triggered. The flare starts in the corona and slams down into the chromosphere in the form of high-speed electrons.

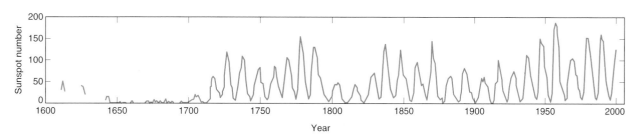

Figure 12.18 Sunspot cycles from 1600 to 2000, plotted in terms of the number of observed sunspots monthly during a year. The peaks (sunspot maxima) come roughly 11 years apart. Note how few sunspots were seen prior to 1700, compared to modern times. (Adapted from a diagram by J. Eddy.)

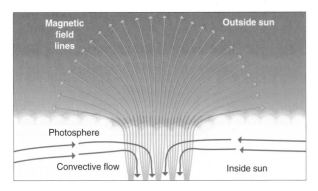

Figure 12.19 One model of the distribution of magnetic field lines in a sunspot driven by the convective upflow and downflow below the photosphere. The convective upflow of the sun's plasma pushes the magnetic field lines together. (Adapted from a figure by E. N. Parker.)

Flares blast energetic particles, mostly protons and electrons, into space. A majority of the flare's energy, however, departs as x-ray and ultraviolet light. Arriving at the earth about 8.3 minutes later, the flare's radiation rips through the upper atmosphere and tears electrons from neutral atoms. The increase in ionization disrupts long-distance radio communication. A few days later, the slower protons and electrons approach the earth, but they are usually trapped by the earth's magnetic field. These particles are called *solar cosmic rays.* They can injure unshielded humans in space (or on the moon). Luckily, the earth's atmosphere and magnetosphere guard us from the hazards of such outbursts.

Occasionally the earth's magnetic reservoirs overflow with charged particles, particularly at times of sunspot maxima when flare activity is at its peak. As the particles spill into the earth's upper atmosphere, the swift electrons bump into atmospheric atoms and excite them. When the atoms de-excite, they emit visible radiation, causing in the sky the glow called the *aurora.*

Coronal Loops and Holes

Because the coronal gas is so hot, it emits low-energy x-rays and has an irregular distribution above and around the sun (Fig. 12.21). The large loop structures indicate where the ionized gas flows along magnetic fields that arch high above the sun's surface and return to it. The hot gas is trapped in these magnetic loops (Fig. 12.22). Solar physicists now view the corona as consisting primarily of such loops.

Some regions of the corona appear dark, especially at the poles and down the middle part of the sun. Here the coronal gas is much less dense and less hot than usual – these regions are called **coronal holes.** The coronal holes at the poles do not appear to change very much, but those above other regions are related to solar activity.

Figure 12.20 Material ejected by an energetic solar flare.

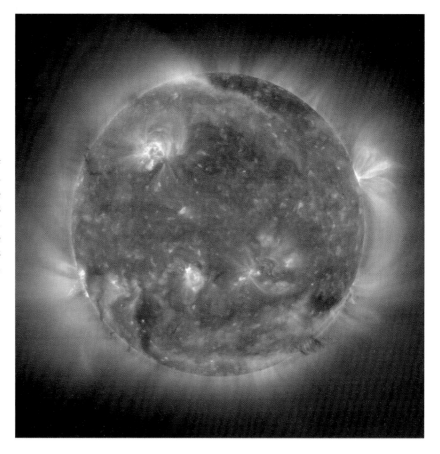

Figure 12.21 False-color SOHO image of the sun showing fine structure in the coronal loops, where the plasma temperature is one million kelvins. The active regions are strung out parallel to the equator. Note how the plasma follows along the magnetic field lines. This composite was made at wavelengths of 171 Å, 195 Å, and 284 Å (red, yellow, blue).

Figure 12.22 False-color TRACE image of hot plasma (temperature about one million kelvins) following along magnetic field loops. Wavelength of observation was 171 Å.

What makes a coronal hole? Solar astronomers believe that coronal holes mark areas in which fields from the sun continue outward into space rather than flowing back to the sun in loops. The coronal gas, not tied down in these regions, can flow away from the sun out of the coronal holes; this flow makes the solar wind.

The coronal gas does not follow the differential rotation of the photosphere. Rather, it rotates at the same angular speed at all latitudes (as the earth does). This fact implies that the bottoms of the magnetic loops are anchored deep below the photosphere, perhaps at the very bottom of the convective zone. There a solar dynamo might generate such fields, where organized flows of electric currents generate magnetic fields.

Now back to why the corona (like the chromosphere) is so hot. The magnetic loops, rising from the photosphere up to 400,000 km into the corona, play the key role. Photospheric motions at its base twist the magnetic field in a loop. Slow twisting generates electric fields that then heat the coronal gas. This doesn't take much energy, because the coronal gas is so thin that it holds little heat.

KEY CONCEPTS

1. The astronomical unit (AU), the average earth–sun distance, is about 150 million km. It's found with precision by bouncing radar signals off Venus and measuring their round-trip travel time when Venus and the earth are separated a known distance in AUs.

1. The astronomical unit (AU), the average 2. With the distance known, the sun's size is found from its angular size, its mass from the earth's orbital period and Newton's law of gravitation and second law (or Newton's version of Kepler's third law), and its low bulk density from its volume and mass. The sun is an ordinary gas throughout that obeys simple laws relating pressure, temperature, and density.

3. To find the sun's luminosity (total radiative energy output per second) requires a knowledge of the earth–sun distance in kilometers and a measurement of the sun's flux at the earth. We measure the flux above the atmosphere with satellite detectors because the atmosphere absorbs some of the incoming light.

4. The sun's surface temperature can be found from its color or the wavelength at which its continuous spectrum hits a peak intensity, if we assume that the sun radiates like a blackbody. (It does, more or less.) Blackbody radiators are hot, opaque bodies whose continuous emission depends only on their temperatures.

5. The solar gases become transparent to light at the sun's surface – its photosphere. The continuous spectrum forms in this region; the absorption lines form in higher, cooler layers. Where the lines form depends on the opacity of the material at different levels.

6. The sun's absorption lines indicate what elements are in the photosphere. The photosphere contains by mass about 74 percent hydrogen, 25 percent helium, and 1 percent all other elements (generally called "metals" by astronomers).

7. Energy flows from the sun's hotter core to its cooler photosphere. This outflow of energy shapes the sun's structure (Table 12.1). For most of the interior, radiation transports the energy. The photosphere has a bubbly structure, indicating that a zone of convection exists beneath it. This convection carries energy to the surface to be radiated into space. The chromosphere above the photosphere is hotter and visible by its emission lines during an eclipse. The corona is even hotter than the chromosphere, as indicated by emission lines of highly ionized atoms.

8. The sun's energy comes from fusion reactions in which four protons are fused into a helium nucleus with the release of energy. Only in the sun's core is the temperature high enough for these fusion reactions to occur. The sun has enough hydrogen to fuel these reactions for gigayears. Gradually, as its hydrogen decreases, the sun's core increases in its percentage of helium.

9. Solar neutrino telescopes have found far fewer neutrinos than predicted by models of the sun and nuclear reactions. This result, if it stands, implies either that our models of the interior are wrong or that we have something new to learn about neutrinos. Current ideas point to the neutrinos as the culprits, not the solar models.

10. The sun undergoes cycles of activity in which active regions – where sunspots and flares tend to be found – become more common and then less so; this cycle lasts an average of 22 years, as indicated by the polarity of sunspots. The magnetic field and activity cycle are driven by the sun's convection and differential rotation.

11. Active regions result from the development of strong – thousands of times higher than the average – local magnetic fields, which produce sunspots (dense concentrations of magnetic field lines) and solar flares, which emit electromagnetic radiation (especially x-ray, ultraviolet, and radio) and high-speed particles that cause auroras when they reach the earth. The solar wind leaves the sun through coronal holes, cool regions in the corona from which the sun's magnetic field streams into space.

12. The sun serves as the local model on which we base our understanding of other stars. In particular, we expect them to be massive, gaseous bodies (made mostly of hydrogen and helium) whose radiance comes from fusion reactions in their dense, hot cores.

Table 12.1 Summary of the Sun's Structure

Region	Properties
Core	Site of fusion reactions Temperature 8×10^6 to 16×10^6 K Energy transport by radiation
Convection zone	Energy transport by convection Temperature below 500,000 K Outer 0.3 of the sun's radius, below photosphere
Photosphere	Origin of continuous and absorption-line spectra Temperature from 6400 to 4200 K Energy transport by radiation
Chromosphere	Energy transport by radiation and magnetic fields Temperature from 4200 to 10^6 K Just above photosphere
Corona	Energy transport by radiation and magnetic fields Temperature from 1 to 2×10^6 K Source of the solar wind

STUDY EXERCISES

1. Why do astronomers bounce radar signals off Venus to find the distance to the sun? Why not just bounce them off the sun? (Learning Outcome 12-1)
2. What measurements must be made to find out the sun's luminosity? Which of these is least accurate? (Learning Outcome 12-2)
3. Suppose you examined sunlight with a spectroscope. Describe, in general, what you would see. (Learning Outcome 12-5)
4. Explain the appearance of the sun's spectrum by simple atomic processes. (Learning Outcome 12-5)
5. Why do you not see helium absorption lines in the sun's visible spectrum, even though helium is the sun's second most abundant element? (Learning Outcome 12-5)
6. Should the chemical composition of the sun's core differ from that of its photosphere? Why or why not? (Learning Outcomes 12-6, 12-8, and 12-10)
7. Describe how to estimate the sun's surface temperature from its continuous spectrum. (Learning Outcome 12-2)
8. Use the concept of opacity to explain why it takes photons hundreds of thousands of years to emerge from the sun's core to the surface. (Learning Outcome 12-4)
9. In one sentence, describe the source of the sun's energy. (Learning Outcomes 12-7 and 12-8)
10. Describe a method by which you could infer the temperature of a sunspot from its continuous spectrum. (Learning Outcomes 12-3 and 12-11)
11. We can see spectral lines of helium from the sun's chromosphere but not the photosphere. Why? (Learning Outcomes 12-5 and 12-6)
12. How do we know that the photosphere marks the top of the sun's convective zone? (Learning Outcomes 12-4 and 12-10)
13. In what sense do neutrinos allow us to "see" the sun's core directly? What have been the results of the solar neutrino experiments to date? (Learning Outcome 12-9)
14. The deeper you go into the sun, the higher the density and temperature. What do you expect, since the sun is a gas, to happen to the pressure? (Learning Outcome 12-12)
15. Consider an ideal gas. Suppose you double its density and triple its temperature. By how much will its pressure change? (Learning Outcome 12-12)
16. Why doesn't conduction play a major role in the energy transport within the sun? (Learning Outcome 12-10)
17. How do we know the sun's
    a. diameter?
    b. mass?
    c. flux at the earth?
    (Outcome 12-2)

PROBLEMS AND ACTIVITIES

1. The sun's luminosity comes from fusion reactions in the core that convert matter to energy. Calculate the number of kilograms converted each second.
2. The sun has been using fusion reactions for close to 5 Gy. If its luminosity has been constant over this time, calculate the amount of matter converted to date.
3. Solar energy heats the earth, and the earth's surface radiates this energy back to space. The earth acts like a blackbody with a temperature close to 300 K. Calculate the peak wavelength of emission and the flux emitted from the earth.

4. Your body also emits like a blackbody. Estimate its temperature and calculate the peak of emission and the energy emitted per square meter. Compare that to the energy per square meter emitted by the sun.

5. What is the wavelength at which the continuous emission from a sunspot has a peak in the flux? *Hint:* Assume that a sunspot radiates like a blackbody!

6. Calculate the fraction of the sun's luminosity that is intercepted by the earth. *Hint:* Consider the earth a disk, whose area is πR^2, where R is the earth's radius.

7. Knowing the sun's flux at the earth and the earth–sun distance in meters, calculate the sun's luminosity.

13 The Stars as Suns

CENTRAL CONCEPT

Astronomers determine the physical properties of stars by finding their distances and analyzing the light received from them. Their properties can be summarized in a mass–luminosity diagram. Like the sun, stars are naturally controlled thermonuclear reactors.

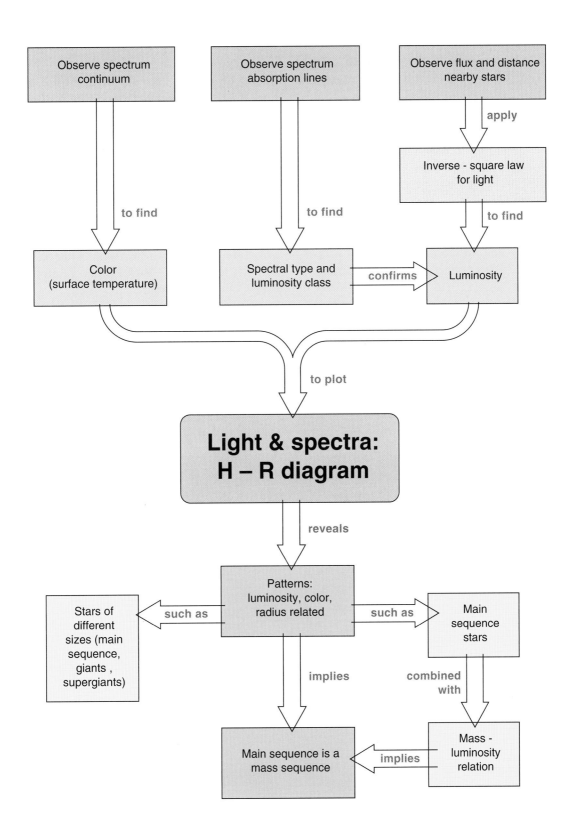

Astronomers know what the stars are: enormous fusion reactors in space. That view is recent, developed in full in the twentieth century. Until the end of the nineteenth century, astronomers really didn't pay much attention to the inner workings of stars. The vast distances of stars from the sun kept them aloof from human understanding.

With the development of spectroscopes and atomic theory (Chapter 5), we finally acquired the crucial tools to analyze starlight and compare the stars to the sun. By finding distances to nearby stars, we were able to infer basic physical properties, which we then applied to more distant stars. These revelations reinforced the connection between the sun and the stars, the essential link to the rest of the cosmos. The stars are suns; the sun is a star. These aspects mark both sides of the solar-stellar connection.

13.1 SOME MESSAGES OF STARLIGHT

Go out on a clear January night in the northern hemisphere. Face south. Orion immediately catches your eye. Two stars in Orion shine the brightest: Betelgeuse, which looks reddish, and Rigel, blazing bluish white. To the south and east of Orion lies Canis Major, the Great Dog, Orion's hunting companion. You can easily notice Sirius, the jewel of Canis Major, as the brightest star in the sky.

The differences in these stars typify the astronomers' dilemma: What can we know about these distant stars? How do these stars compare with our sun? Are they larger? More luminous? Hotter? How to find out these physical properties? In what ways do these stars resemble the sun? How do they differ?

Starlight carries information about the physical properties of stars (Fig. 13.1). By deciphering starlight and comparing it to sunlight, we can infer the nature

Figure 13.1 Various colors of stars: white, reddish, yellowish, and bluish. This view shows the constellation Scorpius with the Milky Way. The bright reddish star at right center is Antares.

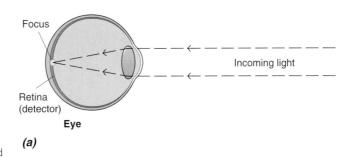

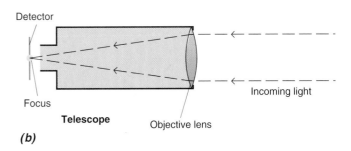

Figure 13.2 Comparison of the optics of the human eye and a telescope. Both focus an image on a detector. In the case of the eye, the retina serves as the detector surface.

of stars. Their physical properties include chemical composition, surface temperature, radius, luminosity, and mass. For some of these properties, we do *not* need to know the distance to the stars; for others, we do. In all cases, we observe, measure, and analyze the light from the stars as we did for the sun.

Brightness and Flux
(Learning Outcome 13-4)

If you have even casually looked at the night sky, you know that stars do not have the same brightness. You make this judgment with your eyes, which work essentially like a refracting telescope (Section 6.2): the lens of each eye focuses light onto its retina, which detects the light by transforming the radiative energy into electrochemical impulses that travel to your brain. How bright a star appears to you depends on how much energy, each second, strikes your retina. The more energy received, the greater brightness you perceive.

Similarly, the objective of a telescope (Fig. 13.2) gathers light to a focus, and a detector senses the radiative energy. How bright a star appears in a telescope depends on how much energy, each second, arrives at the earth from the star and how large the objective of the telescope is. Just to determine the brightness of the star, then, we need to get rid of the dependence on aperture size. We do this trick by

considering only the flux crossing the telescope's objective.

Recall that *flux* is the name given to the amount of energy passing through each unit of area in a second. Since **watts** (Appendix A) are a measure of energy per second, flux can have units such as watts per square meter. For example, the flux at the earth from the sun is 1370 watts per square meter. Compared to the sun, the brightest star, Sirius, sends a tiny flux to the earth – a mere ten-millionth of a watt per square meter!

> **CAUTION.** Brightness actually means flux. "Brightness" is a more common word but lacks the specific physical meaning of flux. (Astronomers have a special way to measure flux at the earth; see Enrichment Focus 13.1.) Also, do not confuse the flux emitted at the surface of a star with that measured at the earth!

Flux and Luminosity
(Learning Outcomes 13-1, 13-5)

Remember how to work out the sun's luminosity (Section 12.3)? Measure the sun's flux, find the earth–sun distance in kilometers, construct an imaginary sphere with a 1-AU radius, and total up all the energy hitting that sphere.

ENRICHMENT FOCUS 13.1

Flux and Magnitude

For reasons of history and convention, astronomers generally use a quirky way to talk about stellar brightness. It's called **apparent magnitude.** Flux and apparent magnitude are two different methods used to describe the same property. I do not use magnitudes in this book because I think they are confusing to the novice and unnecessary for a conceptual understanding of the material. However, you may come across the concept elsewhere, so I'll discuss it briefly here.

The magnitude scale on which stars are rated evolved from a convention established by Hipparchus (160–127 B.C.) and has now become traditional. In his catalog of stars, Hipparchus classified apparent magnitudes by rating the brightest star he could see as magnitude 1 and the faintest as magnitude 6. As this system was used, some stars were found to be brighter than magnitude 1; for example, Vega is magnitude 0, and Sirius is magnitude –1.4. The first peculiarity to note about the magnitude scale is that the larger the magnitude of a star (the more positive), the fainter it is; but the smaller the magnitude (the more negative), the brighter it is. That is, a star of magnitude 6.5 is fainter than one of magnitude 4.2, and one of magnitude –1.3 is fainter than one of magnitude –2.1. A star of magnitude 12.5 is brighter than one of magnitude 17.9, and one of magnitude –1.8 is brighter than one of magnitude –0.9. Confusing? Perhaps it helps to think of magnitude as measuring the amount of faintness; larger numbers mean fainter stars (Fig. F.14).

Most star catalogs or finding charts give the apparent magnitudes of the objects. Here "visual" refers historically to the magnitude as sensed by the eye. Astronomers today use a special filter in the middle of the visual spectrum.

On Hipparchus' scale, stars of first magnitude were about a hundred times brighter than stars of sixth magnitude. The modern system therefore defines a difference of 5 magnitudes as corresponding to a brightness ratio of 100. A difference of one magnitude then amounts to a brightness ratio of 2.512. This strange number pops up because $2.512 \times 2.512 \times 2.512 \times 2.512 \times 2.512 = 2.512^5$ = 100, another way of stating that a difference of 5 magnitudes equals a ratio of 100 in brightness. Table F.2 will help you keep straight the magnitude differences and brightness ratios. Note that differences in apparent magni-

TABLE F.2 Magnitude Differences and Flux Ratios

A Magnitude Difference of:	Equals a Flux Ratio of:
0.0	1.0
0.2	1.2
1.0	2.5
1.5	4.0
2.0	6.3
2.5	10.0
4.0	40.0
5.0	100.0
7.5	1,000.0
10.0	10,000.0

tude provide a way of comparing fluxes if you convert to brightness ratios.

Astronomers also use a system called **absolute magnitude** to talk about luminosities in the form of magnitudes. Astronomers do compare the luminosities of stars, in an imaginary way, by setting them at a standard distance of 10 **parsecs** (pc), which is 32.6 light years.

Imagine that you could transport the stars in the sky, including the sun, to 10 pc from the earth. Stars that are, in reality, closer than this distance would appear fainter; those that are farther would seem brighter, as expected from the inverse-square law. The term *absolute magnitude* refers to how bright a star would appear at 10 pc from us.

Compare the absolute magnitudes of these stars: the sun, 4.8; Sirius, 1.4; Polaris, –4.6. Note that Polaris is actually the most luminous, followed by Sirius, with the sun last.

When we know a star's absolute magnitude, we can compute its luminosity by comparing its absolute magnitude to that of the sun. The trick here is to convert magnitude differences into brightness ratios (Table F.2). Let's compare the sun with Sirius. In the visual range, the sun's absolute magnitude is 4.83, that of Sirius 1.41. The difference is roughly 3.4 magnitudes, so the brightness ratio is about $(2.512)^{3.4} \approx 22$.

If you know a star's distance and apparent magnitude, you can work out its absolute magnitude. In fact, if you know any two of the three properties, you can find the third. Algebraically, the relationship is

$$m - M = 5 \log d - 5$$

where m is the apparent magnitude, M the absolute magnitude, and d the distance in parsecs.

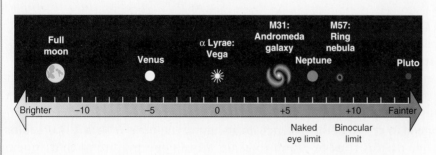

Figure F.14 Astronomical magnitude scale.

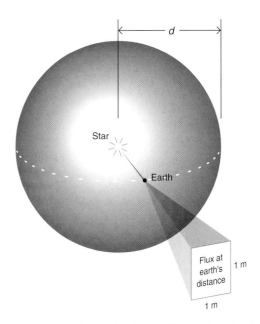

Figure 13.3 Determining the luminosity of a star from its distance and the flux at the earth. When it reaches the earth, a star's light has spread out over a sphere having a radius that is the distance between the earth and the star; its area is $4\pi d^2$. The flux received at the earth is a small fraction of the star's luminosity because the light has spread over a huge area in space.

We can find a star's luminosity by the same procedure (Fig. 13.3). Measure the star's flux at the earth. Then, knowing its distance, find the area of a sphere surrounding the star, and add up the flux over the total area. Simple enough – but we need to know the distance to the star, and that may be tough.

Why worry about a star's luminosity? Because it tells how much radiative energy the star emits each second, and that is most fundamental to understanding stars as suns.

CAUTION. To find a star's total luminosity requires the measurement of its flux over the complete range of the electromagnetic spectrum. That might not be possible. First, our detector may work over only a limited range. Second, the earth's atmosphere absorbs some of the energy. Third, matter in interstellar space may also do some absorbing. If the flux is measured over a limited range of the spectrum, say the visual, astronomers report this fact by using the term *visual flux*. Then the luminosity calculated from it is only the *visual luminosity* of the star. Even sunlike stars emit only 40 to 50 percent of their light in the visual part of the spectrum, so their visual luminosity is only about half their total luminosity. The luminosities for stars given in this book are almost all visual luminosities to have a consistent basis of comparison. But beware that these visual luminosities will be *less* than the total luminosity over all wavelengths.

The Inverse-Square Law for Light (Learning Outcomes 13-1, 13-4, 13-5)

Distance plays a key role in relating flux to luminosity. The brightness of a light source relates in a very specific way to your distance from it.

Consider the following experiment. Put a bare lightbulb in a socket and turn it on. Take a light meter – a device used to measure the intensity of light – and place it 1 meter from the bulb (Fig. 13.4). Note the reading on the light meter and call that one unit. Now move the light meter 2 meters from the bulb; its reading will be one-fourth of that at the 1-meter position. Move the light meter to a distance of 3 meters; the reading will now be one-ninth of that at 1 meter. Note that the light intensity decreases as the inverse square of the distance. The flux is *inversely proportional* to the *distance squared.* (What would the reading be at 4 m? Right: 1/16 of that at 1 m.)

How does this inverse-square relation happen? Imagine a bulb placed in the center of two transparent spherical shells, one having twice the radius of the other (Fig. 13.5). Light the bulb briefly so that it emits a pulse of light. As the light moves away from the bulb, it expands in a sphere in all directions, but the total amount of energy in the pulse remains the same, no matter how large the sphere. As the light passes through the first sphere, it covers a certain area. As it goes through the second, it covers a larger area. How much larger? The area of a sphere is directly related to the radius squared; so the larger sphere has four times the area of the smaller one, and the light spreads out four times as much.

Now pick any square meter of surface for both spheres. Because of the radiation's dilution over a larger area, the small patch you select on the larger sphere has only one-fourth as much light striking it as the patch on the smaller sphere.

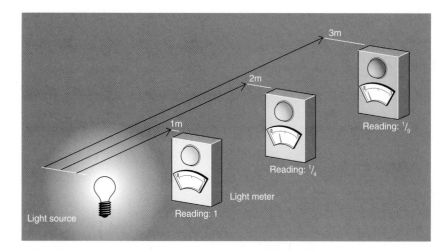

Figure 13.4 Flux and distance. As a light meter is moved away from a lightbulb, the measured flux decreases. The greater the distance, the lower the reading, so that at 2 m, the flux is only ¼ of what it is at 1 m; at 3 m, it is only ⅑ as much.

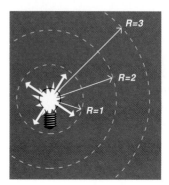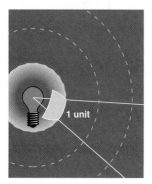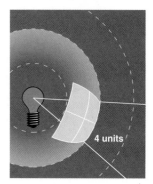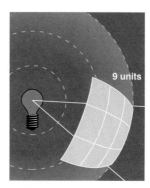

Figure 13.5 Geometry for the inverse-square law for light. Place a lightbulb in a series of concentric, transparent shells. Because each larger shell has a greater area than its predecessor does, light that covers 1 unit of area in the first shell covers 4 units in the second and 9 in the third. Thus, the flux at the second shell is ¼ of the first; at the third shell, it is ⅑ of the first. The spreading out of the light dilutes the flux.

If the ratio of radii were increased to 3, the decrease in brightness would be by 9; if increased to 4, the decrease would be by 16. This is the **inverse-square law for light** (or any electromagnetic radiation) – a crucial concept for astronomy.

Suppose you observe that the flux of one star is a hundred times that of another. If you assume that both stars have the same luminosity, how do their distances compare? According to the inverse-square law, the brighter one must be ten times as close as the fainter one.

13.2 STELLAR DISTANCES: PARALLAXES
(Learning Outcome 13-6)

The key to finding a star's luminosity from its flux is its distance. For nearby stars, we have a direct method to find distances: triangulation, similar to the procedure used by surveyors on the earth. Stellar triangulation is called **heliocentric** or **trigonometric parallax.**

You actually observe parallaxes all the time, although you are usually unaware of it. Put your hand out at arm's length with a pencil held upright (Fig. 13.6). Now alternately open and close each eye. The pencil will appear to jump back and forth relative to the distant background. This angular shift in the pencil's position is the parallax of the pencil. Now if you measure the amount (in angular measure) by which the pencil appears to have shifted and also the distance between your eyes, you can calculate how far the pencil is from your head.

Imagine that your pencil is a nearby star, the background more distant stars, and your eyes the sighting positions of the earth in orbit around the sun separated by a time of 6 months. From these two positions in the earth's orbit (separated by 2 AU), the nearby star appears to shift in position

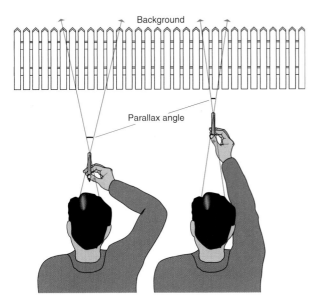

Background

Parallax angle

Figure 13.6 Observing the parallax of a pencil held at arm's length. As you alternately open and close each eye, the pencil appears to jump in position relative to the background markers. The angle of the apparent movement is the parallax angle. The farther away you hold the pencil, the smaller the parallax angle. The baseline for the shift is the distance between your eyes.

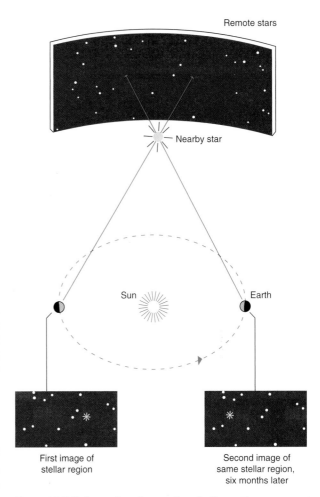

Remote stars

Nearby star

Sun   Earth

First image of
stellar region

Second image of
same stellar region,
six months later

Figure 13.7 Heliocentric stellar parallax. As the earth goes around the sun, the positions of nearby stars shift relative to the positions of more distant stars in six months. The nearby star returns to its original relative position in a year. The closer the star, the greater the observable shift. The shift has a one-year cycle, the earth's period of revolution. The baseline for the shift is the diameter of the earth's orbit. Note that a larger baseline results in a larger shift for a star at the same distance.

(its parallax) compared with the more distant stars (Fig. 13.7). Measure the angular size of that shift. Half of this angular shift is defined as the astronomical parallax of the star. Because you know the diameter of the earth's orbit, you can calculate the distance to the star. Note that the farther away a star is, the smaller its parallax will be. (Enrichment Focus 13.2 shows this concept in more detail.)

You perceive parallax, then, when you view a relatively close object from each of two ends of a baseline. As the earth moves from one place in its orbit to another, a nearby star's position seems to change relative to the more distant stars. The maximum shift occurs when you view the star 6 months apart – that is, when you are sighting from opposite sides of the earth's orbit. Measure the amount of that shift. Then, because you know the diameter of the earth's orbit (2 AU), you can calculate the distance to the star. In essence, the parallax angle gives the size-to-distance ratio, with "size" in this case the size of the baseline and "distance" the distance to the star in units of the baseline.

The parallax of Sirius is 0.38 arcsec. The distance to Sirius is about 546,000 times the distance to the sun (546,000 AU), or a distance of 8.6 ly. From this

information, we can calculate the luminosity of Sirius compared to the sun's luminosity from Sirius' flux. We measure the flux from Sirius at the earth. With the star's distance known, the inverse-square law gives us the rate at which we would receive energy from the star if it were 1 AU from us. We compare that figure to the amount we do receive from the sun. From the comparison, we know how luminous the star is compared with the sun. For example, if we could move Sirius up to the sun, it would appear 23 times brighter than the sun, so Sirius is 23 times more luminous than the sun.

ENRICHMENT FOCUS 13.2
Heliocentric (Trigonometric) Stellar Parallax

Suppose you travel out into space and look back at the earth–sun separation (1 AU). You keep going until the angular size of the AU is 1 arcsec. Now you station a star at this point and hurry back to the earth. You observe this star for one year and measure its shift. Half of the total shift for one year would equal 1 arcsec. We say the star is 1 **parsec** from the sun – that is, the distance at which the parallax is one second of arc – a distance roughly equal to 3.26 light years. Suppose the star were twice as far away; it would have half the parallax, 0.5 arcsec. At half the distance, its parallax would be double, that is, 2 arcsec. Note this simple inverse relationship of parallax and distance. Algebraically,

$$d \text{ (pc)} = \frac{1}{p \text{ (arcsec)}}$$

where d is the distance and p is the parallax.

An example: the star 40 Eridani has a parallax of 0.2 arcsec. How far is the star from the sun? Since the star's parallax is one-fifth that for a star 1 parsec away, 40 Eridani must be 5 pc distant. Explicitly,

$$d = \frac{1}{0.2 \text{ arcsec}} = 5 \text{ pc}$$

Here's a simple geometric explanation for the parallax formula. Travel out in space to some star. Then draw an imaginary circle (Fig. F.15) centered on the star and through the sun (S) so that the circle's radius is d. Note that the parallax angle, p (in degrees), is some fraction of the total circle (360°). Also, R, the earth–sun distance (1 AU), corresponds closely to an arc on the circumference of the circle; its fraction of the circumference is the same as the fraction p is of 360°. So

$$\frac{R}{2\pi d} = \frac{p \,(°)}{360°}$$

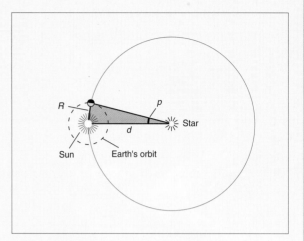

Figure F.15 Geometry for heliocentric parallax.

or

$$d = \frac{360° \times 4}{2\pi p \,(°)}$$

Convert 360° to seconds of arc:

$$d = \frac{(360 \times 60 \times 60 \times R)}{2\pi} = \frac{206,265 \, R}{p \text{ (arcsec)}}$$

Now R is 1 AU; define 1 pc as 206,265 AU. Then

$$d \text{ (pc)} = 206,265 \frac{(1/206,265 \text{ pc})}{p \text{ (arcsec)}}$$

or

$$d \text{ (pc)} = \frac{1}{p \text{ (arcsec)}}$$

Using light years as the unit of distance, we have

$$d \text{ (ly)} = \frac{3.26}{p \text{ (arcsec)}}$$

Heliocentric parallax works accurately only for close stars. Note that the parallax of Sirius, a very close star, is less than 1 arcsec, about 0.0005 the angular diameter of the moon! That's the size of a U.S. quarter (25-mm diameter) at a distance of a little more than 5 km. Because parallaxes are so small, it was not until 1838 that the German astronomer Friedrich W. Bessel (1784–1846) succeeded in measuring the first stellar parallax.

Earth-based observers have a natural limitation in finding parallaxes: the size of the AU. A larger AU would make it possible, given the same techniques, to measure more distant stars reliably. An astronomer on Mars, for instance, has a 50 percent larger baseline to use and could go 50 percent farther out in distance. (Of course, the measurements would take longer because Mars has a larger orbital period than the earth.) People who measure parallaxes with ground-based telescopes have refined their techniques to an accuracy of about 0.001 arc-sec, so they can get accurate distances out to about 300 ly.

Many stars lie more distant than 300 ly. How are their distances measured? By ingenious, indirect methods. A spectroscopic procedure comes later in this chapter (Section 13.5). A special space tele-scope, called Hipparcos, has measured parallaxes of hundreds of thousands of stars with an accuracy about ten times better than can be achieved from the ground today.

13.3 STELLAR COLORS, TEMPERATURES, AND SIZES

Now return to the stars in the winter sky. I have dealt with one observable property: their fluxes. Another property you can observe is color. Betel-geuse looks flaming reddish, Rigel and Sirius appear bluish white, and Capella shines yellowish white. What can we learn from these differences?

Color and Temperature (Learning Outcomes 13-1, 13-2)

The different colors in Betelgeuse, Rigel, Sirius, and Capella suggest different **stellar surface tem-peratures.** (Recall Section 12.3 about the sun's surface temperature.) Rigel is the hottest of the four; it is so hot that it emits more blue light than any of the other visible colors. In contrast, Betel-geuse is the coolest, for it radiates mostly red light. The colors of stars are a direct function of their surface temperatures.

> **CAUTION.** When we look at starlight with our eyes, we see only the visible part of its entire spectrum. Stars, in fact, radiate very much like blackbodies (Enrichment Focus 12.2), so their continuous spectra, from short-est to longest wavelengths, have the charac-teristic blackbody shape, called a *Planck curve.* The visible range covers but a small portion of a blackbody's spectrum.

One way to measure a star's temperature is to find the peak of its Planck curve (Fig. 13.8). The hot-ter the star, the shorter the wavelength of the peak (Enrichment Focus 13.1). Although simple, this tech-

nique has two drawbacks. First, we need to measure a wide range of the spectrum to find the peak. Sec-ond, at the ground, the earth's atmosphere may absorb the radiation at the wavelength of the peak. For example, the hottest stars have spectra that peak in the ultraviolet, which doesn't make it through the atmosphere to a ground-based telescope.

The temperature of a blackbody can also be worked out by measuring the relative brightness at any two wavelengths – the astronomical meaning of *color* (Fig. 13.9). A reddish star does not emit red light only, but the relative amount of the longer (red) wavelengths is greater than that at the shorter (blue) wavelengths. All stars emit light over the entire visible range, but each shows relatively differ-ent amounts of red and blue light.

Temperature and Radius (Learning Outcomes 13-1, 13-3)

One reason to focus on color is that the stellar tem-peratures obtained from the colors are related to the sizes of stars, if the stars radiate at least somewhat like blackbodies.

How can this estimate be done? In the constel-lation Scorpius, you can find the bright reddish star Antares. If you view Antares through a telescope, you'll find it has a faint bluish-white companion. Antares is a binary star (more on this in Section 13.6); it and the bluish-white companion revolve around each other, bound by gravity. The red star is called Antares A, the bluish-white companion Antares B. Their surface temperatures turn out to be roughly 3000 and 15,000 K.

We receive about forty times more flux from Antares A than Antares B. Now both stars lie at the same distance from the earth, so the difference in flux cannot arise from different distances; it must come from differences in the radiative output at the stars' surfaces. So we conclude that Antares A is forty times more luminous than Antares B. How can a cool star be so much more luminous than a hotter one (especially in light of the fact that hotter stars emit more power per surface areas than cooler ones)?

Recall (Enrichment Focus 12.2) that the amount of power emitted by each unit of area of a black-body's surface depends only on its temperature – in fact, on the fourth power of the temperature.

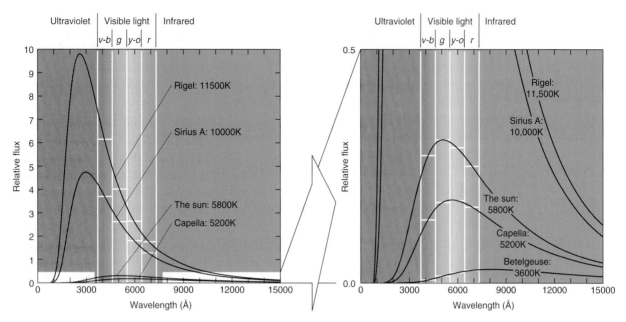

Figure 13.8 Continuous blackbody spectra. The fluxes of blackbodies with temperatures equivalent to those of Rigel, Sirius, the sun, Capella, and Betelgeuse compared in the color bands violet-blue (v-b), green (g), yellow-orange (y-o), and red (r). These are compared to continuous spectra of different temperatures that closely match that of each star. The blackbody spectra have the distinctive shape known as *Planck curves*. The visible range is only a small part of the total spectrum; the color of a star depends on the eye's perception of a star's continuous spectrum in the visible range. The relative flux scale is much larger for the part of the figure on the left.

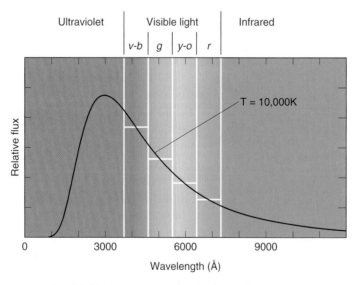

Figure 13.9 A 10,000 K blackbody curve. Note that the flux in the blue range is greater than that in the red. The *ratio* of these fluxes, measured accurately, defines the color of the star. This measurement can be applied to other stars of different temperatures.

For example, Antares B is about five times hotter than Antares A, so each square meter of Antares B emits 5^4, or 625 times the power of Antares A. If Antares A had 625 times the surface area of Antares B, both stars would have the same luminosity. Because Antares A has forty times the luminosity of Antares B, it must have 40×625, or 25,000 times the surface area.

> **TO SUM UP.** A star's luminosity is related to its surface temperature (which determines how much power each square meter emits) and its surface area (which determines the total number of square meters doing the emitting).

Now you have a way to infer the radius of a star from its luminosity and surface temperature. As an example, take the star Capella. Its surface temperature is 5200 K, almost the same as the sun's, but its luminosity is 130 times that of the sun. If Capella radiates like a blackbody, its radius is about 14 times larger than that of our sun.

Direct Measurement of Diameters
(Learning Outcome 13-1)

If we could measure a star's angular diameter and its distance, we could calculate its actual diameter directly. Because stars are so far away, however, their disks have very small angular diameters – on the order of a few milliarcseconds (one milliarcsecond is 10^{-3} arcsec). Stellar images rarely have angular diameters less than about one arcsecond, even during excellent seeing conditions. Ground-based optical interferometers soon will be able to achieve such resolutions.

One star measured is Aldebaran, the eye of Taurus the Bull. The results: Aldebaran's radius is 3.3×10^7 km, or about 50 times larger than the sun's radius. That's about 20 percent smaller than the size inferred from temperature and luminosity (and the assumption that the star radiates like a blackbody). The angular measurement is probably the better value because it is a direct measurement.

13.4 SPECTRAL CLASSIFICATION OF STARS

The absorption (or dark-line) spectra of many stars resemble the spectrum of the sun. From the sun's spectrum, astronomers infer its photospheric com-

position. The same procedure can be applied to stars. But there's one twist: a star's spectrum is affected by its surface temperature as well as by its composition.

Temperature and the Balmer Lines
(Learning Outcomes 13-1, 13-7)

We now know that most stars have pretty much the same composition, but their spectra are not all identical. Consider, for example, the Balmer lines of hydrogen. Stars cooler than the sun have spectra with weaker (less intense) Balmer lines. Stars somewhat hotter than the sun show stronger (darker) Balmer lines. Stars much hotter than the sun also have spectra with weak Balmer lines, however.

How to understand this? Recall (Section 5.3) that the Balmer series arises from transitions between the second energy level of hydrogen and any level above it; these transitions run upward for absorption and downward for emission. To absorb a Balmer line, a hydrogen atom must be excited to level 2. In the sun, only one of every 10^8 atoms is excited. Not enough energetic collisions occur to kick up many electrons from the ground level. So the Balmer lines in the sun's spectrum are not very strong. In a cooler star, fewer energetic collisions occur, and fewer hydrogen atoms are excited. The Balmer lines are even weaker. In a hotter star with more energetic collisions, more atoms are excited, about one out of every million, so the Balmer lines are more intense (darker) in such a star. If the star is sufficiently hot, collisions are so violent that many electrons are knocked out of the atom entirely, leaving hydrogen ions (protons) behind. With fewer excited atoms, the Balmer lines are weaker (less dark).

The key point: Balmer lines of hydrogen are produced by atoms with an electron in the second energy level – *the atom must be already excited.* If for some reason a star does not have very many hydrogen atoms excited to the second level, the Balmer lines in that star will be weak. This happens if the star has a very high temperature, so that virtually all its hydrogen is ionized, or if the star has a relatively low temperature, so that even though there is much neutral hydrogen, there are very few excited atoms in the second level.

Spectral Classification (Learning Outcome 13-7)

At the turn of the last century, workers at Harvard College Observatory in Massachusetts, U.S.A., classified stellar spectra by using absorption lines, especially hydrogen Balmer lines. The results were codified in the *Henry Draper Catalog*. This assignment first fell to Williamnia Fleming (1857–1910), who selected more than 10,000 stars for the classification scheme and supervised a staff of women to bring the catalog into final form.

Much of the final task of developing an expanded catalog in 1924 was done by Annie Jump Cannon (1863–1941; Fig. 13.10), who single-handedly classified the spectra of more than 250,000 stars! The original classification scheme was set up strictly on the basis of the strength of various lines, well before there was any understanding of the effects produced by different temperatures. The Balmer lines played a vital role in this scheme: stars with the strongest Balmer lines were called class A, those with slightly weaker lines class B, and so on. Some classes were later dropped because they contained too few stars or only very peculiar ones, and the order was rearranged to one of decreasing temperature, once the explanation of the line intensity in terms of temperature was understood.

Figure 13.10 Annie Jump Cannon (left) and Henrietta Swan Leavitt (right) did fundamental work in spectroscopic astrophysics at Harvard College Observatory.

The **stellar spectra sequence** from hotter to cooler in the Harvard classification now runs O-B-A-F-G-K-M (Fig. 13.11). Almost all stellar spectra fit into this sequence. The O stars have spectra with weak Balmer lines of hydrogen and lines of ionized helium. The A stars have the strongest Balmer lines. In F stars, the Balmer lines fade and many other lines appear, mostly of metals. The sequence from type O to M, looking at the continuous spectra from the stars, is also a color sequence. O stars appear bluish white, G stars yellowish, and M stars reddish. Astronomers further divide each class into subclasses from hotter to cooler, labeled from 0 to 9 (for example: G0, G1, G2, G3, G4, G5, G6, G7, G8, and G9). Each subclass is distinguished by slightly different intensities of specific absorption lines.

The strengths of the Balmer lines suggest that the differences in stellar spectra reflect primarily differences in temperatures, *not* in abundances. These temperature differences result in different degrees of ionization and excitation of the atoms in the star. The strength of the atom's spectral lines is determined by the number of atoms that are excited and the number that are ionized. This revelation in twentieth-century astrophysics occurred because of the work of an Indian physicist, Meghnad N. Saha (1893–1956), who proposed in 1920 that differences in temperature dramatically controlled the ionization conditions for elements in stellar atmospheres. In 1925, Cecilia Payne-Gaposhkin (1900–1979) applied Saha's ideas to the spectra in the Harvard collection. She came up with the basic temperature classification sequence (described above) from O to M stars, thereby confirming that temperature plays the key role in the spectra. Payne-Gaposhkin determined that stars were made mostly of hydrogen and helium – contrary to the notions of the time that stars were made of stuff like the earth.

Consider first O stars, which have the hottest surface temperatures, 30,000 K and higher. At these high temperatures, atoms collide violently. The energies in such collisions can rip the electrons from hydrogen atoms so that most of the hydrogen is ionized. Very few neutral atoms remain to absorb at wavelengths corresponding to the Balmer series. Because most of the hydrogen is ionized in O stars, the Balmer lines are weak. This is also true for B stars, whose temperatures range from 11,000 to 30,000 K (Fig. 13.12).

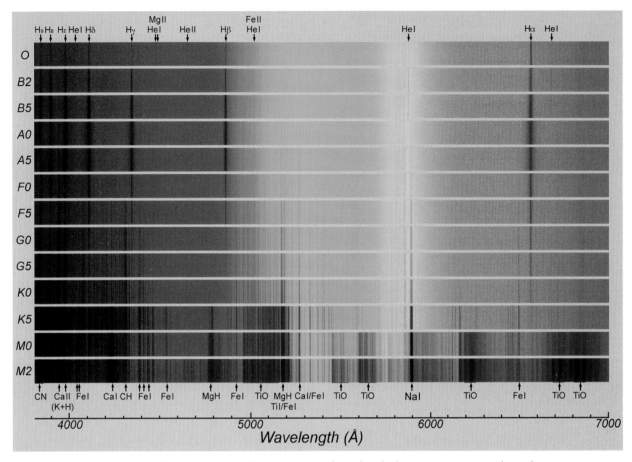

Figure 13.11 Stellar spectral types. This computer simulation has the hottest stars at top, coolest at bottom. Note how the strengths of the absorption lines differ, especially the hydrogen Balmer series, which is strongest in A stars.

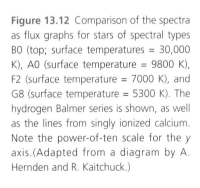

Figure 13.12 Comparison of the spectra as flux graphs for stars of spectral types B0 (top; surface temperatures = 30,000 K), A0 (surface temperature = 9800 K), F2 (surface temperature = 7000 K), and G8 (surface temperature = 5300 K). The hydrogen Balmer series is shown, as well as the lines from singly ionized calcium. Note the power-of-ten scale for the y axis. (Adapted from a diagram by A. Hernden and R. Kaitchuck.)

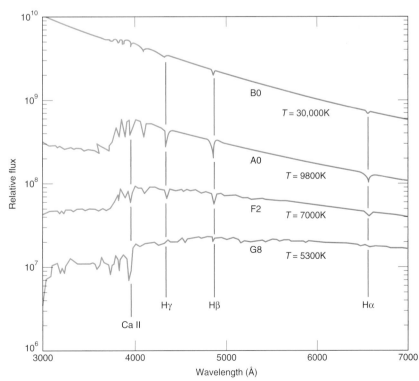

Table 13.1 Features of the Stellar Spectral Classes

Spectral Class	Color	Surface Temperature (K)	Principal Features	Examples
O	Bluish-white	30,000	Relatively few absorption lines. Lines of ionized helium and other lines of highly ionized atoms. Hydrogen lines are weak	Naos
B	Bluish-white	11,000–30,000	Lines of neutral helium. Hydrogen lines stronger than in O-type stars.	Rigel, Spica
A	Bluish-white	7,500–11,000	Strong lines of hydrogen. Also lines of singly ionized magnesium, iron, calcium, and others. Lines of some neutral metals show weakly.	Sirius, Vega
F	Bluish-white to white	6,000–7,500	Hydrogen lines are weaker than in A-type star but still conspicuous. Lines of singly ionized metals are present, as are lines of other neutral metals.	Canopus, Procyon
G	White to yellowish-white	5,000–6,000	Lines of ionized calcium are the most conspicuous spectral features. Many lines of ionized and neutral metals are present. Hydrogen lines are weaker than in F-type stars.	Sun, Capella
K	Yellowish orange	3,500–5,000	Lines of neutral metals predominate.	Arcturus, Aldebaran
M	Reddish	3,500 and lower	Strong lines of neutral metals and molecules.	Betelgeuse, Antares

Now turn to A stars (Fig. 13.12), whose surface temperatures range from 8000 to 11,000 K. Atoms and ions collide less violently, and most of the hydrogen is neutral. To absorb Balmer lines, the hydrogen atom must have its electron in the second energy level. Although the collisions are not strong enough to ionize the hydrogen, they do possess the energy to excite the electrons out of the lowest energy level. In A stars, many hydrogen atoms are excited by collisions with other atoms, so their electrons are in the second level. These excited atoms readily absorb light at the Balmer wavelengths, and the lines appear strongest.

In F and G stars (Fig. 13.12), surface temperatures are still lower, typically 6000 K. Very few hydrogen atoms are ionized. In addition, the impacts between atoms do not have enough energy to excite very many hydrogen atoms, so most have their electrons in the lowest energy state. Such atoms cannot absorb at Balmer wavelengths. As a result, the Balmer lines disappear almost completely. However, the lines from singly ionized calcium become stronger.

The variation in Balmer line intensities arises from collisions that excite and ionize atoms. How much the collisions ionize or excite depends on temperature. Each spectral type corresponds to a restricted range of surface temperatures, which are listed in Table 13.1. Other lines from other elements can be analyzed in a fashion similar to that for the Balmer lines.

Try not to be discouraged by all this detail about spectral lines. Keep in mind the key concept: *the spectral sequence represents a sequence of the surface temperatures of stars.* I will focus on surface temperatures and colors in this book.

The observation of stellar spectra coupled with an understanding of the atom gives astronomers information about the physical conditions in the atmosphere of a star. An analysis of spectral lines based on atomic theory also provides information about the abundance of elements in stars in the same way as for the sun. Astronomers have found that, just like the sun, stars consist mostly of hydrogen and helium. (Stars *do* differ somewhat in composition relative to their abundance of heavy elements, but hydrogen and helium still make up the bulk of their mass.)

13.5 THE HERTZSPRUNG–RUSSELL DIAGRAM

Consider a star-studded sky. You can sense fluxes, which depend on their luminosities and distances. And you can see some colors. How to find any patterns? How are these properties related to the internal anatomy of the stars? Again we face the astronomer's dilemma: how to find out vital information from points of light in the sky. The solution lies in the spectroscopy of these stars. (See the H–R Diagram Celestial Navigator™.)

Temperature versus Luminosity (Learning Outcomes 13-7, 13-8)

Consider a plot: a temperature–luminosity diagram (Fig. 13.13) based on the spectral types of the stars. Such diagrams were independently set up by Ejnar Hertzsprung (1873–1967) and Henry N. Russell (1877–1957). In their honor, such a plot is called a **Hertzsprung–Russell diagram** (commonly abbreviated as *H–R diagram*). Note that astronomers follow the strange convention of plotting temperature so that it increases to the *left* on the horizontal (temperature) axis of the H–R diagram.

Consider the H–R diagram (Fig. 13.13), one for the brightest stars you can see in the sky. Almost all these stars have a much higher luminosity than the sun. Many of them are also much hotter stars (of spectral classes O and B). Any patterns in this plot? Your eye might be attracted to the sloping group ranging from Beta Centauri at the upper left to the

sun at the lower right. Otherwise, the stars appear to be scattered about in the upper right of the plot.

Now inspect a different H–R diagram, one for the nearest stars, all those within 20 ly of the sun (Fig. 13.14). Notice that most of the stars are less luminous and cooler than the sun. (In fact, we can see them only because they are so close to us.) The star Alpha Centauri A has almost the same luminosity and temperature as the sun. This star is the sun's twin and is also one star nearest to us. (Alpha Centauri A is one member of a triple-star system with Alpha Centauri B and Proxima Centauri and in this sense is unlike the sun.) Only a handful of stars appear in both diagrams.

Finally, a distinct pattern! Most stars' properties clearly do not fall in a random scatter. Rather, there is a trend; if you draw a line through the points from luminous, hot Sirius A to the coolest, faintest star in the lower right-hand corner, you have identified the **main sequence.** Most nearby stars fall on the tapered strip of the main sequence in the H–R diagram. The few stars in the lower left-hand corner, Sirius B included, have very high surface temperatures but low luminosities, so they must be very small. Such a peculiar star is called a **white dwarf** star.

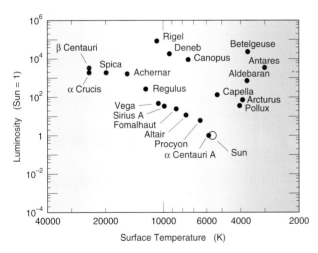

Figure 13.13 A Hertzsprung–Russell (luminosity–temperature) diagram for the brightest stars in the sky. The vertical axis is luminosity (increasing upward) in solar units; the horizontal axis is surface temperature, which increases to the left (rather than to the right as you might expect). The sun and Alpha Centauri A have almost the same luminosity and surface temperature, and so they overlap on this diagram. The stars split into two groups: one rises to the left and the other to the right. Rigel and Betelgeuse, both in Orion, fall at the tops of these ranges. Note the power-of-ten scale for both axes.

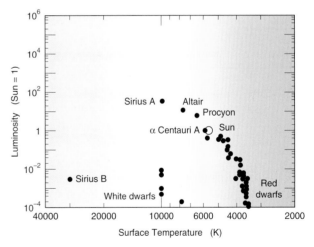

Figure 13.14 A Hertzsprung–Russell (luminosity–temperature) diagram for some stars nearest to the sun. Note that the vertical axis is luminosity in solar units; the sun's luminosity is 1.0. The horizontal axis is surface temperature in kelvins. Except for the white dwarfs, the stars' points fall along a narrow band called the *main sequence*. Note the power-of-ten scale for both axes.

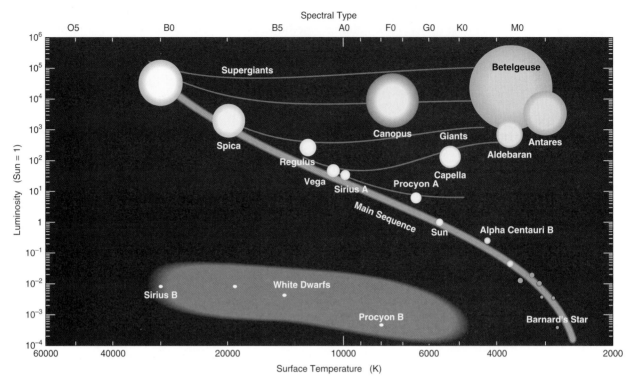

Figure 13.15 True-color representation of the Hertzsprung–Russell diagram. The horizontal axis is temperature; the vertical one, luminosity, in units of the sun's luminosity. The solid white lines show where stars of different luminosity classes fall on the diagram: Ia supergiants at the very top, Ib supergiants below, III giants just below them, and finally V main-sequence stars. The relative sizes of the stars are shown correctly within each luminosity class, but not between them. The colors are those as perceived by the eye looking at these stars through a telescope. Note the power-of-ten scale for the both axes.

What are the physical differences among these stars? Take Betelgeuse, whose properties put it in the upper right-hand corner of the H–R diagram. Here is a star whose surface is much cooler than the sun's, so if Betelgeuse were the same size as the sun, it would be much less luminous. Betelgeuse has a luminosity about 40,000 times that of the sun, however. To be so much cooler and more luminous than the sun, Betelgeuse must be very much larger (Section 13.3), some 1700 times the size of the sun – a star so big that it could swallow up Mars if it replaced the sun of our solar system! Astronomers call Betelgeuse a **supergiant** star.

Here is the reason that the H–R diagram for the nearest stars differs from that for the brightest stars: Figure 13.14 contains ordinary stars with sizes like that of the sun. No giants or supergiants are among the nearest stars, for they are very rare. The sun is a main-sequence star. Most of the stars in the sun's vicinity are also main-sequence stars, of spectral class M. They are cool (surface temperatures about 3000 to 3500 K) and not very luminous, and they have lower mass. Figure 13.13 contains many **giant** and supergiant stars, still visible among the closer stars because of their high luminosity, even though they are scattered widely through space.

Now piece these two diagrams together. Most stars fall on the gentle curve of the main sequence (as in Fig. 13.15). A scattering of stars cuts across the tip of the diagram; these are the very luminous supergiants. A group of luminous stars extends off the main sequence: these are the giants. Finally, note the white dwarf stars at the bottom of the diagram.

You might have heard that the sun is a typical or average star. In some sense this is true, for the sun is neither the largest nor the smallest kind of star. However, it is *not* true that most stars in the Galaxy are like the sun. In the solar neighborhood, G stars like the sun are relatively rare. The most common kind of star in our immediate vicinity, and in fact in

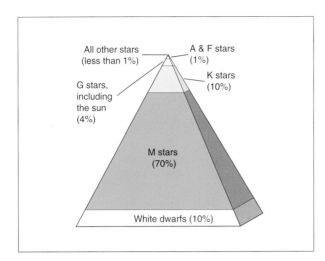

Figure 13.16 Relative populations of stars in the Milky Way Galaxy. This pyramid, which shows the relative numbers of common stars, illustrates that the sun is not an "average" star.

the Galaxy in general, is a main-sequence M star, a cool, reddish star of very low luminosity, on the lower end of the main sequence (Fig. 13.16).

Spectroscopic Distances
(Learning Outcome 13-8)

Once we have an H–R diagram for many stars, we can use it to infer approximate distances to stars. How? First, determine the spectral type of a star.

Suppose it's an M star. Then look at the H–R diagram to find the luminosity. Measure the flux, and use the inverse-square law for light (Section 13.1) to find the distance from the luminosity and the flux.

A problem arises, though: you can see that M stars have a wide range of luminosities, from 10^{-5} solar luminosity for main-sequence M stars to about 10^{5} solar luminosities for supergiant ones. How do we decide what luminosity an M star has over a great range of 10^{10}?

Fortunately, we can tell from a star's spectrum. Recall that the strengths of dark lines relate to a star's temperature. The energy of the collisions depends on the temperature in a star's atmosphere. For gases at the same temperature, however, the rate of collisions depends on the density of the gases: in a denser gas, collisions are more frequent than in a less dense gas at the same temperature.

Giant stars, because they are so huge, have atmospheres of lower density than main-sequence stars. The more frequent collisions in main-sequence stars make certain absorption lines in their spectra appear to be broader than the same lines in spectra of giant or supergiant stars (Fig. 13.17). So a star's size, and hence its luminosity, is given indirectly by the widths of certain absorption lines when comparing the spectra of stars of the same spectral type. (Different lines are used for different spectral types.)

Figure 13.17 Luminosity class and spectra of stars all having the *same* surface temperature (spectral class A0; about 9700 K). The star names are listed to the left of each spectrum. The top three spectra are those of supergiant stars. (Class Iab lies between types Ia and Ib.) Below these is a giant (class III). The next is a peculiar type of A star. Below this one is a subgiant (class IV), and on the bottom a main-sequence star. Note the different intensities of the dark lines, especially those of the hydrogen Balmer series. These lines become broader and stronger from top to bottom and so allow astronomers to infer the luminosity class of the stars.

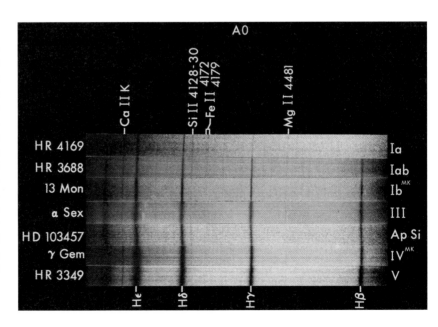

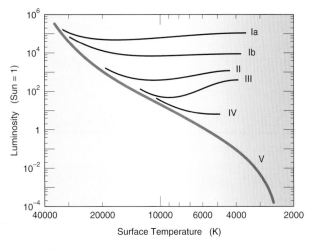

Figure 13.18 Luminosity classes of stars on the H–R diagram. The classes run from I, the largest supergiants, to V, stars like the sun. Classes Ia and Ib are supergiants; class II, luminous giants; class III, normal giants; class IV, subgiants; and class V, main-sequence stars. The lines in this diagram represent the center of the range for each class. Note the power-of-ten scale for both axes.

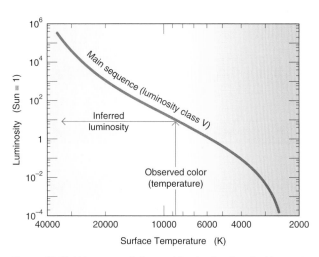

Figure 13.19 Using spectral class and luminosity class (in this case V, main sequence) to infer the luminosity and so the distance to a star, if its flux is known. From a star's color (or spectral type), estimate its temperature on the horizontal axis. Move straight up until you hit the main sequence, and then move horizontally to the left to find the star's luminosity. Comparing the luminosity to the measured flux allows an inference of the distance because the flux varies as the inverse-square law for light. Note the power-of-ten scale for both axes.

Such an analysis reveals that stars fall into **luminosity classes.** The recognized luminosity classes (Fig. 13.18) are Ia, most luminous supergiants; Ib, less luminous supergiants; II, luminous giants; III, normal giants; IV, subgiants; and V, main-sequence stars. The sun falls into luminosity class V. (White dwarfs do not have luminosity classes because they are all essentially the same size.)

A star's spectrum allows it to be classified by spectral type (surface temperature!) and luminosity class. For a given spectral type, you can estimate the luminosity (Fig. 13.19) within a range of probable error (the width of a luminosity class on the H–R diagram). If you know a star's luminosity and its flux, you can calculate its distance from the inverse-square law for light. The procedure for working out distances from spectra is called **spectroscopic distances.**

IN SUMMARY. To make an H–R diagram, first find stars close enough to the sun to measure their distances reliably by parallax. Then calculate their luminosities from their distances and fluxes. Next, take spectra of the stars to find out their spectral class. From the spectra, you determine how hot the stars are.

Then plot the luminosity and spectral type. The result is a calibrated H–R diagram – calibrated in the sense that you have the luminosities of the stars plotted against their surface temperatures. The H–R diagram graphically summarizes some of the important physical properties of stars. It also serves as a visual sorting aid to bring to light different classes of stars, and it can be used as a tool for obtaining the distances of other stars.

13.6 WEIGHING AND SIZING STARS: BINARY SYSTEMS

You have seen how stars differ in properties such as luminosity and size. What about mass? Some stars are larger than the sun, some smaller. But sizes do not tell us directly whether a star is more or less massive than the sun. How to find a star's mass?

Binary Stars (Learning Outcome 13-9)

We have no direct way of knowing the mass of an isolated star. To find masses, we examine the gravitational effects of one object on another. Recall how we find the sun's mass (Section 12.1): we look at the acceleration of the earth as it orbits the sun. Similarly, we use the accelerations of two stars orbiting one another to find their masses. Two stars bound by their mutual gravity and revolving around a common center of mass are called **binary stars.**

If both stars are visible, we can trace out their orbital motion by observing them over a long time, which gives us the angular size of the orbit and the orbital period. But that's not enough to find their masses! In addition, we need to find the distance to the binary system so that we can convert their angular separation into a physical one. Next, it is likely that the plane of a star's orbit is tilted from a direct face-on view; this orbital tilt needs to be accounted for. Then we have enough information to find the sum of the masses from Newton's form of Kepler's third law (Enrichment Focus 4.3). To find the individual masses, we need one more piece of information: we must know how far each star is from the center of mass of the system. (The center of mass is the balancing point between the stars, as if they were on two ends of a seesaw. Refer back to Section 4.5.)

In a binary system, each star orbits the center of mass at a distance *inversely proportional* to its mass. The more massive star lies closer to the center of mass. For instance, in Figure 13.20, M_2 lies 4.5 times farther from the center of mass than does M_1. That means that M_1 has 4.5 times the mass of M_2.

As the system travels through space, and so across our line of sight, the center of mass traces a straight line (Fig. 13.21), while the two stars spiral around it. This corkscrew motion identifies the stars as a binary system and locates the center of mass.

Remarkably enough, most stars are in binary or multiple star systems. For nearby sunlike stars, about 45 percent are probably single, while some 55 percent are known to be double, triple, and even quadruple! For example, Alpha Centauri is a triple system. Two of the stars, Alpha Centauri A and B, orbit each other with a separation of about 20 AU. The third star, called Alpha Centauri C (or Proxima Centauri), orbits a few thousand AU from them.

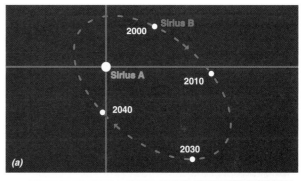

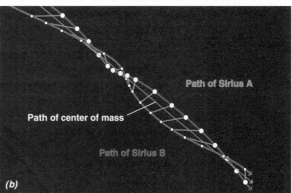

Figure 13.21 Sirius binary system. (a) Orbit of Sirius B relative to Sirius A, plotted with Sirius A as fixed (not moving around the center of mass). The orbital period is about 50 years. (b) Motions of the binary star system consisting of Sirius A and its companion, Sirius B, showing the motions of Sirius A and its companion about their center of mass. It moves on a straight line relative to background stars. The two stars trace a corkscrew path against the sky as they revolve around the center of mass.

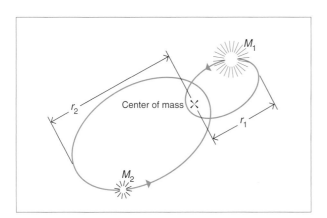

Figure 13.20 Center of mass in a binary star system. Both stars ($M1$ and M_2) move in elliptical orbits (r_1 and r_2) around the center of mass. The more massive star is the one closer to the center of mass.

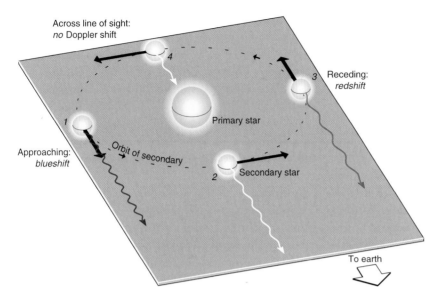

Figure 13.22 Cycle of Doppler shifts from a spectroscopic binary system. Consider the primary star to be fixed (not revolving around the center of mass). When the secondary is at position 1, it is moving toward the earth (negative radial velocity), so its spectral lines appear blueshifted. At 2, the secondary moves across our line of sight, so we see no shift. The same is true at 4. Finally, at position 3, the secondary moves away from us (positive radial velocity), so we see a redshift. Thus, from 1 to 2 we see a decreasing blueshift, from 2 to 3 an increasing redshift, from 3 to 4 a decreasing redshift, and so on.

Spectroscopic Binary Systems (Learning Outcome 13-9)

All binary systems are physically the same, but we observe them in different ways. Suppose two stars are so close together that we cannot distinguish them individually with a telescope, and their orbital periods are so short that the stars move quickly in their orbits. We can identify this binary by looking for two sets of lines in the spectrum (one from each star) and measuring the Doppler shifts (review Enrichment Focus 10.1) produced by the orbital motion. This pair of stars is a **spectroscopic binary.** Note that we can observe the Doppler shifts as long as the orbit is *not* face-on as we view it.

Now to apply the Doppler shift to binary stars. Imagine the more massive star to be stationary, with the secondary revolving about it. As the secondary recedes from the earth, you see its spectral lines redshifted compared with those of the primary; as the secondary approaches, you see its lines blueshifted (Fig. 13.22). At the intermediate points, when the secondary travels across the line of sight, you see no shift.

If the two stars do not differ greatly in luminosity, both spectra can be observed, especially the cycle of shifts of the secondary with respect to the primary. (The smaller-mass star will have the higher velocity, because it is farther from the center of mass.) Sometimes the spectrum of the secondary is too faint to be seen. We then use Doppler shifts in the primary's spectrum alone (but we get less information).

These wavelength shifts can be turned directly into radial velocity shifts by using the Doppler effect (Enrichment Focus 10.1). We use the radial velocities and the period to get the circumference of the orbit. Then we work out the radius of the orbit and so the separation of the two stars. So a spectroscopic binary gives us direct information on the system's orbit.

Note that we can measure the actual velocity of the stars in kilometers per second; we get the actual radius of the orbit, in kilometers, for each star, relative to the center of mass. We can use this information, along with Newton's version of Kepler's third law, to determine the stars' individual masses.

The Mass–Luminosity Relation for Main-Sequence Stars (Learning Outcome 13-9)

Binary systems act like scales that allow us to weigh the two stars. In most cases, the luminosities of the two stars can also be determined. When the luminosities of main-sequence stars are plotted against the stars' masses, the points fall into a definite pattern (Fig. 13.23). For main-sequence stars, the mass determines the luminosity, and the resulting correlation is called the **mass–luminosity relation** (sometimes abbreviated M–L relation).

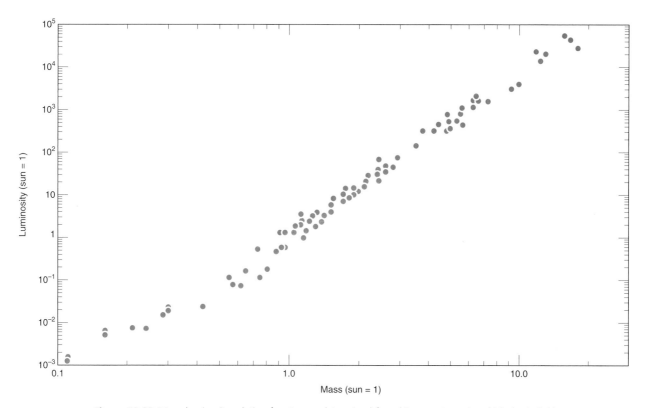

Figure 13.23 Mass–luminosity relation for stars as determined from binary systems, in which the individual masses can be determined. These stars are also those whose distances can be measured, which means that their luminosities can be found. Note the power-of-ten scale for both axes. (Adapted from a diagram by R. C. Smith.)

Basically, the mass–luminosity relation shows that a star's luminosity is *roughly* proportional to the fourth power of its mass for stars with a mass greater than 0.4 solar mass. For example, a star with a mass 10^1 times that of the sun has about 10^4 times the sun's luminosity. Main-sequence stars follow the mass–luminosity relation fairly well; hence, the upward swing in luminosity of the main sequence from M to O stars reflects an increase in the stars' masses. Main-sequence O stars are more massive than main-sequence M stars (Table 13.2). Astrophysicists predicted the mass–luminosity relation theoretically, and its confirmation came from investigation of binary stars. (In Chapter 16 you will see the importance of the mass–luminosity relation for stellar evolution theory.) Giant and supergiant stars also follow mass–luminosity laws, but they differ from that for main-sequence stars.

The mass–luminosity diagram tells us that the masses of other stars do not differ widely from the sun's mass. According to theoretical models, stars with masses greater than about 100 solar masses are unstable, and bodies with masses less than roughly 0.1 solar mass cannot become hot enough to start nuclear reactions and become stars. The predicted narrow range of stellar masses is borne out by the H–R diagram and the mass–luminosity relation.

Stellar Densities (Learning Outcome 13-1)

The mass–luminosity relation says that the masses of stars do not vary over a severely wide range. Yet their sizes do, as indicated by the existence of supergiant and giant stars, so stellar densities vary widely. Remember that density of an object is its mass divided by its volume.

The sun's average density is 1400 kg/m³. Let's compare the sun to Sirius B, a white dwarf. Sirius B is about 3 million times denser than water. This fact tells you that Sirius B – and other white dwarfs – cannot be ordinary stars and cannot have an internal structure like that of the sun.

Stellar Lifetimes
(Learning Outcome 13-10)

The mass–luminosity law immediately offers a way of comparing **stellar lifetimes.** The argument goes like this. The total amount of energy available to a star from the conversion of hydrogen to helium is directly proportional to its mass (its fuel supply). So a star with more mass can produce energy for a longer time than one with less mass, if both stars give off energy at the same rate. The rate at which a star loses energy is given by its luminosity. A greater luminosity means that a star is producing energy faster, using up its mass more quickly. Because luminosity increases as the fourth power of the mass, more massive stars use up their mass faster and have shorter lifetimes. For example, a star of 10 solar masses has a lifetime of about 20 million years, in contrast to 10 Gy for the sun (the representative G star in Table 13.2).

WARNING. Do not confuse the *age* of a star with its *lifetime!* The lifetime is the total span of active life from fusion reactions. The age is the amount of time that has elapsed since fusion reactions began. The sun's expected lifetime is about 10 Gy (true for all solar-mass stars), while its estimated age is 5 Gy. Other solar-mass stars can have different ages. When we say a star is "young" or "old," we are comparing it to stars with similar lifetimes.

Table 13.2 Typical Properties of Main-Sequence Stars

Spectral Class	Surface Temperature (kelvins)	Mass (M_{sun})	Luminosity (L_{sun})	Radius (R_{sun})	Approximate Lifetime (years)
O5	45,000	60.0	800,000	12	8×10^5
B5	15,400	6.0	830	4.0	7×10^7
A5	8,100	2.0	40	1.7	5×10^8
F5	6,500	1.3	17	1.3	8×10^8
G5	5,800	0.92	0.79	0.92	12×10^9
K5	4,600	0.67	0.15	0.72	45×10^9
M5	3,200	0.21	0.011	0.27	20×10^{11}

KEY CONCEPTS

1. Stars are fusion reactors, more or less like our sun (Table 13.2), with similar physical properties, which we can infer by analyzing their light and measuring (or guessing) their distances. The most important physical properties are mass, luminosity, size, surface temperature, and chemical composition. The latter two can be found from spectroscopic analysis, which does not require that we know the distances.

2. Astronomers measure the brightness of stars as seen in the sky by their flux, or the amount of energy from a star that reaches the earth each second over a given area (such as a square meter). A telescope is used to measure the fluxes of stars, which are much smaller than the flux from the sun.

3. Light intensity changes with distance; the flux changes as the inverse-square of the distance. This law applies to electromagnetic waves of all kinds. Hence, distance plays a key role in the fluxes we receive at the earth from stars.

4. Only the distances of the closest stars can be measured directly by a triangulation technique called *heliocentric parallax;* the closer the star, the larger its parallax angle. We use the earth's orbit as the baseline for this measurement.

5. If a star's distance is known, we can find its luminosity from its flux by applying the inverse-square law for light. A star's luminosity tells us how much energy it emits per second.

6. Assuming that stars radiate like blackbodies, their colors indicate their surface temperatures. Hotter stars are bluer, and cooler stars are redder. We can also infer their sizes from their surface temperatures and luminosities, which allow an estimate of surface areas.

7. The angular diameters of stars are generally too small to measure from the earth with current instruments. In a few cases, specialized equipment or techniques allow us to measure stellar angular diameters. Then, if we know the distances, we can calculate the actual diameters.

8. The Hertzsprung–Russell diagram is a graph of the surface temperatures and luminosities of stars. On it, stars fall into distinct groups: main sequence, giants, supergiants, and white dwarfs. Sizes of stars are indicated on the H–R diagram by their luminosity classes. Once an H–R diagram has been made for a large number of stars, we can use it to estimate distances by spectral types and luminosity classes.

9. Binary stars are two stars locked by gravity in orbits around a common center of mass. We can apply Kepler's third law to find the masses of the stars if we know their distance, their orbital period, and the location of each star relative to the center of mass.

10. Binary stars provide the only means of directly finding the masses of stars. For main-sequence stars, we find that the masses and luminosities are related (the more massive stars are more luminous): the luminosities are directly proportional to the fourth power of the mass. We infer that more massive stars have shorter lives than less massive ones.

STUDY EXERCISES

1. In the winter sky, you see the following stars: Capella (yellowish), Betelgeuse (reddish), and Sirius (bluish). List these stars in order of increasing surface temperature. (Learning Outcomes 13-1 and 13-2)
2. Consider stars with the following spectral types: MI, G III, and A V. Which star is the largest? Which the most luminous? (Learning Outcomes 13-3 and 13-10)
3. What limits the accuracy of ground-based heliocentric parallax measurements? (Learning Outcome 13-6)
4. Refer to Figure 13.15 to answer the following questions. (a) Capella and the sun have roughly the same surface temperature. Which star is larger? (b) Regulus and Capella have about the same luminosity. Which star is larger? (c) Vega and Sirius have about the same surface temperature. Which star is more luminous? (d) Which star would appear redder, Vega or Pollux? (Learning Outcomes 13-2 and 13-8)
5. What kinds of stars have very high surface temperatures yet low luminosities? What can you infer about the sizes of these stars? (Learning Outcomes 13-1 and 13-7)
6. What procedure does an astronomer follow to find out a star's density? *Hint:* Divide mass by volume to get density. (Learning Outcomes 13-1, 13-8, and 13-9)
7. Consider a binary star system that does not eclipse and in which one star is much brighter than the other. Then the absorption lines from the fainter star do not appear in the spectrum, but those of the brighter one do. Describe how the Doppler shift would appear from the orbital motion of the stars. (Learning Outcome 13-9)
8. The star Regulus has a mass about five times that of the sun. Use Figure 13.27 to estimate the luminosity of Regulus. (Learning Outcome 13-10)
9. Jupiter is about 5 times as far from the sun as the earth is (5 AUs compared to 1 AU). By how much less is the sun's flux at Jupiter compared to that of the earth? (Learning Outcomes 13-4 and 13-5)
10. Imagine that you observe two stars with the same color. What can you infer about their surface temperatures? (Learning Outcome 13-2)
11. In what region of the H–R diagram do you find supergiant stars? (Learning Outcome 13-7)
12. How do we know that giant and supergiant stars have larger diameters than main-sequence stars of the same temperature? (Learning Outcomes 13-1 and 13-8)

PROBLEMS AND ACTIVITIES

1. Imagine that the sun were moved to one parsec from the earth. What would its flux be then? *Hint:* Enrichment Focus 13.2 will tell you how many AUs make up a parsec.
2. Consider a star whose flux at the earth is one unit of stellar flux. Imagine that the star is moved to five times its distance. How would the flux change?
3. Imagine measuring the parallax of Sirius A from Mars. What size would the parallax angle be?
4. Suppose the sun were placed at a distance of 10 ly from the earth. What would its angular diameter then be?
5. Compare the energy per second per square meter of photosphere emitted by the sun to that of Sirius A; to Betelgeuse. *Hint:* Assume that these stars radiate like blackbodies.
6. For the sun, Sirius, and Betelgeuse, compare the wavelengths at which each body's continuum emission peaks. *Hint:* Same as Problem 5!
7. Sirius has a mass 2.1 times that of the sun. According to the mass–luminosity relation, how luminous should Sirius be compared to the sun?

8. Recall from Enrichment Focus 13.2 that the surface flux of a blackbody (power emitted per unit area) depends only on the temperature. Then the total power or luminosity of a star is simply its surface flux times its surface area. If Sirius has a surface temperature of 10,000 K, what is its surface flux? If its radius is 1.7 solar radii, what is its luminosity? *Hint:* The area of a sphere is $4\pi R^2$, where R is the radius.

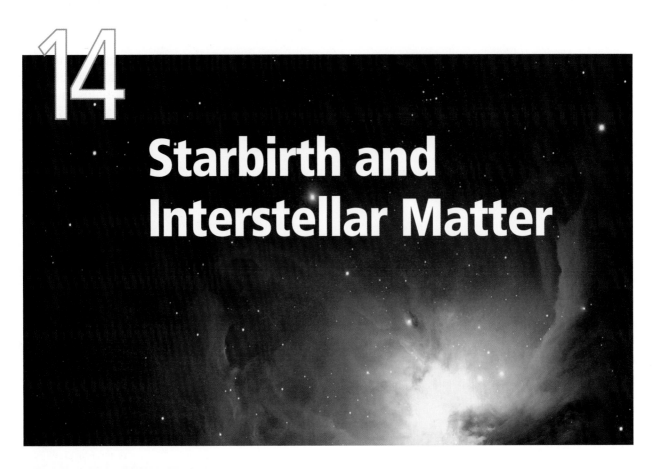

14

Starbirth and Interstellar Matter

LEARNING OUTCOMES

After studying this chapter, you should be able to:

14-1 Present observational evidence for the presence of gas and dust between the stars.

14-2 Compare and contrast the different forms in which the interstellar gas is found and tell how each form is observed.

14-3 Describe three observable effects of interstellar dust on starlight.

14-4 Describe possible physical properties of interstellar dust, such as size and chemical composition.

14-5 Indicate how interstellar molecules and dust might form and suggest the link between them.

14-6 Describe the basic physical ideas of gravitational collapse and contraction for rotating and nonrotating clouds of material.

14-7 Sketch a scenario for the formation of massive stars from molecular clouds and indicate what observations support this model.

14-8 Outline possible processes for the birth of stars like the sun and provide observational evidence to support them.

14-9 Sketch a model for the formation of massive stars and contrast it to that for solar-mass stars.

14-10 Argue that starbirth is occurring now in our Galaxy, with a focus on infrared and radio observations.

14-11 Describe the environment of star-forming regions, with special attention to outflows of gas as a signature.

14-12 Argue that the general process of gravitational collapse can result in planetary systems and brown dwarfs.

14-13 Explain the observational evidence to date for the existence of extrasolar planets around normal stars.

CENTRAL CONCEPT

Stars are born out of the material in the space between the stars. This material consists of gas (in a variety of forms) and dust, mostly collected in clouds.

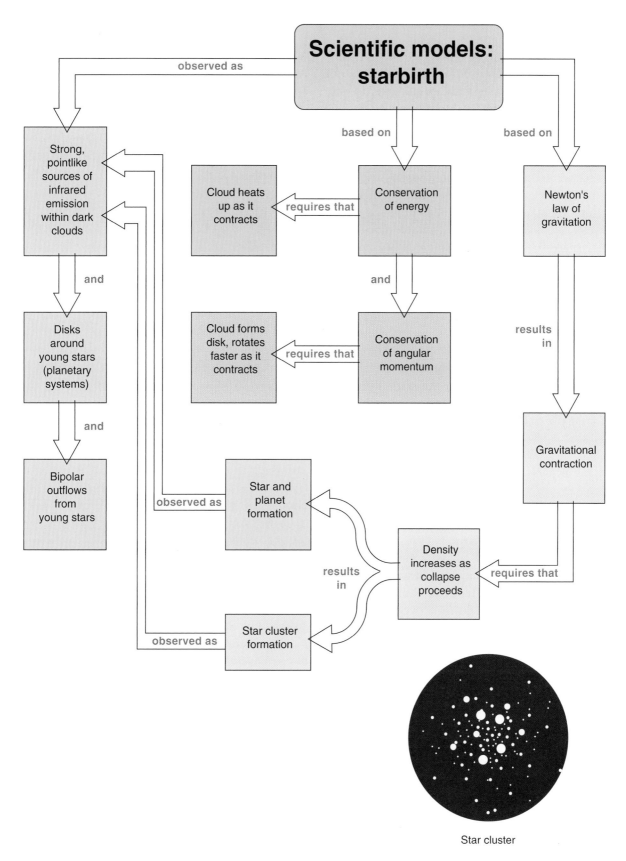

Scientific models: starbirth

observed as

based on

based on

Strong, pointlike sources of infrared emission within dark clouds

Cloud heats up as it contracts

requires that

Conservation of energy

Newton's law of gravitation

and

and

results in

Disks around young stars (planetary systems)

Cloud forms disk, rotates faster as it contracts

requires that

Conservation of angular momentum

and

Gravitational contraction

Bipolar outflows from young stars

Star and planet formation

observed as

Density increases as collapse proceeds

requires that

results in

Star cluster formation

observed as

Star cluster

Stars have finite lives – long by human standards, but limited nevertheless. How long a star lives depends on its mass. A star like the sun will survive for some 10 Gy. More massive stars live scant millions of years. The fact that we observe stars more massive than the sun means not only that star-birth occurred in the past but also that it is going on now. Otherwise we would see fewer stars, especially the massive ones, which die off relatively quickly.

How are stars born? In clouds of gas and dust between the stars. Contrary to first impressions, interstellar space is not empty. It contains gas and dust thinly spread out and in clumps. Hydrogen makes up most of the gas, which typically outnumbers the dust particles enormously. The interstellar gas contains more than neutral atoms: some of it is ionized and some is bound up in molecules. A large fraction of the interstellar gas is locked up in short-lived clouds. From these interstellar clouds, stars are born.

The gas and dust between the stars, called the *interstellar medium,* marks the locus of star-forming action in the Galaxy. Here, stars form in vast stellar nurseries. Some will become massive stars, some sunlike; there will be many binaries, and some will perhaps develop with planetary systems.

14.1 THE INTERSTELLAR MEDIUM: GAS

The **interstellar medium** is the gas and dust between the stars. What do we know, in general, about the **interstellar gas?** First, it is made mostly of hydrogen. Second, it tends to clump in clouds. Third, a hot, diluted gas subsists between the clouds. Fourth, the gas in different locations contains neutral atoms, ionized atoms, free electrons, and molecules. Fifth, on the average, the interstellar gas is very thin. The distance between interstellar atoms is roughly 100 million times larger than the size of the atoms themselves. If a proportional distance relative to their size separated two people, they would be about 100 million meters apart, about the distance between the earth and moon! Sparse as it is, the interstellar gas clumps and forms stars.

Bright Nebulas
(Learning Outcomes 14-1, 14-2)

On a winter night, you can easily spot the constellation Orion. Dangling from Orion's belt is a short sword. If you look closely, the middle star appears fuzzy. A small telescope pointed at this fuzzy patch shows you a diffuse, convoluted cloud surrounding a small cluster of stars. This bright cloud is called the Orion Nebula. (Nebula is the Latin word for "cloud," and the use of the word is an astronomical holdover from the nineteenth century.) Another name for the Orion Nebula is Messier 42 (or M42), the forty-second object in the catalog compiled by the French astronomer Charles Messier (1730–1817) in the eighteenth century.

Only 1500 ly from the earth, the Orion Nebula, roughly 20 ly in diameter, is typical of **emission** or **bright nebulas** (Fig. 14.1), which emit light. At the end of the nineteenth century, spectroscopic analysis demonstrated that these bright nebulas show spectra of emission lines – they consist of hot gas. Recall from Kirchhoff's rules for spectra (Section 5.2) that an emission-line spectrum indicates a hot, diffuse gas. The hot Orion Nebula has lines of hydrogen, helium, and oxygen predominating in its bright-line spectrum.

Nebulas such as Orion do not shine by their own light. Instead, they borrow the energy from hot stars embedded in them. The basic physical process is this: the gas absorbs high-energy ultraviolet photons given off by the central star (or stars) and gives off photons in emission lines at lower energies. For a hydrogen gas, the visible light comes out mainly in the red region of the spectrum. (Enrichment Focus 14.1 has more on this emission process.)

The star (or stars) in a bright nebula is quite hot, about 30,000 K, and emits many photons with enough energy to ionize hydrogen. Most photons are absorbed by the gas surrounding the star so that, out to a considerable distance (a few tens of light years), the gas is almost totally ionized. This zone of ionized hydrogen is called an **H II region** (Fig. 14.2). H I stands for neutral hydrogen and H II for ionized hydrogen. In general, a neutral atom is labeled I, an atom with one electron removed is labeled II, with two electrons removed, III, and so on.

Interstellar Atoms

Bright nebulas are composed almost entirely of hydrogen, so the neutral gas ionized to form them must also have contained mostly hydrogen, as a neutral atom (designated **H I region**). Astronomers surmised that hydrogen atoms populated interstellar space but did not observe the H I gas until 1951.

Figure 14.1 The Orion Nebula (Messier 42), a typical emission nebula of ionized gas. The colors of the emission depend on the elements that have been excited by collisions. The dark regions within the nebula are those bearing concentrations of dust. This image has been processed to bring out the fine detail in the structure of the nebula. The filaments and sheets of gas hint that the nebula is a dynamic place.

21-cm Emission from Atomic Hydrogen (Learning Outcomes 14-1, 14-2)

Recall (Section 5.1) that the hydrogen atom has one proton in the nucleus and one electron in orbit around it. Both the proton and the electron have angular momentum. You can imagine them spinning like miniature tops. According to the rules of quantum physics, the electron and the proton can be oriented in the atom so that the two spins either align or oppose each other. If the spins oppose, the total energy of the atom is just a bit less than if the spins align. As usual, the atom prefers to be in the lower energy state. Suppose that the spins are aligned. Eventually the electron flips over and emits a low-energy photon – energy that corresponds to a wavelength of 21.11 cm (Fig. 14.3).

How are the protons and electrons in hydrogen atoms aligned in the first place? By collisions with electrons and other atoms. The gas in interstellar space is very sparse, and collisions between two atoms occur only once every few million years. On the other hand, once the spins in a hydrogen atom have become aligned, about 10 million years, on the average, pass before the proton flips and the atom drops to its lowest energy state. It emits a 21-cm photon. This is a rare event for any one atom. But because so many hydrogen atoms exist in interstellar space, enough are emitting 21-cm radiation at any given time that the interstellar gas radiates strongly at this wavelength and detected with radio telescopes (Fig. 14.4).

ENRICHMENT FOCUS 14.1

Emission Nebulas: Forbidden Lines

Let me describe in more detail the process by which a bright nebula emits light. Virtually all the atoms in an H II region are ionized. Most of these atoms are hydrogen, but other elements, such as oxygen and nitrogen, are also in the gas and are ionized. In the low-density conditions of H II regions, some ions emit unusual lines called **forbidden lines.** Let's see how.

The ordinary electron transitions described so far (such as the transition from level 3 to level 2 in hydrogen that produces the H-alpha line) occur very quickly, in about 10^{-8} s. Some excited states, though, have much longer durations – as long as a few minutes. Now, under most conditions, such as in the sun's photosphere, an ion in a long-lived excited state is a sitting duck for a collision with another particle in the gas. This collision will knock the electron out of the excited state before it can drop down to a lower one. No photon will be emitted.

In a bright emission nebula, however, gas densities are quite low – roughly 10^9 particles in a cubic meter. Collisions are relatively rare, so an electron in a long-lived excited state can fall to lower energy levels before a collision. Lines from such states are called *forbidden lines* because the electrons stay in the excited states for a very long time compared to the usual 10^{-8} s; some last as long as minutes!

Singly ionized oxygen (O II) has two possible forbidden transitions. This ion has two such sublevels above the ground level (Fig. F.16). A collision can bump an electron into either level. If it goes up to the higher level and then drops back down to the ground state, it emits a photon at 3726 Å; from the lower level, deexcitation results in a 3729-Å photon. Such lines are in the ultraviolet region of the visible spectrum. Other lines in other regions of the spectrum are emitted by other ions. For example, doubly ionized oxygen (O III) can emit two lines at 4959 and 5007 Å, in the green region. Emission lines from H II regions are a mixture of forbidden lines and ordinary emission lines, such as the H-alpha line.

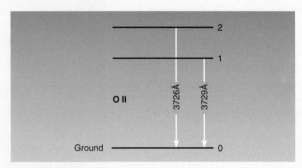

Figure F.16 Energy-level diagram for the forbidden-line emission from singly ionized oxygen (O II).

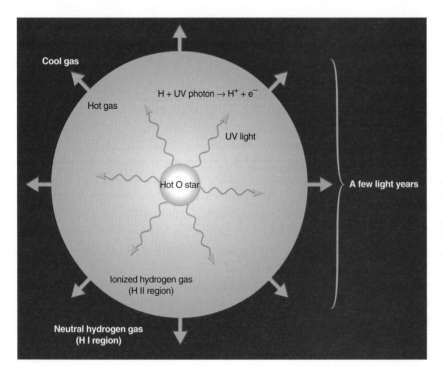

Figure 14.2 Schematic diagram of an idealized H II region. A hot O star (or stars) emits ultraviolet light that can ionize hydrogen for a few light years across. The absorption of the ultraviolet heats the ionized gas to a temperature of about 10,000 K. The hot, ionized gas expands into the cooler, neutral gas surrounding it. Real H II regions are not perfectly spherical, as drawn here.

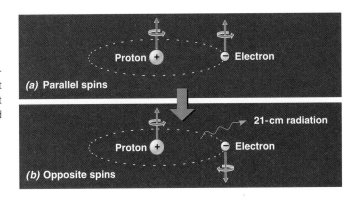

Figure 14.3 21-cm emission from hydrogen atoms. (a) Collisions can line up the spins of the proton and electron so that they are parallel. (b) After some time, if the atom does not hit another, the electron flips so that the spins are opposed and the atom emits a 21-cm photon.

Surveys at 21 cm find that most neutral hydrogen is concentrated in the plane of the Milky Way. On the average, the hydrogen atoms have a temperature of 70 K and a density of 3×10^5 in a cubic meter. In a volume of space equivalent to the volume of your body, you'd find only 3×10^4 hydrogen atoms, whereas your body contains some 10^{27} atoms!

Other gases also dwell in interstellar space. Even before atomic hydrogen was observed with radio telescopes, optical observations had revealed the presence of atoms of several other kinds (and the first molecules). Superimposed on the spectra of some stars, astronomers found sharp, dark lines of elements such as sodium. These narrow absorption lines are produced when starlight passes through cool regions of the interstellar gas, as expected from Kirchhoff's rules (Section 5.2). Such optical observations uncovered atoms of sodium, potassium, ionized calcium, and iron in the interstellar medium.

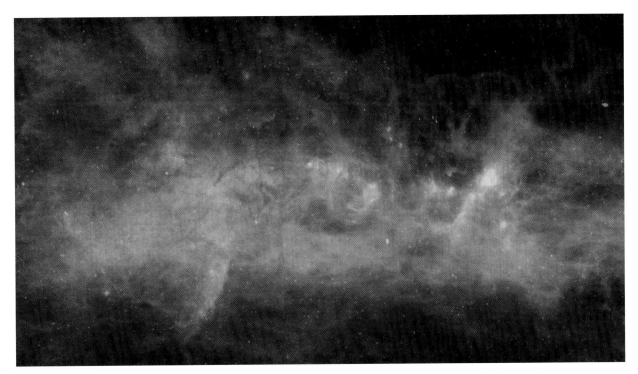

Figure 14.4 Material between stars in the Perseus spiral arm of the Milky Way Galaxy. The 21-cm emission from hydrogen gas clouds is displayed in blue, orange, yellow, and green (depending on their velocities). Infrared Astronomical Satellite (IRAS) images of dust are overlaid on this picture; they appear pink. The image spans about 2000 ly.

By far the most abundant atom in the interstellar gas is hydrogen. Because space is so empty and the temperature of the gas is so low, most hydrogen atoms remain in their lowest energy state, from which they can absorb only ultraviolet light. Because ultraviolet cannot penetrate the earth's atmosphere, these absorption lines cannot be observed with ground-based telescopes. With the advent of earth-orbiting ultraviolet telescopes, we have finally observed cold hydrogen.

Clouds and Intercloud Gas
(Learning Outcomes 14-1, 14-2)

The interstellar absorption lines were the first direct indication of the existence of a pervasive interstellar medium. They led to the idea that cool regions in the medium exist in the form of small clouds. Ultraviolet and radio observations show that the neutral hydrogen gas has a very patchy distribution in clouds with diameters from tenths to tens of light years. The average density of neutral hydrogen is somewhat less than 1 million atoms per cubic meter. However, the observations also show that in certain directions in space, the neutral hydrogen density is ten times less. (For comparison, the density of air at the earth's surface is about 10^{25} particles per cubic meter.)

The space between interstellar clouds also contains gas, and, as expected, it consists mostly of hydrogen. Is it ionized or neutral? Well, it seems to be both – a thin, neutral gas and even thinner, hotter (about 10,000 K) ionized gas between the clouds.

Ultraviolet observations provide direct evidence for a very hot gas of oxygen, stripped of five electrons, permeating the intercloud regions. (The symbol for five-times-ionized oxygen is O VI.) To rip so many electrons from an oxygen atom requires a very high temperature – about 1 million kelvins. Because this hot gas has about the same temperature as the sun's corona, it is called the **coronal interstellar gas.** Evidence for it comes also from the x-rays it emits.

Interstellar Molecules
(Learning Outcomes 14-1, 14-2)

So much for hot stuff. What about cold material? You might expect it to be in the form of simple molecules (such as water) because atoms tend to combine into molecules at cold temperatures. Optical astronomers in the 1930s made the first discoveries of some molecules (such as CH and CN).

The optical part of the spectrum is not the most fruitful region to search, however. A molecule consists of atoms linked together in particular arrangements by electron bonds. It can have different energy states according to how the atoms vibrate or the molecule spins. As with changes in electronic states in an atom, when a molecule changes its vibrational or rotational state, it can emit or absorb a photon. For changes in vibrational states, the photons are infrared ones; for rotational states, radio ones. In the cold regions where molecules can exist in interstellar space, occasional collisions between molecules (or perhaps with atoms) kick the molecules and get them spinning. These excited molecules emit radio photons that can be observed as an emission line, generally at millimeter wavelengths.

The radio search for molecules made of many atoms began in earnest in the 1960s. More than 80 molecules have been found so far. Table 14.1 lists some key ones. Carbon monoxide (CO) is one of the most common molecules in space (Fig. 14.5).

Table 14.1 Selected Interstellar Molecules

Complexity	Molecule Symbol	Molecule Name
Two atoms	H_2	Hydrogen
	OH	Hydroxyl radical
	CO	Carbon monoxide
Three atoms	H_2O	Water
	HCN	Hydrogen cyanide
	H_2S	Hydrogen sulfide
Four atoms	NH_3	Ammonia
	H_2CO	Formaldehyde
	HC_2H	Acetylene
Five atoms	CH_4	Methane
	HCOOH	Formic acid
	HC_3N	Cyanoacetylene
Six atoms	CH_3OH	Methyl alcohol
	CH_3CN	Methyl cyanide
Seven atoms	CH_3NH_2	Methylamine
	CH_3C_2H	Methylacetylene
Eight atoms	$HCOOCH_3$	Methyl formate
	CH_3C_3N	Methyl cyanoacetylene
Nine atoms	CH_3CH_2OH	Ethyl alcohol
	CH_3CH_2CN	Ethyl cyanide

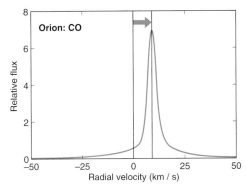

Figure 14.5 Microwave emission line from carbon monoxide (CO) from the region of the Orion Nebula. The molecular emission comes from a cloud behind the nebula as we see it in the sky. Note that the emission line has a width. This widening occurs because different regions of the gas are moving with different radial velocities (horizontal axis). In fact, the central peak lies at a velocity of about 10 km/s. The vertical axis gives the relative flux as measured by the radio telescope. (Adapted from a diagram by John Bally and Charles J. Lada.)

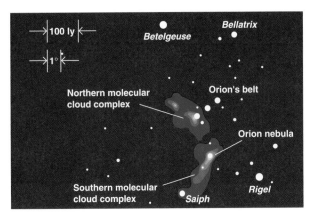

Figure 14.6 Molecular clouds and regions of starbirth in Orion. A carbon monoxide flux map of the emission from a giant molecular cloud associated with the Orion Nebula is superimposed on the stars. This map covers about 15 degrees on the sky. The scale for one degree and 100 ly on this map is shown in the upper left. The hottest, densest part of the cloud lies near the Orion Nebula; it is called *OMC1* for Orion Molecular Cloud 1. It is part of the Southern Molecular Cloud Complex. Another complex lies just to the north and east of the last star in Orion's Belt.

Also common are ethyl alcohol and water. Note that many of the molecules are organic – compounds in which carbon plays a central role. The most abundant atoms in these molecules (carbon, hydrogen, nitrogen, and oxygen) are also the most abundant in living creatures on the earth. Carbon has the ability to form ring-shaped molecules essential to life. After years of fruitless searches, astronomers have finally detected one such carbon ring (benzene) in interstellar space.

By far the most abundant molecule is molecular hydrogen, but molecular hydrogen does not emit or absorb at radio wavelengths. Instead, the hydrogen molecule absorbs and emits ultraviolet and infrared wavelengths. Infrared emission lines of molecular hydrogen have been observed from heated interstellar clouds (at a temperature of about 2000 K, the hydrogen molecules are excited and emit infrared lines), and ultraviolet absorption lines from cool clouds in front of hot stars have been detected in the spectra of these stars. Overall, about half the hydrogen in the Milky Way appears to be in the form of molecules.

Molecular Clouds
(Learning Outcomes 14-1, 14-2)

Most interstellar molecules are localized in dark, dense, cold conglomerates called **molecular clouds.** These clouds often lie near H II regions; one

of the closest molecular clouds sits behind the Orion Nebula, as we view it from earth (Fig. 14.6). The molecular cloud here consists of two parts: a large, low-density cloud surrounding a dense, small core. The low-density cloud has an enormous extent: it is at least 30 ly across and contains at least 10^4 solar masses of material. The core is only 0.5 ly in size and has a mass of only 5 solar masses.

The Orion region presents an excellent example of a **giant molecular cloud.** Observations so far indicate that the bulk of the material of the interstellar medium is bound up in complexes of giant molecular clouds. These immense globs of molecules, held together by gravity, have the following typical properties.

1. They consist mostly of molecular hydrogen; many other molecules are present, but these make up only a small fraction of the mass.
2. The cloud complexes have average densities of a few hundred million molecules per cubic meter; the individual clouds are slightly denser.
3. They have sizes of a few tens of light years.
4. The total masses of the complexes range from 10^4 to 10^7 solar masses; 10^5 solar masses is typical. Masses of individual clouds are about 10^3 solar masses.

Figure 14.7 False-color infrared image of Messier 17, the Omega Nebula. The dark region to the right and above the nebula are dense molecular clouds, out of which the nebula has formed.

The cores of these clouds are unusual places compared with the average interstellar medium. Here the temperatures are a frigid 10 K, and the densities get as high as 10^{12} molecules per cubic meter. That's an immense concentration by interstellar standards, yet it is only 10^{-13} times the density of molecules in the air at the earth's surface!

Giant H II regions, which surround young, massive stars, often conjoin molecular cloud complexes (Fig. 14.7). This proximity suggests that giant molecular clouds play a key role in the process of star formation. Chemically, clouds with active star formation appear to be very different from those in which none is now taking place.

14.2 THE INTERSTELLAR MEDIUM: DUST

Dust also occupies interstellar space. There's not much out there; on the average, only one dust particle in every million cubic meters – that's roughly a cube with sides one soccer field in size! The dust amounts to about one percent of the total mass of interstellar matter, but it can cut out light from distant objects or from those shrouded in dense clouds. Piercing the veil of **interstellar dust** has been an important goal of radio and infrared astronomers. That breakthrough has been critical in revealing the process of starbirth.

Cosmic Dust
(Learning Outcomes 14-1, 14-3)

Direct observations hint at dust between the stars. Dark nebulas, such as the Horsehead Nebula in Orion (Fig. 14.8), display dramatic cutoffs of light due to dust. Astronomers give the name **dark cloud** to an interstellar cloud that contains so much dust that it blots out the light of stars within it and behind it. The dark rifts and lanes in the Milky Way, once attributed to the lack of stars,

Figure 14.8 The Horsehead Nebula in Orion in visible light. The red emission nebula (from hot, hydrogen gas) contrasts with the famous dark, dusty cloud called the Horsehead because of its silhouette.

are actually regions heavily obscured by dust. In some regions of the sky, dusty, swirling clouds blot out the light of the stars and H II regions behind them.

Some bright nebulas are not emission regions but clouds of dust reflecting light from nearby stars. An example is the nebula that surrounds the Pleiades (Fig. 14.9). The spectrum of this nebula does not exhibit the bright emission lines. It shows simply the absorption-line spectrum of the Pleiades stars – light reflected by dust. A bright nebula that arises from the reflection of starlight by dust is called a **reflection nebula,** and it has a bluish color. Why?

Figure 14.9 Reflection nebula in Orion (NGC 1977). Most reflection nebulas appear blue because the dust particles that reflect starlight are tiny, and they scatter blue light rather than other colors. The dust and gas surround a cluster of young stars.

Generally, interstellar dust makes itself known in two ways: by **extinction,** the dimming of starlight, and by **reddening,** the scattering of the blue wavelengths more than the longer wavelengths. Let's look at each of these processes by recalling what happens to sunlight when the sun is near the horizon (Fig. 14.10). One, the sun is dimmer. Two, the sun appears redder when near the horizon because of the preferential scattering of blue light through much more air than when the sun is overhead.

Figure 14.10 Close to the horizon, the scattering of blue light makes the sun appear redder than when higher. Dust or smog in the air enhances the reddening effect. Dust in the interstellar medium has the same reddening effect on visible light passing through it. Here we see the sun setting behind the Gemini 8-meter telescope on Mauna Kea.

Imagine starlight traveling through a dust cloud. The particles can absorb some of this light as it comes through. The dust particles can also scatter the starlight, so it goes off in a different direction from the original one. In either case, less light exits the dust cloud than enters it. Astronomers call this dimming of starlight *extinction.*

When starlight is quenched, blue light is more strongly scattered than red, so red light penetrates the dust cloud more readily than blue. When you observe a star through the dust cloud, more of its red light reaches your eye than does blue. The star appears redder than it actually is; astronomers call this process *reddening.* What about the blue light? The part that is scattered bounces around the dust cloud until it finally exits. So the cloud, a reflection nebula, appears bluish.

Reddening, since it is a color effect, is much easier to measure than extinction. We can estimate the quantity of the dust from the amount of reddening. From the H–R diagram, we know that a certain spectral classification of a star corresponds to a certain color. A star's spectral class can be determined (from the strength of certain absorption lines), even if its light is reddened. We measure the star's color compared with that expected for its spectral class. The difference in color, the reddening, tells how much dust lies along the line of sight to the star.

Dust and Infrared Observations
(Learning Outcome 14-1)

Interstellar dust blocks out visible light. This fact makes optical astronomers unhappy, for it limits their view of distant stars and galaxies. Infrared astronomers fare better, for infrared radiation penetrates dust. In fact, some infrared radiation from space comes from the dust itself. Infrared astronomers can both *see* dust and see *through* dust!

How does dust emit infrared? The dust is made of small, solid particles called *grains*. Basically, dust grains act roughly like (very small) blackbody radiators. If the grain has a temperature of around 100 K, its emission will peak in the infrared.

The Orion Nebula marks a region studded with strong infrared sources. Optically, the core of the nebula is the densest part: hot gas (mostly hydrogen) is ionized and excited by the O stars there. The brightest stars form a trapezoid figure (easily seen in a small telescope) called the *Trapezium cluster.*

Now let's look at an infrared map (Fig. 14.11) of the core region. Quite a difference! What we are seeing is the infrared emission from cold dust (about 70 K) located somewhere at or near the center of the molecular cloud – dust heated by something capable of putting out 70,000 solar luminosities. The visible Orion Nebula, illuminated and sustained by the Trapezium stars, lies in front of the molecular cloud like a hot bubble. This model provides some key pieces to the puzzle of starbirth.

The Nature of Interstellar Dust
(Learning Outcome 14-4)

What is the interstellar dust made of? One indirect clue comes from the cosmic abundances of certain elements: only elements that make up an appreciable fraction of the interstellar material can contribute in a large part to the dust grains. These include (in order) hydrogen, oxygen, carbon, nitrogen, and silicon.

Figure 14.11 False-color infrared image of the young Trapezium star cluster in Orion. Here, newborn stars ("free-floating planets") and many (perhaps a hundred) brown dwarfs, or failed stars, have formed out of the molecular cloud.

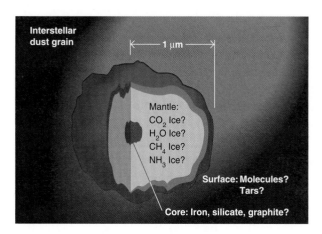

Interstellar dust grain

1 μm

Mantle:
CO_2 Ice?
H_2O Ice?
CH_4 Ice?
NH_3 Ice?

Surface: Molecules?
Tars?

Core: Iron, silicate, graphite?

Figure 14.12 Simplified model of an interstellar dust grain; the composition is not really known. The dust may consist of any of the materials listed or some combination of them. A small core covered with a larger mantle is the main feature. The surface may be coated with a tarry substance.

The most abundant elements make up rather common substances: hydrogen and oxygen for water, carbon and hydrogen for methane, carbon and oxygen for carbon dioxide, nitrogen and hydrogen for ammonia, silicon and oxygen plus metals for silicates (compounds of Si and O commonly found in earth rocks). Compounds like water, methane, and carbon dioxide are loosely called *icy materials* because they are solids at temperatures below about 100 K. (Recall that they make up the bulk of the nucleus of a comet: Section 11.2.)

Now the dust grains themselves. To account for the properties of interstellar extinction, astronomers have developed models featuring **core–mantle grains** (Fig. 14.12). The small core, about 0.05 μm in radius, can consist of silicates, iron, or graphite; silicates are most likely. The mantles, about 0.5 μm in radius, are made of icy materials, likely some mixture of them all. When grains drift into hot regions, such as an H II region, their mantles evaporate, leaving behind a bare core. Infrared observations bolster the idea that silicates and ices (at least water ice) form parts of interstellar grains. Their spectra show absorption bands like those of silicates and water ice.

The icy materials in the mantles may be processed into organic compounds. Laboratory simulations of the conditions in interstellar space show that dirty ice mantles can absorb ultraviolet light (which permeates the dust-free regions of interstel-

lar space). Ultraviolet photons have enough energy to break up chemical compounds and promote the creation of new ones, such as formaldehyde (H_2CO). That should sound familiar, for it is one of the molecules found in the interstellar gas. The experiments have also created residues that can survive much higher temperatures than ordinary ices – up to 500 K. The compositions of these organic materials are unclear, but one of them, a yellow tar, probably makes up a large part of processed grain mantles.

TO SUM UP. Interstellar dust contains elements common to the interstellar gas in the general forms of ices, silicates, graphite, and metals (such as iron). Some of the mantles of grains contain organic compounds.

Dust and the Formation of Molecules (Learning Outcome 14-5)

Dust and molecules are intimately allied in space: wherever you find a molecular cloud, you usually find lots of dust. The reverse is also true. If you see a dark nebula, you have also likely found a molecular cloud. This association implies that grains play a role in the formation of molecules.

Consider the problem of forming an interstellar molecule. You have to get widely separated atoms together and chemically bound – no easy task in the dilute gas of interstellar clouds. Solution? Use cold dust grains as the sticking and forming surfaces. If a hydrogen atom hits a cold grain, it will stick. Add another hydrogen atom to the same grain, and a hydrogen molecule forms, which does not stick as well as atomic hydrogen. It eventually pops off into space. Grain surface formation seems to work for hydrogen. For other small molecules, chemical reactions in the gas alone seem to explain the formation of molecules up to those containing four atoms.

What about more complex molecules? They may form in the icy mantles of grains, where ultraviolet light provides the energy for chemical processing of dirty ice to more complex molecules. Some of these compounds are unstable and explode, hurling the molecules into the interstellar gas.

TO SUM UP. Hydrogen molecules form on interstellar grains, and molecules of up to four atoms form in the interstellar gas. The formation of more complex molecules may be the product of chemical reactions, driven by ultraviolet light, in grain mantles.

Formation of Cosmic Dust (Learning Outcome 14-5)

Where does the dust come from? Interstellar grains have cores of dense solids (such as silicates and graphite), which solidify at a few thousand kelvins (as in the solar nebula).

How and where are these grains made? The denser grains are probably made in the atmospheres of cool supergiant stars. Such stars blow mass into space at rates of about 10^{-5} solar mass per year. As gaseous material streams outward, its temperature drops, and solids can condense out of the vapor. In fact, spectra of some supergiant stars show silicates, indicating that such dust exists around them. In a rarer class of stars, in which carbon is somewhat more abundant than oxygen, graphite particles and particles made of silicon carbide form in the outflowing material.

Cool giant stars, which are more common than supergiants, also lose mass – at about 10^{-6} solar mass per year. These giants add dust to the interstellar medium. For both giants and supergiants, some cool stars (3000 K or so) are enveloped in so-called circumstellar material, which contains both gas and dust. These stars have the highest mass loss rates of all, as great as 10^{-4} solar mass a year. They are large contributors to interstellar dust.

The ices that make up grain mantles likely condense on cores in the deep interiors of dense molecular clouds where they are protected from ultraviolet radiation. Here the temperatures are low and the gas densities high, so bare grains can grow crusts of ices. A core grows a new mantle once every 10^8 years or so, since grains will lose their icy mantles in an environment where temperatures range above a few hundred kelvins.

14.3 STARBIRTH: THEORETICAL IDEAS

Let's deal now with how stars form. Here's the big picture: stars are born out of interstellar molecular clouds by **gravitational collapse.** Because these clouds contain many times the mass of a single star, they fragment into much smaller pieces during the process of star formation. To try to cast a clear light on our present understanding of starbirth, I have divided the topic into two broad parts: theoretical ideas (this section) and observational clues (Section 14.4). (See the Starbirth Celestial Navigator™.)

Collapse Models (Learning Outcome 14-6)

Newton first recognized the basic process of star formation: gravitational collapse. A cloud with enough mass and a low temperature will naturally contract from its own gravity. As gravitational potential energy becomes kinetic energy, the material in the cloud heats up (Section 11.5). Eventually the temperature and density build up to a point that the outward pressure brings down the rapid collapse to a slower contraction. Greater pressure halts the collapse. The temperature in the center finally reaches the kindling temperature of fusion reactions. At that moment, a star is born. Prior to the ignition of fusion reactions, the cloud is called a **protostar.** As a protostar evolves, its luminosity, size, and surface temperature change.

Protostars of different masses evolve in physically different ways. Despite differences because of mass, theoretical models display some common stages.

1. The collapse occurs fast because it starts out in *free-fall,* that is, controlled only by gravity.
2. The central regions collapse more rapidly than the outer parts, and a small condensation forms at the center; this core will become a star.
3. Once the core has formed, it accretes material from the infalling envelope of material surrounding the core.
4. The star becomes visible either by accreting all the surrounding material onto itself or by somehow dissipating the shroud of dust.

With these general features in mind, let's look in a little detail at the formation of protostars.

Protostar Formation (Learning Outcomes 14-6, 14-7)

Imagine a nonrotating interstellar cloud of gas, mostly molecular hydrogen, and dust. Assume that this cloud has sufficient density to contract gravitationally. Its initial diameter is a few light years.

As the collapse proceeds, material at the cloud's center increases in density faster than at the edge. Because of the increase in density, the collapse at the center speeds up (as expected from Newton's law of gravitation). It collapses faster, grows denser, and so collapses still faster. The rest of the cloud's mass is left behind in a more slowly contracting envelope.

With the rapid fall of material in the core, at some point the density of the core reaches a critical value at which the cloud becomes opaque. The core heats up to a few hundred kelvins, its pressure increases, and the core's collapse slows to a contraction and then eventually stops. A protostar forms (Fig. 14.13).

Meanwhile, the envelope continues merrily falling inward, showering mass on the core. The protostar slowly contracts; its size is twice that of the sun, its luminosity, a few times the sun's. The total time from the start of collapse to this stage is about one million years. The infant star hides in its womb, with dust cutting out the protostar's light. However, the light absorbed by the dust heats it so that it gives off infrared radiation. An observational sign of protostars should be small, intense sources of infrared radiation in or near known dense clouds of gas and dust.

Eventually, the star rids itself of its cloaking cloud. It may fall entirely onto the young star, or it may be blown away by a strong stellar wind. As the cloud dissipates, we see a **pre-main-sequence star,** one that is larger and cooler than it will be on the main sequence. The total time elapsed from the onset of collapse to reaching the main sequence is about 50 million years for a solar-mass star.

Collapse with Rotation (Learning Outcome 14-6)

The models just described lack at least one fact of astronomical life: rotation. It is likely that interstellar clouds rotate at least a bit. Any rotating mass has angular momentum (review Enrichment Focus 11.1). An isolated, rotating mass must conserve angular momentum, so as a spinning, spherical mass collapses gravitationally, it must spin faster. It will eventually collapse along its rotation axis into a disk, as described in Section 11.6 for the formation of the planets.

The addition of spin to theoretical models of protostar collapse makes the calculations much tougher and the results less conclusive. To give you the flavor of results: in some cases, a ring or bar of material results. These rings and bars turn out to be unstable in some instances; they break up into two or three blobs, which sometimes coalesce. The smallest mass of a fragment is about 0.01 solar mass; even these small fragments end up rotating rapidly, although the cloud originally may have had very slow rotation.

Three-dimensional computer models (Fig. 14.14) probe this process more extensively. The cloud begins with a simple pancake structure (Fig. 14.14a), but its rapid rotation throws off two spiral arms (Fig. 14.14b). The central region does not split cleanly into two pieces; instead, the fragments crash together to form a bar. Most of the material in the spiral arms wraps up into a ring (Fig. 14.14c).

The important point to remember is this: with

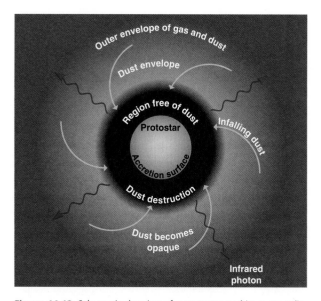

Figure 14.13 Schematic drawing of a protostar and its surrounding environment. Once the central mass has stabilized, material from the envelope continues to fall inward to accrete on the star's surface. Just around the star is a zone free of dust, which has been vaporized by the high temperatures there. The dusty shell emits infrared photons where the dust is thick enough to be opaque to photons emitted by the protostar.

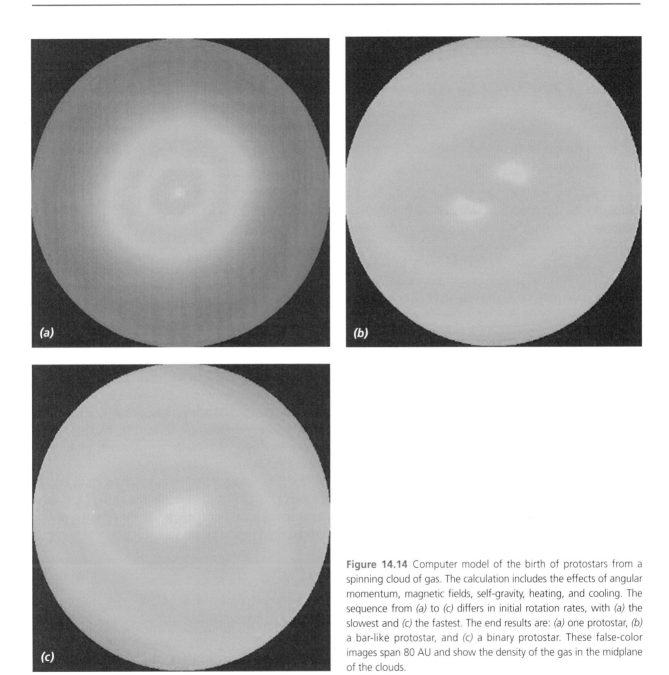

Figure 14.14 Computer model of the birth of protostars from a spinning cloud of gas. The calculation includes the effects of angular momentum, magnetic fields, self-gravity, heating, and cooling. The sequence from (a) to (c) differs in initial rotation rates, with (a) the slowest and (c) the fastest. The end results are: (a) one protostar, (b) a bar-like protostar, and (c) a binary protostar. These false-color images span 80 AU and show the density of the gas in the midplane of the clouds.

spin added to the clouds, the calculations develop rings or bars that fragment into a few blobs. If each blob finally becomes a star, we then have a natural explanation for the finding that most stars in the Milky Way are in binary or multiple systems. If the masses are much less and end up in rings, they can result in planetary systems.

14.4 STARBIRTH: OBSERVATIONAL CLUES

Let's look at the real world to see how models relate to observations. We have uncovered more information about massive starbirth than about the birth of solar-mass stars. There's a good reason: massive protostars have greater luminosities than solar-mass ones, and, when they reach "stardom," massive stars ionize the gas around them. Radio telescopes can then detect the ionized gas. Since all this action is cloaked by dust, infrared and radio observations permit us to inspect stellar wombs.

Signposts for the Birth of Massive Stars (Learning Outcome 14-7)

Let me outline what models predict should be the hallmarks, in the radio and infrared, of the birth of a massive star. With this outline as a guide, we'll then look at the observations.

First, because stars condense from molecular clouds, we need to find a molecular cloud; molecules emit at millimeter wavelengths. Second, the initial free-fall collapse warms the dust to roughly 30 K. This dust emits infrared radiation that peaks at roughly 100 µm. Third, as the protostar forms, the interior dust reaches about 300 K and so emits with a peak at 10 µm. Fourth, as the protostar reaches the main sequence, it ionizes the hydrogen gas, and a compact H II region develops, observable at centimeter wavelengths. Fifth, as the hot, ionized gas expands, the radio and infrared intensity decreases. Finally, the H II region expands enough to blow off its dusty cloak, and the star appears to optical view.

With this scenario in mind, let's return to our old friend the Orion Nebula (Fig. 14.1). The H II region around the Trapezium marks the oldest (most evolved in an evolutionary sense) part of the region. The Trapezium cluster consists of a few hundred stars. The hot stars here, which are no more than a million years old, ionize the gas. The distance between the Trapezium stars and the front edge of the molecular cloud is roughly one light year. In an evolutionary sense, the molecular cloud core that lies behind the Orion Nebula is the youngest (least evolved) part of the region. Where is starbirth happening?

Infrared images reveal that the core of the Orion Nebula contains a dense cluster of young stars (Fig. 14.15). More than 500 stars appear, most visible only in the infrared; these make up the densest young cluster known. Radio observations show that the gas here hits a density of 10^{14} molecules per cubic meter. Infrared observations of hot hydrogen molecules in this region show that the

gas moves along at high speeds – some 100 to 150 km/s – probably powered by a wind from one or more protostars. So the star formation here now occurs in the densest regions, stirred by the swift outflows from the newborn stars.

The Birth of Massive Stars (Learning Outcomes 14-7, 14-9, 14-10)

The foregoing observations add up to a picture of a sequence of massive-star formation from giant molecular clouds. The hot stars have heated the gas around them, ionizing it and destroying the molecules there. The hot gas of the H II region slowly expands and runs into the cold, dense molecular cloud. Here a shock wave forms. The shock wave, moving at about 10 km/s, prompts gravitational collapse and star formation out of the molecular cloud.

Moving more into the fragmented molecular cloud, the shock wave triggers more star formation, and each group – of hot stars – will develop an H II region and another shock wave (Fig. 14.16). So the molecular cloud will finally self-destruct in an orgy of star formation.

Once the star formation has started at an end of a molecular cloud, it propagates through it in a chain reaction. What triggers the initial burst of star formation? We do not yet know.

The Birth of Solar-Mass Stars (Learning Outcomes 14-8, 14-10)

The current observational evidence for the formation of stars like the sun is skimpy and doesn't hold together in an inviting way. One point is clear: like massive stars, solar-mass stars are born from molecular clouds. The questions are in which clouds, and how?

Clusters of young stars are close to gas and dust in the form of dark clouds. *Dark clouds* are one type of small molecular cloud and appear dark because they are dusty. They typically have temperatures of 10 K, densities of 10^8 atoms in a cubic meter, and masses up to a few hundred solar masses.

A typical dark cloud with active star formation shows a central region that emits in the far infra-

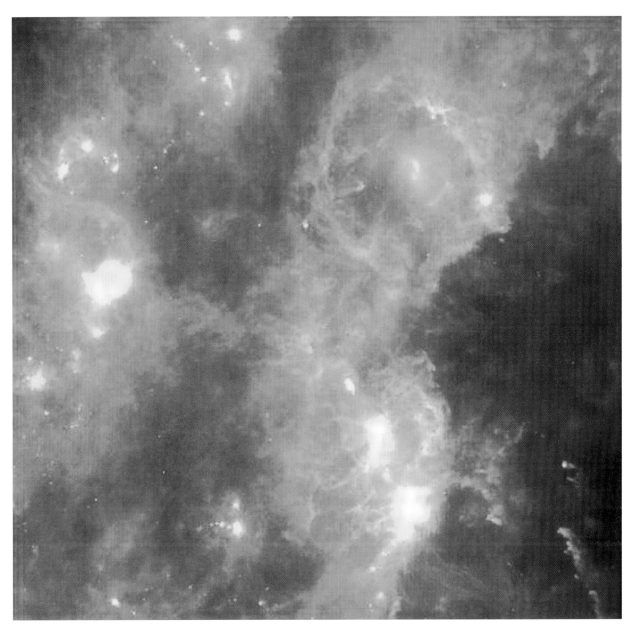

Figure 14.15 Young stars in the Orion region. This infrared image shows hundreds of new-formed stars, most of which are invisible at optical wavelengths. In the false-color coding, the hottest sources (stars) are blue and interstellar dust is green or red. The bright yellow region in the lower right is the Orion Nebula. The Rosette Nebula is the brightest object (left).

red. Near-infrared observations show many possible **young stellar objects** (YSOs), the generic name for all the stellar objects in early stages of formation. The basic observational evidence for solar-mass starbirth is infrared pointlike (starlike) sources in dark clouds.

According to one view, then, solar-mass stars are born in fairly massive dark clouds or perhaps in giant molecular clouds along with massive stars. They form from fragments throughout the cloud, rather than at the edges as massive stars do. The birth of the massive stars sweeps away the gas and

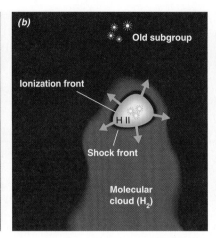
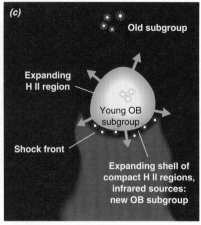

Figure 14.16 Schematic drawing for a sequential model of massive-star formation from a giant molecular cloud. (a) The formation of a small cluster of hot stars in a molecular cloud (b) creates an expanding H II region and shockwave that (c) drives the collapse of more of the molecular cloud to make another small cluster of stars. Each episode of star formation leaves behind an OB subgroup as the formation process moves through the elongated shape of a giant molecular cloud.

dust to reveal the stars. In this picture, most star-birth takes place in dark, massive clouds, out of which OB groups form. The sun may have been born in an OB group, such as the one described in Orion.

From observations so far, we suspect that in general, stars are born from molecular clouds, massive stars from massive clouds, and less massive stars (the majority of those in the Galaxy) from less massive molecular clouds. Infrared observations appear to support this general notion (look back at Fig. 14.15). Many warm (temperatures from 70 to 200 K) point sources of infrared emission lie in the cores of dark, molecular clouds. These may be protostars, heating up the encircling dust. Whatever the details, we estimate that about ten new stars are born each year in the Milky Way.

Molecular Outflows and Starbirth (Learning Outcomes 14-10, 14-11)

Observations of molecular emission around protostars have discovered high-speed (up to 100 km/s; typically 50 km/s) flows of gas. Doppler shift measurements show that these flows tend to be bipolar: two streams moving in oppo-site directions (one has blueshifts, the other redshifts). The flows carry considerable mass – many times that of the sun – and can span a few light years; so enormous amounts of energy push them along (some 10^{40} J). Such **bipolar outflows** appear to be associated with the birth of a star. They probably last only a short time – no more than 10,000 years.

A simple model for the outflows envisions a young massive star putting out a strong stellar wind. Surrounding the star is a dense disk of gas and dust. This disk would naturally channel the flow of the stellar wind, causing it to stream out along the thin axis of the disk, making two streams. When these two streams push enough material outward, two opposing lobes of gas should form. The early stages of this process are buried in gas and dust, but later stages are visible to optical and radio telescopes. HST observations have confirmed this basic model (Fig. 14.17).

Models for these bipolar flows basically involve a disk of material around the protostar, threaded with a magnetic field (Fig. 14.18). The disk forms from gravitational contraction with angular momentum. The magnetic field resides in the sur-

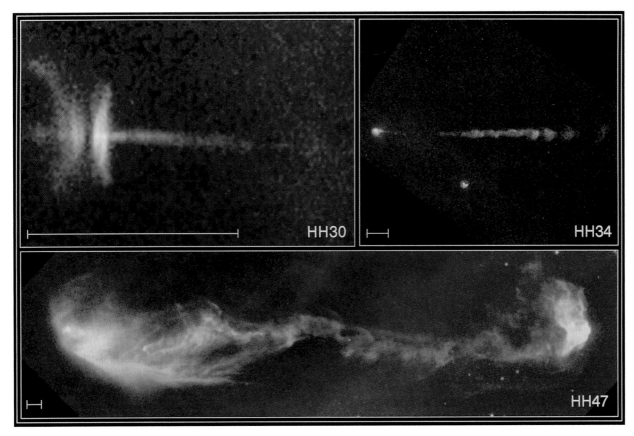

Figure 14.17 HST images of disks and jets associated with young stars (scale at the bottom left indicates 1000 AUs). Upper left: An edge-on disk encircling a protostar (hidden in the dust). Note the jet (a gas outflow) extending to the right. Upper right: This jet shows a beaded structure that indicates that the outflow is episodic. Bottom: This jet collides with the interstellar medium, causing the gas there to glow. The protostar is located in the dust cloud at the left end of the image.

rounding molecular cloud. Ionized gas that falls into the disk drags the magnetic field with it. The hot disk generates both a molecular flow (from the outer regions) and an ionized flow from its inner parts. The ionized outflow is channeled by the magnetic field. Essentially, the rotating disk acts like a flywheel that stores energy that is released in the outflow.

The bipolar flow stage marks a very brief act in the process of starbirth and the evolution of YSOs. The sequence goes like this: a star forms in the gravitational collapse of part of a rotating molecular cloud (Fig. 14.19a); the central part gathers into the protostar, which appears as an infrared source surrounded by a dense disk (Fig. 14.19b). Next, the

star develops a powerful wind that breaks through the disk in opposite directions (Fig. 14.19c). The wind carries along clumps of molecular material and strikes the cloud, producing shock waves. Cavities carved by the outward flow enlarge and push away more of the dark cloud (Fig. 14.19d), eventually revealing the star.

Planetary Systems?
(Learning Outcomes 14-12, 14-13)

The discovery of the bipolar flows strongly hints that disks of material typically form around massive stars during their formation. From such disks, planetary systems might form. So we have a clue that the neb-

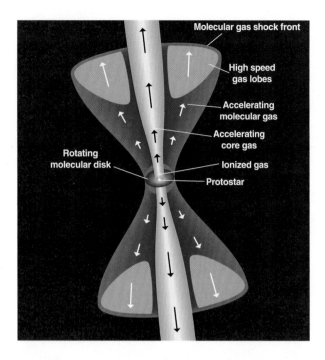

Figure 14.18 Schematic model for the source of the bipolar out-flows from a young stellar object. The key feature here is the presence around the star of a disk of material formed by accretion. It is hot enough to produce ultraviolet photons to ionize some of the gas nearby. If the star has formed with a magnetic field, the ionized outflow follows along the magnetic field lines. The outflow creates a shock front that backs up the material in two lobes. (Adapted from a diagram by R. E. Pudritz and C. A. Norman.)

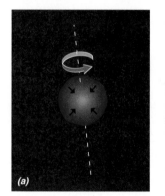 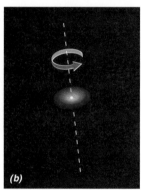  

Figure 14.19 Possible evolutionary sequence in the formation of a bipolar outflow. (a) Gravitational collapse of a rotating interstellar cloud. (b) Formation of a disk with a protostar at the center. (c) Start of the outflow along the axis of the disk. (d) Development of two large lobes from the outflow.

ular model (Chapter 11) for planetary formation might actually operate elsewhere in the Galaxy.

In fact, we are just finding observational clues that planetary formation may be happening now. An image of a star called Beta Pictoris (Fig. 14.20) shows a possible planetary disk. At a distance of 50 ly from the sun, Beta Pictoris is a fairly nearby star. The disk of dusty material extends 60×10^9 km from the star. It shows up edge-on, and calculations indicate that planets may already have formed here. Spectroscopic observations show Doppler shifts that hint at a rapidly rotating, clumpy cloud of gas making up the inner regions of the disk.

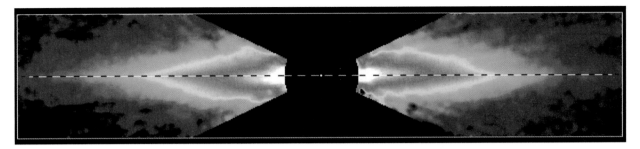

Figure 14.20 Circumstellar disk around the star Beta Pictoris, taken by HST. The disk around the star appears nearly edge-on and extends out some 400 AU.

Observations of bipolar outflows from YSOs, which have about the same mass as the sun, imply that many of these bodies have disks of materials with sizes on the order of 100 AU – just what you would expect for nebulas out of which planetary systems can form.

HST has taken images in the Orion Nebula region that show disks around some YSOs (Fig. 14.21). These disks appear as thick disks with a hole in the middle where the star is located. The radiation from the hot star boils off material from the disk's surface. The material is then blown into a cometlike tail by the wind from the YSO. The formation of such disks may commonly accompany the birth of stars.

Brown Dwarfs: Failed Stars (Learning Outcome14-12)

What becomes of objects born with too little mass to ignite fusion reactions? Theoretical models indicate that such masses develop into huge, Jovian-type bodies with masses less than 75 times that of Jupiter. Gravitational energy released by contraction provides their power. They have the promise of protostars but never

Figure 14.21 Protoplanetary disks around young stellar objects in the Orion Nebula. These HST images show newly formed stars, less than about a million years old. The original spin of the star-forming cloud has spread the material out into a broad disk. Each disk has a hole in the center where the star is located. Material boils off the disk's surface and is blown by stellar winds.

reach stardom. Neither do they make acceptable planets. Astronomers call such masses **brown dwarfs** – a misnomer because they are not really brown but red. Perhaps they are best called "masses below the minimum hydrogen-burning limit," but that phrase is awkward. I will stick with brown dwarfs, though these are really failed stars.

Such masses have very cool (and decreasing) surface temperatures – less than 3000 K – so that most of their emission lies in the red and infrared. These cool temperatures and expected low luminosities (roughly a hundred-thousandths of the sun's luminosity) make brown dwarfs hard to detect. Infrared observations permit detection, and searches have been conducted in regions of dark clouds with young stellar objects. Astronomers also examine nearby stars for faint, low-mass companions.

With such a tough hunt, astronomers did not confirm a brown dwarf until 1995, when one was discovered as a companion to a cool, low-mass star (Fig. 14.22). Called Gliese 229B, this brown dwarf orbits about 6×10^9 km from its much larger companion. It has a mass of 30 to 40 times that of Jupiter and a surface temperature below 1000 K. With a luminosity a million times fainter than the sun, Gliese 229B inhabits the dim zone between real stars and regular planets.

14.5 NEIGHBORING PLANETARY SYSTEMS: FOUND!

What evidence do we have of other planetary systems? Because a planet shines by reflected light from its parent star, because it is small, and because it lies very close to its local sun, as seen from the earth, a planet's gleam would be lost in the stellar glare. We cannot directly observe other planets outside the solar system with earth-based telescopes.

Center-of-Mass Motions
(Learning Outcome 14-13)

We can hunt for the motion around the center of mass of the planet-star system (Section 4.5). As a result of this seesaw effect, the visible star wobbles from side to side about the center of mass if a massive planet orbits it. From the observed stellar wobble and an estimate of the stellar mass, we can estimate the mass of the invisible planetary companion by the same method used to measure binary star masses. In these cases, the center of mass usually lies outside the stars.

For planetary systems, the center of mass falls within the parent star, and so it only sways a little. The wiggles are only about 0.001 arcsec, or about one one-hundredth the size of a star's

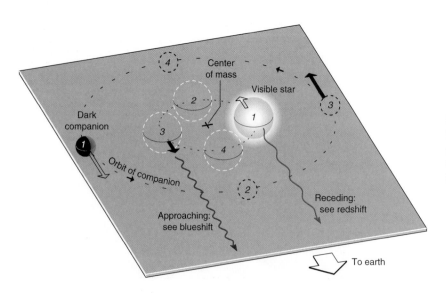

Figure 14.22 Doppler shift from an invisible companion orbiting the center of mass with a visible companion. This motion causes a periodic Doppler shift in the spectrum of the visible star.

image observed from the ground. Such minuscule changes are dramatically affected by changes in the telescopes themselves. No "discovery" of a planet by center-of-mass motions has panned out yet.

Doppler Shift Detections
(Learning Outcome 14-13)

Thirty-one planets and counting! That has marked the exhilarating success of Doppler-shift detections of extrasolar planets around normal stars since 1995. Surprises: most found to date have masses less than two Jupiters (two about that of Saturn), and many have highly eccentric orbits. No earthlike planets have been spotted yet, because the current technology cannot reliably detect such subtle Doppler shifts.

We can examine the spectrum of a star to look for the slight Doppler shift caused by the tug of a planet on its parent star (Fig. 14.22). From these observations, and an estimate of the stellar mass, we can estimate the mass of the invisible planetary companion by the same methods used to measure binary star masses (Section 13.6). Because we generally do not know the orbital tilt, we end up with the *minimum* mass of the planet (Table 14.2).

New techniques give high-precision radial velocity measurements of stars and so enable us to detect tiny variations in velocity from low-mass companions. How small? About 3 m/s – bicycling speed! Such precision allows reliable detection of jovianlike planets (Fig. 14.23). Jupiter's pull on the sun results in a 12 m/s Doppler shift.

A most curious discovery from these surveys is that of an extrasolar planetary *system* around the star called Upsilon Andromedae. The parent star has at least three planets (called prosaically b, c, and d): one at 0.06 AU, the second at 0.82 AU, and the outermost at 2.5 AU (Fig. 14.24). The innermost planet sizzles around in just 4.6 days. The outer two have highly eccentric orbits. All, though, revolve in a zone where Jovian planets would not be expected to form around a 6300-K star.

What do these extraordinary discoveries mean? First, planetary systems are profuse in the Milky Way. Second, Jovian planets appear common – the solar system may be the weird bird with its 50-50 division of Jovian and terrestrial planets. Third, Jovian planets may control the early turmoil of planetary formation by flinging out or absorbing smaller masses. Fourth, our basic models for star and planet birth are legitimate.

Table 14.2 Selected Extrasolar Planets Orbiting Normal Stars

Star Name	Spectral Type	Distance (ly)	Minimum Mass (Jupiter = 1)	Orbital Period (days)	Distance (AU)	Orbital Eccentricity (circle = 0.0)
51 Pegasi	G2	50	0.44	4.2	0.05	0.01
47 Ursae Majoris	G0s	43	2.4	1100	2.1	0.10
70 Virginis	G5	72	7.4	117	0.47	0.40
16 Cygni B	G2	70	1.7	800	1.7	0.68
Upsilon Andromedae b	F8	44	0.69	4.62	0.059	0.04
Upsilon Andromedae c	F8	44	2.0	241	0.82	0.23
Upsilon Andromedae d	F8	44	4.1	1280	2.4	0.31
HD 209458	G0	150	0.63 (actual mass)	3.52	0.045	0.0

Adapted from a table by Geoff Marcy and Paul Butler.

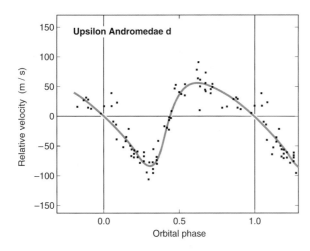

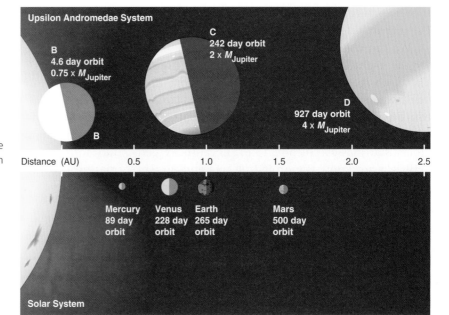

Figure 14.23 Doppler-shift observations of the motions caused by the "planet" Upsilon Andromedae d.

Figure 14.24 Artist's conception of the Upsilon Andromedae planetary system compared to the inner solar system.

KEY CONCEPTS

1. The interstellar medium contains both gas and dust. The gas, mostly hydrogen, comes in a variety of forms: molecules, atoms, and ions. The dust is in the form of small grains. Interstellar dust is far less abundant than the gas but does make up about 1 percent of the total mass of the interstellar medium.

2. The gas clumps in clouds of various sizes, ranging from small clouds of atoms to the giant molecular clouds; between the clouds is a hotter intercloud gas. A wide variety of molecules has been found in molecular clouds; some of these are organic compounds, but the most common element is hydrogen, as atoms, ions, and molecules. Much of the interstellar dust is in molecular clouds.

3. Dust makes itself known by the reddening and extinction of light and also by infrared emission when it heats up. The dust is made of grains about a micrometer in size; the core of the grains contains silicates, graphite, or iron. The mantles are made of icy materials and organic compounds. The grains are formed in the outflow of material from cool, supergiant stars. Dust grains are associated with molecular clouds and aid the formation of simple molecules.

4. Stars are born out of molecular clouds by the process of gravitational collapse and contraction. A protostar forming in a cloud gets its energy from the conversion of gravitational potential energy to kinetic energy. The process of starbirth is hidden from our direct view, but we can infer it by infrared and radio observations. Somehow the surrounding material is dissipated to reveal the star.

5. Interstellar clouds are likely to have initial spin and so angular momentum. That angular momentum must be conserved. As clouds collapse, they naturally form a disk with a few large blobs of material. The gravitational collapse of rotating clouds results in the formation of planetary or multiple star systems.

6. Infrared and radio observations imply that massive stars are formed in small groups (about ten) out of giant molecular clouds in a chain-reaction sequence in which one group triggers the birth of the next. The process may be started by a supernova remnant hitting the giant molecular cloud. Starbirth for massive stars occurs quickly.

7. For solar-mass stars, starbirth happens more slowly than for massive stars. Solar-mass stars may be formed out of small molecular clouds or as a spin-off of the birth of massive stars from giant molecular clouds.

8. The formation of stars involves a bipolar outflow of gas (both molecular and ionized), aligned in part by a magnetized disk or ring around the young star. Many young stellar objects display such outflows, which blow away the surrounding cloud. This outflow marks a very brief phase in the making of a star.

9. Starbirth is taking place now in the Galaxy, as gravity shapes interstellar material into stars. The rate is estimated at about ten stars per year, mostly less massive ones. The observational evidence is the detection of pointlike sources of infrared emission in molecular clouds.

10. Some starbirth results in multiple-star systems, planetary systems, and brown dwarfs. The detection of extrasolar planets confirms our essential concepts about star formation and implies a multitude of such worlds in the Milky Way – and beyond!

STUDY EXERCISES

1. Describe one way in which astronomers observe each of the following: (a) interstellar H I, (b) interstellar H II, (c) the coronal interstellar gas, and (d) interstellar molecules. (Learning Outcomes 14-1 and 14-2)

2. Outline two ways in which astronomers "see" interstellar dust. (Learning Outcomes 14-1 and 14-3)

3. What, if any, evidence do we have that dense materials make up part of the interstellar dust? That icy materials do? (Learning Outcome 14-4)

4. List the observational evidence that leads to the argument for the formation of massive stars in small groups from giant molecular clouds. (Learning Outcome 14-7)

5. What powers a protostar? (Learning Outcomes 14-6 and 14-9)

6. In a theoretical picture for the formation of a massive star, how is it that we never see a massive star directly (optically) until it reaches the main sequence? (Learning Outcome 14-9)

7. Once it has reached the main sequence, how does a massive star influence its parent cloud? (Learning Outcome 14-11)

8. What observational evidence do we have that solar-mass stars form from dark clouds? (Learning Outcomes 14-8 and 14-12)

9. What kind of observation tells us that the outflows from star-forming regions take place in two opposite directions? (Learning Outcome 14-11)

10. In one sentence, argue that starbirth must be occurring now. (Learning Outcome 14-10)

11. In the sequential model for massive-star formation from giant molecular clouds, what mechanism promotes the formation of a new group of stars, once the process has begun? (Learning Outcome 14-9)

12. What is one possible source of interstellar dust grains? (Learning Outcome 14-5)

13. What is the most abundant element in the interstellar gas? (Learning Outcome 14-2)

14. How do we know that extrasolar planets exist? (Learning Outcomes 14-2 and 14-13)

PROBLEMS AND ACTIVITIES

1. A protostar can have a temperature of 600 K and a luminosity 1000 times that of the sun. At what wavelength is its peak emission? What is its radius? *Hint:* Assume that the protostar radiates like a blackbody.

2. Compare the pressure in the core of a molecular cloud to that in an H II region. Which is greater? *Note:* The pressure is directly proportional to the number density and the temperature.

3. A fast bipolar outflow moves at about 100 km/s. How long would it take such a stream of gas to cover a distance of 1 AU?

4. Assume that your body is made of water molecules. Estimate the number of water molecules in the mass of your body. If these were spread out in a density equivalent to that at the core of a molecular cloud, how much volume would they fill? Assume that the cloud is spherical!

5. Imagine a brown dwarf with a temperature of 1500 K and a radius a tenth of the sun's radius. If it radiates like a blackbody, what would its luminosity be?

6. Imagine that you are observing the solar system from a nearby planetary system. At a wavelength of 5000 Å, how much of a Doppler shift do you expect to observe for Jupiter's gravitational effect on the sun?

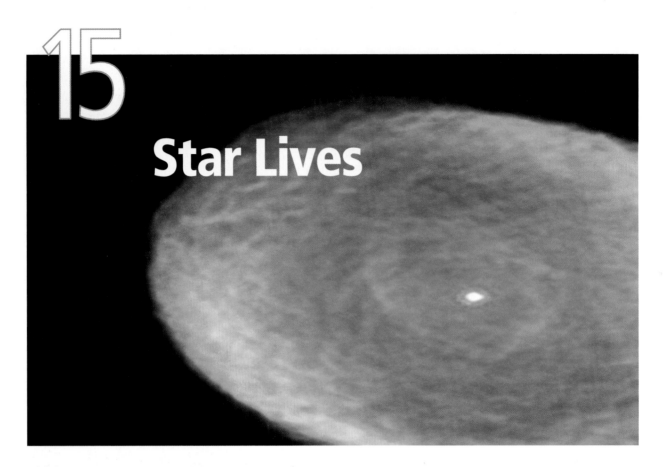

15 Star Lives

LEARNING OUTCOMES

After studying this chapter, you should be able to:

15-1 Show how the Hertzsprung–Russell (H–R) diagram for many stars provides clues about the evolution of individual stars.

15-2 Describe the physical basis of a theoretical model of a star, that is, the physical concepts that go into building a star model.

15-3 Trace the evolution of a 1-solar-mass star on an H–R diagram, describing the physical changes of the star that result from changes in the star's core.

15-4 State in one sentence why stars must evolve.

15-5 Compare the evolutionary tracks of a 1-solar-mass star and a 5-solar-mass star on an H–R diagram.

15-6 Describe the evolution, on an H–R diagram, of a cluster of stars of different masses.

15-7 Back up theoretical ideas of stellar evolution with observational evidence, with special emphasis on star clusters and their H–R diagrams.

15-8 List the sequence of thermonuclear energy generation reactions in stars of different masses.

15-9 Compare and contrast galactic (open) star clusters to globular ones in terms of both their physical properties and their H–R diagrams.

15-10 Indicate how mass and chemical composition affect stellar evolution.

15-11 Describe how fusion reactions in stars during their normal lives result in the manufacture of some heavy elements, and indicate how these processed materials may be recycled to the interstellar medium.

15-12 Trace the flow of energy from the core of a star to its surface for stars of different mass and at different stages of evolution.

CENTRAL CONCEPT

Stars evolve; their physical properties change as they go through their normal lives. The main agent in how and how fast a star evolves is its mass.

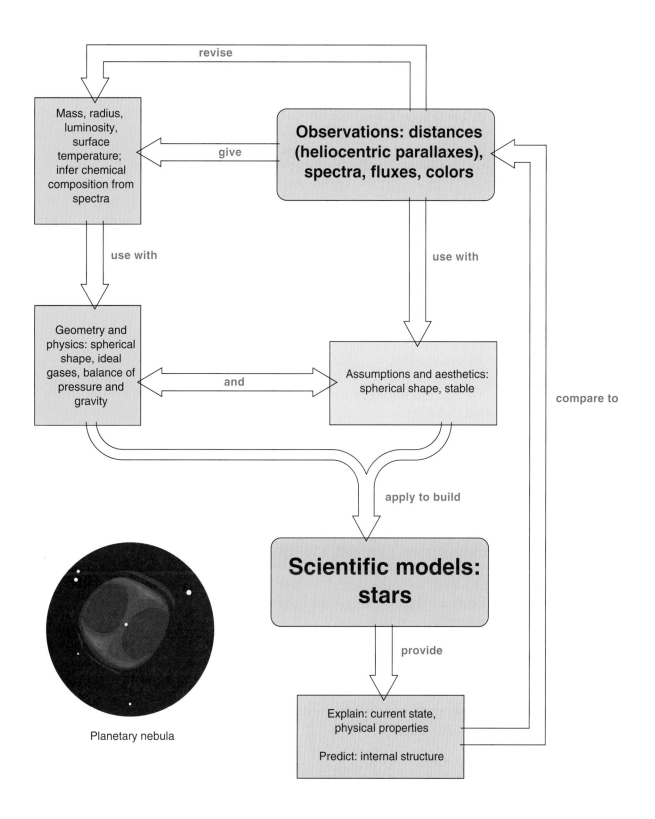

revise

Mass, radius, luminosity, surface temperature; infer chemical composition from spectra

give

Observations: distances (heliocentric parallaxes), spectra, fluxes, colors

use with

use with

Geometry and physics: spherical shape, ideal gases, balance of pressure and gravity

and

Assumptions and aesthetics: spherical shape, stable

compare to

apply to build

Scientific models: stars

provide

Planetary nebula

Explain: current state, physical properties

Predict: internal structure

ravity controls the history of newborn stars. A star survives as long as it can counteract the relentless gravitational crunch. The story of this battle against gravity runs like the aging of a person from birth to death, but it takes millions to thousands of millions of years. As long as a star shines, it lives. When its radiance fades, it dies.

To put this span into a human perspective, imagine time speeded up so that one year passes in one-fifth of a second. Then the sun would live only 65 years or so. The sun's birth would be quick; only 4 months would pass from the start of the collapse of the sun's embryonic cloud to its establishment as an immature but full-fledged star. For about the next 60 years, the sun would shine calmly as it passed through middle age. Old age would gradually fossilize the energy production of the sun. In about 5 years, the elderly sun would slowly expand to almost a hundred times its present size; it would become a bloated red giant. Then a sudden burst would blow off the sun's atmosphere, leaving behind a hot core that would cool quickly.

How a star lives depends mostly on how much mass it has at birth. The important stages that mark its life and the duration of each stage are related directly to a star's mass. Stars that are much more massive than the sun have short, frenetic lives.

This chapter presents, first, some theories of stellar evolution, contrasting the lives of stars like the sun to those of more massive stars. For most stars, the themes are the same; rapid birth, long middle age fusing hydrogen to helium in the core, and a slow aging to a red giant, violence, and death. Second, it offers some observational evidence in support of these theoretical concoctions. You will see how stars burn, falter, and flame out. Our sun, too, will eventually die – and the earth with it.

15.1 STELLAR EVOLUTION AND THE HERTZSPRUNG–RUSSELL DIAGRAM

Stars evolve because they shine. As a star loses energy to space, it must change. While it lives, a star does not cool off. Most objects cool off (their temperature goes down) when they lose energy to their environment. It's natural to think that a star does, too, but it doesn't. Nuclear reactions generally replenish the energy a star radiates into space. When one fuel in the core (say hydrogen) runs out, another (say helium) can ignite, at a temperature made higher by the compression of matter by gravity. Only when a star cannot fuse matter any more will it cool off. Then the star has died (as you will see in Chapter 16).

I will approach stellar evolution from both theoretical and observational points of view. This chapter presents mostly theoretical ideas, because observations of a star's evolution do not come easily. Why? The sun's anticipated lifetime is more than 100 million human lifetimes, so there is no way you could watch a single star, like the sun, evolve. You can see many different stars at one time, however, and organize these stars on the Hertzsprung–Russell diagram by their luminosities and spectral types. Then you can use the H–R diagram of many stars to guess at the evolution of one star. Let's see how.

Classifying Objects (Learning Outcome 15-1)

Suppose you asked 18- and 19-year-old adults in the United States for their weight and height and plotted the data (Fig. 15.1). Note the trend: the points tend to fall along a line that shows that weight generally increases with height. That result should not surprise you.

Suppose you recorded the height and weight of every person you encountered randomly. Plot the data again as weight versus height (Fig. 15.2). You still have a general trend, rather than a random scatter of points, but you also have other groups that do not follow the first set. Why the difference? The first graph includes people of the same age; the second, people at different ages.

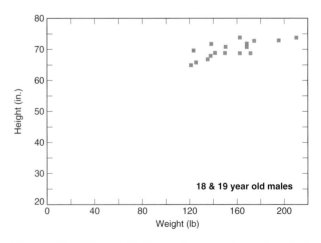

Figure 15.1 Height-weight diagram for a sample of 18- and 19-year-old adults from an introductory astronomy class at the University of California, Berkeley, U.S.A. Units are inches and pounds.

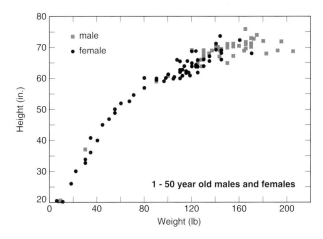

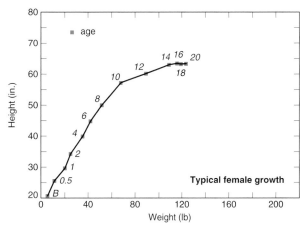

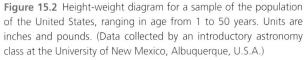

Figure 15.2 Height-weight diagram for a sample of the population of the United States, ranging in age from 1 to 50 years. Units are inches and pounds. (Data collected by an introductory astronomy class at the University of New Mexico, Albuquerque, U.S.A.)

Figure 15.3 Height-weight-time diagram for a typical U.S. female from birth (B) to 18 years of age. Units are inches and pounds. (Data from the U.S. National Center for Health Statistics Growth Chart.)

Now here's a third graph (Fig. 15.3). It is a plot of weight versus height for an average U.S. female at different times in her life, as indicated by points representing the person's age from birth to 18 years. The line connecting the points shows how a *single* person's height and weight change. You can correctly interpret the graph for many people in evolutionary terms if, and only if, you know how one person evolves. Time was implicit in the second graph (Fig. 15.2).

In the same way, time and age implicitly play a role in an H–R diagram, in which are plotted two essential properties of stars: surface temperature and luminosity. These external changes displayed by stars with aging are analogous to height and weight in human beings.

Time and the H–R Diagram
(Learning Outcome 15-1)

How does the H–R diagram tell you about an evolutionary sequence of a star? Imagine that you have a large family. They have gathered together, and you take a snapshot of the group, from the newest-born to the great-grandparents. Most of the people in the picture are in their middle age (20 to 60 years old); you have a few infants, some children and teenagers, and a few old people. You have so many middle-aged people because most of your life you will be middle-aged; you spend relatively less time as an infant, teenager, or old person. The relative numbers of peo-

ple in these stages of their lives reflect the relative duration of each stage. Each person, of course, has his or her unique life. This snapshot just happens to have caught all the family members together at one time.

Now suppose that you have a collection of any objects that evolve in a specific evolutionary sequence. Each follows its own special history but passes through similar stages. You can estimate the relative time spent in any evolutionary stage by the relative numbers you find at that stage compared with others. (This argument holds true only if birth and death go on at a constant rate. If no more people are born, eventually you will see only old people – and then none.)

Recall the H–R diagram for many stars, which shows the luminosity and surface temperature of the stars. Most of the stars fall on the main sequence, so stars found here are going through the longest, most stable stage in their evolution. The main sequence represents a sequence from higher to lower masses. Here's the evolutionary meaning of the main sequence: it marks stars at the stage of converting hydrogen to helium in their cores; stars remain at this stage for the greatest part of their lives. Any star fusing hydrogen to helium in its core is a main-sequence star; it falls on the main sequence of the H–R diagram. That's why we see so many main-sequence stars now. Other stages, such as becoming a red giant, must be shorter. How do we know? Because we see far fewer red giants now than main-sequence stars.

A star's mass (and thus the pressure and temperature in its core) determines how the star will evolve. So you must examine the H–R diagram to find how stars of different masses evolve. What is the correct interpretation of the H–R diagram for the evolution of stars? You need some hints from the physical nature of stars and theoretical calculations to make up the star models.

15.2 STELLAR ANATOMY

What is a star? A huge, hot ball of gas, mostly hydrogen, heated by thermonuclear reactions in its core. You can imagine a star as a naturally controlled hydrogen bomb. Why doesn't it blow up? Because gravity persistently pulls it together. All its life, a star must withstand the inward squeeze of gravity. How does it do it? By producing an outward pressure, usually from fusion reactions.

Pressure and Energy Balance
(Learning Outcome 15-2)

For most of its life, the outward pressure exerted by a star results because the star is hot. A star consists of gas; a hot gas has a high pressure, and its outward force balances the inward gravitational force. This balance must hold true at every level throughout the star (Fig. 15.4); otherwise it would be unstable. The first physical requirement for a model of a stable star is that it be in balance, neither expanding nor contracting.

Second, the star must generate energy internally. For most of a star's life, thermonuclear fusion reactions operate as the internal furnace (details to come in Section 15.4). In general, the rate of energy produced inside the star equals the rate at which the energy radiates away at its surface. This balance holds not only overall but also at each layer within the star. Otherwise the star will be unstable and will expand or contract.

As an analogy, consider an assembly line that carries cars as they pass through different assembly stages. At each stage, the rate at which the cars come in must be equal to the rate at which they go out. Otherwise the cars will pile up! If the flow of heat through a star were uneven, the temperature of various layers would change. These temperature changes would result in pressure changes capable of causing the star to expand or contract. Generally, a

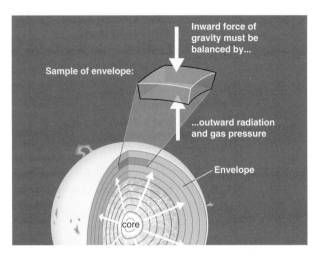

Figure 15.4 Balance of gas pressure and gravity in a star. For the star to be stable, the outward pressure from the internal heat (generated by the fusion reactions in the core) must balance the inward pull of gravity. This balance must hold at every level in the star; otherwise, the star will either expand (pressure is higher) or contract (gravity is greater)

star meets the condition of energy loss equaling energy production, and its energy flows from hotter to cooler regions.

Energy Transport
(Learning Outcomes 15-2, 15-4, 15-12)

Third, we need to look at how a star transports energy from its core to the surface. Basically, three methods are available: conduction, convection, and radiation. Which of these occurs depends on the star's **opacity** (Section 12.3). The opacity of a star's material directly affects its radiative energy transport. The more opaque a star's material, the slower the flow of radiation through it.

You can think of opacity as acting like insulation in a house in winter. The furnace is generating heat. The greater the house's insulation, the slower the flow of the heat to the cold outside. A slow flow keeps the exterior of the house cold and the inside hot. In a poorly insulated house, the outside is warmer and the inside colder than in a well-insulated house.

Likewise, if the opacity of a star were suddenly lowered, radiation would escape more easily and the star would grow more luminous, but its inside would become cooler. The internal pressure would drop, and the star would contract until balance was regained. The opposite occurs if the opacity increases: the radiation is dammed up, the star becomes hotter, and it expands.

Whether convection occurs at any region in a star depends on the opacity of the material there. If the material is transparent, energy will flow more easily by radiation. If it is so opaque that the radiative energy flow gets bottled up, convective transport will take command and operate instead. Generally, a star's opacity depends on its chemical composition, density, and temperature. A greater density or abundance of metals increases the opacity; a greater temperature decreases it.

Gravity keeps a star from expanding and just balances the outward pressure of the hot gases. At the same time, the outflow of energy is just balanced by its production. This balance cannot continue forever. A star loses energy to space – it shines! Thermonuclear fusion reactions supply the energy to keep it hot. Eventually the star runs out of fuel, and the pressure can no longer balance its gravity. Then it contracts. A star can survive as long as it can find a means to produce energy, to stay hot, hence to withstand gravity. When it fails to do this, the star dies. How stars survive marks the central theme of this chapter.

15.3 STAR MODELS

A theoretical model of a star incorporates all the physical conditions just described. A star must produce energy; gravity and pressure must be in balance; and energy must be transported evenly from core to surface. In addition, we must know how the matter in a star behaves. Luckily, for most stars for most of their lives, that behavior is the same as for an ideal gas (Section 12.2). In some massive stars, internal pressure from radiation adds to the internal pressure from the gas. Remember that light has particle properties and so, in a sense, behaves like the physical particles in a gas. It exerts radiation pressure. (See the Star Models Celestial Navigator.™)

We need to know exactly what thermonuclear processes produce energy, as well as the mass and chemical composition of the model star. Complicated equations formulate the problem, and these are solved in a consistent way to find the physical conditions in the star (temperature, pressure, and density, for instance) from the center to the surface. This catalog of values for important physical proper-

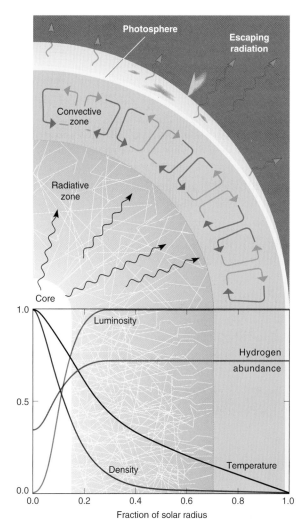

Figure 15.5 Model for the sun's interior at its current age (4.6 Gy). Shown are the temperature (compared to the central temperature), density (compared to the central density), luminosity (compared to the sun's luminosity), and hydrogen abundance (by fraction of total mass) as these change with radius. (Based on calculations by J. Bahcall and R. Ulrich.)

ties for a specified mass and composition is called a **star model.**

In general, we expect density, temperature, and pressure to decrease as we go from the center of the star to its surface (Fig. 15.5). One pair of numbers of this catalog, the luminosity and temperature at the surface, specifies a point on an H–R diagram, the theoretical point for a model star of given mass and chemical composition. This point on the H–R diagram depends on the specifics of the stellar model. The evolutionary sequence of one star involves a series of different models for different stages in a star's life.

Goals of Models
(Learning Outcomes 15-2, 15-4)

Why all this fuss? Because the study of stellar evolution rests on the construction of physically reasonable models of stellar interiors. We rely on the models because we cannot directly observe the interiors of stars and we cannot wait millions of years to watch individual stars evolve. *Models reveal how time enters into the H–R diagram for real stars.* Almost everything said in this chapter rides on the validity of stellar models.

General Results
(Learning Outcomes 15-2, 15-12)

What do star models show? They indicate that as a star evolves, its radius, temperature, and luminosity change in complicated ways. Such changes result in the change in a star's position on the H–R diagram. A series of these points on an H–R diagram makes up a star's **evolutionary track,** just as the graph of height versus weight entails an evolutionary track for a person (Fig. 15.3).

As a simple example, if a star's surface temperature increases (but not its luminosity), the star's position on the H–R diagram moves horizontally from right to left, with no vertical change (line *A* in Fig. 15.6). If the luminosity increases (but not the surface temperature), the star's point on the H–R diagram moves vertically from bottom to top, with no horizontal change (line *B* in Fig. 15.6). Both motions on the H–R diagram represent changes in the physical properties of the star. The path is the evolutionary track. Each star model of a different pair of physical characteristics (such as mass and chemical composition) traces out a different evolutionary track over time.

Note in the first case that to keep the luminosity constant at a higher temperature requires that the star's surface area, and so its radius, decrease. In the second case, with no change in temperature, the star's luminosity can go up only if its surface area increases, and so its radius. You see that a star's surface temperature, luminosity, and radius are all related and change with time.

CAUTION. What moves around on the H–R diagram is *not* the star in space but a point representing the changes in luminosity and temperature of the star.

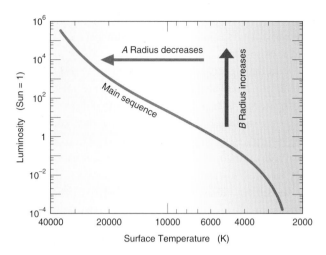

Figure 15.6 Path of a star's point on an H–R diagram as its size changes. A star that stays at a constant luminosity but increases in temperature must be decreasing in size (path *A*). If a star stays at a constant surface temperature but increases in luminosity, its size must increase (path *B*). Generally, a star's changes in luminosity, temperature, and radius are more complex than shown here for these two special cases. Note the power-of-ten scale for both axes.

The goal of studying stellar evolution is to understand the physical causes behind a star's evolutionary track on the H–R diagram.

15.4 ENERGY GENERATION AND THE CHEMICAL COMPOSITIONS OF STARS

What makes a star evolve? The answer lies in the heart of the star. Here thermonuclear reactions cook lightweight elements into more complex ones by fusion. This continual change in chemical composition and its effects on a star's structure mark a secondary theme of stellar evolution.

Hydrogen Burning
(Learning Outcomes 15-8, 15-11)

The sun generates energy by the proton–proton (PP) reaction (Section 12.6). Four hydrogen nuclei combine to form one helium nucleus and release a certain amount of energy. A minor part of the sun's energy comes from another reaction sequence, called the *carbon–nitrogen–oxygen cycle* (CNO cycle). In more massive stars, this cycle becomes more important than the PP cycle. If a star's central temperature is greater than about 20 million K, the

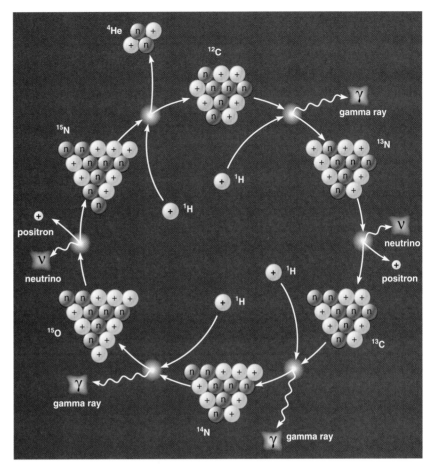

Figure 15.7 Carbon–nitrogen–oxygen (CNO) cycle, a hydrogen-burning fusion process. Note that the net result is the conversion of four hydrogen nuclei to one helium nucleus; the carbon entering the first step returns in the last.

CNO cycle produces more energy than the PP reaction. (Both can go on at the same time.) The net result of the CNO cycle is the same as that of the PP reaction: four hydrogen nuclei are converted to one helium nucleus, with the release of energy. The CNO cycle predominates in stars with masses greater than about 1.5 solar masses.

Let's look at the CNO cycle in detail (Fig. 15.7). The complete cycle has six reaction steps. First, a proton (^{1}H) collides with a carbon nucleus (^{12}C) to convert it to a radioactive nitrogen isotope (^{13}N). The nitrogen nucleus emits a positron and a neutrino to become a carbon isotope (^{13}C). A proton blasts into this particle and, with the emission of a gamma ray, turns it into stable nitrogen (^{14}N). Then a proton bangs into the nitrogen nucleus to form the radioactive oxygen isotope (^{15}O) and a gamma ray. The oxygen isotope decays into the nitrogen isotope (^{15}N), a positron, and a neutrino. Finally, the nitrogen isotope gets hit by a proton and splits into ordinary carbon (^{12}C) and helium (^{4}He). The carbon comes back unscathed; it acts only as a catalyst, helping to glue four protons together to make helium. Total energy released to the star is 4.0×10^{-12} J, some as neutrinos.

These reactions take place in the star's core, where temperatures are high enough to keep them going. Because both involve the fusion of hydrogen as a fuel, they are given the generic name of hydrogen burning. Because a star has only a limited amount of hydrogen to burn, the core is eventually converted almost entirely to helium, and the CNO and PP reactions cease. Fusion reactions no longer supply energy to keep the temperature and pressure high.

Helium and Carbon Burning
(Learning Outcomes 15-8, 15-11)

What next? The core contracts and heats up (remember, the self-contraction converts gravitational potential energy to kinetic energy). When the core's temperature rises to roughly 100 million K, another reaction can take place: the **triple-alpha reaction,** so named because three helium nuclei (also known as *alpha particles*) fuse to form one carbon nucleus, with the release of energy (Fig. 15.8). This process is generally called *helium burning.*

What happens when the helium runs out? The core contracts and heats up again. If the temperature increases enough, carbon can be fused into heavier elements in the processes of carbon burning. Such processes require extreme temperatures, at least 600 million K. Iron, the most stable of all nuclei, ends the sequence of nuclear fusion. To form elements heavier than iron by fusion reactions, energy must be added and absorbed. The steady climb from hydrogen to iron in fusion reactions in stellar cores is one type of **nucleosynthesis,** the nuclear cementing process that makes heavy elements in a universe that otherwise would consist only of hydrogen and helium made in its Big Bang origin (to come in Chapter 20).

You may be wondering why a star's temperature goes *up* when fusion reactions *stop.* The cause is gravitational contraction. A star is always losing energy to space by radiation. If that energy is not replaced by nuclear reactions, it must come from somewhere else – namely, from gravitational contraction. As the star contracts, some of the gravitational potential energy of its particles transforms into kinetic energy and some is radiated away (Section 11.5). So the temperature goes up until the ignition temperature of the next set of fusion reactions is reached. Fusion turns on again, the pressure increases, and the core contraction stops.

During a star's life, short periods of gravitational contraction alternate with long spells of fusion burning. (Note that just the *core* needs to contract, not the star as a whole.)

15.5 THEORETICAL EVOLUTION OF A 1-SOLAR-MASS STAR

Gravity instigates the birth of a star. The shape of the subsequent evolution is controlled by the star's mass. Let's look at the evolution of a star like the sun, a 1-solar-mass star, having the sun's chemical composition.

Evolution to the Main Sequence (Learning Outcomes 15-3, 15-12)

Recall a 1-solar-mass star's pre-main-sequence evolution (Section 14.3). *The key point is that a protostar gets its energy from gravitational energy, not fusion reactions.*

The protostar forms by the gravitational contraction of an interstellar cloud. Once the dense core of the cloud has formed, the rest of the cloud accretes upon it (points 1 through 2 in Fig. 15.9). For a while, a protostar has a larger radius than it will have as a main-sequence star, and the surface temperature is lower. However, the protostar has a higher luminosity than it will have when it reaches the main sequence, about thirty times larger after 100,000 years has elapsed. How so, if it is cooler? Because it is larger and so has more surface area to radiate energy.

At this stage, the star's temperature is so low that its opacity is relatively high (even though its density is low). Convection transports the energy outward, so a protostar is completely convective – a huge, bubbling ball of gas. The efficient transport of energy makes the star very luminous.

The core continues to heat up. Eventually the core hits eight million kelvins, high enough to start thermonuclear reactions. When the protostar gets most of its energy from thermonuclear reactions (PP reactions in the case of the sun) rather than gravitational contraction, it achieves full-fledged stardom (point 3 in Fig. 15.9). The star is now called a **zero-age main-sequence (ZAMS)** star. The total time elapsed from initial collapse to arrival as a star on the main sequence is only 50 million years.

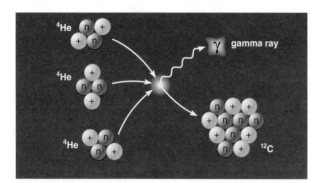

Figure 15.8 Triple-alpha process. This helium-burning fusion reaction converts three helium nuclei into one carbon nucleus and produces energy.

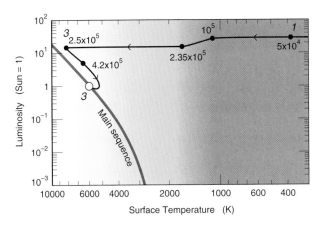

Figure 15.9 The evolutionary track on an H–R diagram of a pre-main-sequence, solar-mass star. The labels indicate the approximate time in years since the formation of a core of an interstellar cloud. The zero-age main sequence (ZAMS) is indicated at the left. Note the power-of-ten scale for both axes. (Adapted from a diagram by K.-H. Winkler and M. J. Norman.)

Evolution on the Main Sequence
(Learning Outcomes 15-3, 15-12)

Where the star ends up on the main sequence depends mainly on its mass. The more massive the star, the hotter and more luminous it is; the less massive the star, the cooler and less luminous it is. The main sequence, on the H–R diagram, is a series of stars of decreasing mass (but similar chemical composition), from the upper left-hand corner (hot stars with high mass) to the lower right-hand corner (cool stars with low mass). A sunlike star spends about 80 percent of its total lifetime on the main sequence, a period called its **main-sequence lifetime.** For this star, energy is transported through its interior mostly by radiation. Convection labors in a zone just below the surface.

How the star evolves further also depends on its mass. Because massive stars have higher luminosities, the hydrogen in them burns faster than in low-mass stars. For example, main-sequence lifetime for a 15-solar-mass star lasts only 5 million years. Why? Because of its greater mass, the 15-solar-mass star has a greater luminosity, about 30,000 times greater than the sun. (Recall the mass–luminosity law, Section 13.6.) Its fusion reactions must therefore go on at a much faster rate than in the sun – almost 30,000 times faster. Even though the 15-solar-mass star has more hydrogen to fuse, it does so at an incredibly fast rate.

The main-sequence phase ends when almost all the hydrogen in the core has been converted to helium. During this time, the temperature in the core increases gradually. This results in a greater flow of energy to the surface, and the star's luminosity increases (point 1 to point 2 on the evolutionary track in Fig. 15.10). The star is now poised to become a red giant.

Figure 15.10 Theoretical evolutionary track on an H–R diagram for a one-solar-mass star. The evolution begins here at the ZAMS (point 1) and ends with the formation of a white dwarf from the core of a red giant star. As hydrogen is depleted in the core, the star's luminosity increases (1 to 2). When the core runs out of hydrogen, the star burns hydrogen in a shell around the core (2 to 4); it becomes a red giant. Gravitational contraction heats the core until the temperature becomes high enough to start the ignition of helium in the core (4); for a short time, helium fusion occurs in the core (5). When the star's core is depleted of helium, the star burns both hydrogen and helium in a shell and becomes a red giant again (5 to 7). Finally, the star throws off its outer layers (7) to expose its core (8) and becomes a white dwarf (9). Note the power-of-ten scale for both axes.

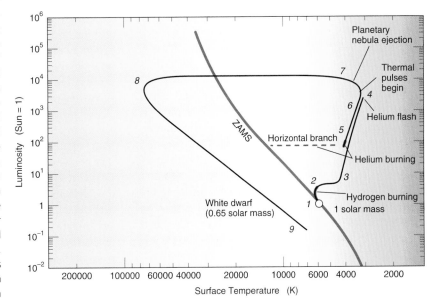

Evolution off the Main Sequence (Learning Outcomes 15-3, 15-12)

When the hydrogen in the core is used up, the thermonuclear reactions cease there. However, they keep going in a shell around the core, where fresh hydrogen still exists. At the end of fusion reactions in the core, gravity takes over and the core contracts. This heats up the shell of burning hydrogen, so the reactions go faster and produce more energy. The luminosity increases. The radius of the star increases, and its surface temperature decreases. The temperature drop increases the opacity, and convection carries the energy outward in the star's envelope.

The star now becomes a red giant as it expands in size. Its position on the H–R diagram moves to the region of lower surface temperatures and higher luminosities (point 3 to point 4 in Fig. 15.10).

Gravity has compressed the red giant core to such a high density that it no longer behaves like an ordinary gas. The core becomes a **degenerate electron gas** (Enrichment Focus 15.1). In this state, the electrons produce a **degenerate gas pressure,** which depends only on density, not temperature, and this enables the core to attain and preserve a balance even though no fusion reactions are going on. Degenerate gases are also good conductors of heat, so they have the same temperature throughout. (You will encounter degenerate gases again in Chapter 16.)

As the bloated star attains its red giant status (point 4 in Fig. 15.10), the core temperature – which has been steadily increasing as the core contracts – hits the minimum to start helium burning by the triple-alpha process. When part of the degenerate core ignites, the heat generated by the fusion spreads rapidly throughout the core. The rest of the

---

ENRICHMENT FOCUS 15.1
Degenerate Gases

When matter is packed to very high densities (greater than 10^8 kg/m³), it no longer behaves in ordinary ways. In a normal gas, the particles are widely separated and rush helter-skelter into one another and rebound away (Section 13.2). In highly compressed material, little space exists between particles. The matter is so jammed together that the electrons on the outside of the atoms are, in a sense, touching one another. The nuclei can no longer hold electrons in their usual energy levels, and the electrons move among the nuclei with little space for moving about.

Electrons abide by a quantum property called the *Pauli exclusion principle*. It states that no two electrons can be together in exactly the same energy state. Picture a small box containing one electron in some energy state. Imagine adding another electron to our box. The electron already in it probably occupies the lowest energy state possible, so the next electron must occupy the second level available, and so on. In contrast to a low-density state, in which energy levels are available for occupation, a very dense gas has far fewer levels, which get filled up quickly. (This is a simplified explanation. Because electrons have spin, two electrons, one with its spin axis up and the other with its spin axis down, can occupy the same energy state.)

An analogy: imagine students filling up a classroom. If they sit randomly wherever they want, that is like an ordinary gas. If they sit so that first the front row of seats is filled, then the second, and so on, then that is like a degenerate gas.

What happens if we try to cool this dense gas by letting electrons give up kinetic energy? They cannot lose very much kinetic energy, for only high energy levels are open. No heat can be extracted from these electrons (in a sense, their temperature is zero), yet they exert a great pressure because they move with high speeds.

A gas in this state is called a *degenerate electron gas,* and the pressure from the uncertainty principle in action is called the **degenerate gas pressure.** Unlike an ordinary gas, for which the pressure is directly proportional to temperature, degenerate gas pressure is nearly *independent* of the temperature.

Electrons become degenerate at densities of about 10^8 kg/m³. Such densities occur in stellar cores after main-sequence hydrogen burning, and also in white dwarfs. When electrons are degenerate, they conduct heat very efficiently, and temperature variations are quickly smoothed out. (In a sense, a degenerate gas acts like a metal; it can conduct current and heat well.) Degenerate cores have the same temperature throughout. A high enough temperature can relieve the electrons of their degenerate condition: "high enough" is at least a few hundred million kelvins for electrons.

core quickly ignites. The increased temperature runs up the rate of the triple-alpha process, generating more energy, further increasing the temperature, and so on. This out-of-control process in the core is called the **helium flash.** The whole process of helium core ignition in the helium flash takes place in a very short time – perhaps only a few minutes! We will never see this helium flash in a star, for the action takes place deep in the core.

The star adjusts to this event by decreasing its radius and luminosity a bit. The star quietly burns helium in the core and hydrogen in a layer around the core (point 5 in Fig. 15.10). This phase is the helium-core-burning analogue to the star's main-sequence phase (in which the hydrogen core burns).

Evolution to the End
(Learning Outcomes 15-3, 15-12)

Eventually the triple-alpha process converts the core to carbon. The reaction stops in the core but continues in a shell around it. This situation – core shut down but the thermonuclear reactions going on in a shell – resembles that of the star when it first evolved off the main sequence. The burning shell makes the star expand. It again becomes a red giant (point 6 in Fig. 15.10).

Because the rate of the triple-alpha reaction changes greatly in response to small changes in temperature, the helium-burning shell causes the star to become unstable. Bursts of triple-alpha energy production cause thermonuclear explosions in the shell. They have the prosaic name of **thermal pulses.** The explosions occur about every few thousand years and cause the luminosity of the star to rise and dip rapidly by up to 50 percent. "Rapidly" here means in a few years or tens of years! Each blast generates an assault of energy. To move it out efficiently, the region becomes convective. The bubbling gases carry outward the heavy elements fused in each explosion.

Meanwhile, the star has developed a very strong outflow of mass from its surface, sometimes called a **superwind** to distinguish it from the normal stellar wind of a red giant. The superwind blows in gusts that, in about 1000 years, rip off the envelope of the star (point 7 in Fig. 15.10). A hot core is left behind. The expelled material forms an expanding shell of gas heated by the hot core. Astronomers call this a

planetary nebula for historic reasons. (It looks like a planetary disk when viewed with a small telescope. William Herschel, who discovered Uranus, first gave this description.) The hot core appears as the central star of a planetary nebula (Fig. 15.11). A star may shed its envelope in a series of bursts, each releasing a shell of material. This gas expands at a typical speed of 20 km/s.

The nebula keeps expanding until it dissipates in the interstellar medium in only a few thousand to a few hundreds of thousands of years. What happens to the core? For a star of roughly 1 solar mass or less, the core never reaches the ignition temperature of carbon burning. Why not? Because the core has become degenerate and cannot contract and heat up to ignite carbon burning. In about 75,000 years it forms a white dwarf star, composed mostly of carbon (point 8 to point 9 in Fig. 15.10).

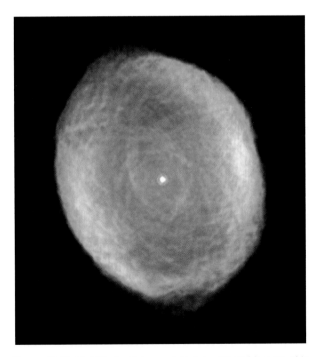

Figure 15.11 IC 418, the Spirograph Nebula, imaged by HST. This planetary nebula represents the final stage in the evolution of a star. There was once a red giant at its center that ejected its outer layers to form the nebula. Ultraviolet radiation from the still-hot core causes the gas of the nebula to disperse gradually, and this will cool the star. After billions of years as a white dwarf, it will fade away. Note the power-of-ten scale for both axes.

Without energy sources, the white dwarf cools to a **black dwarf,** the dark culmination of a 10-Gy biography. The cooling down occurs over thousands of millions of years while the degenerate gas pressure continues to support the star.

The Fate of the Earth

What will happen to the earth when the sun dies? The sun will expand to a size of about 1.1 AU, so that the earth will orbit in its outer atmosphere. The earth's orbital velocity of 30 km/s corresponds to a supersonic speed within a red giant envelope. The earth's air will be ripped off and the mantle vaporized. The drag will cause the earth's orbital radius to decrease, and in less than 200 years, the earth will fall into the sun's core to vaporize completely. Zap – our planet will be gone!

This prediction gives an astronomical twist to the notion of "global warming."

Lower-Mass Stars
(Learning Outcome 15-10)

The evolution of stars of lower mass than the sun resembles that of the sun, with one exception. Few stars less massive than the sun have had time to evolve off the main sequence. The universe simply isn't old enough. A 0.74-solar-mass star, for example, has a main-sequence lifetime of 20 Gy, longer than most estimates for the age of the universe. Below 0.08 solar mass, the object never becomes hot enough to ignite fusion reactions – it becomes a Jovian planet or brown dwarf.

Chemical Composition and Evolution
(Learning Outcome 15-10)

A star's mass most strongly directs its evolution. Its chemical composition plays a secondary role. If a star has a much smaller percentage of heavy elements than the sun, 0.01 percent instead of 1 percent, would its evolutionary track make much different gyrations?

The answer is *no*. The general trends stay the same: the star becomes a red giant, undergoes a helium flash, sits for a while during helium core burning, returns to red giant status, undergoes a series of thermal pulses, and finally expels its outer layers to make a planetary nebula, leaving a white dwarf.

However, low-metal stars do differ significantly with respect to where they bide their time on the H–R diagram during their stage of burning helium in the core. Basically, stars with smaller percentages of heavy elements and less mass than the sun make a horizontal line on the H–R diagram, called the **horizontal branch** (Fig. 15.10), during the phase of burning helium in their cores. These stars have a lower mass than when they were main-sequence stars. They have lost some mass as a result of stellar winds during their first red giant phase (before the helium flash).

TO SUM UP. The evolution of low-mass stars like the sun proceeds through the stages of protostar, pre-main-sequence star, main-sequence star, red giant, planetary nebula, white dwarf, and finally black dwarf.

15.6 THEORETICAL EVOLUTION OF MASSIVE STARS

Now to sketch the history of stars much more massive than the sun but with the same chemical composition. Regulus, the bright star that marks the heart of Leo the Lion, is a 5-solar-mass star. The evolution of stars such as Regulus shows the trends for middle-mass stars – those that range from about 5 to 10 solar masses. Such stars differ in their evolution from less massive stars because they can reach higher temperatures in their cores and throughout their interiors. Radiation pressure plays a larger role in providing the internal pressure supporting the stars. (A massive star cannot exceed about 100 solar masses because radiation pressure blows it apart.) The greater temperatures have other notable outcomes for the general evolutionary trends:

1. While on the main sequence, the star burns hydrogen by the CNO cycle.
2. The star's main-sequence lifetime is shorter.
3. The higher temperatures kindle fusion of carbon and heavier elements in the core.
4. The helium-rich core does not become degenerate.

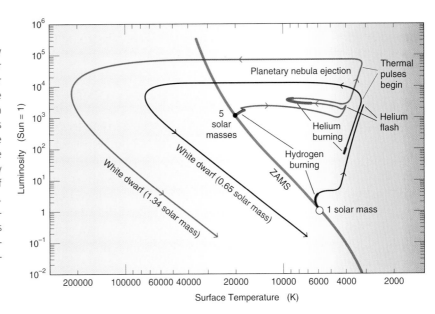

Figure 15.12 Theoretical evolutionary tracks off the main sequence (ZAMS) for stars of 1 and 5 solar masses. The thicker parts of the evolutionary tracks indicate phases of long-term fusion reactions in the cores. This diagram shows both stars going through a planetary nebula phase before they become white dwarfs; note the large percentage of mass lost by each. Note the overall similarity of the sequence of evolutionary stages. The thermal pulses occur in a helium-burning shell after helium burning has ceased in the core. Note the power-of-ten scale for both axes. (Based on a diagram by I. Iben, Jr.)

5. Energy is transported through the interior mostly by convection rather than radiation.

All this evolution happens about a hundred times faster than for a solar-mass star.

Evolution of a 5-Solar-Mass Star
(Learning Outcomes 15-5, 15-12)

The PP reaction first ignites in the core as the star reaches the ZAMS. When the core's temperature rises to about 20 million K, the CNO cycle produces more energy each second than the PP reaction. The CNO cycle uses up the core's hydrogen quickly. Note that a 5-solar-mass star on the main sequence is more luminous and hotter than a 1-solar-mass star (Fig. 15.12).

When the central hydrogen fusion fires are exhausted, the star contracts. New hydrogen falls to the inner regions and ignites in a shell around the burnt-out core. The luminosity increases, but the radius expands, so the surface temperature drops. The star becomes a red giant (Fig. 15.12).

Meanwhile the core contracts until it gets hot enough to ignite the triple-alpha process. In contrast to a 1-solar-mass star, the core has not become degenerate, so no helium flash occurs in the core, just a relatively gentle triple-alpha ignition. Stars

with masses greater than 2.25 solar masses do not develop degenerate helium cores because helium ignition takes place before the stars can acquire sufficient density.

The star burns helium in its core for about 10 million years (Fig. 15.12). When the helium runs out, the star again contracts. Now fresh helium falls into the core to make a thin, helium-burning shell around the core. The star becomes a red giant again. (Note that whenever fusion reactions take place in a shell rather than in the core, the star expands.) During the second time as a red giant, thermal pulses ignite in a shell around the core (Fig. 15.12), driven by the triple-alpha process. The star flashes and pulsates.

What next? We are really not sure. One possibility is that superwind develops and blows off the outer layers to leave behind a hot core surrounded by a planetary nebula (look back at Fig. 15.11). The core becomes a white dwarf.

More Massive Stars
(Learning Outcome 15-10)

A somewhat more massive star can develop a degenerate carbon core of about 1.4 solar masses. When the core gets hot enough to ignite carbon burning, it should do so in a flash fire. This reaction may generate so much energy that the star blows apart. Such a cataclysmic explosion may result in a supernova. The outer layers of the star blast into space. The core is crushed to immense densities.

We are not certain what main-sequence mass marks the transition to stars that certainly will explode as supernovas. Such a demise seems sure for high-mass stars – those with masses of 20 solar masses and greater. Their evolutionary tracks off the main sequence are very complex, in part because these stars shed mass with strong stellar winds. These stars turn into supergiants after they leave the main sequence, and soon afterward they self-destruct in a fierce supernova.

15.7 OBSERVATIONAL EVIDENCE FOR STELLAR EVOLUTION

The description of stellar evolution given in the preceding sections is based on theoretical calculations. How well do these ideas connect with the real astronomical world? To find out, we look for actual stars at various stages in their life cycles, and with different masses and compositions, that are like the stars predicted by our models of stars. The way we do so is to examine stars in clusters.

Stars in Groups
(Learning Outcomes 15-6, 15-7, 15-9)

Most of the stars in the Milky Way are in multiple systems, which are isolated from other stars by great distances, but we do find that some stars come in groups held together by their own gravity, at least for some time. These groups are called *star clusters*. The Milky Way contains star systems of two contrasting types: **open clusters** (sometimes called **galactic clusters**) and **globular clusters.** How do they differ?

The Pleiades in the constellation Taurus is a good example of an open cluster with a loose array of stars. The Pleiades lie about 400 ly away. The cluster contains about 100 stars within a diameter of 10 ly. Its average density is about 0.1 star in a cubic light year. These statistics are pretty typical of open clusters: they contain from fewer than a hundred up to a thousand stars in a space a few tens of light years in size. Their star densities are not more than a few stars per cubic light year. Astronomers have cataloged some thousand open clusters to date, and perhaps 20,000 inhabit the Milky Way (Fig. 15.13).

Figure 15.13 Pleiades open cluster (Messier 45). A dusty cloud encompasses this young galactic star cluster; it creates a bluish reflection nebula.

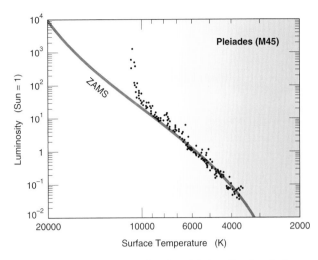

Figure 15.14 H–R diagram for the Pleiades. The line indicates the zero-age main sequence; the dots are the observed points for the stars. The luminosities are given for the visual range of the spectrum, with the sun's visual luminosity equal to 1.0. Note the power-of-ten scale for both axes. (From *An Atlas of Open Cluster Colour-Magnitude Diagrams* by G. L. Hagen, David Dunlop Observatory.)

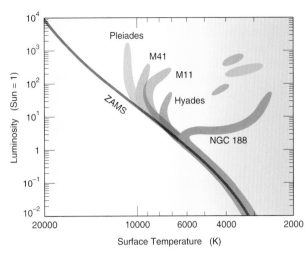

Figure 15.15 Schematic H–R diagram for selected open clusters whose distances are known. Each cluster turns off the main sequence at a different point. The stars above the turnoff point have evolved off the main sequence to become red giants. All clusters have stars that occupy the lower part of the main sequence below the turnoff points. Note that NGC 188 is relatively old and the Pleiades cluster is relatively young. Note the power-of-ten scale for both axes.

One key characteristic of open clusters is their H–R diagrams. The one for the Pleiades is pretty typical (Fig. 15.14). Note that the stars below a surface temperature of 10,000 K fall squarely on the main sequence. Above 10,000 K, the stars lie above and to the right of the main sequence. Most open clusters show the same properties: the lower-mass stars fall on the main sequence, but at some point the higher-mass stars turn off it (Fig. 15.15).

Globular Clusters
(Learning Outcome 15-9)

Globular clusters contrast dramatically with open clusters (Table 15.1). As the name implies, a globular cluster has a distinct spherical shape (Fig. 15.16). Binoculars show it as a miniature, fuzzy sphere, like a cotton ball. A small telescope reveals some of the brightest stars.

Table 15.1 Comparison of Star Groups

Characteristic	Open Clusters	Globular Clusters
Mass (solar masses)	10^2–10^3	10^4–10^5
Diameter (ly)	6–50	60–300
Color of brightest stars	Reddish to bluish-white	Reddish
Density of stars (solar masses/ly³)	0.1–10	1.0–100
Examples	Pleiades	Hercules (M13)

Figure 15.16 Globular cluster Messier 15 as imaged by the HST in true colors. Compare the distribution of stars to the Pleiades cluster; note the great concentration in the center. The brightest stars are red giants; they appear orangish. Less luminous stars are hotter and look bluish.

The stars at the center of a globular cluster are crammed together as densely as a hundred per cubic light year! If the earth orbited a star in the core of a globular cluster, its nearest neighbors would be a few light months away. You would see thousands of bright stars scattered over the sky. In all, a globular cluster contains up to a million stars of roughly 1 solar mass or less jammed into a space only 100 ly or so in diameter.

An H–R diagram of a globular cluster (Fig. 15.17) clearly shows its major difference from an open cluster in terms of its stellar type. The main sequence turns off to the red giant branch,

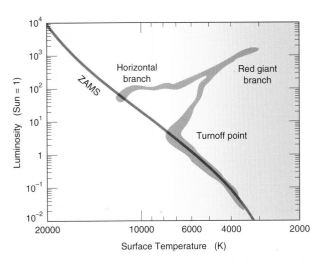

Figure 15.17 Schematic H–R diagram for a globular cluster. The stars along the lower part of the main sequence have masses less than 1 solar mass. The horizontal branch indicates the points for stars that have metal abundances less than that of the sun; on the left of the branch, the masses are smaller. Note the well-defined red giant branch. Note the power-of-ten scale for both axes.

and the upper end of the main sequence has disappeared. The more massive stars have evolved off the main sequence. A horizontal branch of stars returns from the giant region to the region of the absent upper main sequence. This slash across the H–R diagram, called the *horizontal branch,* is the special signature of a globular cluster. These features indicate that globular clusters are very old, with an age of 13 Gy!

Recall that a horizontal branch contains low-mass stars, burning helium in their cores and having chemical compositions that differ significantly from that of the sun. The observation that globular clusters have a horizontal branch strongly implies that the stars there are less massive and have fewer heavy elements than the sun. They are poor in metals because they are so old, and the gas and dust from which they were born contained only a small percentage of heavy elements. However, that means that at least some heavy elements were in the cloud from which the cluster formed, so it seems that every globular cluster survived at least one supernova explosion very early in its life.

Stellar Populations
(Learning Outcome 15-10)

The striking difference between the H–R diagrams of open and globular clusters implies that the stellar types in the two kinds of cluster are quite different. Astronomers call stars of the kinds found in open clusters **Population I stars** and those in globulars **Population II stars.** The brightest Population I stars are bluish-white; the brightest Population II stars, reddish. Also, the brightest Population I stars (O stars) have about one hundred times the luminosity of the brightest Population II stars (red giants). The luminous Population I stars must be relatively young, since they are O and B supergiants.

A really crucial distinction cannot be seen directly in the H–R diagram: chemical composition. Spectroscopic observations show that Population I stars have essentially the same chemical composition as the sun – 1 to 2 percent, by mass, of heavy elements (which are all the elements more massive than hydrogen and helium). Population II stars contain about 0.01 this amount, only 0.01 to 0.02 percent of the mass.

From the description of stellar evolution so far, you should see that Population I stars must be younger than Population II stars, in the sense that they were born later. According to the Big Bang model for the origin of the cosmos (coming in Chapter 20), the original stuff of the cosmos was mostly hydrogen and helium – essentially no heavy elements. The heavy elements must be made in stars. When a massive star dies, it spews a lot of material back into the interstellar medium. This blown-off material has been enriched in heavy elements; from it, new stars will be born. So we know that Population II stars are older (formed earlier) than Population I stars because they have fewer heavy elements. Population I stars have been formed out of enriched, recycled material.

CAUTION. Not all Population I stars are luminous bluish-white stars. In fact, many Population I stars are stars like the sun, and the vast majority are faint red dwarf stars (the lower right-hand end of the main sequence). Bluish-white Population I stars are the most luminous and so the easiest to spot. Also, not all red giants are Population II stars; but in a group of Population II stars, the most luminous ones will be red giants. Population I stars have a range of ages, from a few tens of millions of years old to perhaps 10 Gy old.

Comparison with the H–R Diagram of Clusters (Learning Outcome 15-9)

When a star cluster forms (galactic or globular), its stars are born at essentially the same time with the same chemical composition, but the masses vary. The more massive stars evolve more rapidly than the less massive stars. The more luminous stars evolve more quickly into red giants. Each star of different mass follows a different evolutionary track. As a cluster ages, stars of lower mass will evolve off the main sequence.

At the beginning of its life, a galactic cluster's H–R diagram resembles that of a young open cluster (Fig. 15.18a); a little later, that of a middle-aged open cluster, such as the Pleiades (Fig. 15.18b). Much later, the H–R diagram is similar to that of an old open cluster (Fig. 15.18c). Note that the **turnoff point** away from the main sequence moves down to lower-mass stars as the cluster ages; a cluster's turnoff point indicates its age.

An example: the old galactic cluster Messier 67.

Figure 15.19 shows an H–R diagram for the stars in M67. Note the theoretical line that corresponds to stars with masses ranging from 0.7 to 3 solar masses after an evolution of 5×10^9 years. The evolutionary line fits the turnoff point, so we infer that the cluster has this age.

The stars in the galactic clusters do not peel off the main sequence at the same point. In contrast to galactic clusters, all globular clusters have remarkably similar H–R diagrams, with roughly the *same* turnoff points.

What are the implications of this comparison for stellar evolution? First, it says that globular clusters are older than open clusters. Second, it implies that the stars in globular clusters are roughly the same age. Calibrated with theoretical models, the turnoff points for globular clusters indicate that they are 13 Gy old, give or take about 2 Gy. The stars in the globular clusters are the oldest known. Third, open clusters have a large range of ages. The ages of open clusters can also be estimated from their turnoff points by comparing them to those found from theoretical calculations. For example, M67 has an approximate age of 5 Gy, and the Pleiades, 100 million years.

The turnoff point acts like the "year hand" of the stellar evolution clock. Where it points along the main sequence gives the age of the cluster as calibrated by computer models.

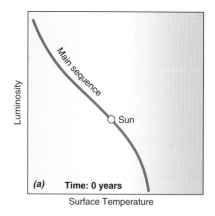
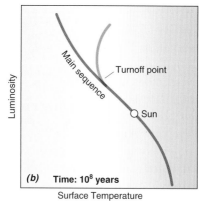
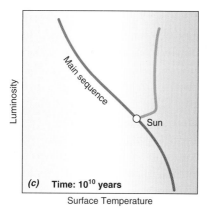

Figure 15.18 Theoretical H–R diagrams of a cluster of stars with the same chemical composition, born at the same time (a), but having different ages. "Sun" indicates the position of a solar-mass star. Note how the turnoff point moves down the main sequence as the cluster ages (b and c).

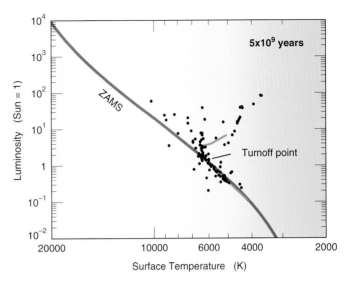

Figure 15.19 Comparison of theoretical models (solid line) and observations of the stars in the cluster Messier 67 (solid dots) to find the turnoff point from the ZAMS and so the age of the cluster (roughly 5 Gy, which means that it is a very old open cluster). Note the power-of-ten scale for both axes. (Based on calculations by D. A. VandenBerg.)

TO SUM UP. As a cluster ages, the main sequence will gradually shorten as the stars peel off, in order of mass, and evolve into the red giant region. This turnoff point gives the age of a cluster, when compared to the evolutionary tracks of theoretical models. The comparison also confirms the general validity of the star models.

Variable Stars

A star's luminosity varies little and slowly during its main-sequence sojourn. Later in life, a star's luminosity varies dramatically, over short periods of a few hours to a few years. Some of these **variable stars** (so called because their luminosities change rapidly with time) lie above the main sequence on the H–R diagram (Fig. 15.20). What is their evolutionary status? We know from model calculations that these variables are post-main-sequence stars and that those that vary regularly are in the helium-core-burning stage.

There is a staggering array of variable stars. I will mention just two: RR Lyrae variables and cepheid variables (Fig. 15.20). The RR Lyrae stars and Population I and II cepheids are **periodic** or **regular variables;** their change in luminosity with time follows a regular cycle over a certain period.

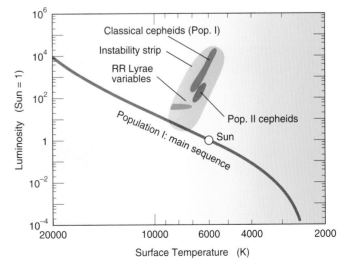

Figure 15.20 Positions of selected variable stars on the H–R diagram. The Population I and II cepheids and the RR Lyrae stars occupy a region known as the *instability strip;* here stars are burning helium in their cores and are giants. Note the power-of-ten scale for both axes. (Adapted from a diagram by J. P. Cox.)

RR Lyrae stars (named after their prototype, RR Lyrae, whose period is 13.6 hours) vary in luminosity with periods of typically 12 hours. They are Population II stars and have about one hundred times the sun's luminosity. About 5000 RR Lyrae stars are known. **Cepheid variables** (named after their prototype Delta Cephei) fall into two groups: Population I, called *classical cepheids,* and Population II. Classical cepheids have periods of light variation typically 5 to 10 days in duration. Population II cepheids vary in periods from 12 to 20 days.

These variables have a key physical characteristic in common: they pulsate. Doppler shift observations of their spectra show that as they vary, these stars expand and contract. (When the star is expanding, its surface moves toward us, producing a blueshift in the lines of the spectrum; when it contracts, the surface moves away, producing a redshift in the lines of its spectrum.) Population I cepheids, for instance, expand and contract at speeds of about 30 km/s.

Cepheids and RR Lyrae stars lie in a region of the H–R diagram called the **instability strip.** Evolutionary tracks of low-mass (0.5 to 0.7 solar mass) Population II stars and high-mass (3 to 20 solar masses) Population I stars traverse this strip during their helium-core-burning phase. During this phase, a star's outer layers become unstable and pulsate with a regular period.

Theoretical calculations indicate where on the H–R diagram we expect instability and variations. Observations validate the theoretical work and confirm the basics of the models.

Central Stars of Planetary Nebulas (Learning Outcome 15-7)

In the scenario for the evolution of a 1-solar-mass star, when a red giant blows off its outer layers, it leaves behind a hot, dense core. This cinder cools to form a white dwarf. We expect the same to be true of low-mass and some middle-mass stars.

If this picture is correct, you would expect the central stars of planetary nebulas to fall along the evolutionary track after the red giant stage. Well, they do (Fig. 15.21). Some central stars are extremely hot and luminous. One measured by the HST suggests a surface temperature of 200,000 K – the hottest star known to date! Others are not so hot. Their positions on the H–R diagram fall neatly above and to the left of those for white dwarfs, so the stars of planetary nebulas mark a transition between the core of a red giant and a white dwarf. This observation nicely confirms that stars evolve from red giants to white dwarfs.

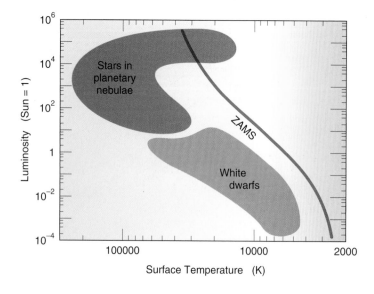

Figure 15.21 Schematic diagram for the location of central stars of planetary nebulas on the H–R diagram. These stars are the hottest known; note the very high surface temperatures. The close fit of the region of planetary nebula central stars to the region for white dwarfs supports the idea that the two are linked in evolutionary sequence. By the time the star has cooled to the white dwarf region, its planetary nebula has expanded and faded away. Note the power-of-ten scale for the both axes.

Table 15.2 Stages of Thermonuclear Energy Generation in Stars

Process	Fuel	Major Products	Approximate Temperature (K)	Approximate Minimum Mass (solar masses)
Hydrogen burning	Hydrogen	Helium	2×10^7	0.1
Helium burning	Helium	Carbon, oxygen	2×10^8	1
Carbon burning	Carbon	Oxygen, neon, sodium, magnesium	8×10^8	1.4
Neon burning	Neon	Oxygen, magnesium	1.5×10^9	5
Oxygen burning	Oxygen	Magnesium to sulfur	2×10^9	10
Silicon burning	Magnesium to sulfur	Elements near iron	3×10^9	20

Adapted from a table by A. G. W. Cameron.

15.8 THE SYNTHESIS OF ELEMENTS IN STARS

Gravitational potential energy furnishes the initial heat to get fusion reactions going. To survive, a star must fuse lighter elements into heavier ones to generate energy. The more mass a star has, the greater the central temperature produced by gravitational contraction and the heavier the elements it can eventually fuse. From the ignition temperatures needed for fusion reactions, we can set limits on the heaviest elements that a star of a certain mass can fuse (Table 15.2). For example, our sun can burn helium to carbon in its core but will never get hot enough to fuse carbon.

Table 15.2 summarizes the principal stages of nuclear energy generation and nucleosynthesis in stars. Note that the products (or ashes) of one set of reactions usually become the fuel for the next set of reactions. What a beautiful scheme for energy production in the universe!

Also note that only very massive stars can produce elements heavier than oxygen, neon, sodium, and magnesium in their cores. Few stars have this much mass, so most stars come to the end of their nuclear evolution without having manufactured some important elements. This fact emphasizes the importance of massive stars in the scheme of cosmic evolution – they create heavy elements and throw some back into the interstellar medium.

Nucleosynthesis in Red Giant Stars (Learning Outcomes 15-8, 15-11, 15-12)

Red giant stars play a vital role in the scheme of cosmic nucleosynthesis. That view, based on theoretical models, sees the thermal pulses in a helium-burning shell as the site for producing certain isotopes, especially those that are rich in neutrons. (A neutron-rich isotope is one that has more neutrons than protons in the nucleus.) This process occurs differently for stars of low and middle mass.

One episode takes place when a low-mass star first becomes a red giant. Thermal pulses, sparked by the triple-alpha process, burn helium to carbon, which is then transformed to oxygen. The convective zone that develops as a result of the pulses reaches down to the star's core and pulls up elements that have been made with hydrogen burning.

For medium-mass stars, such as one of 5 solar masses, another episode goes on after a star has burned the helium in its core. Thermal pulses can then convert helium to carbon, carbon to oxygen, nitrogen to magnesium, and iron to certain neutron-rich isotopes of heavier elements. The convective zone brings these to the surface.

These processes would have no effect on the rest of the cosmos except for one crucial fact: red giant stars have strong stellar winds. They blow off material from the surfaces of the stars, with the result that the processed material is sent out into the interstellar medium, from which future stars will form.

You now see where part of the periodic table of the elements (Appendix F) comes from. What about the rest? You will find out in Chapter 16.

KEY CONCEPTS

1. A Hertzsprung–Russell diagram provides a picture of many different stars at different stages of their lives. Time is implicit in an H–R diagram; neither axis is time. We make time explicit by constructing theoretical models of stars at different stages of their lives. A series of stages for any one star traces out its evolutionary track on an H–R diagram.

2. Stars maintain a balance between gravity and internal pressure generated by the heat from fusion reactions. To strike this balance, a star must produce energy inside. The loss of this energy to space requires that the star evolve. If stars did not shine, they would not evolve.

3. Fusion reactions in cores normally generate a star's energy. When fusion reactions stop, gravitational contraction of the core can produce heat and light and raise the temperature high enough to ignite the next stage of fusion reaction. As fusion reactions use up fuel, the ashes produced can become the next fuel, if the temperature gets high enough. Fusion products are the sources of new fuel in the sequence of stellar nucleosynthesis.

4. Star models are based on the physical properties of stars. These models assume that a star's internal pressure balances its gravity; that an energy source (usually fusion) is in the core; that energy flows out from hotter to cooler regions by radiation or convection; that a star is a gas throughout; and that equal amounts of energy are radiated from the interior and pro-

duced within it. Specific models depend on the initial mass and chemical composition.

5. A newly born star shines from energy produced by gravitational contraction. When fusion fires ignite in the core, the star finally achieves the main-sequence stage of its life. The star first appears on the zero-age main sequence. Stars on the main sequence in the H–R diagram are fusing hydrogen to helium in their cores.

6. More massive stars (upper part of the main sequence) use up their fuel very fast, hence evolve faster than less massive stars (lower part of the main sequence). The more massive stars have higher fuel requirements because they use the CNO cycle and because they have higher core temperatures than stars of lower mass.

7. When fusion reactions are occurring in a star's shell (or shells) around the core and not in the core, the star expands in size. This shell burning usually means that the star becomes a red giant. Our sun (like other low-mass stars) will evolve to a red giant and will blow off its outer layers to make a planetary nebula. The former red giant core will become a white dwarf, then cool to a black dwarf. Some middle-mass stars follow the same sequence.

8. Other medium-mass stars become red giants a number of times, then die in a supernova explosion. High-mass stars (greater than 10 to 20 solar masses) become supergiants after their main-sequence phase. They also die in supernovas.

9. A comparison of the H–R diagrams for clusters of stars confirms our basic ideas about stellar evolution because these stars were formed about the same time, with the same initial chemical compositions. Only their masses differ. The more massive stars evolve first off the main sequence, so the main-sequence turnoff point indicates the age of a star cluster. Theoretical models of stars calibrate it.

10. Massive stars can fuse elements up to iron in their cores in their normal lives. Red giants can fuse some heavy elements in convective helium-burning shells. These are brought to the star's surface by convection and blown off by stellar winds into the interstellar medium, so the percentage of heavy elements in the interstellar medium grows as time passes.

11. A star's mass determines how it develops and how fast it lives (Table 15.3); more massive stars have shorter lives than less massive ones. Initial chemical composition plays a secondary role in determining the history of a star.

Table 15.3 Comparison of the Evolution of Stars of Different Masses

Mass	Evolutionary Sequence
Low (< 1 solar mass)	Protostar → pre-main sequence → main sequence → red giant → planetary nebula → white dwarf → black dwarf
Middle (≈ 5 to 10 solar masses)	Protostar → main sequence → red giant → planetary nebula or supernova
High (> 20 solar masses)	Protostar → main sequence → supergiant → supernova

STUDY EXERCISES

1. A star like the sun consists completely of an ordinary gas. Why doesn't it suddenly collapse gravitationally? (Learning Outcome 15-2)

2. Present calculations indicate that a solar-mass protostar is much more luminous than the sun, yet it's much cooler at the surface. How can the protosun be much cooler and yet more luminous than the present sun? (Learning Outcome 15-4)

3. How can you tell from an H–R diagram that the stars in the Pleiades cluster are younger than those in the Hyades? (Learning Outcome 15-6)

4. Why are massive stars able to fuse heavier elements than less massive stars? (Learning Outcome 15-11)

5. What evidence do we have that red giant stars become white dwarfs? (Learning Outcome 15-7)

6. What is a main difference between the evolution of a 1-solar-mass star and a 5-solar-mass star? (Learning Outcomes 15-5 and 15-10)

7. Compare and contrast the stellar types in open and globular clusters. (Learning Outcomes 15-7 and 15-9)

8. What kind of star is our sun likely to become, and what kind of corpse will it eventually leave? (Learning Outcome 15-3)

9. Outline the evolution of a cluster of stars containing half 5-solar-mass stars and half 1-solar-mass stars. (Learning Outcome 15-6)

10. What kind of fusion reaction can produce carbon in a star? What minimum mass is needed for this reaction to occur? (Learning Outcomes 15-8 and 15-11)

11. In what sense is time implicit in an H–R diagram? How do astronomers make it explicit? (Learning Outcomes 15-1 and 15-9)

12. Why do stars evolve? How do we know that stars evolve? (Learning Outcomes 15-4 and 15-7)

13. How does the method of energy transport differ in high-mass stars compared to solar-mass ones? (Learning Outcome 15-12)

PROBLEMS AND ACTIVITIES

1. How many kilograms of matter are converted to energy each second in a 5-solar-mass, main-sequence star?

2. When the sun was a protostar, it had a maximum luminosity about 1000 times its present luminosity and a surface temperature of roughly 1000 K. What was its radius? *Hint:* Assume that it radiates like a blackbody.

3. Let's draw a few points of a small range of the sun's evolutionary track on an H–R diagram. Use a piece of graph paper with the horizontal axis for temperature (increasing to the left) and the vertical axis for luminosity. For a luminosity equal to 1.0 and a surface temperature of roughly 6000 K, place a point for the sun now. The radius equals 1.0. Label this point with an age of 4.6 Gy. Then add points from the table below. Note you will need to calculate the surface temperatures at different times, assuming a blackbody radiator.

Time (Gy)	Luminosity (sun [now] =1.0)	Radius (sun [now] = 1.0)	Surface Temperature (kelvins)
3.0	0.95	0.95	
6.6	1.2	1.08	
7.7	1.3	1.14	
9.8	1.8	1.36	

These calculations pretty much give the sun's evolution from the pre-main-sequence phase to the end of its main-sequence phase. (Table from calculations by S. Turck-Chièze, S. Cahen, M. Cassé, and C. Doom.)

4. A typical planetary nebula will have an angular diameter of 20 arcsec. What is its size-to-distance ratio? A typical distance to a planetary nebula is 500 ly. What is the nebula's physical diameter?

5. The smallest angular diameter of a galactic cluster is 0.5 arcmin. The largest angular diameter is 50 arcmin. What is the size-to-distance ratio of each? Assume that all galactic clusters have the same physical diameter. Then what is the range of distance for such clusters?

16

Stardeath

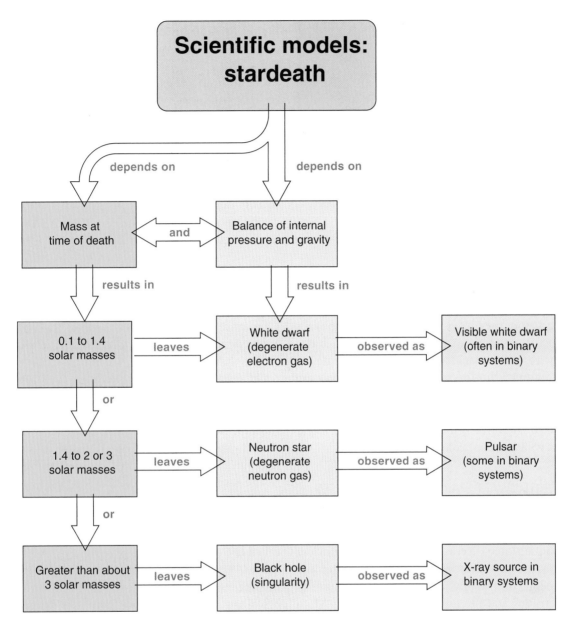

Chapter 16
Celestial Navigator™

Many stars die violently! Imagine that one year equals one-fifth of a second; then the sun would live 65 years. In this speeded-up time, about twenty-five stars in the Milky Way wink out every second. These deaths are signaled by blasts of the superwinds that eject a star's outer layers. The discarded shells replenish the interstellar medium, which has been depleted by the formation of stars and planets. And they add heavy elements to it.

The dead star's remnant core cools. Locked tight by gravity, it forms a cinder in space. In some instances, the core develops into a neutron star, a smooth, spinning sphere of nuclear matter. In others, the core disappears through a warp in space-time as a black hole – where gravity has gone bad. In still others, the burned-out core becomes a white dwarf star, a solid carbon crystal – the eventual fate of our sun.

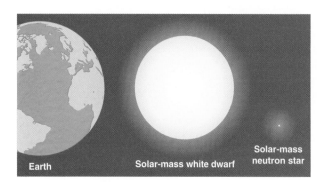

Figure 16.1 Comparison of the relative sizes of the earth, a solar-mass white dwarf, and a solar-mass neutron star drawn to the same relative size scale.

Observations imply that almost all stars throw off mass before they meet their ends. A supernova is the most destructive example of mass loss, but a supernova is constructive too – in its immense explosion, many heavy elements of the universe are made and thrown to the currents of space. Supernovas spice the interstellar medium and provide the impulse to the birth of new stars. Gravity is a star killer as well as a star maker.

16.1 WHITE DWARF STARS: COMMON CORPSES

The evolution of a 1-solar-mass star (Section 15.5) illustrates the constant battle between pressure and gravitational forces. Because, unlike thermonuclear reactions, gravity never lapses, the final state of any star depends on the physical properties of matter at high densities and the total mass of the star. When all the thermonuclear reactions cease, what pressure can support the star? What will the star become at the time of its death? (See the Stardeath Celestial Navigator™.)

Physics of Dense Gases
(Learning Outcomes 16-1, 16-2)

The conventional model of the atom surrounds the nucleus with a cloud of electrons. The distance from the nucleus to the first electron shell is about a thousand times the diameter of the nucleus, so an atom is mostly empty space. At the high temperatures in stars, atoms are ionized, and the electrons

run around free of the nuclei. As a star is crushed to higher densities in its evolution, the electrons form a degenerate electron gas (explained in Enrichment Focus 15.1).

In an ordinary gas, the pressure depends directly on the temperature, but in a degenerate gas, the pressure is not related to the temperature; it depends only on density. The density at which a gas switches to a degenerate state is about 10^8 kg/m^3. A degenerate star of 1 solar mass has an average density of about 10^9 kg/m^3. If the sun were compressed to this density, it would be roughly 7000 km in radius, about the size of the earth (Fig. 16.1).

White Dwarfs in Theory
(Learning Outcomes 16-1, 16-2)

In 1935 the Indian astronomer Subrahmanyan Chandrasekhar (1910–1999) applied the physics of a degenerate electron gas to the model of a star. He found that the pressure exerted by the electrons could resist the force of gravity only for stars of less than 1.4 solar masses and that such stars would have a density of about 10^9 kg/m^3. Such a star, at the end of its thermonuclear history, is a **white dwarf.** No heavier elements are fused, and no new energy produced. The stored thermal energy flows to the surface by conduction because of the great density.

How does a white dwarf fend off gravity? By the outward pressure from the degenerate electron gas, called the *electron degeneracy pressure*. What about the nuclei? They form a crystal structure embedded in the degenerate gas. If the expulsion of the outer layers of the star left a carbon core, the white dwarf, once it had cooled enough, would be a solid carbon crystal. Many white dwarfs contain a mixture of largely carbon and oxygen.

Chandrasekhar also discovered that the *more massive* the white dwarf, the *smaller* its radius. (In contrast, the more mass a main-sequence star has, the larger it is.) How? More mass means more gravity. To balance gravity requires internal pressure. In a white dwarf, the pressure arises from the nature of a degenerate gas, where greater pressures are a response to greater density, so the more massive a white dwarf, the smaller its size.

A crucial point arrives when the mass of the white dwarf is about 1.4 solar masses. Such a star has the highest density and smallest radius possible. Add a bit more mass and the gravitational forces overwhelm the degenerate electron gas pressure. The star collapses; it cannot be stable. This amount of mass, 1.4 solar masses, is called the **Chandrasekhar limit** and signals the point at which degenerate electron matter is crushed by gravity.

Observations of White Dwarfs (Learning Outcome 16-1)

In 1862 the American optician Alvan Clark (1804–1887) observed Sirius B, the faint companion to Sirius. Later this star was found to be a white dwarf with an average density of about 3×10^9 kilograms per cubic meter! On the earth, a spoonful of white dwarf stuff would weigh about as much as a car.

Some white dwarfs are actually white, but a few are yellowish, and some are reddish with surface temperatures 4000 K and lower. A few have very high surface temperatures – up to 70,000 K! These hotter ones (temperatures from 25,000 K and up) emit x-rays. Their luminosities simply result from the outflow of internal thermal energy left over from nuclear reactions in the past. Very slowly, over gigayears, their stored internal energy radiates into space. Ultimately, a white dwarf becomes a black dwarf (not to be confused with a black hole!).

To date, about 1500 stars have been identified as white dwarfs. Because of their low luminosities, white dwarfs are hard to see. This makes it hard to estimate the total number of white dwarfs in the Galaxy, but they may constitute up to 10 percent (or more!) of all stars.

> **TO SUM UP.** A white dwarf is a stellar corpse with roughly the mass of the sun and the size of the earth. Fusion reactions have halted. A degenerate electron gas supports it against the crush of gravity. Our sun will become a white dwarf, then a *black dwarf* – a cold corpse in space.

16.2 NEUTRON STARS: COMPACT CORPSES

What happens to stars that have more than 1.4 solar masses of core material at the end of their evolution? Strange things! Degenerate electron gas pressure cannot support them, and self-gravity crushes them to higher and higher densities.

Degenerate Neutron Gases (Learning Outcomes 16-1, 16-2)

At about 10^{13} kg/m^3, the inward pressure forces **inverse beta decay** to occur (Fig. 16.2). In this process, an electron (sometimes called a *beta particle*) and a proton join together to form a neutron and a neutrino. This happens both to free protons and to protons in the nucleus of a heavy element. At around 10^{15} kg/m^3, the neutrons are no longer bound to the nuclei and begin to form a separate gas. At 10^{17} kg/m^3, the nuclei suddenly fall apart into a gas with 80 percent neutrons. A spoonful of this material on the earth would weigh about a thousand million tons.

At this density, the neutrons become degenerate in the same manner that electrons do at white dwarf densities. The neutrons provide a degenerate gas pressure, called the *neutron degeneracy pressure*, and so balance the inward pull of gravity. This pressure allows the formation of a stable **neutron star,** a star composed mainly of neutrons. Its diameter will be about 10 to 20 km, depending on its mass. (As with white dwarfs, the greater the mass of a neutron star, the smaller its radius, because it is made of a degenerate gas.)

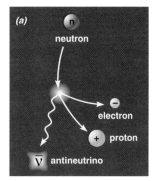 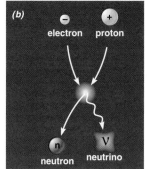

Figure 16.2 Beta decay and its reverse process, inverse beta decay. (An electron is sometimes called a *beta particle for historic reasons.*) *(a)* In beta decay, a neutron decomposes into a proton, an electron, and an antineutrino (an antimatter neutrino). *(b)* In inverse beta decay, a proton plus an electron transforms into a neutron and a neutrino.

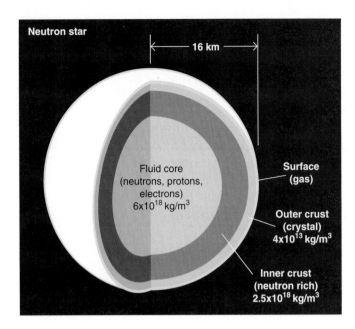

Figure 16.3 Theoretical model for the cross section of a 1.4-solar-mass neutron star. Note that the model shows three distinct interior layers. The exact details of any model depend on the known and assumed properties of the behavior of matter at high densities. (Adapted from a diagram by F. K. Lamb.)

Physical Properties
(Learning Outcomes 16-1, 16-2)

A neutron star is an eerie beast compared to an ordinary star. The exact structure of a model depends on assumptions made about the behavior of matter at very high densities. In one model for a 1.4-solar-mass neutron star, the radius is about 16 km (Fig. 16.3). The inner 11 km is a fluid core, mostly of neutrons. The next 4 km out makes an inner crust, a neutron-rich fluid or perhaps a solid lattice. The outer crust, about 1 km thick, is a crystalline solid similar to the interior structure of a white dwarf. In the outer few meters, where the density falls quickly, the neutron star has an atmosphere of atoms, electrons, and protons. The atoms are mostly iron.

Because a neutron star is so dense, it has an enormous surface gravity. For example, a 1-solar-mass neutron star with a radius of 12 km has a surface gravity 10^{11} times greater than that at the earth's surface! This enormous pull means that there are no mountains on a neutron star – minimountains reach a few centimeters at most. This intense gravitational field also results in a huge escape speed, as much as about 80 percent the speed of light. Objects falling onto a neutron star from a great distance have at least the escape speed when they hit. That means that even a small mass carries a fantastic amount of kinetic energy. For

example, a marshmallow dropped onto a neutron star from a few AU out will clobber the surface with megatons of kinetic energy!

An ordinary star, having at the end of its evolution a mass equal to or greater than 1.4 solar masses, probably ends up as a neutron star. Theoretically, a stable neutron star with a mass of less than 1.4 solar masses can also form. These low-mass neutron stars could be made in the pile-driver compression of a supernova explosion in which the interior of the star is crushed to the very high densities of neutron stars. The lowest mass possible is about 0.1 solar mass.

In an analogy to the Chandrasekhar limit, a mass limit for neutron stars arises when the gravitational forces overwhelm the degenerate neutron gas pressure. This limit – not known exactly, but estimated at about 2 to 3 solar masses – signals the next crushing point of matter by gravity.

TO SUM UP. A star with a mass between 1.4 and (roughly) 3 solar masses at the time of its death naturally forms into a degenerate neutron gas; hence the name "neutron star." Neutron stars lack any fusion reactions.

16.3 NOVAS: MILD STELLAR EXPLOSIONS

A new star, or *stella nova,* a Latin phrase usually contracted to simply **nova,** seems like the sudden

Figure 16.4 Firework Nebula (GK Per): a nova remnant near the constellation Perseus, which exploded in 1901 and lies about 1500 ly distant. It is a binary system with an orbital period of almost two days. The material expands at up to 1200 km/s. This view spans 0.7 ly.

appearance of a star where none had been visible before. But appearances are deceiving. Astronomers now consider that a nova is not a new star but rather the sudden eruption of light from an existing star (Fig. 16.4). Also, astronomers recognize that some of these outbursts take place with extraordinary violence. These special flare-ups are now called **supernovas** to distinguish them from ordinary novas.

Both novas and supernovas designate explosions of stars. For ordinary novas, only the outer layers of the star participate in the explosion. For supernovas, the interior regions are also involved.

Ordinary Novas (Learning Outcome 16-4)

In a typical nova outburst, a star in just a few days increases in brightness up to 50,000 times (sometimes up to a million). It stays at peak brilliance for several hours. Then slowly, in a few hundred days, the nova's light declines to an inconspicuous level – usually brighter, however, than the star's prenova level. A plot of a nova's rise and fall in brightness (or luminosity) is called its **light curve** (Fig. 16.5). Most novas have the same general shape for their light curves: a sharp rise and a gradual decline.

A typical nova hits a peak at greater than 100,000 solar luminosities. All told, a nova emits during its flare-up and demise some 10^{38} J; thus, in a few hundred days, a nova emits as much energy as the sun generates in about 100,000 years! We estimate that about 100 novas take place each year in the Milky Way.

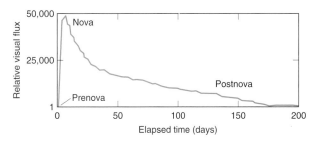

Figure 16.5 Light curve for a typical nova. Note the sharp rise from the prenova luminosity and the more gradual decline over a span of a few hundred days.

A nova's spectrum reveals the evolution of the outburst. First, the star's photosphere expands dramatically in size, from 100 to 300 solar radii. Second, the photosphere then collapses back onto the star. Third, a shell of material is blown off the star and rapidly expands away from it, as measured by the Doppler shift. Overall, a nova spurts off about 10^{26} kg of material, about 10^{-4} solar mass, only a small fraction of the total.

A Nova Model (Learning Outcome 16-5)

What prompts a nova to explode? One major clue: observational studies indicate that almost all novas occur in close binary systems, that is, binary stars with short orbital periods – typically about 4 hours. In such systems, the two stars are so close together (about 1 solar diameter) that matter may flow between them. A nova occurs on the white dwarf companion of about one solar mass in such binary systems.

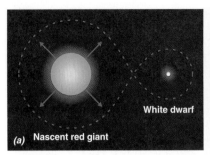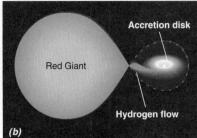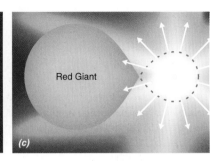

Figure 16.6 One model for the evolution of a binary system to produce a nova outburst. *(a)* One star has evolved to a white dwarf; its companion expands as it evolves off the main sequence to become a red giant. *(b)* The companion is now a red giant and matter flows through the Roche lobe into an accretion disk around the white dwarf. *(c)* The fresh hydrogen falls onto the white dwarf and fuels an explosion within the surface layers of the white dwarf.

Here is a scenario. Consider a binary star that originally consists of a main-sequence star and a less massive companion. At the end of its life, the main-sequence star has become a white dwarf (Fig. 16.6). In a close binary star system, hydrogen can come from the companion star to the white dwarf. How?

Around each star lies a region of space in which its gravitational force dominates. This region is called the **Roche lobe** (Fig. 16.6a). Any matter within the Roche lobe is gravitationally bound to that star and cannot escape to the other star. Whereas in a binary system, the gravitational pull from one star just cancels that from the other, the two Roche lobes touch and join. This means that a gravitational highway opens for the flow of matter between the stars. For the mass exchange to happen, material from the companion star has to get out to the Roche lobe. How? When it becomes a red giant (Fig. 16.6b), its atmosphere swells up and reaches its Roche lobe boundary. Material then streams from the red giant to the white dwarf.

The donor star does not have to be a red giant. In very close binary systems, the Roche lobe around a low-mass, main-sequence companion star can be about the size of the star. The white dwarf tidally distorts its companion, and mass transfer takes place.

As the matter falls toward the white dwarf, the gas forms into an **accretion disk** (Fig. 16.6b). The material gathers in a disk from the conservation of angular momentum (discussed in Enrichment Focus 11.1). The matter spirals into the white dwarf's sur-face from the accretion disk. Fresh hydrogen gradually accretes on the white dwarf's surface, forming a new envelope of fuel. Additional material piles on, compressing and heating the disk.

When the temperature at the bottom of the accreted layer has reached a few million kelvins, hydrogen fusion reactions ignite. Because the gas here is degenerate, the ignition is explosive (Fig. 16.6c). The runaway fusion reactions heat the entire layer to some hundred million kelvins. Then the layer expands and rapidly results in an envelope about the size of a red giant! This material then blows into space at speeds of hundreds of kilometers per second.

In any case, the ejected material will have more heavy elements from the thermonuclear processing. Observations of the expanding nebular material around novas do show enhanced (compared to the sun) abundances of elements such as carbon, nitrogen, oxygen, and neon.

In summary, novas occur when material accretes onto a white dwarf. This infall heats the material, igniting the runaway thermonuclear reactions that blow off the outer layers. Most novas are members of short-period binary systems, which contain a white dwarf companion.

16.4 SUPERNOVAS: CATACLYSMIC EXPLOSIONS

As violent as novas may appear, they cannot match the fierce destruction of a star in a supernova. These cataclysmic explosions spew out energy in extraor-

Figure 16.7 Supernova 1998S (brightest star just below center) explodes in the spiral galaxy NGC 3877.

Figure 16.8 The Crab Nebula (Messier 1), remnant of a supernova in A.D. 1054. The red tendrils are hot gas emitting H-alpha. A rotating neutron star within the nebula supplies its energy.

dinary amounts, about 10×10^9 times the sun's luminosity at their peak (more than 100,000 times that of a nova). A supernova usually signals the death of a massive star.

The name supernova was coined by Fritz Zwicky (1898–1974) and Walter Baade (1893–1960) for the extraordinary novas discovered in our own and other galaxies. Hundreds of supernovas have been found in other galaxies (Fig. 16.7). One exploded nearby in the Large Magellanic Cloud in February 1987 (see below). Only six have been confirmed in the Milky Way during recorded history. Over the age of the Galaxy, however, hundreds of millions of supernova explosions have occurred.

Chinese astronomers used the term *guest stars* for temporary celestial objects, such as novas or comets. One guest star flared in the sky on July 4, 1054. Close study of Chinese and Japanese accounts of this visitor confirms that the star remained visible to the unaided eye for more than 650 days in the night sky and was visible in daylight for 23 days! The position noted by the Oriental astronomers placed the event in the constellation Taurus. In the twentieth century, the Crab Nebula became the first firmly identified supernova remnant in our Galaxy (Fig. 16.8).

Classifying Supernovas
(Learning Outcome 16-4)

Astronomers classify supernovas into two general categories (Table 16.1) by their spectra and the shape of their light curves (Fig. 16.9). **Type I** light curves exhibit a sharp maximum (reaching about 10×10^9 solar luminosities) and die off gradually. **Type II** have a broader peak at maximum (emitting about 1×10^9 solar luminosities) and die away more rapidly. Studies of other galaxies have revealed that Type II supernovas occur in association with Population I stars. In confusing contrast, Type I supernovas occur in association with both Population I and II stars. Recall that Population I stars tend to be younger and richer in metals than Population II stars.

Table 16.1 General Properties of Supernovas

Property	Type I	Type II
Origin	Binary system with one star a white dwarf	Massive star near end of its life
Ejected mass (solar masses)	≈ 1	≈ 5
Velocity of ejected mass (km/s)	10,000	5,000
Total kinetic energy (J)	$\approx 10^{44}$	$\approx 10^{44}$
Visual radiated energy (J)	$\approx$ few $\times 10^{42}$	$\approx 10^{42}$
Spectra	No hydrogen lines	Hydrogen lines
Corpse	None?	Neutron star
Frequency	1 in 60 years	1 in 40 years

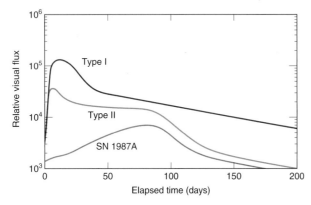

Figure 16.9 Typical light curves for Type I and Type II supernovas. Both have a sharp rise to maximum, but their declines in brightness differ in character: Type I supernovas generally have higher peaks and are brighter overall during the declining stage. The light curve for supernova 1987A is shown for comparison. Note the power-of-ten scale for the *y* axis.

The spectra of Type I and II supernovas can contain both emission and absorption lines, which change as the properties of the ejected material evolve. The basic difference is that Type II spectra show strong hydrogen lines and those of Type I do not. This indicates that Type I explosions involve stars that lack hydrogen – bodies that are highly evolved. The features of the spectra are the best way to differentiate the type of supernova explosion.

The total energy output (excluding neutrinos) of a supernova is stupendous: 10^{44} J, or roughly the entire energy output of the sun during its projected lifetime of 10 Gy. At its brightest, a supernova shines with a light of 10^{10} suns – about that of a typical galaxy! Most of the energy actually is liberated in the form of neutrinos (10^{46} J!) rather than light – an undercover way to die.

How often this kind of cosmic violence occurs is still debated. The rate of occurrence in any one galaxy is low in human terms, but the vast number of visible galaxies ensures that a few supernovas will be observed every year. From the rate in other galaxies, we estimate that a Type I supernova bursts forth in a galaxy, on average, once in 60 years. Type II supernovas may be more frequent: one explosion in a galaxy roughly every 40 years. Some astronomers argue that the true frequency is greater because we do not observe all supernova events.

The Origin of Supernovas (Learning Outcome 16-6)

What kinds of stars become supernovas? Because Type I supernovas often occur in Population II stars, these stars may have a mass about that of the sun. In contrast, Type II supernovas are thought to be stars much more massive than the sun – stars that live their normal lives as O and B stars. These stars probably explode once they have evolved off the main sequence to become red supergiants and have formed an iron core (the next section gives details). They blow off a few solar masses of material and leave a remnant mass behind.

Type I supernovas are more puzzling. How can a 1-solar-mass star detonate as violently as a supernova? A widely held idea resembles that for binary novas (Fig. 16.6). Imagine a binary system containing a white dwarf and a red giant star. Assume that the white dwarf has a mass close to the Chandrasekhar limit (1.4 solar masses). As material flows

Figure 16.10 Supernova 1987A (arrow) in the Large Magellanic Cloud. To the upper left of the supernova is the Tarantula Nebula, a large, complex H II region in the LMC.

onto the white dwarf, it builds up enough to push the star's mass over the Chandrasekhar limit. The star collapses because it cannot support the increased gravitational forces. The collapse heats the material and ignites the carbon of the white dwarf. The carbon burns swiftly into nickel, cobalt, and iron – elements that are seen in the spectra of Type I supernovas. These stars probably blow themselves to bits and leave no remnant mass.

Supernova 1987A (Learning Outcome 16-6)

It finally happened – the astronomer's dream. On the night of February 24, 1987, SN 1987A burst into view in the Large Magellanic Cloud (Fig. 16.10) – the brightest supernova since Kepler's in A.D. 1604. ("A" is for the first supernova discovered in 1987.)

The supernova in the Large Magellanic Cloud

(LMC) exploded some 170,000 years ago, at the time when glaciers enveloped the earth. When its flash finally reached us, current astronomical instruments permitted the first modern opportunity to study a close-by supernova in detail. These observations have confirmed our basic model of a Type II supernova.

The foremost observation was the discovery of neutrinos from the explosion – a prediction first made for supernovas in 1966 by Stirling Colgate and collaborators. In Kamioka, Japan, a joint project of the United States and Japan, called the Kamiokande II, detected a burst of neutrinos about one day before the supernova burst into visibility. Neutrino events were also observed from detectors in Ohio, U.S.A., by the Irvine-Michigan-Brookhaven experiment (Fig. 16.11). The detection of neutrinos strongly denotes a Type II supernova – the death of a massive star and birth of a neutron star!

In the models for Type II supernovas, a star with a mass of 10 or more solar masses develops an iron core. When the core has grown to a mass greater than 1.4 solar masses, it suddenly collapses. In less than one second, the electrons in the core are forced to merge with protons to form neutrons. The infalling material bounces off this neutron core, sending out a shock wave that creates the optical supernova explosion, which takes a day or so to become visible. In contrast, the neutrinos that are produced (one for each combination of an electron and proton) zip right out of the core. These neutrinos, in fact, carry away most of the energy, some 10^{46} J, but only a few (about 20!) were detected of the estimated 10^{15} per square meter in the burst that arrived at the earth from SN 1987A.

Most observations backed up the idea that this was a Type II explosion, but with a twist. The supernova did not get as bright as expected. It was about a hundred times fainter than a Type II should be at maximum (see the light curve in Fig. 16.9). Initial spectra showed strong, broad hydrogen lines, as expected from the hydrogen in the envelope around the exploding core. Doppler shift measurements indicated an expansion rate of about 17,000 km/s, as expected for a Type II event. The evolution of the spectrum was typical. It first showed very strong lines of hydrogen alone. Later, strong emission lines of heavier elements (some formed in the explosion) dominated the spectrum.

The star that blew up was a blue super-giant, which was located in a region of active star forma-tion in the LMC. The star was estimated to have had a mass of 20 solar masses, a luminosity of 10^5 times the sun's, a surface temperature of 16,000 K, and a radius about 50 times that of the sun.

Supernova 1987A marked the supernova of a lifetime – the first time we have seen the core collapse of a massive star. It provided a number of crucial tests for supernova models – tests that were passed with flying colors. We can now even observe some of the material expelled in the explosion and some discharged before the explosion (Fig. 16.12).

Supernova Remnants (Learning Outcomes 16-6, 16-8)

A supernova bangs out a blast of material into the interstellar medium. Traveling at supersonic velocities, the shell of material creates a shock wave that plows through the interstellar gas and dust. Its collisions with the cool clouds of the interstellar medium can excite the interstellar material, making it glow and produce emission lines. This luminous material marks a **supernova remnant.** Such remains look spherical – a shell produced by the interaction between the interstellar medium and a supernova shock wave. These shocks stir up the interstellar medium, creating turbulence in the gas.

Radio astronomers have one up on optical astronomers in the hunt for galactic supernova remnants: they can observe low-density excited gas that has no detectable optical emission. Radio astron-

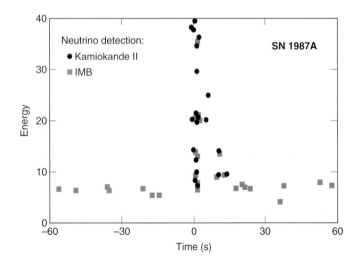

Figure 16.11 Detection of neutrinos from SN 1987A. Time zero on the horizontal axis indicates the moment of core collapse. Two sets of data are plotted here: one from the Kamiokande II experiment and the other from the Irvine-Michigan-Brookhaven (IMB) detector. The vertical axis gives the energy of the neutrinos in units of millions of electron volts (MeV), where 1 eV = 1.6×10^{-19} J. The points near and below 8 MeV are background noise. The neutrinos from the supernova all arrived within about 12 seconds (peak).

omers recognize supernova remnants by a special property of their radio emission. A plot of flux versus frequency displays a nonthermal spectrum (Enrichment Focus 16.1). If a radio source is observed at a variety of frequencies (Fig. 16.13), the shape of its spectrum distinguishes between a possible supernova remnant (*nonthermal*) and an ordinary H II region (*thermal;* a hot gas). The Planck curve of a blackbody radiator is the prototype of thermal emission.

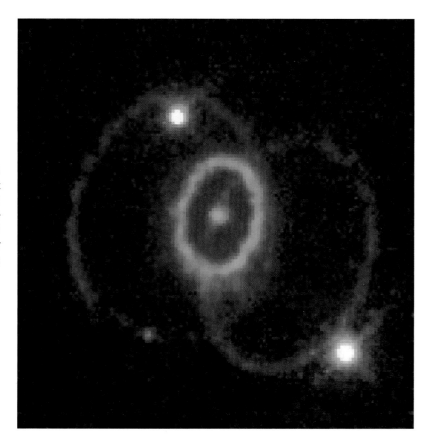

Figure 16.12 False-color HST image of SN 1987A. The material thrown off by the blast is the central, reddish region. It is about 0.1 ly across, or 100 times the diameter of the solar system. The outer yellowish ring is 1.3 ly across. It is made of material discharged by the star prior to the explosion by a strong stellar wind.

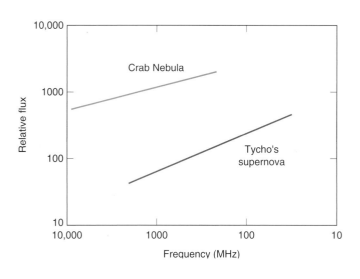

Figure 16.13 Radio spectra of two supernova remnants. Plotted here are the overall spectra at frequencies from 10 to 10,000 MHz of the Crab Nebula and Tycho's supernova. The flux decreases at higher frequencies (shorter wavelengths) – a typical trend for the nonthermal spectra of supernova remnants. Note the power-of-ten scale for both axes.

ENRICHMENT FOCUS 16.1
Thermal and Nonthermal (Synchrotron) Emission

The spectrum of blackbody radiation has a characteristic shape. The energy output increases from longer to shorter wavelengths, peaks at a wavelength that depends on the blackbody's temperature, and then decreases at shorter wavelengths until it hits zero. The spectrum of blackbody emission is the archetype of thermal emission, which arises basically from the motions of the particles involved. The greater the motions, the higher the output of radiation (and the hotter the source).

Nonthermal emission does not follow the characteristic signature of blackbody radiation. In general, the nonthermal spectrum increases in intensity at longer wavelengths (Fig. F.17). Synchrotron emission is a frequently found example of nonthermal emission; it arises from the acceleration of charged particles (usually electrons) in a magnetic field. Moving charged particles interact with magnetic fields so that they spiral around the magnetic field lines rather than traveling across them (Fig. F.18). The spiral paths are curved, so the particle is continually accelerated and thus emits electromagnetic radiation. The frequency of emission is directly related to how fast the particle spirals; the faster the spiral, the higher the frequency. Increasing the magnetic field strength tightens the spiral and so increases the frequency.

The velocity of the charged particle also affects the frequency directly, so more energetic particles can produce higher-frequency emission, but they also require strong fields to keep them in a tight spiral. As the particle radiates, it loses energy and generates lower-energy (longer-wavelength) radiation. So a synchrotron source needs a continually replenished supply of electrons to keep emitting at relatively short wavelengths (high-energy photons).

One important point about synchrotron emission: it is polarized. Thermal emission, such as that from the sun, is not polarized. Synchrotron-emitting electrons, when viewed side-on in their spiral motion, appear to be moving back and forth along almost straight lines. Their synchrotron emission has its waves more or less aligned in the same plane. Synchrotron radiation can

be polarized and, at visible wavelengths, can be observed with Polaroid filters.

We use the term *synchrotron emission* because such radiation was first observed from the General Electric synchrotron, in which magnetic fields contained electrons that were accelerated to high energies.

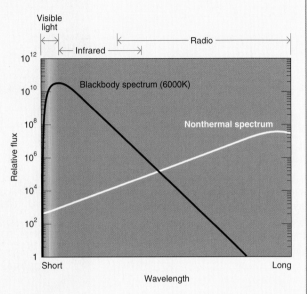

Figure F.17 Comparison of the spectra of a thermal (blackbody) source at 6000 K and a nonthermal (synchrotron) source. Note that the wavelengths run from short to long on the horizontal axis. Note the power-of-ten scale for both axes.

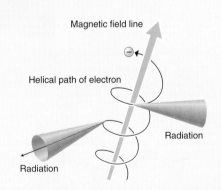

Figure F.18 Schematic diagram of synchrotron emission from an electron spiraling along a magnetic field line. The emission occurs in a narrow beam in the plane of the electron's spiral path.

A nonthermal spectrum indicates that the radio emission comes from very-high-speed electrons accelerating in magnetic fields, producing **synchrotron radiation.** The flux and wavelength range of this radiation depend on the intensity of the magnetic field and the kinetic energy of the electrons. Because supernova remnants display nonthermal spectra, we know that the synchrotron process gen-

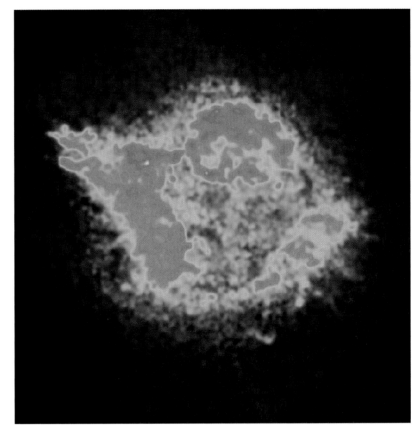

Figure 16.14 False-color x-ray image of the supernova remnant Cassiopeia A clearly shows one of the elements ejected by the explosion: silicon ions.

erates their radio emission, which means that the remnants contain magnetic fields and some high-energy particles.

X-ray astronomers can observe young supernova remnants directly. The huge shock waves plow through the interstellar medium at speeds of hundreds of kilometers per second. They compress and heat the interstellar gas to temperatures of a few million kelvins. This hot gas shows up in x-ray images (Fig. 16.14).

The Crab Nebula: A Supernova Remnant
(Learning Outcomes 16-6, 16-7, 16-8)

Eleventh-century Chinese and Japanese astronomers carefully watched the supernova that produced the Crab Nebula. The star that exploded may have had a mass of 9 solar masses. The material blown off in the explosion is still expanding. Indeed, Doppler shift measurements of the optically visible filaments in the Crab Nebula demonstrate that the gas is expanding. The Doppler shift of the expanding filaments implies – if the expansion rate has been constant at 1450 km/s – that they began expanding

at around A.D. 1132. That's pretty close to the actual date of the explosion observed from earth. (The supernova's distance is 6500 ly, so the explosion actually preceded human observation by thousands of years.)

What has powered the nebula since the explosion expelled the outer layers of the star? In 1953 the Ukrainian astronomer Josef Shklovsky (1916–1985) solved the source of the Crab Nebula's radio and optical emission when he suggested that the synchrotron process produced it (Enrichment Focus 16.1). Shklovsky's argument was clinched when astronomers found that the optical emission was strongly *polarized* (Fig. 16.15). Later observations showed that the x-ray emission was also polarized.

What does it mean for electromagnetic radiation to be polarized? You already know if you wear polarizing sunglasses (with Polaroid lenses). Look at sunlight reflected off a road or a car through such sunglasses in their normal orientation. Then twist them 90°. You'll see that the brightness of the reflected sunlight changes. That's because reflected light is polarized, and the polarizing lenses when correctly oriented, cut down the glare from the reflected light.

Figure 16.15 False-color image of the Crab Nebula, viewed through polarizing filters, confirms that synchrotron radiation is the source of most of the emission.

Light has wave properties and vibrates up and down. Waves undergo **polarization** if the planes of their vibrating motion tend to be oriented in some direction – for instance, the plane of this page of paper. Light that is unpolarized has no preferred orientation: the planes of wave vibration occur in all directions in equal amounts. Synchrotron emission is usually strongly polarized.

The solution to one puzzle posed another even more vexing: what is the source of the energetic electrons? As these electrons spiral through the magnetic field emitting synchrotron radiation, they lose energy rapidly. For the electrons producing the optical emission, half their energy would be drained off in only 70 years. Somehow the supply of electrons must be continuously replenished. The problem of the electrons' source became even more acute when x-ray emission was discovered in 1963. The electrons that produce synchrotron x-ray emission have higher energies than those that produce optical emission. They also deplete their energy faster, losing half in only a few years. The Crab Nebula emits about 100 times more energy in the form of x-rays (Fig. 16.16) than as radio or optical emission, so a large amount of energy must be added to the nebula over a period of only a few years.

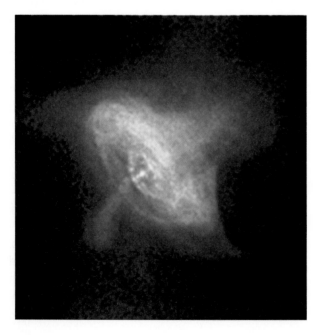

Figure 16.16 X-ray image of the Crab Nebula. The bright patch at the center is a pulsar. Note the bipolar outflows from the pulsar. The diffused glow around it is synchrotron emission from the expanding material.

The energy problem disappeared in 1968 with the discovery of a pulsar in the Crab Nebula (to come in Section 16.6).

16.5 THE MANUFACTURE OF HEAVY ELEMENTS

A massive star dies in a supernova. The blast can leave behind a neutron star or a black hole. A massive star digs its own grave, but it also blasts newly synthesized elements into interstellar space.

Nucleosynthesis in Stars
(Learning Outcome 16-10)

Massive stars (those of 8 solar masses or greater) burn heavier elements in their sequence of thermonuclear energy generation. Each step up in the fusion chain requires higher temperatures to overcome the greater repulsive electrical force of nuclei with more protons. These fusion reactions can fire only in the cores of massive stars. Compared to the main-sequence hydrogen burning, they have very short durations – a thousand years or less.

The end to the fusion chain comes with iron, the most tightly bound of the normal nuclei. To split iron into lighter elements takes an input of energy. To fuse it to heavier ones also requires additional energy. So nuclear reactions naturally stop at iron; no other rearrangement of nuclei can generate any more energy.

The elements that we know do not end at iron (Appendix F); many are heavier. Where and when are they manufactured in the course of cosmic evolution? A supernova explosion acts as nature's special workshop for forging some elements heavier than iron. In the most important process, high-energy neutrons bombard various nuclei. Because they have no charge, the neutrons have an easy time penetrating a nucleus. This nuclear capture of neutrons leads to a buildup of heavier nuclei.

Two processes take part in this buildup. A neutron under normal conditions of low density will disintegrate (in about 1000 s) into a proton, an electron, and an antineutrino – **beta decay** (look back at Fig. 16.2). The rate of beta decay naturally divides neutron nucleosynthesis into two processes. In one process, neutrons are captured at a rate faster than the beta-decay rate, so neutron-rich nuclei are formed. This process is called the **rapid process** (*r-process* for short). In the other, nuclei capture neutrons at a rate slower than the beta-decay rate, so proton-rich nuclei are made. This process is called the **slow process** (or *s-process*). A combination of both these processes leads to the manufacture of most of the elements and isotopes heavier than iron. In supernovas, the time scales are short, and it is mainly the r-process that is effective.

Nucleosynthesis in a Supernova
(Learning Outcomes 16-6, 16-10)

Here's one model of how a Type II supernova happens (Fig. 16.17a). Imagine a very massive star with a core of iron. Its interior temperature decreases from the core outward, so the star's interior is layered like an onion. Around the iron core is a silicon layer; here temperatures are not high enough to fuse silicon to iron. Around that layer is one of oxygen; here temperatures are too low to fuse oxygen to silicon. The star has taken all its life to get to this stage, but at an accelerating pace. For a 25-solar-mass star, for instance, the hydrogen-burning phase in the core lasts 7 million years; helium, 500,000 years; carbon, 600 years; neon, one year; oxygen, 6 months; and silicon, one day – a rapid dive to disaster!

When the core ends up as iron, its fusion reactions stop. Relentlessly, gravity squeezes the core to higher temperatures and densities. When the core's temperature hits about 5×10^9 K, the photons there have so much energy that they penetrate the iron nuclei and break them down into helium. As the iron disintegrates into helium, large amounts of heat are used up. Photons that normally would provide the radiation pressure to support the star are splitting iron nuclei instead. The pressure in the core drops, and it caves in suddenly. The gravitational collapse rapidly pumps heat into the material. This collapse is fast – speeds in the outer part of the core hit 70,000 km/s. In one second, a core the size of the earth crumbles to a radius of only 50 km and its central density rises to several times that of an atom's nucleus!

Three important events mark this destruction. First, protons and neutrons released by the disintegration of nuclei in the core pelt and penetrate remaining nuclei. If these particles capture neutrons, they can be transformed to heavier elements. Second, the layers above the core plummet inward toward the core and heat up. Abruptly, ignition temperatures of many fusion reactions are reached. They turn on explosively; neutrinos are produced.

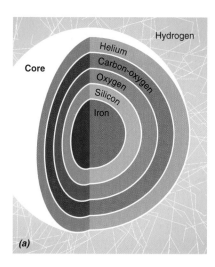
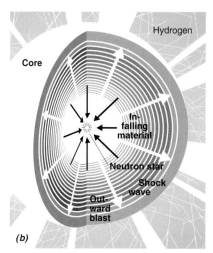

Figure 16.17 One model of the interior of a massive star during a supernova explosion. Here we are looking at the core of a Population I star at the end of its life. Fusion reactions in shells give layers of different elements *(a)*. At high enough temperatures, the iron core disintegrates into helium. This process soaks up heat from the core, and it collapses rapidly *(b)*, sparking explosive ignition of fusion reactions. This explosion produces a burst of neutrinos. (Adapted from a diagram by J. C. Wheeler.)

Third, the inner core's collapse creates a neutron star; its degenerate pressure halts the collapse, and the material rebounds outward. As the infalling matter from the outer core crashes into the rebounding inner core, a shock wave forms; it bullies its way outward from the core in just tens of milliseconds (Fig. 16.17b). It carries material with it into the interstellar medium – material enriched with heavy elements. About 99 percent of the energy released in this explosion comes out in the form of neutrinos.

Other Sites of Nucleosynthesis

Because I have emphasized nucleosynthesis in supernovas, you might be thinking that *only* supernovas make elements heavier than iron. That's not the case. Red giants also manufacture some heavy elements by the s-process. While a red giant undergoes helium shell burning and thermal pulses, neutrons are produced and added slowly to iron (and other elements) to fuse heavier elements in isotopes that are rich in neutrons (Section 15.8). Although this process is indeed slow, the red giant stage with thermal pulses lasts long enough to synthesize an appreciable amount of heavy elements. These are then dredged up to the surface by convection. A red giant can blow off s-process materials by its stellar wind.

In general, elements made in red giants complement those made in supernovas to fill up the periodic table. Nucleosynthesis in red giants makes many of the elements lighter than lead but heavier than iron. Supernovas synthesize elements heavier than lead, such as uranium and thorium. (When you use electricity generated by a nuclear power plant, you are turning on your light by the fossil remains of a massive star!) Old stars and dying stars spice up the composition of the universe. From this enriched material, new stars and planetary systems form. We live on a planet rich in metals from stars.

16.6 PULSARS: NEUTRON STARS IN ROTATION

Models of supernovas suggest that a neutron star may remain as the corpse of the exploded star. A neutron star found in a supernova remnant would clinch this argument, but how would a neutron star be visible? In a way not anticipated by astronomers: as a **pulsar,** a member of a category of objects accidentally discovered in the summer of 1967 by an English radio astronomy group headed by Anthony Hewish.

The Hewish group was mapping the sky at radio wavelengths to observe quasars. Jocelyn Bell Burnell, then a graduate student in charge of the data analysis, noticed a strange signal that suddenly disappeared, only to reappear 3 months later. The Hewish group focused on this strange signal and found radio pulses occurring at a very regular rate, once every 1.33730113 seconds. Flushed with excitement, they searched the sky for any similar signals and discovered three more objects emitting radio bursts at different rates. The Hewish group concluded that the objects must be natural phenomena and named them pulsars.

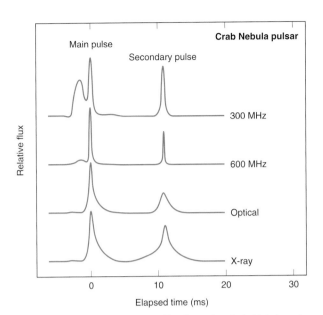

Figure 16.18 Average pulse profiles from the Crab Nebula pulsar observed at a variety of frequencies from radio (300 and 600 MHz) to optical and x-rays. This pulsar, called PSR 0531 + 21, emits a strong main pulse and a weaker secondary one. Each pulse from a pulsar has a special shape at a given frequency of observation. Averaged over many pulses at that frequency, every pulsar has a characteristic shape to its pulses. Note how these differ at different frequencies. (Adapted from a diagram by F. G. Smith.)

Observed Characteristics (Learning Outcome 16-3)

To date, more than 600 pulsars have been discovered. For a given pulsar, the period between pulses repeats with very high accuracy, better than 1 part in 100 million. The amount of energy in a pulse, however, varies considerably; sometimes complete pulses are missing from the sequence. Although the flux and shape vary from pulse to pulse, the average of many pulses from the same pulsar defines a unique shape (Fig. 16.18). The average pulse typically lasts for a few tens of milliseconds. (A *millisecond* is 10^{-3} second.)

For the well-studied pulsars, the periods (time interval between pulses) range from milliseconds to a few seconds, with an average value of 0.65 second. When accurate radio observations have been made, periods have been seen to increase. The rates of slowing down have typical values of about 10^{-8} s/y. A digital wristwatch is off by about 100 seconds in a year.

These observed general properties of pulsars provide clues to the physical properties of these precise cosmic clocks. The pulse duration indicates the largest size of the bodies emitting the radiation. How? Suppose that you could switch off the sun. Because of the finite speed of light, it would take slightly more than 2 seconds for the entire solar disk to appear dark, and then another 2 seconds to regain its original brightness if quickly turned on again. The part of the sun that we see when we look at the edge of the visible disk is at a greater distance (halfway around the sun) than the part at the center of the disk. So if the sun turned off, we would not know that the edge was dark until about 2 seconds after the center became dark.

Clearly an object the size of the sun, or any ordinary star, is too large to be a pulsar, which has pulse durations amounting to a few hundred milliseconds or less. The typical pulse durations of a few tens of milliseconds imply sizes of roughly 3000 km or less. What stellar objects exist at this size or smaller? And what objects can contain the large quantities of energy that pulsars emit? A dense, rapidly-rotating object acts as a storehouse of kinetic energy. This clue points to degenerate stars as the candidates for pulsars.

Clock Mechanism (Learning Outcome 16-3)

Here's the basic issue: pulsar models must account for the precise clock mechanism of pulsars, that is, the extremely regular repetition of pulses. Basically, one clock mechanism works the best: *rotation*.

Consider one rotation period equal to the period between pulses. Then what kinds of objects are rotating? If a spherical mass rotates too rapidly, its gravity will not be able to hold it together. Mass will fly off tangent to its equator. Neutron stars have average densities of about 10^{17} kg/m^3, so they can rotate once every millisecond without losing mass – fast enough for even the fastest pulsar. (White dwarfs cannot do so.) Of course, they can rotate more slowly, too.

TO CONCLUDE. A rotating neutron star provides the clock mechanism for pulsars.

Pulsars and Supernovas

Are supernovas or their remnants connected to pulsars? Yes! One excellent example: a pulsar in the Crab Nebula.

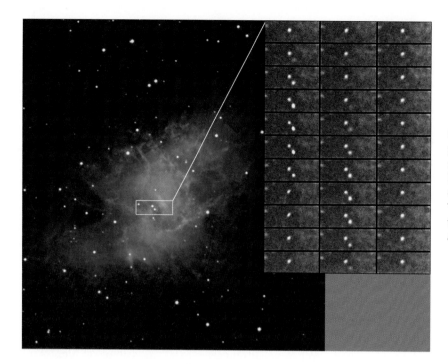

Figure 16.19 Time sequence for the Crab pulsar. The sequence of 33 photos at the right runs from top to bottom and from left to right. It shows the light variation over one complete cycle of 33 milliseconds. The brighter primary pulse is visible in the first column and a weaker secondary pulse appears in the second column.

The Crab Nebula pulsar (Fig. 16.19) is called PSR 0531 + l21. (PSR stands for pulsar, and the numbers give the pulsar's position in the sky.) PSR 0531 + l21 has a fast period: 0.033 second, or 30 pulses per second! Two key features of the Crab pulsar: first, it pulses not only at radio and optical wavelengths, but also in the infrared, x-ray, and gamma-ray regions of the spectrum. The total power emitted in the pulses is about 10^{28} W, which is about 100 times more than the total luminosity of the sun. Second, the Crab pulsar was one of the first to exhibit a definite slowdown in pulse period, at a rate of about 10^{-5} s/y.

The discovery of the Crab pulsar solves the energy problem of the Crab Nebula. At all wavelengths, the Crab Nebula emits about 10^{31} W, or about 10^5 times the sun's luminosity. What is the source of this energy? If the pulsar is a rotating neutron star, its slowdown in period gives a change in rotational energy of about 5×10^{31} W.

That's enough to power the nebula – if the rotational kinetic energy of the neutron star can somehow be converted to the kinetic and radiative energy of the nebula. In other words, the light we see now from the nebula ultimately derives from the pulsar.

A Lighthouse Model for Pulsars (Learning Outcome 16-3)

Now to tie these observations together in the accepted basic model for pulsars – a rotating, magnetic neutron star – otherwise known as the **lighthouse model.** The model has two key components: the neutron star, whose great density and fast rotation ensure a large amount of rotational energy, and a dipole magnetic field that transforms the rotational energy to electromagnetic energy. Needed is a very strong magnetic field, some 10^8 T. (The strongest magnet on earth was 60 T.)

The region close to the neutron star, where the magnetic field directly and strongly affects the motions of charged particles, is called the pulsar's magnetosphere (in analogy to planetary magnetospheres). Here all the energy conversion takes place. One model pictures the magnetic axis as tilted with respect to the rotational axis (Fig. 16.20). As the pulsar spins, its strong magnetic field spins with it, and this spin induces an enormous electric field at its surface like an electric dynamo. This electric field pulls charged particles – mostly electrons – off the solid crust of iron nuclei and electrons. The electrons flow into the magnetosphere, where the rotating magnetic

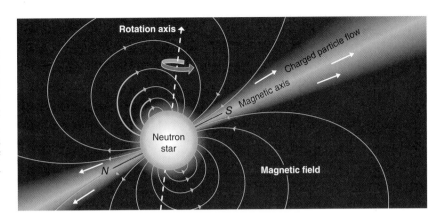

Figure 16.20 Lighthouse model of a pulsar. The magnetic axis is tilted with respect to the spin axis. Electrons from the neutron star's surface flow out along the magnetic field lines and so escape mostly at the north and south magnetic poles. These emit synchrotron radiation, which we see as pulses when a pole spins across our line of sight. The rotation of the neutron star provides the clock mechanism for the pulsar.

field lines accelerate them. The accelerated electrons emit synchrotron radiation (Enrichment Focus 16.1) in a tight beam more or less along the field lines.

You can now see how a pulsar pulses *without* actually pulsating. (Recall that the cepheid variables, in contrast, actually expand and contract.) If the magnetic axis falls within our line of sight, each time a pole swings around to our view (like the spinning light of a lighthouse), we see a burst of synchrotron emission. The time between pulses is the rotation period. The duration of the pulses depends on the size of the radiating region. As the pulsar generates electromagnetic radiation, the torque (Enrichment Focus 11.1) from accelerating particles in its magnetic field slows its rotation. This slowdown is observed.

For us to see a pulsar in the lighthouse model, the neutron star's magnetic axis must be oriented just right for the pulses to beam at the earth and give a blinking effect. The Milky Way contains many pulsars that we cannot see because of unfavorable orientations.

Binary Radio Pulsars

Most stars in the Milky Way are members of binary or multiple star systems. Even after one member of a binary goes supernova, the system usually remains intact, so we expect to find radio pulsars in binary systems.

The first one found, called PSR 1913 + 16, has a pulse period of only 0.059 second. Surprisingly, its period goes through a large cyclic change in only 7.75 hours. Such regular changes would naturally come about in a binary system of the pulsar and a companion with an orbital period of 7.75 hours. What is seen is a Doppler shift in the signal produced by the orbital motion of the system. When the pulsar is moving away from us, its pulses are spread out and come at longer intervals. When it is moving toward us, the pulses are pushed together and come at shorter intervals.

From radio observations alone, we know that the pulsar and its companion have an orbital semimajor axis of only a few times the sun's radius! Their combined masses are 2.8 solar masses of about equal mass, so perhaps both are neutron stars.

PSR 1913 + 16 neatly confirms a prediction that Einstein made in 1915 from his general theory: an accelerating mass should radiate energy in the form of gravitational waves. (In analogy, accelerating charged particles emit electromagnetic waves.) Such gravitational waves interact feebly with matter, so it is hard to detect them directly. PSR 1913 + 16 serves as a laboratory for an indirect test: the pair's accelerations should produce gravitational radiation. If so, the orbit of the pulsar should decrease in size. Indeed, such a decrease has been accurately measured using the pulses. The orbit decreases about 3.5 meters per year, decreasing the orbital period, at a rate just about that predicted by general relativity.

Very Fast Pulsars

While investigating a peculiar radio source in the constellation of Vulpecula, radio astronomers homed in on an extremely fast pulsar – a period of 1.56 milliseconds (Fig. 16.21). If the lighthouse model is applicable to this pulsar, called PSR 1937 + 214, it must spin 642 times per second (twenty times faster than the Crab pulsar) so that its surface rotates at roughly one-tenth the speed of light, very close to its breakup speed. Astronomers call such pulsars **millisecond pulsars.** Some 50 are known to date. Many have been found in globular clusters, which implies that they are very old.

One of the curious features of millisecond pulsars is the great stability of their rotation rates. For instance, PSR 1937 + 214 loses only 3×10^{-12} ms in a year. This even beats out atomic clocks! One explanation is that the millisecond pulsars have very weak magnetic fields, perhaps a thousand times weaker than typical. Then how does such a pulsar become observable?

The scenario is that of a pulsar resurrected in a binary system. Millisecond pulsars may have aged gracefully after formation in a supernova. Then, thousands of millions of years later, their low-mass companions evolved finally to red giants and matter flowed from them into a disk around the dead pulsar. This material makes a rapidly spinning accretion disk around the neutron star. The magnetic field of the pulsar is entwined with the disk; this linkage spins up the pulsar so that it lives again.

Pulsars with Planets

Millisecond pulsars are thought to be old (many gigayears), companions to stars in binary systems. Radio astronomers can time their pulses with great precision and detect even small changes in them.

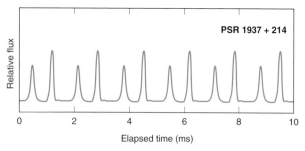

Figure 16.21 Radio pulses from a millisecond pulsar (called PSR 1937 + 214) observed at a frequency of 1412 MHz. Like the Crab pulsar, this one shows a large main pulse and a smaller secondary pulse. Note that the time scale at the bottom spans a mere 10 ms. (Based on observations by D. C. Backer, S. R. Kulkarni, C. Heiles, M. M. Davis, and W. M. Goss.)

One millisecond pulsar, called PSR 1257 + 12, showed two periods in the small variations of its timing compiled over a time span of more than a year: one around 67 days and the other around 98 days. (These changes were so small that they amounted to a 0.7 meter per second wobble in the pulsar.) What could cause them? Small masses orbiting the neutron star!

Radio astronomers Alexander Wolszczan and Dale A. Frail have interpreted this tottering as the effect of three planet-sized masses orbiting a 1.4-solar-mass neutron star. One, which has a mass of about 4 earth masses, swings around the neutron star in 67 days at a distance of 0.36 AU. The other also contains about 4 earth masses and orbits in 98 days at a distance of 0.47 AU. The third and much less massive (0.02 earth mass) orbits at 0.2 AU with a period of only 25 days. In contrast to the orbits of extrasolar planets around normal stars, these companions pursue nearly circular paths.

How might such objects form? Millisecond pulsars are apt to have an accretion disk made of matter from their stellar companion. Somehow planet-sized masses form in this disk, maybe in a manner similar to the formation of the protoplanets in the solar nebula (Section 11.6). For now, this is speculation. But if planets do form under conditions that involve a mass as different from the sun as a neutron star, it may well be plausible that planetary systems are much more common than now believed.

16.7 BLACK HOLES: THE ULTIMATE CORPSES

Many stars in the Milky Way will die with masses much greater than the neutron star limit of about 3 solar masses. Will anything stop the collapse of this material? No, nothing! The crush of gravity overwhelms all outward forces, including the repulsive forces between particles with the same charge and degenerate pressure. No material can withstand this final crushing point of matter. The collapse cannot be halted; the volume of the star will continue to decrease until it reaches zero. The density of the star will increase until it becomes infinite.

Before a mass reaches total collapse, bizarre events occur near it. As its density increases, the paths of light rays emitted from the star are bent more and more away from straight lines going away from the star's surface. Eventually the density reaches such a high value that the light rays wrap

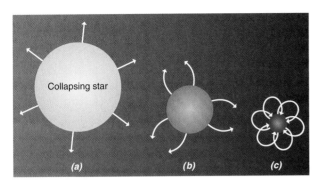

Figure 16.22 Trapping of light by the collapse of a mass into a black hole. Imagine a person standing on the star's surface with a flashlight. The star's mass, when much larger than its black hole size, has a gravity low enough to permit light to leave on straight-line paths *(a)*. As the star contracts, the gravity at the surface increases and light paths are bent. When the star is a little larger than its black hole size *(b)*, all light paths, except for the one straight up, fall back to the surface. When the star is smaller than its black hole radius *(c)*, even the straight-up photon returns.

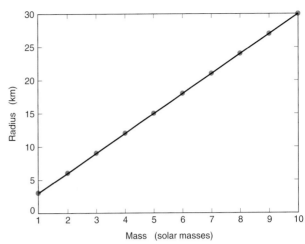

Figure 16.23 Linear relation between the mass of a black hole (in solar masses) and its radius (in km).

around the star and do not depart. The photons are trapped by the intense gravitational field in an orbit around the star. The escape speed from the star is then greater than the speed of light. Any additional photons emitted after the star attains this critical density can never reach an outside observer. The star becomes a **black hole** (Fig. 16.22). Gravity has gripped its ultimate, dark conclusion.

The Schwarzschild Radius
(Learning Outcome 16-9)

Let's consider the meaning of "black" in black holes in terms of escape speed (Section 4.5 and Enrichment Focus 4.4). The escape speed from the earth is about 11 km/s. Imagine squeezing the earth to became smaller and denser. The escape speed from its surface would increase. Imagine the earth compressed until its escape speed equaled the speed of light. Then nothing emitted at its surface, not even light, could escape into space. Nothing could get away, so to an outside observer, the earth would appear to be black.

How small must an object become to be dense enough to trap light? Einstein's general theory of relativity provides an answer. Just after its publication, the German astrophysicist Karl Schwarzschild (1873–1916) calculated this critical size, now called the **Schwarzschild radius.** For the sun, the Schwarzschild radius is about 3 km – smaller than the typical sunspot! The mass of any object gives its Schwarzschild radius directly. For 1 solar mass, it's 3 km; for 2 solar masses, 6 km; for 10 solar masses, 30 km; and so on – a direct proportion (Fig. 16.23). The earth's Schwarzschild radius is a mere centimeter!

How to get a star as small as its Schwarzschild radius? Two ways are possible. First, runaway gravitational collapse. If you put together more mass than about 3 solar masses, it must naturally squeeze itself into a black hole. Nothing we know – not even the hardness of matter itself – can stop this final crunch. Second, a supernova. A star's self-destruction can slam matter into a size smaller than its Schwarzschild radius (and make black holes with masses less than 3 solar masses).

NOTE. An object of any mass can be made into a black hole if a force compresses it enough. But a stellar corpse with a mass greater than the neutron star limit of 3 solar masses *must* become a black hole after thermonuclear reactions have ceased, for there is no known source of pressure that can support it. This limit applies only to the fate of stars, where a black hole is born when a massive enough star dies.

The Singularity
(Learning Outcomes 16-9, 16-11)

Once a black hole has formed, what happens to the matter that makes it? Einstein's general theory predicts that the matter will keep collapsing gravitationally until it has no volume. Yet it still has mass, so its density is infinite. This theoretical end to runaway gravitational collapse is called a **singularity.** Matter has literally squeezed itself so small that it occupies no space. Yet it is still there. What a paradox! How can matter not take up space? The general theory of relativity points to the formation of a singularity, cloaked in the center of a black hole, as the uncivil but natural end of gravitational collapse.

Perilous Journey into a Black Hole
(Learning Outcomes 16-9, 16-11)

Now to focus the *theoretical* properties of a black hole, both inside and outside. A person falling into a black hole meets a fate an outside observer cannot even discover – unless the outsider drops in too. Let's take an imaginary journey into a black hole to follow the adventures of a crazy astronaut who takes the plunge. I will compare this trip with what an outside observer would see of it. You and a friend start out in a spaceship orbiting a few AU from a 10-solar-mass black hole. Nothing peculiar here. The ship orbits the black hole in accordance with Kepler's laws, as it would any ordinary mass. Kepler's third law and the spaceship's orbit permit you to measure the hole's mass.

Your friend volunteers to hop in. She takes with her a laser light and an electronic watch. You and she synchronize watches. Once a second, according to her watch, she will send a laser flash back to you.

Down she goes! For a long time as she falls feet first toward the black hole, nothing strange happens. But as she gets closer, stronger and stronger tidal gravitational forces (Enrichment Focus 9.1) stretch her out from head to toes. Also, another tidal force squeezes her together, mostly at the shoulders. (You feel such tidal forces on the earth, but the forces are so weak that they do not bother you.) Near a black hole, tidal forces grow enormously. An ordinary human being would be ripped apart about 3000 km from a 10-solar-mass black hole. Let's suppose your friend is indestructible, so she can continue her trip. As she drops, the tidal forces rapidly get stronger, but

nothing else seems strange to her. Every second on the dot she sends out a blast of laser light. Peering down, she can just make out a small black region in the sky. (Recall that a 10-solar-mass black hole has a radius of only 30 km.)

Then she crosses the Schwarzschild radius. Nothing new happens to her! No solid stuff, no signs mark the edge of the black hole. No amount of energy can push her out of the black hole, however. She has crossed a one-way gate in spacetime. The trip now swiftly ends for your unfortunate friend. Quickly – in about 10^{-5} second (by her watch) after crossing the Schwarzschild radius – she crashes into a singularity (if it exists!). Crushed to zero volume, she rudely vanishes. Even if a singularity does not lie in the black hole's center, the mass that made the black hole probably does, and your friend would smash into it.

What of your view, back in the spaceship, of your friend's adventure? You would never see the end of her journey. In fact, you would not even see her fall into the black hole. As she dropped closer to the black hole, you would notice that the light from her laser was redshifted, with the shift increasing as she fell closer to the black hole. (The light had to work against gravity to get to you, so it lost energy and increased in wavelength.) Also, you would find that the time between laser flashes increased. What had happened? Compared with your watch, your friend's watch appeared to slow down as she traveled into regions of stronger gravity. So your watch and hers disagreed about how long it took her to reach the black hole.

As she came closer to the Schwarzschild radius, the watches got more and more out of synchronization. The times between your reception of her flashes stretched out. In fact, a laser burst sent out just as she crossed the Schwarzschild radius would take an infinite time to reach you. It also would suffer an infinite redshift. To you, her fall would seem to grow slower and slower as she got closer to the black hole, but she would never appear to fall into it. Your measurement would show time slowing down so much near a black hole that it seemed to be frozen. In addition, the light would get more and more redshifted until you could no longer detect it.

A black hole practices cosmic censorship. It prevents you from seeing your friend even fall into it, and you cannot know what happens to your friend, once she's inside.

CAUTION. Black holes are *not* cosmic vacuum cleaners! Many people think that they have infinitely powerful gravitational fields that suck up everything that gets near them, scouring out the universe. Not so! Suppose you'd never heard anything about the strange properties of black holes. Just thinking about Newtonian gravitation, what do you think would happen to the orbit of the earth around the sun if the sun suddenly shrank down to a ball 6 km across? That's right, nothing! The masses of the sun and earth haven't changed, and neither has the distance from the earth to the center of the sun. So the force of gravity on the earth hasn't changed, and neither has its orbit.

You can think of black holes as gravitational ghosts. The Schwarzschild radius is not a physical object. It is a region of spacetime so severely curved that weird things happen near it. Likewise, the singularity predicted to be in the center of a black hole cannot be a real object. It's a region where space and time end in this universe. What a strange corpse from the death of a massive star!

16.8 OBSERVING BLACK HOLES

Black holes are the big game of the astronomical hunt. How to actually spot a black hole? With difficulty! Light emitted inside cannot get out. Light sent out close by is strongly redshifted, so it's hard to detect. In addition, a black hole is small, only a few kilometers in size. You'll have a tough time seeing an isolated black hole.

A black hole near any mass might be observable, however. Matter falling toward a black hole gains energy and heats up. (It is also squeezed by tidal forces.) Heated enough, the atoms are ionized. Gravity accelerates the ionized gas, and it emits electromagnetic radiation. If heated to a few million kelvins or so, the material gives off x-rays. Before it is trapped in the gravitational gulf, infalling material can send x-rays into space. A black hole passing through an interstellar cloud or close to a star can sweep up material, which can radiate before it crosses the Schwarzschild radius. Hence, x-ray sources are good candidates for locating black holes.

Binary X-Ray Sources (Learning Outcome 16-12)

Why are binary x-ray sources most suspect? Imagine a black hole orbiting a supergiant star. If they are very close together, their orbital period is a few days or so. The star has a huge, distended atmosphere – and material from this atmosphere can fall to the black hole from a Roche lobe (Fig. 16.24).

Falling toward the black hole, the material gains kinetic energy, heats up to a million kelvins or so, and emits x-rays. Only a small region around the black hole gives off x-rays, and since the material may fall in sporadically, you might expect the intensity of the x-rays to vary quickly. Also, imagine the black hole and star with their orbital plane in our line of sight. When the black hole goes behind the star, its x-rays will be cut off. In this case we would see an eclipsing x-ray binary system.

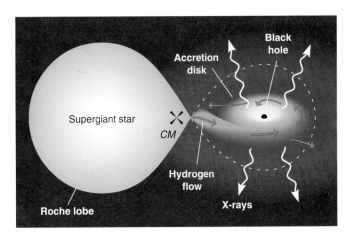

Figure 16.24 One model for a black hole as an x-ray source. In a binary system, the black hole is coupled with a supergiant star (which fills its Roche lobe). Material from the supergiant star flows to the black hole, where it first falls into an accretion disk. As it falls, the gas heats up to about a million kelvins and emits x-rays from the accretion disk before it ultimately enters the black hole. The arrows indicate the flow of material from the supergiant star. "CM" marks the center of mass of the system.

So a sign of a possible black hole is a rapidly varying x-ray source, which may be eclipsed at regular intervals in a binary system. Have we seen such variable x-ray sources? Yes! Each of these objects is believed to be a main-sequence or post-main-sequence star swinging around an x-ray source. The objects have x-ray luminosities in the range of 10^{29} to 10^{31} W. (That's 200 to 20,000 times the luminosity of the sun, and *all* in x-rays!) Some sources have short-period x-ray pulses; they are *x-ray pulsars*.

Other systems exhibit x-ray eclipses – the x-ray source passes behind the normal star as we view the system. Using spectroscopic analysis of the light from the visible star, we can observe the changes in Doppler shift, which permits us to find out the orbital periods – they are typically a few days. These short periods indicate that the orbits are only a few times larger than the primary stars. If we can determine the separation of the two objects, we can ascertain – from Kepler's third law – the sum of the masses (normal star plus x-ray source). If we can get an idea of the mass of the normal star from its luminosity, then we can also determine the mass of the x-ray source. If that mass turns out to be large enough (greater than 2 to 3 solar masses, the upper limit for a neutron star), the x-ray source must be a black hole.

Observing Black Holes (Learning Outcome 16-12)

To prove the reality of black holes we need to observe one. We have only a few solid candidates. A firm one is Cygnus X-1, a strong x-ray source in the constellation Cygnus. Cygnus X-1 emits about 4×10^{30} W in x-rays. Astronomers have identified Cygnus X-1 with a hot supergiant star.

Optical observations show that the dark lines in the spectrum of the blue supergiant go through periodic Doppler shifts in 5.6 days (Fig. 16.25). So the supergiant orbits with the x-ray source around a common center of mass every 5.6 days. The visible star has a massive but optically invisible companion: Cygnus X-1.

Recall that only for binaries can we find the masses of stars directly, but we need to know the separation of the stars, their distances from the center of the mass, and the orbital tilt. In these two regards, the mass of Cygnus X-1 is hard to pin down. We can observe the Doppler shift in the spectrum of the visible companion, but we cannot obtain the velocity of the x-ray source. Because Cygnus X-1 does not eclipse, we cannot clinch its orbital inclination. So we do not have enough information to determine accurately both individual masses.

We can make some reasonable estimates. The mass of a blue supergiant star is typically 15 to 40 solar masses. The orbital period and velocity of the supergiant give us a relation between the masses of the supergiant and the x-ray source; this relation, however, is uncertain by the amount of orbital tilt. X-ray observations imply a tilt of about 50°. These values suggest that Cygnus X-1 has a mass of about 6 to 9 solar masses. If so, and if the limit for a neutron star is 3 solar masses, Cygnus X-1 is a black hole (Fig.16.26a).

Another strong candidate has the prosaic name of A0620-00. It is located in Monoceros, about 3200 ly from earth. This binary system contains a main sequence, K star orbiting the center of mass every 7.75 hours at an orbital speed of 430 km/s (about 14 times the orbital speed of the earth

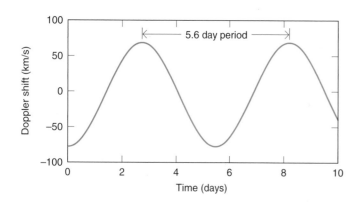

Figure 16.25 The orbital period of Cygnus X-1 inferred from the Doppler shift in the visual spectral lines of the blue supergiant star about which Cygnus X-1 orbits. These lines show a cycle of redshifts and blueshifts that indicate a period of 5.6 days. (Based on observations by C. Bolton.)

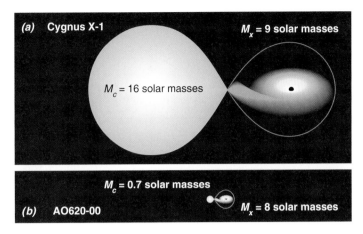

Figure 16.26 Possible black hole binary systems shown schematically. The Roche lobes are drawn, with the companion star (M_c) assumed to fill its Roche lobe. (a) Cygnus X-1, based on a 16-solar-mass companion and a reasonable orbital inclination. (b) A0620-00, also based on a reasonable value for the orbital inclination. (Adapted from a diagram by J. E. McClintock.)

around the sun). The distance of the K star from the center of mass has been measured: it is 0.014 AU, and we can infer its mass as 0.7 solar mass.

What about the companion? It is invisible, but emission lines from the hot gas in an accretion disk surrounding it have been recorded. The Doppler shift in these lines indicates that the companion has at least 3.2 solar masses – just enough to be a black hole. Again, we do not know the orbital tilt of the system; the 3.2-solar-mass result is a minimum mass that comes from assuming that we observe the system edge-on. A plausible estimate for the companion is 8 solar masses (Fig. 16.26b) – a pretty firm case for a black hole.

Not all binary x-ray sources contain black holes. The x-ray emission could arise from accretion disks around neutron stars. The evidence again relies on the mass – less than 3 solar masses – which implies a neutron star. An analogy: the model of radio pulsars as magnetic neutron stars. A flow of material from a companion (from a Roche lobe) forms an accretion disk. The x-ray pulses might arise from accreting matter channeled into the magnetic polar regions by the intense magnetic fields.

16.9 HIGH-ENERGY BURSTERS EVERYWHERE!

The bizarre objects paraded here have one aspect in common: all emit x-rays vigorously at some stage in their lives. Another surprise comes with the pageant – powerful x-ray bursts, increasing luminosities a hundredfold (Fig. 16.27a). These *x-ray bursters* pale compared to the power of *gamma-ray bursters,* whose flashes illuminate deep space.

X-Ray Bursters (Learning Outcome 16-13)

Some **x-ray bursters** flare up regularly every few hours or days. Others fire off in rapid sequence – thousands of bursts per day in a staccato pattern, like a cosmic machine gun. Some bursters reside in globular clusters, which imply that they involve solar mass old stars. Most collect along the plane of the Milky Way, which implies that they are local – within the Galaxy.

What's going on? One established model explains x-ray bursts as thermonuclear flashes from the surfaces of neutron stars. It pictures fresh hydrogen accreting on the neutron star from a companion in a binary system. This fuel ignites explosively as the gravitational energy converts to thermal energy, and the temperature hits the ignition point of hydrogen. Then, bang – an x-ray burst!

Sound familiar? It should: the model mimics that for nova explosions, with the white dwarf companion in a binary system replaced by a neutron star.

Gamma-Ray Bursters: Mystery Solved? (Learning Outcome 16-13)

Gamma-ray bursters flash as beacons more spectacular and bewildering than x-ray bursters. A typical burst lasts only a few seconds (Fig. 16.27b) and outshines all other gamma-ray sources in the sky.

Some gamma-ray satellites detected gamma-ray bursters. Typically, they observed about one burst per day and recorded a few thousands in total. Remarkably, the bursts appear evenly strewn on the sky, which implies that their sources are either very close ("local") or very far away ("cosmological").

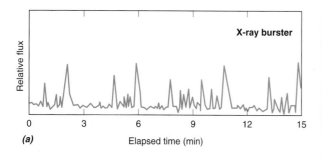

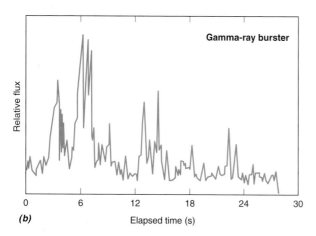

Figure 16.28 First optical observations of a gamma-ray burst (object enclosed by circle) right after the detection of the gamma rays.

Figure 16.27 High-energy celestial bursters. *(a)* x-ray bursts from the rapid x-ray burster called MXB 1730-335. These observations cover a time span of about 15 minutes; the interval between bursts is as short as 10 seconds. (Adapted from a diagram by W. Lewin and colleagues.) *(b)* The Pioneer Venus Orbiter observed a gamma-ray burster in 1978. Note the irregularity in the spacing and flux of the outbursts. (Adapted on a diagram by R. W. Klebesdal, W. D. Evans, E. E. Fenimore, J. G. Laros, and J. Terrell.)

But what are they? The answer largely depends on their distances. Resolving the mystery of gamma-ray bursters loomed as one outstanding problem of twentieth-century astrophysics. We now have tangible clues to solve that mystery.

The case broke open in January 1999 when a special wide-field camera in New Mexico, U.S.A., bagged the first optical images of a gamma-ray burst in progress (Fig. 16.28). Within 22 seconds of the burst detected by gamma-ray satellites, the robotic camera swung into position and caught the visible peak of the explosion. With the position pegged down, other telescopes zoomed in on the fading afterglow. The image seemed most like a supernova in a very distant galaxy. Spectra furnished a redshift of 1.6, which from Hubble's law implied a distance of some 9 Gly. For a few seconds, the burster shone with a power of 20×10^{12} solar luminosities – equivalent to about one solar mass converted wholly to gamma rays!

What process can power these bursts? The contending model pictures an exceptional blowup called a **hypernova** (to top a supernova). Here a rapidly spinning core of a very massive star implodes, as in a supernova. The implosion creates a black hole, and the spinning gives rise to an accretion disk. Jets shoot from the disk along the spin axis and punch through the star, blowing it apart. When they slam into the interstellar medium at close to the speed of light, the jets generate gamma rays. Millions of millions of years later, these gamma rays sweep past the earth. Recent evidence that gamma-ray bursters are associated with host galaxies that show active starbirth support this hypernova notion, as such massive stars would form in these regions.

Though still tentative, a hypernova model gives new meaning to the phrase "a blast from the past."

KEY CONCEPTS

1. The mass of a star at the time of its death determines the corpse it leaves behind (Table 16.2): white dwarf, up to 1.4 solar masses; neutron star, 1.4 to perhaps 3 solar masses; black hole, greater than 3 solar masses. (The actual neutron star limit may be as low as a little more than 2 solar masses.)

2. White dwarfs and neutron stars are supported against gravity by the degenerate gas pressure from electrons in a white dwarf and from neutrons in a neutron star. Gravity will never crush them any smaller; they will exist forever.

3. Novas occur in close binary systems when hydrogen from the companion falls onto the surface of a white dwarf from an accretion disk. This fresh fuel ignites explosively on the surface to produce a nova outburst.

4. Supernovas (Type II) are the explosions of massive stars (greater than 5 to 10 solar masses) that have evolved iron cores. Type I supernovas are the explosions of solar-mass stars in a binary system, perhaps involving white dwarfs close to the Chandrasekhar limit.

5. Supernova explosions create many of the elements heavier than iron and blast them into the interstellar medium (along with elements made in their normal lives). These newly minted nuclei enrich the matter between the stars.

6. The Crab Nebula is a supernova remnant, emitting light now by the synchrotron process. A pulsar in the center powers its current emission; that pulsar was formed in the supernova explo-sion of A.D. 1054, which involved a star of 10 solar masses.

7. Pulsars are rapidly rotating neutron stars. This idea is inferred most strongly from the regular timing of the fastest pulsars; only neutron stars are small enough and dense enough to rotate so rapidly. Pulsar emission comes from high-speed electrons in a neutron star's intense magnetic field – one example of synchrotron emission.

8. A black hole forms when a mass becomes so compacted that its escape speed is greater than the speed of light. Black holes are small: a 1-solar-mass black hole has a radius of 3 km; 2 solar masses, 6 km; 10 solar masses, 30 km; and so on.

9. Einstein's general relativity predicts that time appears frozen near a black hole and that a singularity resides in its center. Black holes can be seen only by their interaction with visible matter; an especially good circumstance would be a black hole–ordinary star binary system.

10. Cygnus X-1, in a binary system, is a good candidate for a black hole. It emits x-rays from a hot accretion disk around the suspected black hole, which has a mass of at least 6 solar masses. Most other binary x-ray sources contain neutron stars. Very few others are candidates for black holes.

11. X-ray bursters probably involve binary systems with at least one neutron star. Gamma-ray bursters may involve the disastrous collapse of massive, spinning stars in very distant galaxies. Whatever their specifics, they create the most extreme blasts in the cosmos.

Table 16.2 Deaths of Stars

Main-Sequence Mass (solar masses)	Normal Life	Death and Final Corpse
0.1–0.5	M stars	White dwarf
0.5–8	K-B0 stars	Planetary nebula, white dwarf
8–20	B0-B5 stars	Supernova, neutron star, or black hole
20–60	O stars	Supernova, black hole

STUDY EXERCISES

1. In a short paragraph, describe the primary characteristics of a white dwarf. (Learning Outcome 16-1)
2. In a short paragraph, describe to a friend who has not studied astronomy the chief features of a neutron star. (Learning Outcome 16-1)
3. What observational evidence do we have for the actual existence of neutron stars and white dwarfs? (Learning Outcomes 16-1, 16-2, and 16-3)
4. Look around you. Of the items you see, what would not be there if supernovas did not occur? (Learning Outcomes 16-6 and 16-10)
5. Assuming no loss of mass, what will be the final form of a
   a. 0.5-solar-mass star
   b. 2-solar-mass star?
   (Learning Outcome 16-1)
6. Make a list of the observational evidence that supports the idea of the Crab Nebula as a supernova remnant. (Learning Outcomes 16-6, 16-7, and 16-8)
7. In what way does a black hole practice censorship? (Learning Outcomes 16-9 and 16-12)
8. If a black hole is really black, how can it be an x-ray source? (Learning Outcome 16-9)
9. Why can't you find out what happens inside a black hole? (Learning Outcome 16-9)
10. What one feature of pulsars links them most strongly to rapidly rotating neutron stars? (Learning Outcome 16-3)
11. What is the source of the electromagnetic radiation in the pulses of pulsars? (Learning Outcome 16-3)
12. What are the main similarities of models of x-ray and gamma-ray bursters compared to those of novas and supernovas? (Learning Outcomes 16-4, 16-5, 16-6, 16-8, and 16-13)
13. How do we know that each type of stellar corpse exists? (Learning Outcomes 16-1 and 16-12)

PROBLEMS AND ACTIVITIES

1. Calculate the escape speed from the surface of a solar-mass neutron star with a radius of 10 km.
2. The Cassiopeia A supernova remnant has an angular diameter of 6.5 arcmin. What is its size-to-distance ratio? Its distance is some 9000 ly. What is its actual diameter?
3. Tycho's supernova expands at a radial speed of about 2000 km/s. This expansion is visible as a motion of the material of 0.2 arcsec/y. What is the distance to the remnant?
4. If our sun were reduced to the size of a neutron star, what would its average density be?
5. If your mass were reduced to its Schwarzschild radius, how large would it be?

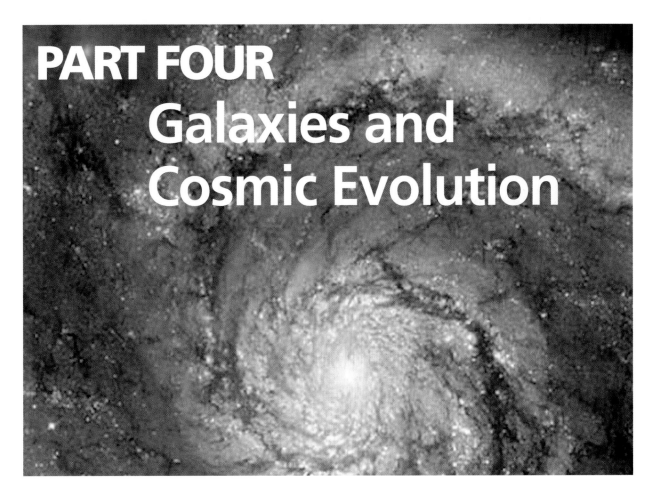

PART FOUR
Galaxies and Cosmic Evolution

PART OUTCOME

To describe the universe of galaxies, the cosmos itself, and to tie together the major themes of cosmic evolution in discussing the origin and history of the cosmos from birth to the present day.

INQUIRY FOCUS

How do we know:

The physical characteristics and layout of the Milky Way Galaxy?

The distances to and physical characteristics of other galaxies?

The layout of galaxies in space?

That the universe is expanding?

That dark matter exists in the cosmos?

The source of violence in the nuclei of galaxies?

That the cosmos began in a Big Bang?

The future fate of the universe?

17

The Evolution of the Galaxy

LEARNING OUTCOMES

After studying this chapter, you should be able to:

17-1 Explain at least one astronomical difficulty in trying to figure out the structure of the Galaxy from our location in it.

17-2 Name the important spiral-arm tracers and state generally how they are used to map spiral structure, with an emphasis on cepheid variable stars.

17-3 Present the observational evidence for the Galaxy's having a spiral structure; that is, describe what specific methods astronomers use to work out the positions of spiral arms.

17-4 Sketch the rotation curve of the Galaxy, describe how to find from it the approximate mass of the Galaxy, and argue that a significant amount of the Galaxy's mass must exist in the halo in an unseen form.

17-5 Describe the sun's location in the Galaxy and its orbit around the galactic center; explain the techniques used to find the distance and speed of this orbit.

17-6 Explain how radio astronomers used 21-cm-line and millimeter-line observations to trace spiral arms; indicate the limita-

tions of their method and explain its advantage over optical observations.

17-7 Describe the contents of a typical spiral arm and explain how these evolve.

17-8 Describe the evolution of spiral arms in terms of the density-wave model for spiral structure.

17-9 Outline a model for the evolution of the disk of the Galaxy.

17-10 Describe the current content of the Galaxy's halo, and outline a model for the evolution of the halo of the Galaxy.

17-11 Make two rough sketches of the entire Galaxy, from a top and a side view, labeling the disk, spiral arms, halo, globular clusters, and nucleus, and indicating the sun's position.

17-12 Describe what information radio, infrared, and x-ray observations provide about the nucleus of the Galaxy and make a case for the possible existence there of a supermassive black hole.

17-13 Speculate on the birth and the future of the Galaxy from current information and models.

CENTRAL CONCEPT

The evolution of the Milky Way, a spiral galaxy, is driven primarily by the evolution of the parts that make up its disk.

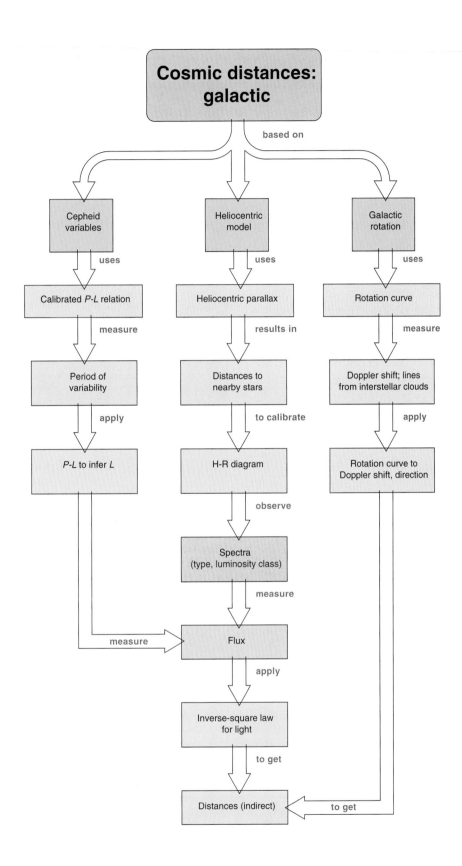

Like a majestic cosmic pinwheel, our Milky Way Galaxy spins slowly in space. Our sun orbits the center of it along with more than 10^{11} other stars. Because the sun is about 30,000 ly from the nucleus, it completes one revolution roughly every 250 million years.

What is the three-dimensional structure of this enormous system of stars? Optical astronomers probe the structure near the sun, and radio astronomers study regions farther away. The details are not yet in, but astronomers have been able to establish the broad outlines of the Galaxy's structure – a remarkable achievement, considering that we are buried within the Galaxy. These investigations show that the Galaxy does have a spiral pattern.

This overall spiral pattern of the Galaxy evolves. Theoretical models see the grand design as driven by waves; it links many of the galactic entities in the disk by the process of galactic evolution. The Galaxy's structure evolves – but at a rate so slow that we will not see any changes in our lifetimes.

The Galaxy looked very different in the past, especially just after its formation. The evolution of stars, interstellar matter, and the nucleus has driven significant changes. This chapter takes a brief excursion into galactic history to see how our spiral Galaxy evolved.

17.1 THE GALAXY'S OVERALL STRUCTURE (Learning Outcomes 17-1, 17-11)

Astronomers study the structure of the Galaxy to infer its past and predict its future. Our current model is pretty good. It pictures the Galaxy in three main parts: a central region called the nuclear bulge, a disk, and a halo (Fig. 17.1). The **disk,** the main body of the Galaxy, has a diameter of some 120,000 ly and a thickness of about 1000 ly. Population I stars, open clusters, and interstellar clouds of gas and dust inhabit the disk, the gas extending out farther than the stars. The sun resides a little above the disk at a distance of approximately 30,000 ly from the Galaxy's center. (See the Galactic Distances Celestial Navigator™.)

The **nuclear bulge** encases the central regions of the Galaxy, including the mysterious **nucleus,** the very heart of the Galaxy. The bulge is about 12,000 ly in diameter and 10,000 ly thick; it contains old Population I stars. Infrared observations from space clearly disclose the nuclear bulge (Fig. 17.2).

The spherical **halo** encircles the nuclear bulge and the disk. Globular clusters, containing Popula-

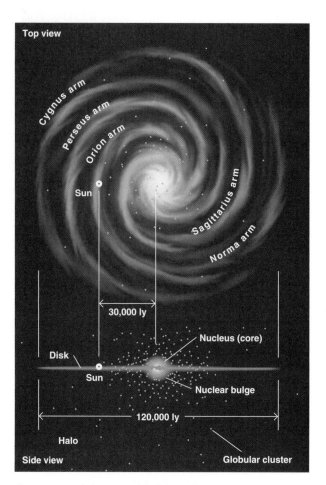

Figure 17.1 Schematic model of the Milky Way Galaxy showing its main features: nucleus, halo, and disk. The sun lies in the disk, about 30,000 ly from the center, on the inner edge of a spiral arm. The realm of the globular clusters defines the halo, shown here only in part. The nuclear bulge in the center surrounds the core.

tion II stars, make up the most obvious material in the halo, which has a diameter of some 300,000 ly.

Stars of different types inhabit each part of the Galaxy: young, metal-rich Population I stars in the disk; old, metal-rich Population I stars in the bulge; and very old, metal-poor Population II stars in the halo. You will find this distribution a key clue about the Galaxy's history.

We are in a bad position to observe the Galaxy's structure, for we reside in its dusty disk. Imagine, for example, that you are watching the half-time show at a U.S. football game. The band has set up an elaborate formation, and from the stands, you can easily see what the formation represents. But suppose you were down on the field instead, at the edge of the formation. It would at first appear to be a jumble! You could eventually figure out the shape

Figure 17.2 Wide-angle view of the central bulge and the disk of the Galaxy. The Cosmic Background Explorer (COBE) took this infrared image. The false colors display the emission at 1.2 μm (blue), 2.2 μm (green), and 3.4 μm (red). Stars in our Galaxy dominate the emission at these wavelengths. The region shown spans 168° across, or almost half the sky. Note the central bulge.

if you could find the distances to all the band players. You could make a map of their positions by plotting the distances and directions of each person. Now suppose that you had to do this mapping in a fog so dense that only the closest people were visible. Then you would need some other method of estimating distances – perhaps by the intensity of the sounds of the instruments that came through the fog. This is what it's like for astronomers in the disk to figure out what the Galaxy looks like!

We believe that the disk of the Galaxy contains a few spiral arms. A **spiral arm** extends for thousands of light years and contains many hot, young stars, gas, and dust with a density higher than in the region between arms. The spiral arms contain most of the gas, dust, and young stars in the Galaxy. Near

the sun, for example, about half the matter is in gas and dust, the rest in stars.

Optical astronomers, who try to find the distances to spiral arms, find their view blocked by interstellar dust. Radio waves get through. Radio astronomers can pick up the centimeter and millimeter radio emission from clouds of gas that mark spiral arms, but they have more difficulty in determining distances than the optical astronomers do. Although the results from the two techniques do not agree in all details, the following structure has been uncovered. We infer that our Galaxy also has two major arms (interior to the sun's orbit) and perhaps four arms (exterior to it) wound around the nucleus. The coherent, spiral pattern is clearest in these outer parts. Other galaxies also exhibit a spiral design (Fig. 17.3).

Figure 17.3 Spiral galaxy NGC 4414 with well-defined spiral structure, taken by the HST. Note the differences between the colors in the nucleus and those in the spiral arms. The yellow nucleus contains old stars; the bluish color of the arms comes from concentrations of young, hot stars. This galaxy lies about 60 Mly distant.

17.2 GALACTIC ROTATION: MATTER IN MOTION
(Learning Outcome 17-5)

Relative to nearby stars, the sun travels at a speed of about 20 km/s. These stars and the sun have roughly similar orbits. They are pretty much in the plane of the Galaxy and almost circular. But how to find out what motion the sun and nearby stars share in relation to the center of the Galaxy? And what about the orbital motions for matter in other parts of the Galaxy?

The Sun's Speed around the Galaxy
(Learning Outcomes 17-1, 17-5)

We use indirect approaches to find out how fast the sun moves around the Galaxy. One method utilizes the motions of globular clusters. They orbit the Galaxy in random orbits, with a roughly spherical distribution around the nucleus (Fig. 17.4). With respect to the nucleus, the average motion of all globular clusters is roughly zero. In other words, the system of globulars has no overall rotation about the galactic center (although individual clusters move rapidly, some in one direction, some in another, following Kepler's laws).

The radial velocities of the globular clusters found by Doppler shifts reveal the sun's motion with respect to the system of globulars. Now, the system of globulars has no rotational motion with respect to the nucleus, so the sun's motion relative to the system of globulars is its rotational motion with respect to the Galaxy's center. Such an analysis shows that the sun revolves at about 250 km/s.

The Sun's Distance from the Center
(Learning Outcome 17-5)

How far is the sun from the Galaxy's center? That's also a tough question to tackle because we cannot see the center optically. However, we can look above and below the galactic plane, where the obscuration is less, to observe objects thought to be symmetrical about the galactic center.

For example, we can use the RR Lyrae variables found in globular clusters. These are Population II stars; they have periods of light variation that typically are about a half-day in duration (Section 15.7). Whatever the period, RR Lyrae stars have essentially the *same* average luminosity: 50 solar luminosities. So once an RR Lyrae star is identified by the shape of its light curve, we can calculate its distance from measuring its flux and by applying the inverse-square law for light. By this technique, the sun's distance from the galactic center is approximately 28,000 ly, within a range of 24,000 to 33,000 ly. I will use 30,000 ly; but you must realize that the measured distance to the center is somewhat uncertain. Other observations denote a distance as small as 23,000 ly.

The American astronomer Harlow Shapley (1885–1972), around 1915, developed this technique of using RR Lyrae variable stars in globular clusters to find the distance of the sun from the Galaxy's center. Shapley's argument rests on the observation of the distribution of globular clusters, which concentrate in the southern sky (Fig. 17.5). From what vantage point do we, whirling around the sun, view these groups of stars? Shapley knew that our Galaxy has the shape of a flattened disk.

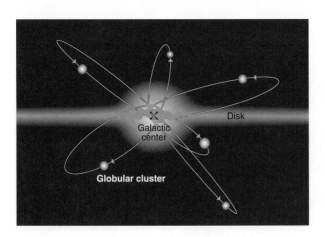

Figure 17.4 Schematic picture of the orbits of globular clusters around the galactic center. Each cluster moves on a highly eccentric orbit, so that most of its time it lies far from the nucleus, as expected from Kepler's Second Law. The far ends of these orbits define the spherical halo of the Galaxy.

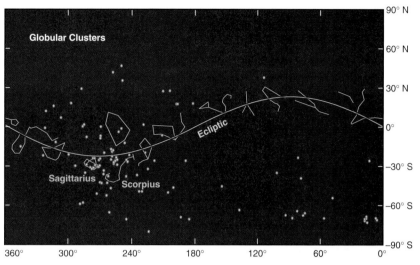

90° N
60° N
30° N
0°
−30° S
−60° S
−90° S

Globular Clusters

Ecliptic

Sagittarius   Scorpius

360°   300°   240°   180°   120°   60°   0°

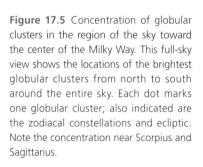

Figure 17.5 Concentration of globular clusters in the region of the sky toward the center of the Milky Way. This full-sky view shows the locations of the brightest globular clusters from north to south around the entire sky. Each dot marks one globular cluster; also indicated are the zodiacal constellations and ecliptic. Note the concentration near Scorpius and Sagittarius.

He assumed that the globular clusters had a uniform distribution around the Galaxy's nucleus in a huge sphere that outlines the halo.

Now to infer the size of the Galaxy and the sun's location. Suppose that the sun were located in the center of the Galaxy (Fig. 17.6a). Trace lines of sight in a number of directions. Because we have assumed a central vantage point in a uniform distribution of objects, every line of sight you choose should intercept the same number of globular clusters. The expected distribution should be uniform over the sky. But this uniformity does *not* exist.

Shapley therefore chose to abandon the idea that the sun was centrally located in the Galaxy (Fig. 17.6b). Now, some lines of sight cut longer distances than others through the globular clusters did, so more clusters are seen in these directions than along other lines of sight. So, instead of being symmetrical, the expected distribution is most

heavily concentrated in the direction of the galactic center. Allowing for the sun to be away from the center, the predicted result matched the observational one.

The distances to the globular clusters are established from RR Lyrae stars. Then the diameter of the sphere of globulars marks the extent of the halo (and so the size of the Galaxy), and the distance of the sun from the center of this sphere indicates the distance of the sun from the Galaxy's center.

Rotation Curve and the Galaxy's Mass

Knowing the sun's velocity and distance, we apply Newton's version of Kepler's third law to deduce the mass of the Galaxy (Enrichment Focus 17.1). The result, about 10^{11} solar masses, refers only to the mass interior to the sun's orbit. What about the mass outside it?

Figure 17.6 Position of the sun in the Galaxy inferred from the observed distribution of globular clusters in the sky. Assume that the clusters have a uniform distribution around the center of the Galaxy. *(a)* The situation if the sun were in the center of the Galaxy: you would see roughly the same number of globular clusters in every direction in the sky. *(b)* The actual situation, with the sun away from the center: more globular clusters are visible in the direction of the center than in other directions.

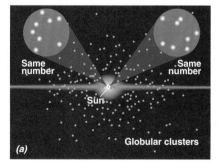
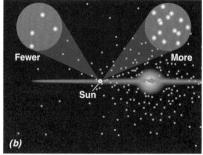

ENRICHMENT FOCUS 17.1

The Mass of the Galaxy

The sun swings around the Galaxy at 220 km/s at a distance of 30,000 ly from the center. Let's use this information to estimate the mass of the Galaxy. Assume that the sun moves in a circular orbit. Apply Kepler's third law in the form used for binary star systems (where $k = 1$) to the Galaxy:

$$M_1 + M_2 = \frac{R^3}{P^2}$$

where R must be in AU and P in years. The mass then comes out in solar masses. At 220 km/s it takes the sun about 2.8×10^8 years to complete a circuit of the

Galaxy. Now R is roughly 30,000 ly and 1 ly = 6.32 $\times 10^4$ AU; so 30,000 ly = 1.9×10^9 AU. Then

$$M_1 + M_2 = \frac{(1.9 \times 10^9)^3}{(2.8 \times 10^8)^2} = \frac{(6.9 \times 10^{27})}{(7.8 \times 10^{16})}$$
$$= 0.9 \times 10^{11} \text{ solar masses}$$

What is $M_1 + M_2$? It is the mass of the sun plus the mass of the Galaxy. The mass of the sun is so small compared with that of the Galaxy (just look at the result!) that we ignore it, so the Galaxy's mass interior to the sun's orbit is about 10^{11} solar masses. The flat rotation curve implies that much more lies beyond the sun's orbit. We need to observe farther out until we see a decline in the rotation curve. Then we can accurately calculate the total mass of the Galaxy.

With the sun's orbital distance and velocity known, we find the **galactic rotation curve** – how fast an object some distance from the galactic center revolves around it. The rotation curve tells us the overall distribution of matter in the Galaxy, because gravity controls those orbital motions.

Imagine that the Galaxy's mass is mostly concentrated in the nucleus – a layout that resembles the solar system. Although the planets exert mutual attraction, each orbits around the sun following Kepler's laws (Fig. 17.7). The stars in the galactic disk *should* revolve around the nucleus of the Galaxy much as the planets revolve around the sun. The stellar motions *should* follow Kepler's laws (Section 3.5), and the orbital speeds of the stars *should* decrease with increasing distance from the Galaxy's center. That's a lot of "shoulds."

In fact, the orbital speeds do *not* follow the

expected Keplerian decline (Fig. 17.8). That exposes an important fact about the Galaxy: the major part of the mass is not concentrated at the center! From close to the center out to 1000 ly, the curve rises steeply, then drops, bottoming out at about 10,000 ly. It then rises slowly out to the position of the sun, then drops again. In the outer parts of the Galaxy, the curve rises beyond the sun's orbit. It then appears to flatten out at a distance of 50,000 ly from the galactic center. If the motions followed Kepler's laws, we would expect the rotation curve to decline beyond the sun's orbit, and it does *not*.

What does this rotation curve say? Much of the Galaxy's material must lie out beyond the sun's orbit. From the rotation curve out to 60,000 ly, the Galaxy's mass is 3.4×10^{11} solar masses. Other estimates give close to 10^{12} solar masses. At least as much mass lies exterior to the sun as interior to it.

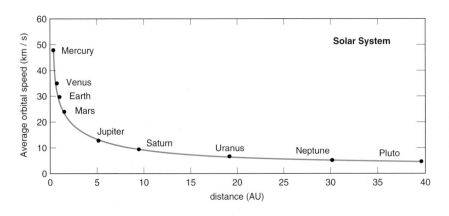

Figure 17.7 Solar system rotation curve for the planets. This is Keplerian rotation, because almost all the mass of the solar system is concentrated in its center – the sun. Note how rapidly the orbital speeds *decrease* with distance.

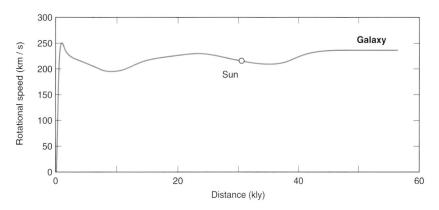

Figure 17.8 Galactic rotation curve, based on a combination of carbon monoxide and hydrogen observations, assuming that the sun's speed is 220 km/s and its distance from the galactic center is 30,000 ly. Note that the rotational speeds do not follow a downward trend outside the sun's orbit but reach a constant value. (Adapted from a diagram by D. P. Clemens.)

17.3 GALACTIC STRUCTURE FROM OPTICAL OBSERVATIONS

Photos in blue light taken of nearby spiral galaxies, such as Messier 31 (the Andromeda Galaxy) show the bluish spiral arms distinctly. This appearance results from bluish supergiants, H II regions, and Population I cepheids that lie in the spiral arms. These objects are called **spiral tracers.**

Spiral Tracers and Spiral Structure (Learning Outcome 17-2)

How to apply these spiral tracers to our Galaxy to determine the layout of its spiral arms near the sun? It's not a simple operation! First, it requires an accurate and reliable technique for measuring the distance to each of the tracers. Second, the blotting out of starlight restricts optical observations by dust: since most of the interstellar dust lies concentrated in the galactic plane, the sun sits in the thick of the interstellar smog. Third, the sun's location in the plane gives us a poor vantage point for seeing the Galaxy's spiral structure because we are forced to observe it edge-on rather than face-on.

Let's see, for example, how cepheid variables have been used to delineate spiral features. The cepheids have the advantage that their distances are easy to determine by the **period–luminosity relationship.** This crucial distance-measuring technique is so important that you should know how it works. (It will appear in later chapters.)

A variable star is one whose luminosity (and so flux) changes with time. A light curve is a graph of the change in a star's flux with time. A star whose light varies in a regular fashion is known as a *peri-*

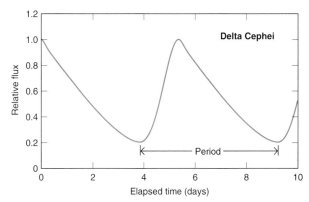

Figure 17.9 Light curve for a typical cepheid variable star – Delta Cephei, the prototype of the class – shows how the star's brightness varies with time. Note that the rise in brightness is steep, but the decline less so. The period of this wavelike brightness variation is about 5.4 days.

odic variable; Delta Cephei sets the standard for one such class of variables – called *cepheid variables* or *cepheids* (Section 15.7) – by the special shape of their light curves (Fig. 17.9). From Doppler shift observations, we know that cepheids expand and contract as they vary in luminosity.

In 1912 the American astronomer Henrietta S. Leavitt (1868–1921) noted a number of cepheid variable stars in the Small Magellanic Cloud, a companion galaxy to the Milky Way. At the time, the distance to this galaxy was not known, so Leavitt could not find out the luminosities of these stars. She did recognize that their periods were related to their average fluxes. Because these stars all lie at the same distance from the earth, their luminosities must be related to their periods. For the cepheid variables, Leavitt found that a special relationship, the **period–luminosity relationship,** connects the luminosity of a cepheid variable to its period (Fig. 17.10). Basically, it states that *the longer the period of light variation of a cepheid, the more luminous it is.*

Here's how to use the relationship to find distances to cepheids.

1. Find a cepheid (identifying it by its light curve).
2. Measure its period of light variation from peak to peak.
3. Find the star's average luminosity from the period–luminosity relationship.
4. Measure the star's flux by telescopic observations.
5. Calculate its distance from the inverse-square law for light. You have the flux and luminosity, so you can find the distance.

Today we know that the stars that used to be lumped together as cepheid variables are actually three types: Type I (classical) cepheids, Type II cepheids, and RR Lyrae stars. Type I cepheids are Population I stars, Type II belong to Population II. RR Lyrae stars, also Population II, are commonly found in globular clusters. (Shapley used these and the technique described here to get the distances to globulars: Section 17.2.)

Optical Maps of Spiral Structure (Learning Outcomes 17-3, 17-4)

The optical maps of spiral structure must be viewed with caution because interstellar dust restricts where and how far we can look from the sun and how well we can estimate distances to optical tracers. Dust decreases the measured flux, and we would estimate a larger distance than the actual distance. Although the outline of spiral structure is likely to be correct, observations of other galaxies show that irregularities commonly occur. It's futile to draw a master map from optical data alone, which works out to distances of only some 15,000 ly.

Despite discords about the details, most optical astronomers concur on at least three major arm segments spaced about 7000 ly apart. The Galaxy appears to have a spiral structure with much irregularity in the general pattern, which may consist of two or four spiral arms – but we cannot tell for certain yet which is correct.

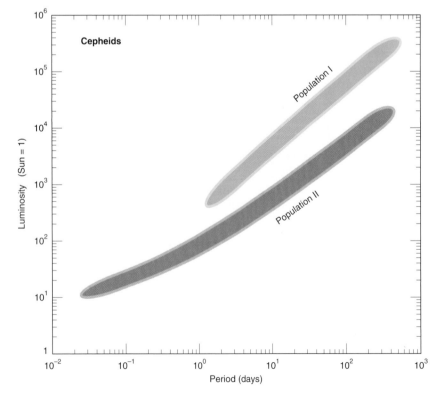

Figure 17.10 Simplified period–luminosity relation for cepheids, which fall into two groups, Types I and II, based on stellar population. In both cases, the general trend is the same: the longer the period, the more luminous the star. By comparing the observed flux to the luminosity estimated from the period, you can use the inverse-square law for light to find the distance to the cepheid.

17.4 EXPLORING GALACTIC STRUCTURE BY RADIO ASTRONOMY
(Learning Outcome 17-6)

The prime drawback to optical mapping of the Galaxy is obscuration from interstellar dust. Radio observations do not have this handicap because dust does not stop radio waves. The 21-cm line from hydrogen (Section 14.1), which comes from the concentrations of neutral hydrogen clouds in the spiral arms, is useful as a spiral tracer. But the best radio tracer seems to be the carbon monoxide millimeter-line emission from giant molecular clouds.

Radial Velocities and Rotation

To distinguish among spiral arms, we look in *different directions* and at *different velocities* in the same direction. The 21-cm radiation from H I clouds arrives at the earth Doppler-shifted to different wavelengths because of the different velocities of the hydrogen gas clouds. These differences in velocities come mostly from the rotation of the Galaxy. So if we know how the Galaxy rotates (and that's the tricky part), we can translate 21-cm observations into a map of spiral structure. The same method applies to emission from carbon monoxide (and most other molecules).

Here's a case looking outward from the sun (Fig. 17.11). Our line of sight intercepts two nearby clouds at successively greater distances. The inner cloud travels faster than the outer one. In its revolution around the Galaxy, the sun is approaching both clouds. So the difference between the sun's velocity and the innermost cloud's velocity is the least, and its Doppler shift is the least. In contrast, the difference between the sun's velocity and velocity of the outer cloud is the most, so it has the greatest Doppler shift. If each of these clouds corresponds to a piece of a spiral arm, how to arrange them at different distances from the sun?

Assume circular orbits. We know the sun's orbital speed around the Galaxy. When we observe 21-cm emission, we know the direction in which we are looking and can measure a radial velocity. From that radial velocity, we infer the rotational speed of the cloud. We then look up this speed on the galactic rotation curve to find the distance to which it corresponds.

In essence, the galactic rotation curve tells us that at a given position in the sky, a certain radial

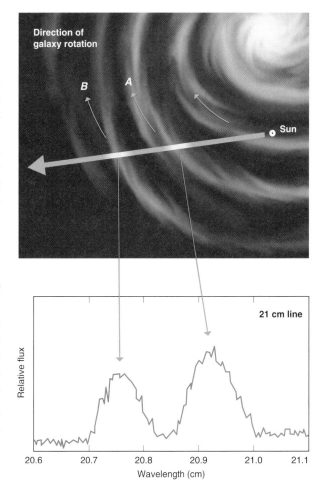

Figure 17.11 Using 21-cm emission to trace out spiral structure. Consider looking out of the Galaxy through two spiral arms (*A* and *B*). The sun is overtaking *A* and *B*. Then the radial velocity difference between us and *A* is smaller than that between us and *B*, and the blueshift from *A* is less than that from *B*. The emission from both is blueshifted to wavelengths shorter than 21.11 cm.

velocity corresponds to a specific distance from the sun. Scanning around the plane of the Galaxy, we make a series of Doppler shift observations. These can be connected to trace a spiral arm and so outline the Galaxy's structure.

Radio Maps of Spiral Structure

Different investigators have drawn conflicting radio maps. The heart of the problem: the gas clouds do *not* follow the simple scheme of circular rotation. In addition to their circular motion, they have their own random motions. Unfortunately, such random motions lead to incorrect distances and so disrupt the unity of the spiral-arm map.

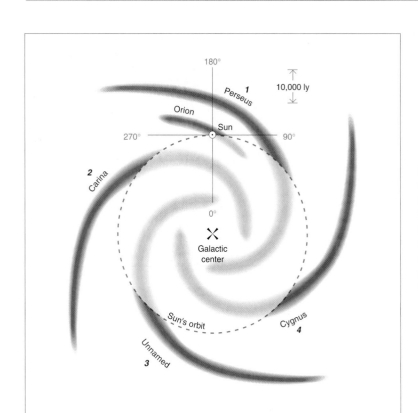

Figure 17.12 One model for the overall spiral structure of the Galaxy. The circle indicates the sun's orbit. The numbers around the cross centered on the sun indicate the galactic longitude as viewed from the sun, with 0° in the direction of the galactic center. Note the pattern of four arms (labeled 1, 2, 3, and 4) and the distance scale. (Adapted from a figure by L. Blitz, M. Fich, and S. Kulkarni.)

Despite such problems, the radio maps do hint at a large-scale spiral structure, with four arms overall (Fig. 17.12). The Carina arm stands out most clearly. It stretches for more than 80,000 ly, with a giant molecular cloud roughly every 2000 ly along the segment. Overall, the maps from the giant molecular clouds probably provide the best portrayal of the Galaxy's spiral pattern.

17.5 THE EVOLUTION OF SPIRAL STRUCTURE
(Learning Outcome 17-7)

For years astronomers supposed that the spiral arms in our Galaxy and others were material arms, a coherent bunch of objects – stars, nebulas, gas, dust – somehow physically held together. Such a viewpoint faced hard questions: What holds the material in an arm together? How does an arm persist for a long time? The answers proved the idea wrong!

The Windup Problem

The persistence of an arm is a puzzle, because it should wind up. Why? Some parts of the arm rotate more slowly than others. After a few rota-

tions of the Galaxy, the arms should have disappeared.

An analogy to this windup problem: imagine that you and two friends are going to run around a track. You station yourself in the middle, one friend a few meters in from you, closer to the track's center, and the other a few meters out from you. You start running, lined up. Now insist that the friend inside run around faster than you and the one outside slower. You can guess what will happen after one or two laps: the lineup will be disrupted. If spiral arms were material arms, the same thing would happen to them after a few rotations.

Astronomers have been struck by the stability of spiral arms. Not from our Galaxy alone; it could be that we are observing at a very special time, right after the formation of the arms. But that cannot be true for all visible galaxies. Of the brightest galaxies in the sky, more than 60 percent have well-defined spiral arms.

How does this tell us that spiral arms endure? Recall that to look out in space is to look back in time. Suppose that a galaxy's spiral structure did in fact disappear by winding up after a few rotations. Assume, too, that all galaxies formed at the same time. Then we should not see spiral structure in nearby galaxies, but we should see it in more distant galaxies. Not the case! Both nearby and distant galaxies exhibit spiral structure. Seeing so many galaxies as spiral in shape implies that the structure lasts for many years or that it is regularly renewed.

The Density-Wave Model
(Learning Outcomes 17-2, 17-7, 17-8)

How to explain this persistence? The current model pictures spiral arms not as material arms at all but rather as the result of spiral waves of higher density moving through the Galaxy's disk. These waves produce all the signposts of a spiral arm – young stars, H II regions, lanes of dust. None of these objects lasts very long. As they die and the density wave moves on, new spiral-arm tracers are born from the interstellar medium at the new location of the wave.

So a spiral arm always contains objects of the *same kinds*, but not the *same objects*. Any particular arm is a fleeting phenomenon. Individual objects revolve at the speed determined by gravitational forces appropriate for their distance from the center, but the wave pattern rotates with a constant angular speed and does not wind up. This approach is called the **density-wave model** of spiral structure. Of the models proposed to date, it best describes the overall scheme.

What's a density wave? A sound wave is a density wave. Clap your hands. You push against air molecules to compress them together. This first group of molecules bangs into adjacent ones in the direction of their motion, which transfers the squeezing to the next bunch of molecules. As this compression (a density wave!) travels outward, it leaves behind a trough of lower density. Two important points here: a sound wave requires a source to start it, and the high-density part of the wave persists even though the specific particles that make it up change at different points in the medium.

An analogy: suppose that you are driving on a heavily traveled mountain road (Fig. 17.13). Everyone moves along happily at the speed limit. Ahead, an overloaded truck can go about half the maximum speed. Cars jam up just behind the slow-moving truck as the drivers wait for a clear road ahead in order to pass. When they do pass, they resume moving along at the speed limit, leaving the poor trucker behind. Imagine that you watched this situation from the air and concentrated on the motion of the cars. You would see a denser region of traffic just behind the truck, where the cars pile up for a short time. (The truck moves down the road at half the average speed of the cars.) You would also note that the jam-up persists, even though it does not contain the same cars. New cars get caught up in it as other cars move out.

You can think of the cars as the stars, the interstellar medium in the galactic plane, and the jam-up as the visible effect of a moving density wave (the truck). The jam-up creates a region in the disk of increased density of stars and gas – a spiral arm.

This density-wave idea assumes that a two- or four-armed spiral density wave sweeps through the galactic plane. Although the origin of this wave is not explained, once formed, it persists for gigayears. The gas in the disk piles up at the back of the wave. The compression squeezes small molecular clouds together to form giant molecular cloud complexes, which in turn form young stars and H II regions. The compression of the interstellar medium by the density wave forms the features associated with a spiral arm.

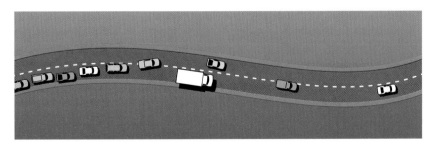

Figure 17.13 A traffic bottleneck as an analogy to a density wave. The slow-moving truck has cars jammed up behind it waiting to pass. The blockage consists of different cars at different times, but it is always behind the truck. Viewed from above, the jammed-up area moves more slowly than the average speed of the cars; it is a density wave.

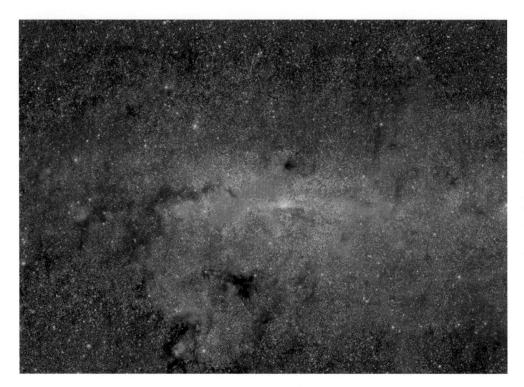

Figure 17.14 Infrared view of the galactic center. The red areas represent thick interstellar and glowing dust and the blue areas represent thinner dust through which distant red giants are visible. Note the bright area marking the nucleus.

During the short lifetimes of the newly formed massive, hot stars, a density wave moves only a short distance, so these stars, while they last, mark the spiral arm clearly. As a density wave moves on, it provokes the formation of more stars. These take the place of the ones that had rapidly faded out. The spiral arms are maintained by density waves. They generate a spiral wave pattern by making regions of higher gravitational forces and so higher densities of matter in the Galaxy's disk.

CAUTION. Just because I have been emphasizing luminous, hot stars (they *are* good tracers of spiral arms), don't be misled into thinking that a density wave initiates the formation of only such stars. It probably prompts the formation of stars with all possible masses. The massive ones die quickly compared with the less massive ones, which are far more numerous. The less massive stars remain between the spiral arms after the density wave has moved on, but they are less conspicuous because of their low luminosities.

Status of the Density-Wave Model (Learning Outcome 17-8)

How well does the density-wave model work? First, it outlines the grand scheme of an overall spiral pattern that we observe in other galaxies and in our own. About half of all spirals show this grand design. Second, it explains the persistence of the spiral arms in spite of galactic rotation. Third, it predicts the general features of a spiral arm. The density-wave model succeeds fairly well in explaining the prominent features of spiral structure.

However, the model so far falls flat on key points. It does not explain the origin of the density waves. Nor does it clearly work out what sustains them. As the density waves ripple through the interstellar medium, they lose energy and should dissipate in about a billion years. As evidenced by the abundance of spiral galaxies, however, the density waves – if they are the correct explanation – last longer than gigayears or are renewed several times as the old waves fade out. Some mechanism keeps supplying energy to maintain them or to trigger a new series of density waves.

TO SUM UP. Observations and computer simulations leave little doubt that spiral density waves exist and indeed are fairly common, but we do not yet know how they come into being. This is a puzzle to be solved in the study of our Galaxy and others.

17.6 THE HEART OF THE GALAXY

Although dust largely obscures the center of the Galaxy, a few regions of low absorption open up optical glimpses of the nuclear region (Fig. 17.14). In addition, we can surmise the nature of the stars in the nucleus from observations of the nuclei of other galaxies. Observations of these two kinds imply that the nucleus contains mostly old Population I stars densely packed together. So jammed are these stars that if you lived in the nucleus, the nighttime sky would be as bright as twilight on the earth. In fact, the total power output of the nucleus is about one billion solar luminosities!

Radio, infrared, and x-ray observations can all probe the nucleus. They show that the heart of the Galaxy is a bizarre and active place. Motions of gas here suggest a high concentration of mass at the very center – perhaps a supermassive black hole. In the very center of the Galaxy lies a radio source less than 13 AU (about 2 light hours) in size.

Radio Observations (Learning Outcome 17-12)

Let's look first at the continuous emission of the galactic center (Fig. 17.15). An intense radio source lies smack in the direction of the center. It is called Sagittarius A (Sgr A for short). Clustered around Sgr A – and all lying more or less along the galactic plane – are radio sources in a string. Some sources appear to have characteristics of H II regions: hot ionized gas around young hot stars. The ultraviolet energy output from the massive stars needed to keep these regions ionized is at least 5 million solar luminosities.

Other radio emission here is from ionized gas, but some also comes from high-energy electrons traveling through a magnetic field – synchrotron emission (review Enrichment Focus 16.1). That implies that these sources are supernova remnants – just what you would expect in a region containing massive stars.

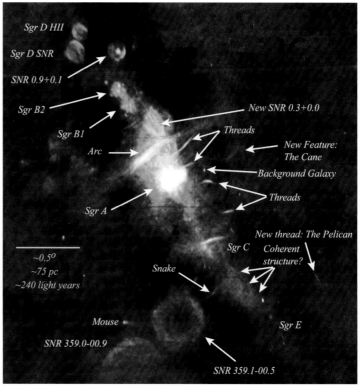

Figure 17.15 Wide-angle radio view of the galactic center at 330 MHz (wavelength 0.90 m). Note the scale at the bottom left. "SNR" indicates a supernova remnant. Objects with numbers indicate their location along the galactic plane and somewhat above or below it. Sagittarius A (Sgr A) appears in the center. The concentration of sources on a diagonal line show the disk of the Galaxy viewed edge-on.

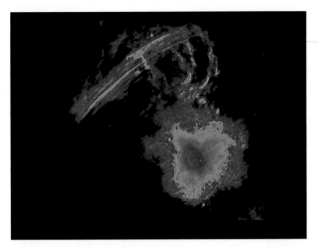

Figure 17.16 Radio observations of Sgr A at a wavelength of 20 cm (1.44 GHz), taken by the VLA telescope near Socorro, New Mexico, U.S.A., from 1982 to 1984. This false-color image shows the halo surrounding Sgr A, about 70 ly across. The many filaments have widths of about 3 ly. Note the spiral-like structure of Sgr A.

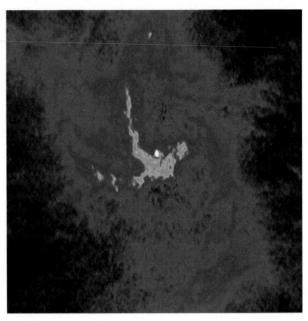

Figure 17.17 Close-up false-color image of the galactic center by the VLA at 6-cm wavelength. This map shows thermal emission. Note the spiraling structures emerging from Sgr A*.

High-resolution radio maps show that Sgr A actually consists of two separate radio sources: one, called Sgr A East, emits by the synchrotron process; the other, Sgr A West, seems to be more like a giant H II region. Sgr A West is associated with an agglomeration of infrared sources. Within Sgr A West lies a compact radio source, smaller than 13 AU, which appears to mark the actual core of the Galaxy; it is called Sgr A*. It shows synchrotron emission with a variable power of some 10^{28} W. Overall, the ionized gas here, which amounts to a few million solar masses of material, rotates at about a few hundred kilometers per second.

Also eye-catching are a filamentary arc, threads of material, and coherent structures with puzzling shapes. The arc may consist of material extending from Sgr A and guided in loops by local magnetic fields. A high-resolution map of the inner 10 ly of the Galaxy (Figs. 17.16 and 17.17) reveals emission that has a spiral shape (not to be confused with spiral arms in the disk!).

Sgr A also gives off radio-line emission from molecules such as carbon monoxide. The observations indicate that this molecular cloud, which may contain as much as a million solar masses of material, is in front of Sgr A East. The gas and dust appear to form a ring about 15 ly in diameter.

Infrared Observations (Learning Outcome 17-12)

Early infrared observations showed that the galactic center region emits strongly at 2.2 μm. The most intense part of this emission coincides with Sgr A. What is the source of this radiation? Simply the combined 2.2-μm emission from all the old Population I stars that inhabit the galactic nucleus. High-resolution observations have revealed many sources packed together around the nucleus.

Some of the heating radiation comes from the old Population I stars. Some also derives from newly formed massive stars. Their combined luminosity, in the range from 2 to 20 μm, is roughly a million times that of the sun. Over the entire infrared range, the galactic center emits about 100 million times the sun's luminosity!

X-Rays (Learning Outcome 17-12)

The galactic center emits x-rays in the form of an extended x-ray source about 2° in size and point-like, short-lived sources very close to the galactic center. Some of these sources give off bursts of x-rays, with the amount of energy in each burst comparable to that from the short-lived sources. To date it is not clear how or indeed whether these sources relate to each other.

X-ray images show the galactic center in its high-energy glory (Fig. 17.18). Within 300 ly of the center, the x-ray emission is modest in strength. It consists of a complex of weak sources (covering some 200 ly by 150 ly) embedded in a halo of weaker, diffuse emission. About a dozen sources are here. One coincides with Sgr A West; it has an x-ray luminosity of 10^{28} W. The other sources lie along the ridge in the same location as the cluster of infrared sources.

The Inner 30 Light Years
(Learning Outcome 17-12)

We now have a pretty good overall idea of the distribution of gas and dust within 30 ly of the Galaxy's center (Fig. 17.19). The bulk of the continuous radio emission comes from Sgr A West in a filamentary structure that shows a western arc and a northern arm centered on the radio point source Sgr A*, which lies very close to or exactly at the galactic center. The far-infrared observations point to a disk of neutral gas and dust some 30 ly across. The inner few light years of this disk have a lower density than its average. The dusty disk lies pretty much in the plane of the Galaxy. The two lobes that mark the inner edge of this disk appear on either side of the ionized region. The disk, which contains hydrogen molecules at some 2000 K heated by shocks, rotates with its axis lined up to that of the general galactic rotation.

Does a Black Hole Lurk in the Core?
(Learning Outcome 17-12)

A puzzle generated by the radio and infrared line observations arises from the rapid rotational motions near the Galaxy's core – the rotational velocities increase closer to the core, as shown by their Doppler shifts.

Why is that a problem? Well, the rotational velocities are so high that a huge concentration of mass is needed to hold the speedy gas together. For example, if the Galaxy's core simply contained a cluster of stars, you would expect the rotational

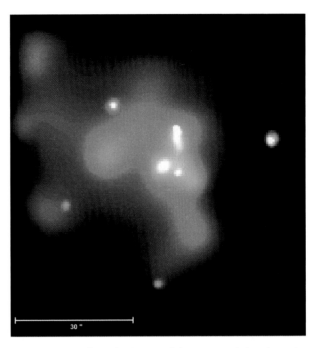

Figure 17.18 False-color image of the x-ray emission from the galactic center. Sgr A* is the largest white dot at the center of the picture. The hot gas cloud surrounding it has been heated by shock waves produced by supernova explosions.

velocities to decrease toward the core, because as you get closer in, you have less and less mass to bind the moving materials gravitationally. To account for the rapid rotation requires a mass in the core of 4 million solar masses – all lumped together in a region only 0.1 ly in diameter!

What form might this mass have? One possibility is that it is locked up in a supermassive black hole. If it were in the form of, say, a cluster of solar-mass stars, these stars would be located, on the average, only 1 to 2 AU from each other. Some million solar masses of material could be jammed in such a star cluster in the nucleus – and the more in the form of stars, the less mass needed in a black hole. The dynamics of the nucleus do not require a supermassive black hole; one of moderate mass, say a thousand solar masses, could also explain the motions there.

The idea that a supermassive black hole lurks in the heart of the Galaxy has yet to be confirmed. Indirect support for the idea comes from observations that some other galaxies do have a similar mass concentration in their nuclear region. (More in Chapter 18.)

TO RECAP. Infrared, radio, and x-ray telescopes can probe the galactic center directly. They have shown that although the nucleus is very small, it emits enormous amounts of energy, about 10 percent of the total from the Galaxy. Within it, material orbits the center at a rapid rate. At the very center lies a massive object, perhaps a black hole. If it does exist, such a supermassive black hole can explain the observed radio and infrared emission and the total luminosity of the galactic center region.

17.7 THE HALO OF THE GALAXY
(Learning Outcomes 17-10, 17-11)

The globular clusters outline the halo around the Galaxy (Fig. 17.20). Little else is visible there. Stray stars are sometimes seen, as well as a bit of hot, ionized gas. The halo may contain as-yet-undetected objects, such as very faint low-mass stars. For sure it embraces dark matter.

Globular Clusters

To review briefly: A globular cluster has a spherical shape (some tens to hundreds of light years in diameter) and contains up to a million Population II stars, each with a little less than 1 solar mass. The globular clusters form a spherical distribution around the Galaxy's center. Some elliptical orbits bring them out to extreme distances of 300,000 ly from the Galaxy's nucleus. The clusters orbit at speeds about 100 km/s, diving into and shooting out of the disk.

The outer halo of the Galaxy has no exact boundary. Observations of the placement of globular clusters indicate that the halo extends out to some 300,000 ly, far beyond the limits of the Galaxy's disk and the Magellanic Clouds, two companion galaxies that are gravitationally bound to the Galaxy.

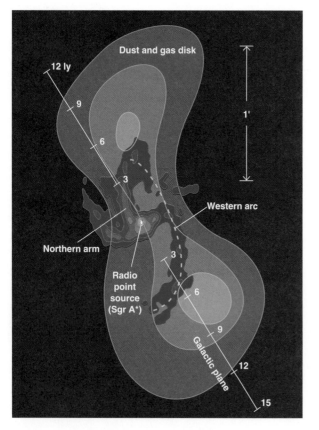

Figure 17.19 Schematic diagram of the inner 3 ly of the Galaxy. The galactic plane is marked by the diagonal line; the scale for 1 arcmin lies vertically at the right. The radio contours show the point source at the nucleus (Sgr A*), the northern arm, and the western arc of Sgr A. The dashed lines outline the disk of gas and dust, which is tilted to our line of sight. (Adapted from a diagram by M. K. Crawford and colleagues.)

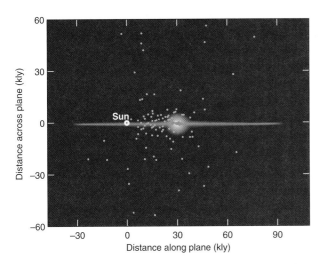

Figure 17.20 Distribution of some globular clusters about the plane of the Galaxy. The view is edge-on. The sun's position is indicated. (Adapted from a diagram by W. E. Harris.)

Other Material in the Halo
(Learning Outcome 17-4)

The rotation curve discloses that the halo contains considerable dark matter, perhaps much more mass than is in the rest of the Galaxy. This conclusion rests on the fact that the rotation curve flattens out at distances far from the center of the Galaxy. More mass is out there, but its form is not obvious. Using other spiral galaxies as a guide, the mass in the halo may be a few times greater than the total mass known so far. What is this stuff?

The halo may contain a large number of low-mass, faint red stars that are difficult to observe directly. Observations of a few other nearby galaxies of the same type as the Milky Way imply that they may have extensive, massive halos of faint red stars.

The halo also contains gas, but much less than the disk of the Galaxy. Observations at 21 cm show hydrogen clouds traveling with high speeds above and below the galactic plane. Most of the halo's gas, however, is probably ionized hydrogen. This gas could come from the disk, blown out by supernova explosions, expanding H II regions, and stellar winds. So the halo may be fairly hot and expanding into intergalactic space.

The halo may also contain other objects, currently unobservable. Astronomers give the generic name **dark matter** to this invisible material. Some might be in the form of tiny stars (such as brown dwarfs) and planetary masses. A generic name for these is *MACHOs – massive compact halo objects*. If a MACHO passes in front of a distant star, gravitational lensing by the MACHO causes the starlight to brighten then dim in a speical way. A few such MACHO events have been seen. They imply that MACHOs make up only 20 percent of the halo's mass – still room for ample dark matter.

Finally, a massive halo can help support the density-wave model for the spiral structure in the disk. The mass in the halo could stabilize the arms gravitationally and ensure that the density waves propagate for long periods of time (gigayears). A massive, invisible halo may well be the necessary ingredient to lock in a well-defined spiral structure in a galaxy.

17.8 A HISTORY OF OUR GALAXY
(Learning Outcome 17-13)

We have a pretty fair idea of the architecture of the Milky Way Galaxy. What clues does this information provide about its birth and evolution? The crucial clues come from two sources: the chemical composition of galactic material and its dynamics.

The process of galactic evolution links the chemistry with the dynamics. This linkage marks an important theme of cosmic evolution, because the evolution of the Galaxy results from the evolution of all the stuff that makes it up.

Populations and Positions
(Learning Outcomes 17-9, 17-10)

Population I and Population II stars differ considerably in their abundances of heavy elements. In general, Population II stars contain about 1 percent of the metal abundance of Population I stars.

In fact, however, we do not find a stark and simple division of metal abundances into just two groups. Rather, we find a continuous range of abundances, from about 3 percent to less than 0.1 percent for the ratio of metals to hydrogen; so, though the division into two populations is a useful tool, a continuous range of populations actually exists.

When these populations are cataloged by metal abundance, we find a striking correlation with average distance from the Galaxy's disk. The lower the metal abundance, the farther the objects are found from the plane of the disk.

To interpret this key observation, we rely on basic concepts of starbirth, stardeath, and the recycling of the interstellar medium (Chapters 14 and 16). First, stars are born from clouds in the interstellar medium. Their atmospheric elemental abundance reflects that of the gas from which they formed. Second, stars inherit the orbital motions about the Galaxy of their parent gas and dust clouds. Third, massive stars evolve quickly and spew back into the interstellar medium material that is enriched by heavy elements. As long as new stars – especially massive ones – are born, the abundance of heavy elements in the interstellar medium gradually increases as the Galaxy ages.

With these basics in mind, let's try to make sense of the observations. They show that the youngest objects (highest heavy-element abundances) hug close to the disk, whereas the oldest objects (lowest heavy-element abundances) range far from the disk. Other objects fall in between these extremes. The halo of the Galaxy is its oldest part, and the spiral arms its youngest.

We can estimate the Galaxy's age by finding the oldest stars in the halo. A comparison of theoretical models for globular cluster stars with H–R diagrams for them (Section 15.7) indicates an age of 13 Gy (with an error range of about 2 Gy). That's when the Galaxy formed. Let's see how it formed.

The Birth of the Galaxy
(Learning Outcome 17-13)

Because globular clusters contain the oldest stars associated with the Galaxy, the halo marks the fossil remains of its birth. Within it, globulars orbit the Galaxy on extremely elongated elliptical paths. Most of the time, the globulars move slowly through the halo at the outer extremes of their orbits; only briefly do they whip in and around the nucleus. These stars exhibit the motions of the cloud from which they were formed. So the Galaxy must have been born from a gas cloud that was initially huge – at least 300,000 ly in radius.

Here's one model of the Galaxy's birth. Imagine a tremendous, ragged cloud of gas roughly twice as big as the Galaxy's halo today (Fig. 17.21). Its density is low. This proto-Galaxy cloud probably is turbulent, swirling around with random churning currents. Slowly at first, the cloud's self-gravity pulls it together, with its central regions getting denser faster than its outer parts. Throughout the cloud, turbulent eddies of different sizes form, break up, and die away. Eventually, the eddies become dense enough to contain sufficient mass to hold themselves together. These might be hundreds of light years in size – incipient globular clusters. Each blob then splits up to form individual stars – all born at about the same time (15 Gy ago).

Meanwhile, the gas contracts more and falls slowly into a disk. (Sound familiar? Recall the formation of the solar nebula and starbirth.) Why a disk? Because the original cloud had a little spin, and the conservation of angular momentum (Enrichment Focus 11.1) requires that it spin faster around its rotational axis as it contracts. The kinetic energy of the cloud slowly decreases, as gas clouds collide and heat is radiated away. The disk rapidly flattens.

As the disk forms, its density increases, and more stars form. Each burst of starbirth leaves behind representative stars at different distances from the present disk. Finally, the remaining gas and dust settle into the narrow layer we see today. Somehow density waves appear and drive the formation of spiral arms.

During this time, massive stars were manufacturing heavy elements and flinging them back into the cloud by supernova explosions. So as stars were born in succession, each later type had more heavy elements. That enrichment continues today in the disk of the Galaxy.

What of the Galaxy's future? Let's speculate a bit. If we assume that no new gas is added from outside the Milky Way, stellar evolution points to a day when most stars become corpses. Matter that once made up the interstellar medium will be locked up for good. The Galaxy will literally run out of gas; starbirth will halt. Even if density waves still endure, they will have little gas to move around. When the disk of the Galaxy stops evolving, the Galaxy will be essentially defunct. Globular clusters will still swing on their leisurely orbits around the core. The supermassive black hole (if there!) will survive forever. Overall, the Galaxy will be quiet and dull, engulfed in dark matter.

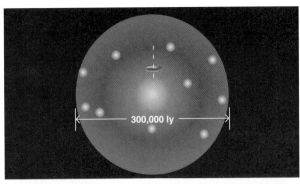

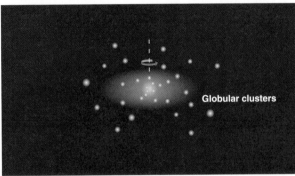

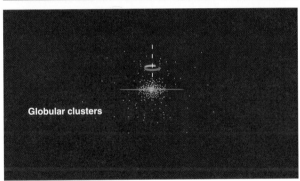

Figure 17.21 Schematic sequence of the collapse and condensation of a large cloud of gas and dust to make globular clusters and the Galaxy's disk. Because of its original spin, the matter eventually makes a disk with a central bulge. The stars in the globular clusters form before the disk has developed. The entire process takes less than 1 Gy. Note how the globular clusters now give the approximate size of the original gas cloud.

KEY CONCEPTS

1. The main parts of the Galaxy are the encircling halo, the flat disk, and the central nuclear bulge, which contains the nucleus.

2. Stars of different types inhabit these regions: the disk contains young, metal-rich (few percent) Population I stars; the nuclear bulge has metal-poor (few tenths of a percent) Population I stars; and the halo, metal-poor Population II stars.

3. The disk contains at least two and possibly four spiral arms, which contain concentrations of young stars, gas, and dust. The sun lies on the inner edge of one arm, and we can see pieces of other arms toward and away from the galactic center.

4. We trace out spiral arms by the use of objects found in them: optically, young massive stars, cepheids, and H II regions; with radio, H I clouds, and molecular clouds. In all cases, the essential problem is to determine the distance to the object observed.

5. The rotation curve shows how fast objects at different distances from the galactic center orbit around it. The failure of the curve to follow Kepler's laws at large distances from the center indicates that a large fraction of the Galaxy's mass lies beyond the sun; this mass has not yet been seen directly.

6. The sun orbits the Galaxy at a distance of some 30,000 ly from the nucleus at a speed of 250 km/s, with some uncertainty in both numbers.

7. Cepheid variable stars show a period–luminosity relationship: the more luminous the cepheid, the longer the period of its light variation. This relationship is a powerful tool for inferring the distances to cepheid variables, along with the inverse-square law for light.

8. Radio astronomers use the Doppler shift in the molecular clouds and clouds of atomic hydrogen to infer the spiral-arm structure of the Galaxy; to do so, they assume that the clouds move along near-circular orbits (they don't) and that the rotation curve has been well observed (it has not).

9. Any model for the evolution of the Galaxy must explain the persistence of spiral arms. The density-wave model does this by having two spiral density waves disrupt the gas of the Galaxy's disk to promote the formation of spiral arms. As the density waves plow through the disk, different material condenses into the spiral arms as the old material dissipates; so the persistence of the original arms is really an illusion.

10. The nucleus of the Galaxy emits intense radio and infrared radiation; it contains supergiant M stars, young massive stars, dust, and gas rotating at high speeds. To account for the motion of the gas requires a concentration of millions of solar masses of material in the inner few light years – perhaps a supermassive black hole.

11. The Galaxy's halo contains globular clusters, some gas, and the invisible objects that make up the mass that shows up in the rotation curve. This invisible material is called *dark matter.*

12. The Galaxy formed some 13 Gy ago from the gravitational contraction of a large, slowly spinning cloud of gas and dust; the halo formed first, then the disk. We can estimate the relative sequence of formation from the heavy-element content of stars: the more metals they contain, the younger they are.

STUDY EXERCISES

1. What limits an optical astronomer's investigation of the Galaxy's structure? (Learning Outcomes 17-1, 17-2, and 17-3)
2. Why are Population I cepheids good optical spiral-arm tracers? (Learning Outcome 17-2)
3. Radio astronomers need the rotation curve of the Galaxy to use 21-cm line observations to establish its spiral structure. Why? (Learning Outcomes 17-4 and 17-6)
4. Argue that a spiral arm cannot be a material arm. (Learning Outcome 17-7)
5. What are the kinds of celestial objects found in spiral arms? (Learning Outcomes 17-7 and 17-8)
6. What characteristics of spiral arms does the density-wave model account for? In what respects is the model at present inadequate? (Learning Outcome 17-8)
7. What observational evidence do we have that a large fraction of the Galaxy's mass is not in the core, nor, in fact, within the radius of the sun's orbit? (Learning Outcome 17-4)
8. Relate the orbits of globulars and their chemical composition to the birth of the Galaxy. (Learning Outcome 17-10)
9. Some of the radio emission from the nucleus of the Galaxy is nonthermal. What does that imply about the physical conditions there? (Learning Outcome 17-12)
10. What observational evidence and physical argument can be used to infer that a supermassive black hole may reside in the Galaxy's core? (Learning Outcome 17-12)
11. What is the sun's distance from the center of the Galaxy? How is the value of the distance determined? How uncertain is its current value? (Learning Outcome 17-5)
12. As the disk of the Galaxy evolves, what do you expect to happen to the percentage of metals found in the interstellar medium? (Learning Outcome 17-9)
13. Which region of the galaxy do the orbits of globular clusters define? (Learning Outcome 17-11)
14. In the future, what do we expect to happen to the molecular clouds that we now see making up so much of the interstellar medium? (Learning Outcome 17-13)

PROBLEMS AND ACTIVITIES

1. Assume that Kepler's laws apply to the rotation of the Galaxy. What would the orbital period be of a star 150,000 ly from the center? *Hint:* Use the right units!
2. What is the orbital period of the sun around the galactic center?
3. At 3 ly from the Galaxy's center, material orbits at speeds of about 100 km/s. What is the gravitational mass interior to this point?
4. A length of 1 ly at the galactic center corresponds to what angle when seen from the sun?
5. A few globular clusters reach maximum distances of 350,000 ly from the center of the Galaxy. What are their orbital periods?

18 The Universe of Galaxies

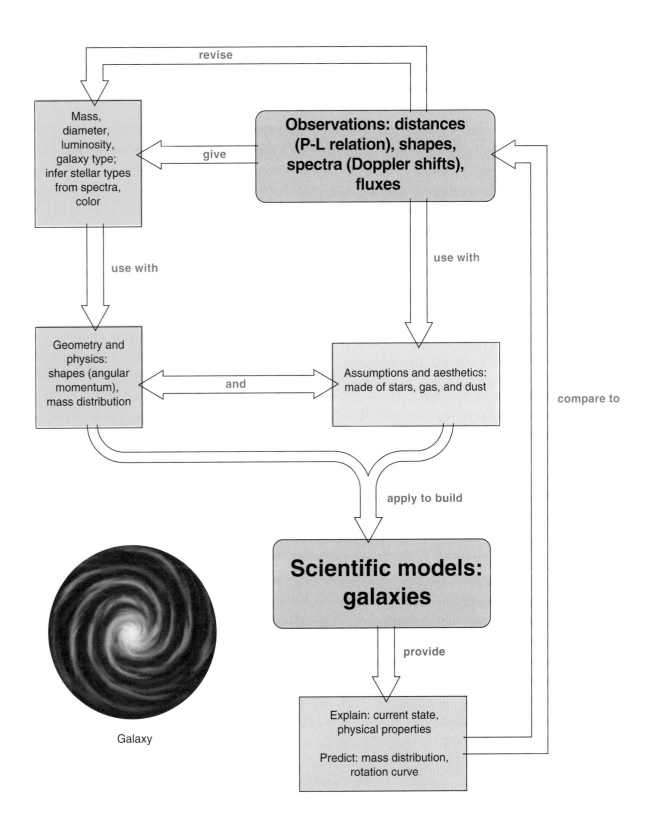

Mass, diameter, luminosity, galaxy type; infer stellar types from spectra, color

Observations: distances (P-L relation), shapes, spectra (Doppler shifts), fluxes

revise

give

use with

use with

Geometry and physics: shapes (angular momentum), mass distribution

and

Assumptions and aesthetics: made of stars, gas, and dust

apply to build

compare to

Galaxy

Scientific models: galaxies

provide

Explain: current state, physical properties

Predict: mass distribution, rotation curve

No celestial object is quite as grand as a galaxy. In a large telescope, a bright spiral galaxy is a stunning sight, a star-bright nucleus and a misty swirl of spiral arms. Thousands of millions of stars are caught in a whirlpool spanning hundreds of thousands of light years.

The galaxies form the basic elements of our modern cosmological vista. Their sheer numbers are beyond our comprehension. The diversity of structure in these galaxies is also astounding. Even more surprising is the fundamental unity found in spite of their wide variety. The fact that galaxies can be sorted into broad divisions hints at a common evolutionary process.

The galaxies form the skeleton of the universe. The crucial problem is the measurement of their distances, for many conclusions of modern cosmology hinge on them. This chapter notes that the notorious difficulties of surveying our own Galaxy are amplified when we try to appraise the vastness of the universe. In spite of present problems, our vision of the universe, underpinned by general relativity, has a coherence that allows us to draw conclusions about the large-scale structure of spacetime.

18.1 THE EXTRAGALACTIC DEBATE
(Learning Outcome 18-4)

For almost two centuries, astronomers hotly debated whether the spiral nebulas they saw with their large telescopes were simply clouds of gas within the Milky Way or were other galaxies like ours but far beyond it (Fig. 18.1). The controversy climaxed in April 1920, when American astronomers Harlow Shapley and Heber D. Curtis (1872–1942) debated the point publicly.

Basic Arguments

Shapley believed that the spiral nebulas were distant parts of the Milky Way. Curtis, who claimed that they were galaxies in their own right, relied on the principle of the **uniformity of nature** – that is, similar astronomical objects are assumed to have similar properties until proven otherwise by observations. (In other words, innocent until proven guilty!)

Curtis argued that the wide range of angular diameters of spirals – approximately 4° (for M31, the nearest) to 10 arcmin and less (for the smallest)

Figure 18.1 Close-up view of the four largest galaxies in the small group of galaxies called Stephan's Quintet. Three of the galaxies have the same redshift, so they are at the same distance (about 300 Mly). These galaxies are colliding, each ripping the others apart by tidal forces. The bluish spiral at lower left is a closer galaxy and not part of the collision.

– required a large range of distances, and so the spirals could not be part of our Galaxy. (Remember that an angular diameter gives you a diameter-to-distance ratio.) Starting from the principle of the uniformity of nature, Curtis assumed that all spirals have roughly the same physical diameter. The range in observed sizes (about 10 to 1) implied that the spirals must be at enormous distances from the Galaxy. For if they were the same in diameter, the range in apparent size would mean that the ones ten times smaller must be ten times farther off, or about ten times the radius of the Galaxy (Fig. 18.2). The more distant spirals could not be members of the Milky Way Galaxy, as Shapley argued.

In addition, Curtis noted the so-called **zone of avoidance,** a region near the plane of the Galaxy where very few spirals are visible (Fig. 18.3). Curtis

argued that interstellar dust in the galactic plane cuts out the light from the spirals; we see fewer in the zone of avoidance because there we have to look through the disk of the Galaxy. If the spirals were actually associated with the Galaxy, they would be found concentrated in the plane – rather than avoiding it – along with the stars, galactic clusters, and H II regions. As evidence for this point of view, Curtis cited photographs of spirals showing dark lanes cutting through their planes (Fig. 18.4). If the Milky Way Galaxy and other spirals were similar in structure, Curtis reasoned, then our Galaxy must also have dusty material collected in the plane.

Finally, Curtis pointed out that the spectra of spirals are not bright-line spectra like those of emission nebulas (such as the Orion Nebula). Rather, they resemble those from a conglomeration of stars. The spectra show faint dark lines against a bright background – the same spectrum the Galaxy would show if viewed from a great distance.

The Debate's Resolution (Learning Outcome 18-3)

In 1924 Edwin Hubble settled the dispute conclusively by discovering cepheid variables (Section 17.3) in the Andromeda galaxy (M31). He derived a distance of 490,000 ly for M31, far beyond the farthest globular clusters that marked the outer limits of the Galaxy. (Hubble's estimate was too small; current work on M31 establishes a distance of 2.2 million light years.)

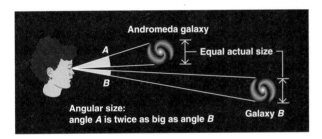

Figure 18.2 Distances to galaxies estimated by their angular sizes. If galaxies have roughly the same physical size, the more distant will have smaller angular sizes. Here, Galaxy B is about twice as far away as the Andromeda Galaxy, and so it is half the angular size. So angular size is inversely proportional to distance.

Figure 18.3 The "zone of avoidance." This plot shows the brightest galaxies in the sky, the zodiacal constellation outlines, and the galactic equator (the plane of the Galaxy) completely around the sky and from north to south. Note how few galaxies are visible above and below the galactic equator: dust blocks the view.

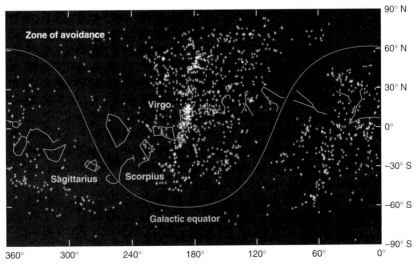

Figure 18.4 Dust in the plane of a spiral galaxy (NGC 891). The dark band is due to dust cutting out starlight. Note the nuclear bulge at the center. Our Milky Way has a similar layer of dust, creating the zone of avoidance.

CAUTION. The principle of the uniformity of nature has great power, but it must be used with care. Sometimes it works; sometimes it leads you astray. For example, novas in M31 are pretty much the same as those in our own Galaxy, because M31 is a spiral galaxy very similar to the Milky Way. But the variable stars in globular clusters are not, it turns out, the same as the cepheids in the disk of our Galaxy, for they belong to different populations of stars. You can assume that things are *uniform* everywhere, and this can lead to new knowledge and insights, but you must always keep looking for new ways to check the assumption and be prepared to abandon it if inconsistencies develop.

NOTE. As indicated before, in naming some galaxies we use the prefix *M;* for others we put *NGC* before the number. The prefixes refer to two different catalogs of galaxies. The Messier (M) catalog, compiled by Charles Messier in the eighteenth century, contains star clusters and nebulas in our Galaxy as well as the brighter galaxies visible from mid-northern latitudes. The New General Catalogue (NGC) was compiled in the nineteenth century with larger telescopes. Two supplements to the NGC, called the first and second Index Catalogues (IC), were published in 1895 and 1908.

18.2 NORMAL GALAXIES: A GALAXIAN ZOO
(Learning Outcome 18-1)

Edwin Hubble pioneered the field of extragalactic astronomy. He recognized that galaxies have different shapes. To catalog the differences in form, Hubble in 1926 proposed his first scheme for the classification of galaxies. Although his initial design is now considered to be too simple, modern classifications still use the fundamental categories of elliptical, spiral, and irregular galaxies. Most galaxies fall into these categories – normal galaxies, whose emission is mostly from starlight. (Chapter 19 describes extraordinary galaxies.) Let's look at these basic galaxy types.

Figure 18.5 A giant elliptical galaxy, Messier 87 (NGC 4486), about 50 Mly distant. Note the symmetry of the shape and the lack of distinct structure; it is an elliptical of type E1 and contains mostly old, yellowish stars.

Ellipticals

An **elliptical galaxy** (Fig. 18.5) exhibits an elliptical shape. Very little gas or dust appears in elliptical galaxies, and OB stars are also absent. The ellipticals generally have a reddish overall color.

Hubble subdivided the ellipticals into classes from E0 to E7, according to how elliptical they appear. Imagine looking at a circular plate face-on; such is the appearance of an E0 galaxy. Now slowly tilt the plate so that it looks more elliptical and less circular. This flattening of shape presents the same views as the sequence from E0 to E7 galaxies. Be warned that Hubble based the classifications on the *appearance* of the galaxy, not on its true shape. For example, an E7 is really a flat elliptical viewed edge-on, but an E0 may be either a truly spherical galaxy or a flattened galaxy seen face-on.

Ellipticals range in size from supergiants to dwarfs. The largest ellipticals, found in clusters of galaxies (ahead in Section 18.6), have diameters of a few million light years! In contrast, the smallest dwarf ellipticals span merely thousands of light years in diameter.

Disks (Spirals)

A **disk galaxy** usually forms a **spiral galaxy.** If not seen edge-on, it displays an obvious spiral structure, usually with two, but sometimes more, spiral arms (Fig. 18.6). One type of spiral has a prominent bar through the nucleus, the spiral arms winding out from the end of the bar (Fig. 18.7).

Figure 18.6 Two spiral galaxies are superimposed in the HST image, with the smaller galaxy in the foreground. It is Hubble type Sc, and one of the dustiest galaxies known. Two spiral arms are visible extending from the nucleus; bluish patches of star-forming regions are visible within them.

Figure 18.7 The barred spiral galaxy NGC 4314, imaged by the HST, reveals clusters of infant stars, born a few million years ago, in a ring around the core.

Figure 18.8 A dwarf irregular galaxy of the Local Group, NGC 6822. Young, hot stars give it a bluish color.

Hubble arranged the spiral forms in sequence according to the sizes of their nuclear region, the tightness of the spiral arms, and the degree to which the arms were resolved into patches of stars. He termed the spirals without a bar *normal* (denoted S) and the others *barred* (denoted SB). Our Galaxy is now known to be a barred spiral.

The normal and barred spirals are subdivided further into categories a, b, and c. These types are judged by how tightly the spiral arms wind around (a, the tightest; c, the most open) and the relative size of the nucleus compared to the disk (a, the largest; c, the smallest). For example, a Hubble Sa is a normal spiral with a large nucleus and tightly coiled arms. A few galaxies appear to have the disk of a spiral but no arms. Hubble dubbed those S0. These are now sometimes called **lenticular galaxies** because of their shape, like a convex lens.

Irregulars

Finally, as a catch-all category, Hubble designated as **irregular galaxies** (denoted Irr) those that were devoid of spiral structure or symmetry but were resolvable into distinct patches of stars (Fig. 18.8). OB stars and H II regions dominate these strange beasts. Usually there are no conspicuous dust clouds. Radio observations of neutral hydrogen gas tend to show a rotating disk of material. In this respect, they resemble spiral galaxies.

Modern classifications have expanded Hubble's format. I will stick with Hubble's three basic categories: spirals, ellipticals, and irregulars. This Hubble scheme does not include all types of galaxies, however. Some galaxies stand out as peculiar in shape and do not fit into the three general Hubble categories. Many of these peculiar galaxies turn out to have evidence of unusual activity. Some are pairs of galaxies close together, interacting gravitationally by tidal forces (Fig. 18.9).

Luminosity Classes

Astronomers have recognized that the same Hubble type of galaxy, say Sb, comes in a range of luminosities. Like stellar luminosity classes (Section 13.5), galaxies also have luminosity classes of I, II, III, IV, and V, with I the most luminous and V the least. An Sc I galaxy, for instance, is a very luminous spiral with a small nucleus and spread-out arms. It turns out that luminosity class I galaxies are larger than class II, and so on. Class I galaxies can be thought of as supergiant galaxies.

What does a census of galaxy types reveal? A complete survey of a region out to 30 Mly showed

Figure 18.9 Interacting galaxies (NGC 2207/IC 2163). The shape of these spiral galaxies is twisted by the tidal interactions with each other.

that 34 percent of the galaxies in this volume are spirals, 12 percent are ellipticals, and 54 percent are irregulars. Hence, the majority of galaxies in the nearby universe are irregulars of fairly low luminosity (just as low-luminosity stars make up most of the stars in the Galaxy).

Remember that a telescope is a time machine, and the relative numbers of galaxies may well change with time. An observation by the HST makes this point for a group of galaxies that lies at a distance of 4 Gly (Fig. 18.10). The image clearly shows that long ago, most of the galaxies were spirals! That's not the case in the recent past or today. Somehow, the spirals slipped from dominance as the cosmos evolved, but we do not know what caused most of them to disappear.

Figure 18.10 HST image of a massive cluster of galaxies (Abell 2218) some 2 Gly from earth. Both spiral and elliptical galaxies are visible. Note the arcs, which are distorted images of very distant galaxies.

To group galaxies by shape marks an initial step toward delving into the far depths of the universe. To probe the physical properties of galaxies – their masses, sizes, and luminosities – ultimately depends on knowledge of their distances from us, distances that are as difficult to survey as they are vast to imagine.

18.3 SURVEYING THE UNIVERSE OF GALAXIES

Less than a century has passed since we learned for certain that galaxies are faraway islands of stars. Yet we know the distances to most galaxies only roughly. For the nearest galaxies, we have measurements that are good to about 10 percent. For the most distant galaxies, the error is larger. It's a hard but essential astronomical business to survey the distances in the huge universe of galaxies. (See the Galaxies Celestial Navigator™.)

Judging Distances
(Learning Outcome 18-3)

The distances to galaxies rely on two essential techniques. The initial indicators are the criteria that brightness means near-ness and smallness means farness. Galaxies with the smallest angular size tend to be the most distant, and faint galaxies also tend to be far away. By applying these simple criteria, you can make rough estimates of the relative distances to galaxies.

For example, if you look at two galaxies, one apparently half as large as the other, and if both are in reality approximately the same size, the apparently smaller galaxy is twice as far away as the apparently larger one. A similar argument applies for relative brightness: if one galaxy is a hundred times fainter than another, it is roughly ten times farther away (recall the inverse-square law for light, Section 13.1).

Refining this rough first approach requires the use of known physical properties of stars and galaxies inferred from theoretical models and careful observations. Each step in surveying the universe applies to certain objects and over a certain range of distances; astronomers work step by step to establish a distance scale.

What are some steps of the scale? First, we need to establish the scale of distances in our Galaxy,

starting with the solar system. Parallax measurements are used: heliocentric and spectroscopic parallaxes. Along with these are distances using the cepheid period–luminosity relationship. So far, so good, for we can then infer the size of the Galaxy.

Distance Indicators
(Learning Outcomes 18-5, 18-6)

To bridge the distances to other galaxies, we assume the uniformity of nature: that the essential character of objects in our Galaxy (such as cepheid variable stars or supernovas) is the same for similar objects in other galaxies. Without this necessary assumption, we couldn't get anywhere. It's not a blind, unsupported assumption, however. Observations made so far are consistent with it. For example, cepheids in other galaxies have the same spectra and light curves of the same shape as cepheids of the same type in our Galaxy and follow the same period–luminosity relationship (Section 17.3).

Then, to find distances to galaxies, we must use identifiable objects (within galaxies) whose luminosities we know. We compare their fluxes with their luminosities to infer their distances. Unfortunately, even the largest telescopes have a limit, and some objects are too faint to be picked up, so we want to choose the most luminous objects in galaxies for which we actually can determine the luminosities.

As an analogy, imagine that you know that all streetlights have the same luminosity, say 500 watts. Then as you look out at night at a city, you can judge the distances to different locations by measuring the flux of the relevant streetlights with a light meter. The inverse-square law for light (Section 13.1) allows you to find the distance from the measured flux and the assumed luminosity.

Starting with close galaxies, we apply the period–luminosity relationship to cepheids in other galaxies. That way we find the luminosities of cepheids observed in other galaxies. The cepheids, however, are hard to detect beyond 20 Mly with ground-based telescopes. The HST extends this range to 100 Mly.

To go beyond this limit requires the establishment of other objects whose luminosity is greater

Figure 18.11 HST image of supernova 1994D, Type Ia.

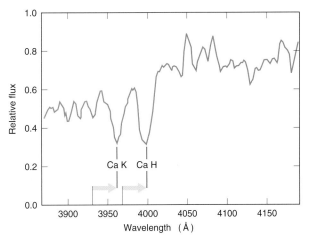

Figure 18.12 Graph of the spectrum of a spiral galaxy (NGC 1357), showing the redshift of the H and K lines of calcium. At bottom is given the unshifted position of the lines (marked *Ca K* at 3944 Å and *Ca H* at 3968 Å). The redshift, indicated by the arrows, is about 28 Å. (Spectrum obtained by Robert C. Kennicut, Jr.)

than that of the cepheids. For instance, supernovas. Recall that supernovas come in two types, distinguished by the shape of their light curves (Section 16.4). Best estimates give, for a subclass of Type I supernovas (called Type Ia), a uniform luminosity at maximum of about 10^{10} solar luminosities. That's some million times the most luminous cepheids! Type Ia seem to occur in a standard model, as a white dwarf in a binary system undergoes a thermonuclear detonation (Chapter 16). By measuring the flux of a Type Ia supernova in another galaxy at maximum, we can infer its distance with the inverse-square law for light (Fig. 18.11). We can see such staggering explosions at great distances, about 1000 times farther than cepheids.

Of course, for this method to work, we need a calibration for the peak luminosity of these Type Ia supernovas. How to get it? By using cepheid variables to find the distances to galaxies in which such explosions appear. HST provided independent distances based on cepheids.

The designation of other distance standards follows the same strategy: select fairly common bright objects, find their luminosities in our own or nearby galaxies, check other galaxies by methods thought to be reliable, and then utilize the standard to the limits of its accuracy. Beyond these limits, we rely on the expansion of the universe as described by Hubble's law.

18.4 HUBBLE'S LAW AND DISTANCES

The universe is expanding, and that expansion is described by Hubble's law. Hubble's law states that the farther a galaxy is from us, the greater its radial velocity of recession. The number that relates the recessional speed and the distance is **Hubble's constant,** *H.* Finding the value of *H* has consumed the energy of astronomers since the construction of large telescopes and was the main goal of HST.

Redshifts and Distances
(Learning Outcomes 18-6, 18-8)

It is relatively easy to measure the redshift of a galaxy, compared with the task of finding its distance. A comparison spectrum made at the same time as the galaxy's spectrum affords a direct measurement of the shift in some prominent spectral lines (such as the absorption lines of calcium; Fig. 18.12). The redshift indicates the radial velocity of a galaxy.

We then need to fix the redshifts and distances from a range of near and far galaxies, such as in Figure 18.13. Look at the galaxies (each in a cluster) and at their respective spectra with redshift indicated. Notice that as the redshifts get larger, the galaxies appear smaller and fainter. Now plot the distances for these galaxies versus their redshifts.

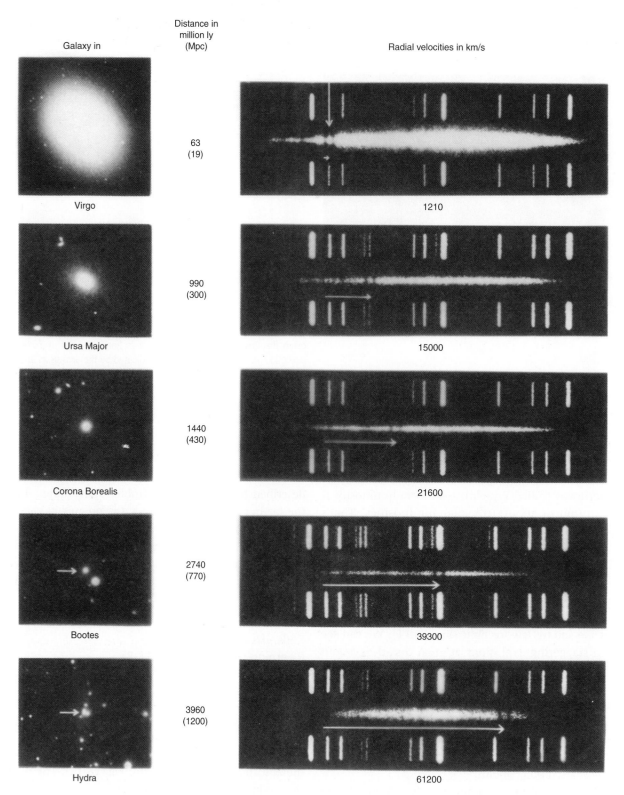

Figure 18.13 Measured redshifts and distances for selected galaxies in order of increasing redshift and distance (from top to bottom) from a galaxy in the Virgo cluster to one in the distant Hydra cluster. The redshifts are visible in the spectra on the right, where a white arrow indicates the size of the shift in the H and K lines of calcium; these are the two darkest lines in the spectra. Below each spectrum is the radial velocity in kilometers per second. Above and below each spectrum are emission-line spectra that serve as wavelength markers for a source at rest. The galaxies (at left) are in clusters named for the constellation in which they appear. Note, in general, that the farther galaxies appear smaller than the closer ones.

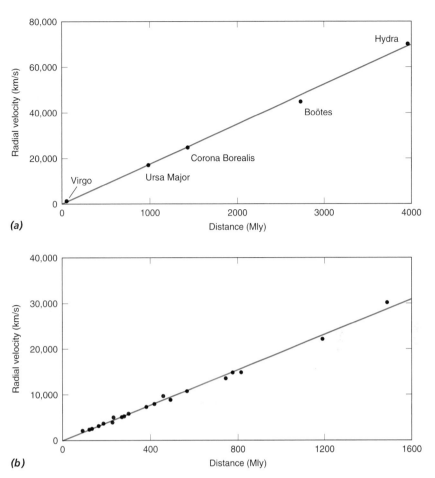

Figure 18.14 Hubble's law. *(a)* Hubble plot using the data from Figure 18.13. The straight line is the "best fit" through the points. Its slope is 15 km/s/Mly, which is the value of the Hubble constant derived from these data. Note that Hubble's law is a direct relationship (radial velocity directly proportional to distance). *(b)* Hubble plot based on distances inferred from Type Ia supernovas. The slope is 21 km/s/Mly. (Adapted from a diagram by Riess, Press, and R. Kirshner.)

You can draw a straight line that represents the trend of these points and so find a value for Hubble's constant, which is the slope of the line (Fig. 18.14a). Once we have a value of Hubble's constant, we can use Hubble's law to find the distances to galaxies beyond the limits described in Section 18.3.

In equation form, the Hubble law is

$$v = H \times d$$

where v is the radial velocity, in kilometers per second, H the Hubble constant, and d the distance, in millions of light years (Mly). Note that the radial velocity is *directly proportional* to the distance.

Now, we reverse this procedure to find distances from redshifts. Suppose you were given the redshift of a galaxy. You find the distance to which this value corresponds by looking it up on the redshift axis of your plot. If the redshift is larger than measured earlier, you can assume that the line you've drawn for the plotted points may be extended farther out and still be valid. Suppose a galaxy's mea-

sured redshift is 40,000 km/s. What is its distance? If H is 20 km/s/Mly, then the plot gives 2×10^9 ly.

CAUTION. This indirect method of distance measurement rests on one crucial fact: that the galaxies with known distances give an accurate value for Hubble's constant! It also depends upon a linear (direct proportion) Hubble law out to great distances. Using Type Ia supernovas, we have reliable distances out to some 1.5 Gly! Some of these supernova data (Fig. 18.14b) indicate a value for H of 21 km/s/Mly; others as low as 19 km/s/Mly.

In contrast, a special HST team, relying mostly on cepheids, obtained a value of 23 km/s/Mly, with a stated uncertainly of only 10 percent. That's a bit higher than the supernova value. But far better when, before HST, the value ranged from 15 to 30 km/s/Mly! I consider the case closed, and for this book (and for a simple number), I use a value of 20 km/s/Mly.

Hubble's legacy finally reached its most important conclusion. And the Copernican legacy continues, because a linear Hubble law fits with the notion that our place in the cosmos is *not* a special one. An observer on another galaxy would find the same linear relationship.

"Age" of the Cosmos (Learning Outcome 18-9)

A specific value for H implies an "age" for the cosmos (Enrichment Focus 18.1). That "age" marks the time the expansion began, and it assumes that the value has not altered with time. The Hubble law, even without a firm value for H, implies that the universe's expansion began at a *finite time* in the past! About when?

Consider driving at a constant speed between two points. If you know the distance and the speed, you can find the elapsed time. For example, if you travel in a car from Boston to New York, a distance of 400 km, at a constant speed of 80 km/h, the trip takes 5 hours (400 km/80 km/h = 5 h). The inverse of the value of H plays the same role as the elapsed time in this example. If it is 20 km/s/Mly, then the "age" is 15 Gy.

That result poses no problem with other ages (after all, you cannot be older than your mother, and the cosmos is the mother of us all). The oldest stars (in globular clusters) are some 14 Gy old. So a value near 20 km/s/Mly (which this book uses) will not cause serious problems with the ages of stars and other dated celestial objects.

ENRICHMENT FOCUS 18.1

Hubble's Constant and the Age of the Universe

You can infer from the measured value of H the time since the **expansion of the universe** began. To do this simply, you need to make the assumption that the expansion rate is uniform. (This assumption does not apply to the real universe. For any realistic model, velocities must decrease with time. However, this assumption allows a simple calculation of the oldest possible age of the universe from the present value of H.)

At a constant velocity, the distance traveled, d, is equal to the velocity, v, multiplied by the elapsed time, t. In algebraic form (Focus 4.1),

$$d = vt$$

Conversely, the travel time is

$$t = \frac{d}{v}$$

Hubble's law written as an equation is

$$v = Hd$$

Here v is the radial velocity of a galaxy, d the distance to the galaxy, and H Hubble's constant. If H has the units kilometers per second per million light years (km/s/Mly) and d is given in millions of light years (Mly), then v comes out in kilometers per second (km/s). Now compare Hubble's law written as

$$t = \frac{d}{Hd}$$
$$= \frac{1}{H}$$

to the trip formula

$$t = \frac{d}{v}$$

The travel time equals $1/H$. If H is 20 km/s/Mly, then $1/H$ equals 1.5×10^{10} years. To get this result we have to put in the conversion factors to get everything in the same units:

$$t = \frac{1}{H}$$
$$= \frac{1}{20 \text{ km/s/Mly}} \times 10^6 \text{ ly/Mly} \times 10^{13} \text{ km/ly}$$
$$= 4.5 \times 10^{17} \text{ s}$$
$$= \frac{4.5 \times 10^{17} \text{ s}}{3.3 \times 10^7 \text{ s/y}}$$
$$= 1.5 \times 10^{10} \text{ y}$$

So the expansion of the universe started about 1.5×10^{10} years ago!

This conclusion is valid *only* if our assumption of uniform expansion is correct. However, even if the expansion were a bit faster in the past than it is now, the value of 1.5×10^{10} years marks an upper limit to the actual time since the expansion began.

An Accelerating Expansion?

To estimate the universe's age, we assumed that H was constant over all time. Now, because the universe contains visible and dark matter, as well as photons that have a mass equivalent ($E = mc^2$), we naturally expect that the expansion slows down – it should decelerate. Again, imagine a mass leaving the earth. As it moves outward, its initial speed (even if greater than escape speed!) decreases from the gravitational tug of the earth. We expect the same to occur in the expanding cosmos: the rate of expansion should decrease with time – a cosmic deceleration.

The Type Ia supernovas are the pivotal players. Two separate research teams have measured several dozen very distant blasts, and both came to a startling conclusion: the expansion is accelerating! In essence, the supernovas appear more distant than expected. Astronomers can concoct various models to explain this effect. Acceleration is one of the simplest. The extra stretching it causes of space, compared to deceleration, puts the supernovas farther away than expected.

Acceleration, if correct, has cosmic consequences. It implies that the cosmos contains a long-range repulsive force, previously unknown, a sort of cosmic antigravity. Recall (Section 7.4) that Einstein's model was static. How so? He imposed a long-range repulsive force to keep gravity from collapsing the universe but later rejected it when the expansion was discovered. Now that notion may come back into play. It would become another triumph of general relativity – Einstein's legacy for this new century.

But the observations really reach the ends of the observable universe and are somewhat iffy. I judge the conclusion about cosmic acceleration to be reasonable but not yet firm. More observations, say in the next ten years, will solidify or evaporate the interpretation. Stay tuned!

18.5 GENERAL CHARACTERISTICS OF GALAXIES (Learning Outcome 18-2)

Let's step back a bit from our cosmological vista and study the characteristics of galaxies (Table 18.1), now that we have methods to find out their distances.

Size

Once you have learned the distance to a galaxy, you can find out its actual size from a measurement of its angular size. The hitch here is that the definition of the edge of a galaxy is more or less arbitrary – and hard to measure well! Different definitions of the edge (which is not simply where the visible stars end) result in different sizes. Also, the galaxy may be tilted to our line of sight.

Despite these problems, we can still make general statements about the sizes of galaxies. Dwarf ellipticals and small irregulars tend to be the smallest galaxies – only 300 to 3000 ly in diameter for some. Giant ellipticals can range up to 200,000 ly in size. To put this into perspective, imagine your height (about 2 m) to be the size of a dwarf galaxy. Then an irregular galaxy would be about twice your size, a spiral ten times your size, and a giant elliptical some twenty times your size!

Table 18.1   General Properties of Galaxies

Property	Disks (Spirals)	Irregulars	Dwarf Ellipticals	Giant Ellipticals
Diameter (ly)	90×10^3	20×10^3	30×10^3	150×10^3
Mass (sun = 1)	10^{11}	10^6	10^5 to 10^7?	10^{13}
Luminosity (Sun = 1)	10^{10}	10^9	10^8	10^{11}
Color	Bluish (disk), reddish (halo and nucleus)	Bluish	Reddish	Reddish
Neutral gas (fraction of mass)	5%	15%	Less than 1%	Less than 1%
Types of star	Young (disk), old (halo and nucleus)	Young	Old	Old

Note: Mass and luminosity are given in solar masses and solar luminosities, respectively.

The very largest galaxies, the supergiant ellipticals, can have sizes up to 3 million ly. That's greater than the distance from our Galaxy to the Andromeda Galaxy (M31)! These **supergiant elliptical galaxies** tend to define the gravitational centers of clusters of galaxies (Section 18.6).

Mass

To find a galaxy's total mass is a tough task. The light from a galaxy comes mostly from its stars, but its material may not emit visible light – or may not emit at all (black holes, for example). This dark matter may inhabit the halos of most (all?) spiral galaxies (as found for the Milky Way Galaxy – Section 17.7).

The most widely used methods of finding a galaxy's mass are rotation curves and binary galaxies. The rotation curve method (which can be used for spiral galaxies only) works as follows. A galaxy's material orbits the nucleus in a specific way, so at every distance the orbital velocity has a certain value. This rotation curve comes from the orbital motions, described by Newton's laws that arise from the distribution of mass within the galaxy. (Recall the rotation curve for the Galaxy, Section 17.2.) If we observe a galaxy's rotation curve and make up a model for that galaxy's mass distribution, we can work out the galaxy's total mass, including dark masses that we can't see directly. We observe the rotation curve by directing a spectroscope's slit across the galaxy and measuring the Doppler shift at a number of points from the center out to the edge (Fig. 18.15). The same technique can be applied to the 21-cm emission from spirals.

Spiral galaxies hit as high as 10^{12} solar masses, although their rotation curves become flat without dropping off to the extent measured so far (Fig. 18.16). American astronomer Vera Rubin and her colleagues have pioneered observations such as these of many spiral galaxies.

For binary galaxies, we again make use of the versatile Doppler shift. Imagine two galaxies orbiting about the center of mass of the binary. Assume that the orbits are stable. Then, just as with visual binary stars (Section 13.6), we could apply Newton's version of Kepler's third law to find the masses, if we knew the distance, the angular size of the orbit, the period, and the position of the center of mass.

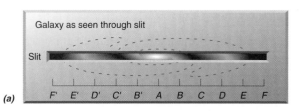
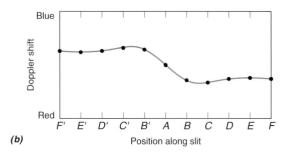

(a)

(b)

Figure 18.15 Observing the rotation curve of a galaxy. *(a)* A spectroscope is attached to a telescope, and its slit is placed over the image of a spiral galaxy; the positions along the slit are marked (F – F'). *(b)* Each position has a different Doppler shift, depending on the radial velocity from the rotation.

However, galaxies revolve too slowly for us to see their actual orbits, periods, and relative centers of mass; so we cannot find the individual masses of binary galaxies. All we can measure are the present radial velocities and the separation; we don't know what part of the orbit the galaxies are on, or what the inclination is, so we don't know what the true orbital velocities are. If we examine a large sample of galaxies and assume that their orbits are nearly circular and randomly oriented to our line of sight, however, we can estimate from these data the average masses of the galaxies sampled. An investigation of a few hundred binary systems, mostly spirals, using the Doppler shift of the 21-cm line to get their velocities, finds an average mass of 10^{12} solar masses for these spirals (for $H = 20$ km/s/Mly).

CAUTION. These two methods – rotation curves and binary galaxies – do *not* give the same results. Masses deduced from rotation curves are generally smaller than those deduced from the binary galaxy method. This difference has reinforced the idea that spiral galaxies have massive halos surrounding their bright, starry disks. Observations show

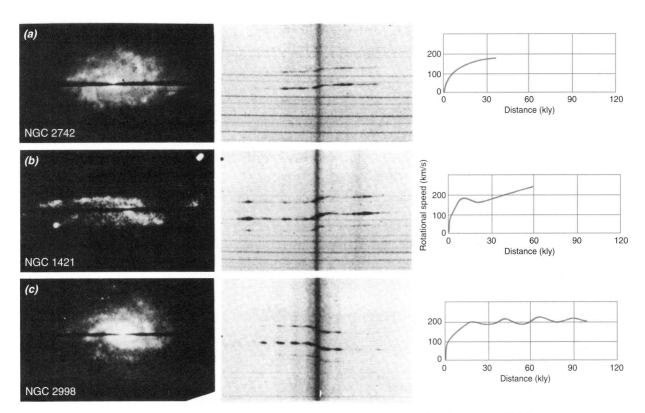

Figure 18.16 Rotation curves of spiral galaxies NGC 2742 *(a)*, NGC 1421 *(b)*, and NGC 2998 *(c)*: left-most column, the galaxies, with the position of the respective spectroscope slits indicated; center column, actual spectra along each of the slits (a negative image, so the bright appears dark); right-most column, the measured rotation curves. Note in the spectra the sharp shift of the slope of the lines across the center of the galaxy.

that the rotation curves for spirals remain flat out to some 300,000 ly from their centers. These flat curves imply directly that these galaxies have massive, invisible halos of dark matter. Otherwise the rotation curves should decline, as expected from Kepler's third law. Our own Galaxy has such a massive halo.

Luminosities

If we know the galaxies' distances and fluxes, we can work out their luminosities using the inverse-square law for light. One trouble here is that it is not easy to measure a galaxy's total flux. Why not? Because a galaxy thins out gradually at its edge, making it hard to be sure that you're catching all the light from the galaxy. In addition, corrections must be applied for light absorption: first, for that due to dust in our Galaxy, and second, for absorption due to dust in the galaxy itself (espe-

cially for spirals). Finally, since we view most galaxies tilted to our line of sight, we must account for the oblique view.

The luminosities of galaxies range from 10^5 solar luminosities for the dwarf ellipticals to 10^{12} solar luminosities for the supergiant ellipticals. The latter types are very rare, however. The Milky Way Galaxy, if we could see it all from space, has a luminosity somewhere near 2.5×10^{10} solar luminosities.

Mass–Luminosity Ratios

Divide the total mass of a galaxy by its luminosity and you have its **mass–luminosity ratio** (abbreviated M/L), an indication of the average mass per unit energy output from the galaxy. (It is usually expressed in units of solar masses and solar luminosities, so the M/L for the sun is 1.) Modern determinations using binary galaxy masses give 35 for the average mass-to-luminosity ratio for spiral galaxies (rotation curves give M/L of about 10 for ordinary spirals), about 100 for giant ellipticals, and 200 for supergiant ellipticals (Fig. 18.17).

For comparison, the M/L ratio for stars in the sun's neighborhood is about 1 or a bit larger; that's because these bodies are indeed solar-type stars. A galaxy made of such stars would have the same value. If it contained only B stars, the M/L would be about 0.01; if all M stars, about 20. Note that dark matter adds to the mass but not to the luminosity, so it makes the ratio larger. Spiral galaxies show a trend of rising M/L as the distance from the center of the galaxy increases. In the nucleus, the value is about 2. At the edge of the visible disk, the value climbs to about 20.

Ellipticals have a larger M/L because they contain a greater percentage of low-mass stars with low luminosities – main-sequence stars of class M. This extra abundance of M stars (and the absence of luminous bluish stars) would mean that ellipticals should be redder overall than spirals (see the following paragraphs). Other objects that may contribute to the mass but not to luminosity are neutron stars, black holes (including perhaps a giant one in the nucleus), brown dwarfs, and dark interstellar matter.

Colors

We can measure the colors of galaxies by comparing the flux at two different wavelengths. The color of a galaxy links directly to the kinds of stars in the galaxy. For example, a galaxy with many OB stars is bluer than a galaxy with few such stars.

A direct connection exists between a galaxy's type and its color. Ellipticals tend to be much redder than spirals, and spirals redder than irregular galaxies. Within the spiral group, the galaxies appear redder as their nuclear bulges grow larger and their spiral arms less extensive. The progression of color from the bluer irregulars to the redder ellipticals reflects a trend in the mix of the galaxy's stellar population. The reddish color means that ellipticals and the nuclei of spirals contain an old Population I. In general terms, then, an old Population I predominates in ellipticals, whereas a much younger Population I stands out in the irregulars. The mixture in the spirals is determined by the size of the nuclear bulge (old Population I) compared with that of the spiral arms (young Population I). Population II, which exists mainly in the globular clusters and galactic halo, probably is a minor contributor in all large galaxies.

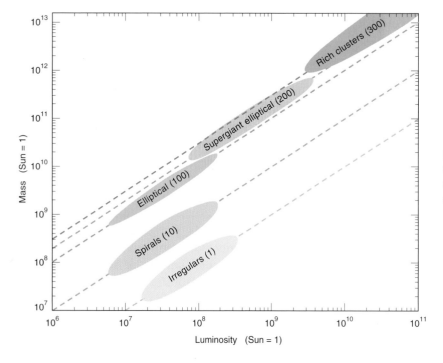

Figure 18.17 Mass–luminosity plot for various types of galaxies and rich clusters of galaxies. Both the mass and luminosity are in solar units.

Here is a way to distinguish galaxy types. Recall that our Galaxy has a halo, a nuclear bulge of reddish stars, and spiral arms of bluish stars. Imagine our Galaxy without the halo and the nuclear bulge. What remains (the arms) is like the stars, gas, and dust found in irregular galaxies. Now imagine our Galaxy stripped of its spiral arms. The remains (nuclear bulge and halo) are typical of the composition of elliptical galaxies. Irregulars have both old and young stars, but ellipticals contain only old stars. Why? Because in irregulars, enough gas and dust remains so that star formation continues; in ellipticals, it halted many years ago because the raw materials were all used up.

As you might suspect, a galaxy's color also is related to its overall content of gas and dust. The reddest galaxies, ellipticals, contain almost no gas and dust. The bluest galaxies, irregulars, contain the greatest percentage of gas and dust relative to their total mass.

Spin

The overall shape of a galaxy depends on its spin (actually, its angular momentum – review Enrichment Focus 11.1). The more spin a galaxy has, the flatter its shape. Elliptical galaxies have a small angular momentum. Spirals have a higher angular momentum in the disk than in the halo. In contrast, irregulars have no shape to speak of; their angular momentum is essentially zero. Since these galaxies of different types have different amounts of angular momenta, they cannot evolve by themselves from one type to another if angular momentum is conserved. (However, interacting galaxies can exchange angular momentum and so change their shapes.)

Note that the sequence from ellipticals to irregulars marks a sequence from galaxies in which starbirth ceased long ago (ellipticals: old Population I and some Population II, no gas and dust) to those in which starbirth is carried on very actively (irregulars: young Population I, large percentage of gas and dust). This trend gives a key clue to understanding the evolution of galaxies.

TO SUM UP. Galaxies contain stars, gas, and dust. The mixture of gas, dust, and stars, as well as the stellar types, are related closely to a galaxy's type and structure. Although the masses of galaxies are difficult to measure for distant systems, we know that, in general, giant ellipticals are the most massive and dwarf ellipticals the least massive.

18.6 CLUSTERS OF GALAXIES (Learning Outcome 18-10)

Take a close look at wide-angle images of the deep sky. On the images of regions in which the stars thin out – away from the Milky Way – you can see the tiny forms of galaxies. If you look at many images, you will notice that if you find one galaxy, you're likely to see others nearby. Galaxies tend to come in clusters! In fact, it may be true that all galaxies belong to clusters – though many of these clusters may be a simple marriage of two galaxies. Our universe contains **clusters of galaxies!**

The Local Group

The cluster of galaxies to which the Milky Way Galaxy belongs is called the **Local Group** of galaxies. This small cluster takes up a volume of space nearly 3 Mly across in its long dimension (Fig. 18.18). Our Galaxy is located near one end of the Local Group, and M31 is near the other.

As the most massive objects in the cluster, the Milky Way Galaxy and M31 dominate its motions and secure the other members gravitationally. In fact, the Galaxy and M31 orbit each other. The other members of the Local Group come along for the ride. The Local Group contains about thirty galaxies; they are mostly dwarf ellipticals, which contrast dramatically with the giant ellipticals found in other clusters. The obscuring matter in the Milky Way probably clouds our sight of other members, especially faint dwarf ellipticals. (Radio observations can find them.) Let's look briefly at some of the more important members of the Local Group.

The Large and Small Magellanic Clouds (Fig. 18.19) lie closest to the Galaxy. The Large Cloud lies at a distance of 170,000 ly; the Small Cloud at 200,000 ly. Both are connected to our Galaxy by a bridge of hydrogen gas. A large but thin envelope of neutral hydrogen and a thread of stars physically connect the two clouds. Tidal interactions with our Galaxy distort both.

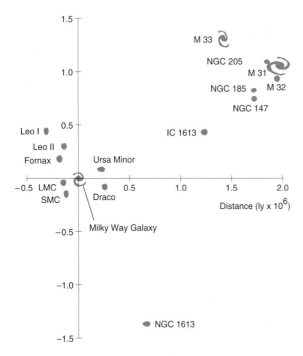

Figure 18.18 The most prominent galaxies in the Local Group, as viewed looking down on the plane of the Milky Way Galaxy: most of these are dwarf elliptical or irregular galaxies. The distance scale is in units of 330,000 ly.

The Large Magellanic Cloud (abbreviated LMC) contains stars totaling about 20×10^9 solar masses. Most of these stars are similar to those in the solar neighborhood, but with more young, hot stars. Radio studies demonstrate that the LMC has large amounts of neutral hydrogen gas, a total of approximately 3×10^9 solar masses. The LMC is a medium-sized galaxy, with a diameter about half that of the Milky Way and a tenth of its total mass. Its shape is irregular, with a hint of barred spiral structure – perhaps as many as six or seven ill-defined arms – but without an obvious nucleus. (Supernova 1987A exploded in the LMC; Section 16.4.)

The Small Magellanic Cloud (SMC) contains a stellar population similar to that of the LMC. The SMC has a total mass of about one-fourth that of the LMC and is only a little smaller – about one-third the diameter of the Galaxy. Like the LMC, the SMC has an irregular shape with a bar but no nucleus at its center.

M31 is a large spiral galaxy easily visible to the unaided eye on a dark, moonless night. With binoculars you find an elliptical, hazy patch of light; the spiral arms are too faint to be seen without a big telescope. The light from M31 arriving now departed from its source 2.2 million years ago!

We can examine directly the nuclear region of M31 (Fig. 18.20), which contains some gas and dust. Infrared observations have revealed that the star formation in the nucleus takes place in a distinct ring of clouds – perhaps similar to the ring in the nuclear region of the Milky Way.

M31 tilts 15° to the line of sight. Because of the tilt, dark lanes of obscuring material are plainly visible along with the spiral arms marked by hot stars.

Figure 18.19 The Large Magellanic Cloud (LMC). This galaxy, along with the Small Magellanic Cloud (SMC), are the two closest galaxies to the Milky Way Galaxy, and they interact with it gravitationally.

Figure 18.20 Double nucleus of Messier 31, the Andromeda Galaxy. The brighter object is the nucleus as seen from earth, but the HST reveals that the dimmer object is the true center of the galaxy. It is possible that the double nucleus is an illusion created by thick dust across the middle, which seems to split it in two.

Figure 18.21 Central region of the dwarf galaxy NGC 205 (Messier 110), Hubble type E5. It is the second small companion, along with Messier 32, to the Andromeda Galaxy, and thus one of our nearest neighbors.

A halo of globular clusters surrounds M31 like bees around a hive, in a distribution like that around our galactic system. The Andromeda Galaxy also has companions – a total of seven dwarf ellipticals that orbit it. One of these is called M32, a dwarf elliptical with a diameter of only 1000 ly containing a mere 400 million stars. HST imaged the nucleus of M32 (Fig. 18.21) and found a surprise: a dense concentration of stars in the center that hints at the presence of a supermassive black hole.

Other Clusters of Galaxies

Other clusters range from compact ones to rather loose arrays of galaxies. The huge **Coma cluster** spreads over at least 20 million ly of space and contains thousands of galaxies. Even observations of just the brightest galaxies show how common clustering is for galaxies. From these observations, we find that a typical cluster contains about a hundred galaxies and is separated some tens of millions of light years from its neighboring clusters.

The **Virgo cluster** stands out as one of the most stupendous in the sky (Fig. 18.22). Of the 205 brightest galaxies in the Virgo cluster, the four brightest are giant ellipticals. The Virgo cluster covers about 10° in the sky (twenty times the diameter of the moon!), which implies that its physical diameter is some 10 Mly at a distance of roughly 50 Mly. The Virgo cluster is so massive and so close that it influences the Local Group gravitationally; we are moving toward the Virgo cluster.

Clusters and the Luminosity of Galaxies (Learning Outcome 18-13)

From the description of the Local Group, you might suspect that very luminous galaxies are few in number compared to low-luminosity ones. Only three local galaxies are very luminous (the Milky Way Galaxy, M31, and M33), whereas most galaxies in the Local Group are dwarf, low-luminosity galaxies.

Figure 18.22 Central region of the Virgo cluster of galaxies. Visible are the giant elliptical galaxies M 86 and M 84 near center and bottom center. Note the shapes of the other galaxies. Can you see which ones are spirals?

Galaxies in clusters have a range of luminosities: there are many faint galaxies and few bright ones in a cluster.

Much of a cluster's visible mass resides in very faint galaxies. It is difficult to estimate the masses of these clusters because not all the material in them can be seen at visual wavelengths; so adding up all the galaxies gives a lower limit to the cluster's mass. On the other hand, if the cluster is bound by self-gravity, the motions of the galaxies within it establish an upper limit on its mass. (The actual value lies between these two limits.) Masses range from 10^9 to 10^{15} solar masses.

You can use the M/L ratio to consider the amount of dark matter in clusters. Take the Coma cluster as an example – two supergiant galaxies lie near the center. Their huge masses collect matter around them. By studying the velocities of the galaxies in the inner 4 Mly, we estimate a mass of almost 10^{15} solar masses in this region. When we count up the luminosities of visible galaxies, we find that the M/L ratio is 300! (The typical range for clusters is 300 to 500.) Given the types of galaxy here, we expect a value of 10! This means that visible stars make up only a few percent of the mass in the Coma cluster. The rest is the so-far mysterious dark matter.

Interacting Galaxies

A remarkable fact about clusters of galaxies is that the spacing of galaxies is pretty close, compared to the sizes of the galaxies themselves. Consider planets and stars in terms of sizes and spacing. In the solar system, the planets are spaced out about 100,000 times their diameters. In the Galaxy, stars are spaced out about a million times their diameters. But in a cluster of galaxies, the spacing amounts to only about a hundred times a typical galaxy's diameter. Astronomically speaking, galaxies in a cluster are very crowded together, so you might wonder if they would ever pass close to one another. They do!

Consider this additional fact: the most massive galaxies (supergiant ellipticals) are at least 10 million times more massive than the least massive ones (the dwarf ellipticals). Then it's not hard to imagine the tidal forces (Enrichment Focus 9.1) of the largest galaxies disrupting the smallest ones strongly enough to rip them apart and then pull the pieces in. Astronomers call this devouring of a smaller galaxy by a larger one **galactic cannibalism.**

A striking name, but what evidence supports it? Observations show that the supergiant elliptical galaxies do have peculiar properties. These include (1) extensive halos, up to 3 Mly in diameter, (2) multiple nuclei near their centers, and (3) locations at the center of clusters. These observations, when combined with the motions of supergiant elliptical galaxies within clusters, suggest that these special galaxies formed from galactic cannibalism. How? By close encounters at the centers of clusters that tidally strip material from other galaxies; then the stripped-away material is harvested to promote the growth of a supergiant elliptical galaxy.

Galaxies do not actually have to merge to show the effects of tidal forces. A modestly close encounter would also have observable consequences. Material would be pulled out in tongues from both sides of each galaxy; and because galaxies rotate, their material after an encounter would flow off in arc-shaped streams. We expect that bridges of material might join two tidally interacting galaxies with tails pointing away from each in opposite directions. Some galaxies with peculiar shapes that do not fall into the standard Hubble categories show indications of tidal interactions (Fig. 18.23). Note a bridge of material between the galaxies and the tadpolelike tails extending from them. Computer simulations show that these forms result naturally from a tidal interaction during a close encounter.

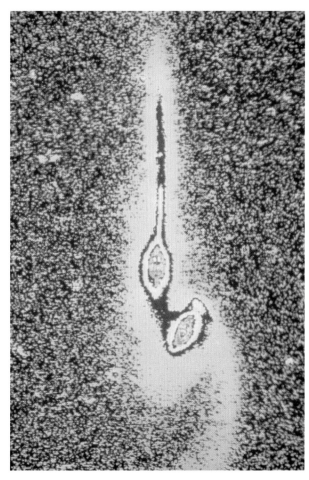

Figure 18.23 The galaxies NGC 4676a and 4676b, a tidally-interacting pair. The false colors in this computer-enhanced photograph bring out the details of the gravitational interaction, such as a bridge of red material between the two galaxies. Note the tadpole tails extending from each galaxy.

Computer simulations also indicate the possibility that collisions can transform spiral galaxies into elliptical ones. The interaction of stars, gas and dust, and a halo of dark matter results in the merger of the two galaxies into a single one (Fig. 18.24). The collision drives vast amounts of gas into the center of the resulting elliptical, which might spur the formation of a supermassive black hole there.

Tidal interactions may well trigger the short, large-scale episodes of star formation seen in many irregular galaxies – sometimes called **starburst galaxies.** In particular, the gas and dust throughout a galaxy respond strongly to the tidal forces of a companion, even without an actual merger of the two, such that the gas becomes unstable and starts to clump.

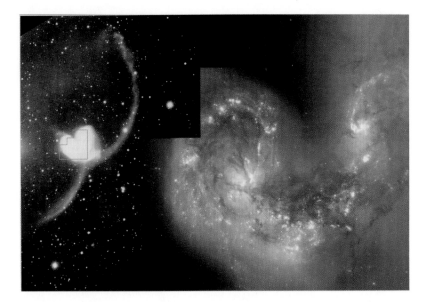

Figure 18.24 Colliding galaxies: NGC 4038 and 4039 in the constellation Corvus, 63 million ly from earth. The blue spirals are young stars born from the collision.

A notable example of such a galaxy is the irregular galaxy M82 (Fig. 18.25), which is thick with dusty clouds of gas in its nuclear region. Infrared observations demonstrate that these are active sites of star formation. Radio data show radio hot spots – giant H II regions – that coincide with the star-forming regions found in the infrared. The companion to M82 is the spiral galaxy M81, which is close enough to have tidal effects.

A few very luminous galaxies emit 95 percent of their energy in the infrared region of the spectrum. HST has imaged one of these, called Arp 220, a super example of a starburst galaxy. The merger of two spiral galaxies may have formed Arp 220.

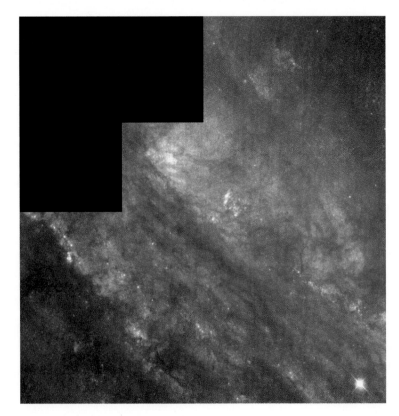

Figure 18.25 HST image of a starburst galaxy (NGC 253). Note the clumpy gas clouds, darkened dust lanes, and young, luminous star clusters.

The HST image reveals gigantic clusters of young stars being produced at a furious rate from the gas and dust of the former galaxies (Fig. 18.25). Other HST images show new star clusters, about the size of globular clusters, in the core of a peculiar galaxy that probably is the head-on collision of two spirals. Eventually, a giant elliptical galaxy may form. Such mergers of spirals may well explain why spirals are less numerous now than in the past.

18.7   SUPERCLUSTERS AND VOIDS
(Learning Outcome 18-12)

Are there clusters of clusters of galaxies? Do **super-clusters** exist? Yes! A **Local Supercluster** has a diameter of roughly 100 Mly. It contains the Local Group as well as the Virgo cluster among others, for a total mass of some 10^{15} solar masses. The Local Supercluster appears to be somewhat flattened, which may mean that it is rotating. The center of mass lies in or near the Virgo cluster; our Galaxy and M31 lie on the outskirts. Most of the supercluster is empty space: 98 percent of the visible galaxies are restricted to 5 percent of the volume. These clouds of galaxies outline a disklike structure six times wider than it is thick – a true cosmic pancake of clusters of galaxies!

Before observations drove home the reality of superclusters, most astronomers imagined that the universe contained more or less spherical clusters of galaxies embedded in a more or less uniform background of noncluster galaxies. That image has changed drastically.

For example, the Hercules supercluster (Fig. 18.26) covers more than 600×10^9 cubic light years of space. (It lies at a distance of 550 Mly, if H is 20 km/s/Mly.) The supercluster itself occupies a broad band, spreading out 250 Mly. In front of the supercluster lies a **void** some 250 Mly deep that separates the supercluster from foreground galaxies.

The superclusters that have been well mapped to date exhibit common features. First, they confirm the existence of superclusters as organized structures composed of multiple clusters of galaxies. Second, and this was a surprise, they contain large areas devoid of visible galaxies. These voids must be an integral part of the process that forms superclusters – a process about which we have only vague ideas right now. Third, streams of galaxies appear to

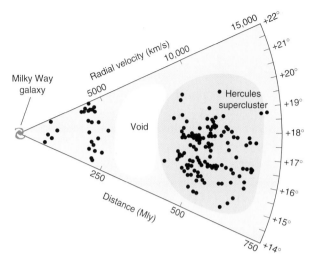

Figure 18.26 A slice of the Hercules supercluster. Shown here are the radial velocities (from the redshifts), distances (for $H = 20$ km/s/Mly), and positions in the sky for the galaxies in the supercluster. The Milky Way lies at the vertex of this wedge diagram. Note the clumping of the galaxies and the void of space in front of the supercluster. (Adapted from a diagram by S. A. Gregory and L. A. Thompson.)

connect the main concentrations in superclusters. Although we have examined only a few percent of the sky in depth, we are beginning to catch a glimpse of the architecture of the cosmos.

Even the apparent voids may not be empty. Several dwarf galaxies have been found in a void in the direction of the constellation of Bootes. This gigantic void covers some 30 million cubic light years of space. The fact that a few, albeit small, galaxies reside within a void has important consequences for any models for the origin of the universe of superclusters.

Finally, keep in mind that we have so far seen just small slices of the cosmos in three dimensions. For instance, the Galaxy and the Local Group appear to be moving at speeds of a few hundred kilometers per second toward the Hydra–Centaurus region of the southern sky. This observation has resulted in the notion of the "Great Attractor" – a conglomeration of galaxies covering about a third of the southern sky. A survey of some 17,000 galaxies here reveals a grouping of galaxies at a redshift of 4000 km/s. Its mass may account for the motion of the Local Supercluster in that direction.

The Cosmic Tapestry

What does the universe of galaxies look like on a grand scale? If we put together a map of the nearby superclusters, we get a view riddled with structures. This map (Fig. 18.27) shows that galaxies cluster in knots and filaments in a hierarchical fashion. These are chains of galaxies and clusters of galaxies (looking like a chain-link fence). Nearly all galaxies are within these clusters, with huge holes between them devoid of luminous matter. Is this clumpiness the character of the cosmos on larger scales?

No. A recent redshift map has expanded our view of the geography of the cosmos out to some 4 Gly (Fig. 18.28). More than 100,000 galaxies dot the map in majestic arcs and filigree patterns that span millions of light years. Visible are giant agglomerations of galaxies – and nothing larger in the big picture. They mark the end to the greatness of structure in the universe. From this zoomed-out view, the cosmos takes on a smooth and featureless character. At least, for the visible matter!

18.8 INTERGALACTIC MEDIUM AND DARK MATTER (Learning Outcomes 18-11, 18-13)

Is intergalactic space empty? Or does an intergalactic medium exist, in analogy to the interstellar medium? We can look for the **intergalactic medium** in two locations: between the clusters of galaxies and within clusters of galaxies.

To get some idea of how much material might be in an intergalactic medium, imagine the following. Take the matter from all the galaxies we can see and spread it out over the entire volume of space that we can observe. This spread-out material would have a density of about 4×10^{-28} kg/m³. (That's about 2 hydrogen atoms every 10 cubic meters.) What is the evidence for such a density of dust, neutral hydrogen, or ionized hydrogen?

Ionized hydrogen (H II) is the most likely candidate for the intergalactic medium. Because intergalactic material does not have a high density, ionized hydrogen would take a very long time to find an electron and recombine. If the gas is hot (in the 10^7 K range), you can search for x-ray or ultraviolet emis-

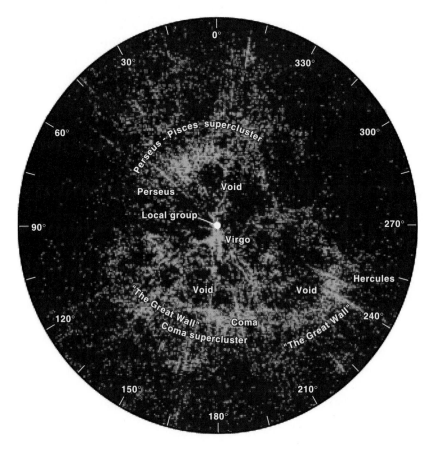

Figure 18.27 The nearby universe of more than 14,000 galaxies shown by the layout of nearby superclusters. Our Galaxy lies in the center of this schematic representation, which spans 500 Mly in space. Each dot represents a galaxy; at the cores of clusters, it is not possible to show the galaxies individually. The view spans 360° around the sky and has a depth of 36°. The Virgo cluster dominates the Local Supercluster. Note the immense voids between the Milky Way and the great Coma supercluster. (Adapted from a diagram by J. Huchra, M. Geller, and R. Marzke of the Harvard–Smithsonian Center for Astrophysics.)

sion. X-ray observations of local superclusters show sources that are contained within superclusters. The sources consist of spots centered on clusters; this implies that a hot gas exists in superclusters, that it is highly clumped, and that little gas exists between clusters. The gas has a temperature of some 10^8 K.

Another good candidate for matter within clusters is (again!) ionized hydrogen gas. X-ray observations back up this idea. Many clusters of galaxies are known to emit x-rays (Fig. 18.29); they tend to lie in the central regions of the clusters. The x-ray luminosities of clusters range from 10^{36} to 10^{38} W. The diameters of the x-ray-emitting regions range up to 5 Mly.

A reasonably confirmed model for this x-ray emission is that it comes from hot, ionized gas. Such a gas tends to settle around the center of mass of a cluster. It has typical temperatures of 10^7 to 10^8 K and densities of about one ion per cubic meter; such properties nicely explain the x-ray observations. In other words, we have reasonable evidence of intergalactic gas in clusters – about equal to the amount of mass in the galaxies themselves. But its density does not appear to be sufficient to bind the cluster gravitationally.

Consider the Coma cluster again. Its x-ray emission indicates that no more than 20 percent of the total mass in the inner 4 Mly is in the form of hot plasma. Recall that the visible galaxies make up just a few percent; the rest is dark matter. The x-ray-emitting material does not solve the dark matter problem for us, nor does it appear to be sufficient to close the universe. We are getting close, though, because the M/L for clusters is about 300, and we need about 700 for a closed universe.

We know that matter resides generally in clusters in a dark form that is undetected so far. That matter may be in the form of elusive particles such as neutrinos (especially if neutrinos have even a *very* small mass). If such dark matter exists in a proportion of about ten to twenty times that of visible matter, then the dark matter controls the structures of clusters and superclusters, with the luminous matter as the visible tip of the large-scale clustering of matter.

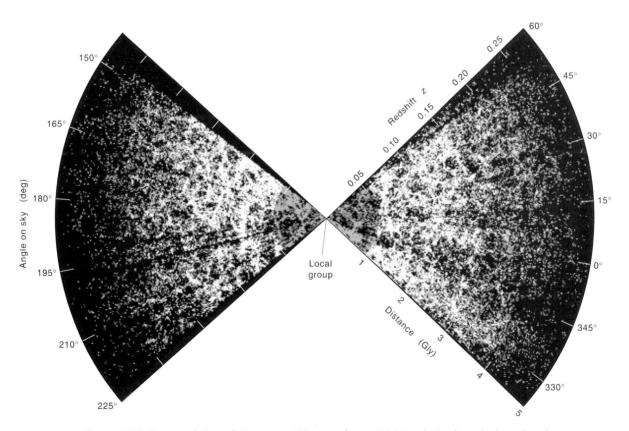

Figure 18.28 Two grand slices of the cosmos. This map of over 100,000 galaxies shows both north and south views out of the Milky Way. The web-like pattern marks the superclusting. At the far reaches, large-scale structure seems to disappear. (Adapted from a diagram by P. Norberg and S. Cole, 2-Degree Field Galaxy Redshift Survey.)

Figure 18.29 A map of X-ray emission from the Hydra A cluster of galaxies (840 Mly distant). The emission comes from a hot (30,000,000 K!) cloud of ionized gas within the cluster. It spans several million light years. Note how the emission is strongest around the central region.

KEY CONCEPTS

1. Distances to galaxies are essential to finding out their properties; they can be estimated in a relative sense from brightness and angular sizes.

2. Galaxies come in three main types, based on shape: ellipticals (dwarf, giant, and supergiant), spiral (normal and barred), and irregulars.

3. Measuring distances to galaxies is difficult and relies on indirect schemes. The basic trick is to find very bright objects whose luminosities can be reasonably estimated, identify these in the target galaxies, compare their flux to their luminosity, and use the inverse-square law for light to infer their distance.

4. Measured redshifts and distances of galaxies result in Hubble's law and Hubble's constant, H. The greatest uncertainty in H (its value lies between 15 and 30 km/s/Mly) arises from the uncertainties in distances.

5. Galaxies differ in terms of size, mass, luminosity, mass–luminosity ratio, spin, and colors – all of which reflect their content of stars, gas, and dust. In general, galaxies have mass–luminosity ratios that are greater than one, an indication of dark matter.

6. Galaxies come in clusters, containing from two to thousands, held together (at least for a while) by the gravity between the galaxies. The mass–luminosity ratios are around 100, an indication of a large fraction of dark matter.

7. The Local Group contains around thirty galaxies, mostly of low mass and luminosity, dominated by the Milky Way and Andromeda galaxies. The Local Group spreads over a volume some 3 Mly in diameter.

8. Clusters are grouped in superclusters; for example, the Local Group is one small piece of the Local Supercluster (which is centered on the Virgo cluster). Superclusters seem to be surrounded by voids and to come in long chains; they are the largest entities in the cosmos. At least a few superclusters are known to have a mass–luminosity ratio of a few hundred.

9. A very thin, very hot gas exists between clusters and superclusters; within clusters, hot, thin, ionized gas exists and is observed by the x-rays it emits. This plasma does *not* account for the dark matter in clusters, whose nature is as yet unknown.

STUDY EXERCISES

1. Which galaxies generally appear redder, ellipticals or spirals? (Learning Outcome 18-1)
2. Which galaxies contain more gas and dust relative to their total mass, spirals or irregulars? (Learning Outcome 18-1)
3. At the same distance from us, would irregular galaxies appear larger than spiral ones? (Learning Outcomes 18-1, 18-2, and 18-3)
4. How do astronomers know that other galaxies are made of stars? (Learning Outcome 18-4)
5. In recent years the value of Hubble's constant has been revised by some investigators from 30 to 20 km/s/Mly. How does this change affect the distances to galaxies inferred from redshift and Hubble's constant? The "age" of the cosmos? (Learning Outcomes 18-6, 18-8, and 18-9)
6. Describe how cepheid variable stars can be used to estimate distances to nearby galaxies. State the assumptions and limits of this method. (Learning Outcomes 18-6 and 18-7)
7. Why must intergalactic gas be both ionized and hot? (Learning Outcome 18-11)
8. Describe the layout of a supercluster of galaxies. (Learning Outcome 18-12)
9. How do astronomers see the depth of superclusters in space? What uncertainty exists in their technique? (Learning Outcomes 18-6 and 18-12)
10. What is the most common type of galaxy found in the regions of space that we can probe? (Learning Outcomes 18-1 and 18-10)
11. What is the basic information that implies that a large fraction of the matter in clusters of galaxies is in the form of dark matter? (Learning Outcome 18-13)
12. The observed rotation curves of most spiral galaxies become flat at large distances from their centers. What can we infer about the distribution of mass in these galaxies? (Learning Outcome 18-2)
13. How do we know that the space between galaxies and clusters of galaxies contains a very small amount of dust? (Learning Outcome 18-11)
14. Imagine that you observe two spiral galaxies through a telescope. The smaller one appears to have an angular diameter that's ¼ that of the larger one. How far away is the smaller galaxy relative to the larger one? Which galaxy would appear fainter? (Learning Outcome 18-3)

PROBLEMS AND ACTIVITIES

1. Consider a galaxy whose redshift is 0.10. What is the radial velocity of the galaxy? What is its distance if H is 20 km/s/Mly?
2. Imagine a dwarf galaxy at the edge of the Local Group. What is its orbital period?
3. Consider a hot, opaque cloud of hydrogen gas at a temperature of 10^6 K. At what wavelength would its spectrum peak?
4. If our Galaxy and M31 make up an orbiting pair, what is their orbital period?
5. The typical speed of a cluster of galaxies within a supercluster is about 300 km/s. How long would it take a cluster to move across a supercluster? How does this compare to the age of the universe?
6. The Coma cluster has a radial velocity from its redshift of about 6650 km/s. What is its distance?
7. The angular diameter of M31 is about 4.5°. What is its diameter-to-distance ratio? What is its diameter?
8. A new spiral galaxy is discovered. Its angular diameter is found to be 10 arcmin. What do you estimate its distance to be?

19

Cosmic Violence

After studying this chapter, you should be able to:

19-1 Outline the observational evidence for violent activity in our Galaxy and other galaxies, with special emphasis on synchrotron radiation, and an overall view of their spectra.

19-2 Compare and contrast the nuclei of active galaxies to ordinary ones (such as the Milky Way Galaxy).

19-3 List important observational characteristics of quasars and specific types of AGNs.

19-4 Sketch a physical model that could account for the observed characteristics of quasars.

19-5 Discuss the significance of redshifts for quasars for both their emission and absorption lines and their estimated distances.

19-6 Summarize the evidence for quasars as the active nuclei of distant galaxies.

19-7 Compare and contrast the nucleus of the Milky Way Galaxy, active galactic nuclei (AGNs), and quasars.

19-8 State the importance of the discovery of extragalactic gravitational lenses.

19-9 Outline a generic model for AGNs and quasars, using a supermassive black hole as the power source.

19-10 Describe the observational properties of jets in the nuclei of galaxies and outline a possible model for them.

CENTRAL CONCEPT

Observations, especially at radio wavelengths, show violent activity in peculiar objects beyond the Milky Way Galaxy. These may be powered by supermassive black holes.

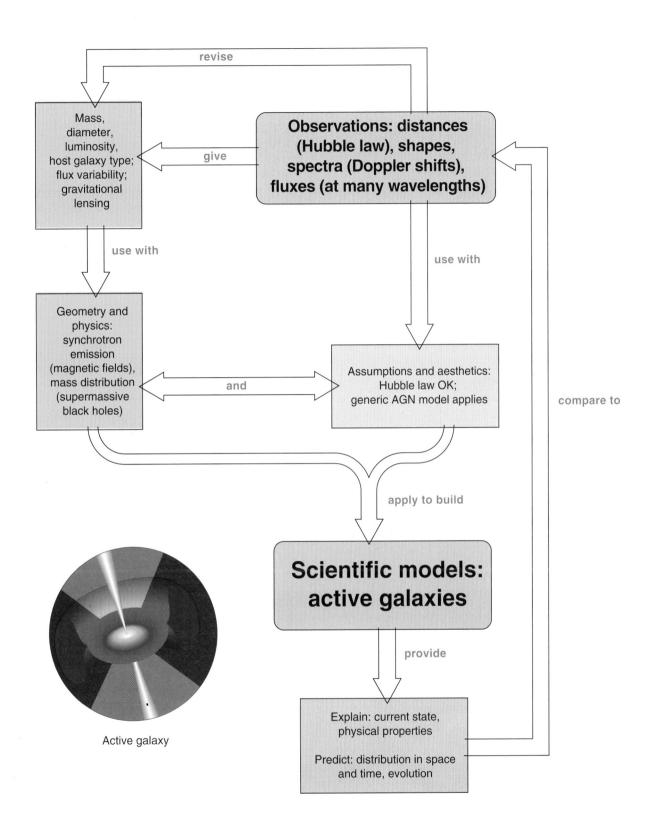

revise

Mass,
diameter,
luminosity,
host galaxy type;
flux variability;
gravitational
lensing

give

**Observations: distances
(Hubble law), shapes,
spectra (Doppler shifts),
fluxes (at many wavelengths)**

use with

use with

Geometry and
physics:
synchrotron
emission
(magnetic fields),
mass distribution
(supermassive
black holes)

and

Assumptions and aesthetics:
Hubble law OK;
generic AGN model applies

compare to

apply to build

**Scientific models:
active galaxies**

provide

Explain: current state,
physical properties

Predict: distribution in space
and time, evolution

Active galaxy

The universe appears calm, caught up in the well-controlled generation of energy in stars and the strict Newtonian dance of matter. Until the middle of the twentieth century, we saw it as a gentle cosmos. Rare outbursts such as supernovas only occasionally shattered the stillness. The advent of radio astronomy ripped the veil from a violent universe. Radio astronomers found the nucleus of our Galaxy a strong radio emitter. They also detected intense radio sources beyond the Galaxy, objects at enormous distances that required a tremendous outpouring of energy. By the 1980s, observational evidence pinned down violent events in the nuclei of galaxies of many types. These are called *active galaxies* for their energetic properties.

Observations of jets emerging from the nuclei of active galaxies have reinforced this view. These jets appear as cosmic umbilical cords, channeling the power from the nucleus out to immense reservoirs of charged particles. Supermassive black holes in the nuclei of such galaxies may power these beamed outflows.

Quasars, starlike-looking objects, top the range of cosmic violence. They have the largest known redshifts of any extragalactic objects. If the redshifts of quasars result from the expansion of the universe, then the quasars are the most energetic bodies in the universe. At the fringes of the cosmos, quasars represent some of the first-formed objects in the visible universe – and the most powerful. How quasars produce their enormous energies seems solved: observations imply that quasars are the nuclei of young, hyperactive galaxies.

19.1 VIOLENT ACTIVITY IN GALAXIES

The nuclei of many galaxies harbor a compact arena of violent events. These have been lumped into the category of **active galaxies,** in contrast to normal galaxies (Chapter 18). The nature of the physical conditions and processes at the heart of an active galaxy signal its special nature. (See the Active Galaxies Celestial Navigator™.)

Evidence of Violence in Our Galaxy (Learning Outcome 19-1)

Our home Galaxy's nucleus shows signs of cosmic violence. The nucleus lies at the center of the strong radio source Sagittarius A, a part of which appears to mark the actual core of the Galaxy – a region smaller than about 10 AU. It is surrounded by clumps of ionized gas that move at speeds of about 100 km/s. This region exhibits a jetlike structure above and below the galactic plane from the nucleus. It is also an intense source of gamma rays (Fig. 19.1). (You will see that such systems typify many active galaxies.)

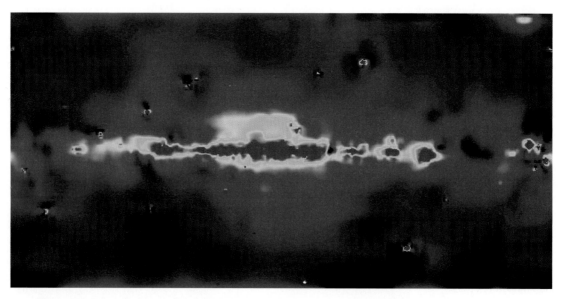

Figure 19.1 False-color image of the Milky Way Galaxy's halo of gamma rays (pale blue), most intense in the plane and at the galactic center. Brown and green regions show known sources of gamma rays.

The spectrum of the nucleus is, in part, nonthermal. That is, some of the emission comes from the *synchrotron process*. So high-energy electrons, moving at speeds close to that of light, spiraling in huge magnetic fields, subsist in the nucleus.

Galaxies with nuclei more active than ours display energy outputs at a much higher level of violence. You can regard the nucleus of the Galaxy as a scaled-down example of the emission from the nuclei of more active galaxies. The key clues are (1) emission over a wide range of wavelengths (usually nonthermal in character), (2) radio output concentrated in a small space, (3) most of the energy coming out in the infrared, (4) variability in the emission, and (5) jetlike structures arising from the nucleus (Fig. 19.2).

We will be dealing with large redshifts in the spectra of the objects discussed in this chapter. Keep in mind that astronomers define the amount of the redshift by the change in wavelength of a spectral line divided by the line's wavelength at rest.

Synchrotron Emission Revisited (Learning Outcome 19-1)

The process of synchrotron emission (review Enrichment Focus 16.1) underlies the basic physics of this chapter so firmly that you must have the basic concept in mind. Whenever charged particles accelerate, they give off electromagnetic radiation. That's basically how a radio transmitter works; a transmitter forces electrons to accelerate back and forth at the frequency of transmission. These accelerated charges emit at that frequency of oscillation.

Now magnetic fields bend the paths of charged particles – electrons are twisted more than protons because electrons have less mass. So as electrons move through magnetic fields, their paths are bent and the electrons accelerated. (Remember that curved motion is accelerated motion according to Newton's laws.) The electrons emit electromagnetic radiation as they spiral along the magnetic field lines. The magnetic fields do not have to be especially strong (the earth's field strength or less) to accomplish this twisting.

In general, the faster electrons travel, the more energetic (shorter in wavelength, higher in frequency) the radiation they emit. As electrons continue to emit, they lose energy and slow down. To keep a synchrotron source powered up, there must

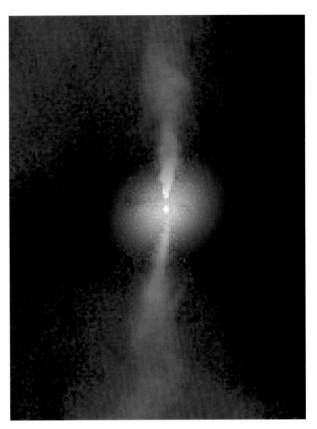

Figure 19.2 Radio emission from the active galaxy M 84 (optical image in Fig. 18.22). This VLA false-color map shows that the emission extends in opposing jets from the nucleus. The nucleus of the galaxy contains the engine that drives these high-speed, bipolar outflows.

be a supply of electrons moving close to the speed of light.

Synchrotron emission turns out to power most of the strange objects presented in this chapter. How do we know? From the shape of the spectrum and the polarization of their energy output.

Active Galaxies (Learning Outcome 19-2)

Radio astronomers have found that many active galaxies have nonthermal spectra. But how active is "active"? I will use the term here in contrast to "normal," which applies to our Galaxy. Basically, an active galaxy's spectrum does *not* look like a collection of stars. It is mostly nonthermal and has infrared, radio, ultraviolet, and x-ray outputs greater than those in the optical (Fig. 19.3). (Our own Galaxy does emit at all these wavelengths, but not nearly as strongly as an active galaxy.) Active galaxies make up only a small percentage of all known galaxies, but they pack the most punch.

Look carefully at the spectra of galaxies (Fig. 19.3). Note that a normal galaxy has a blackbody spectrum that peaks in the optical range – just what you'd expect from a group of stars. In contrast, active galaxies have a strong peak of emission in the far infrared from heated dust. The synchrotron part of the emission crops up most noticeably in the radio. Although the emphasis here is on the synchrotron emission (which can occur at any wavelength), you should not think that it is the *only* source of radiation from active galaxies. Their spectra combines synchrotron as well as the continuous emission from stars and dust (which is radiated in the infrared by dust grains heated to temperatures of a few hundred kelvins from the absorption of other forms of electromagnetic radiation). In addi-

tion to these three forms of continuous emission, we may also find emission lines from thin, hot gases.

Because the activity originates in the nucleus of an active galaxy, astronomers call these **active galactic nuclei,** or **AGNs** for short. The central issue is: what powers AGNs?

19.2 RADIO GALAXIES (Learning Outcomes 19-3, 19-10)

The largest class of active galaxies consists of the **radio galaxies,** which have strong radio emission. Two principal types of radio galaxy have cropped up: **compact** and **extended.** In an "extended" galaxy, the radio emission is larger than an optical image of the galaxy. "Compact" indicates that the radio emission is the same size or smaller.

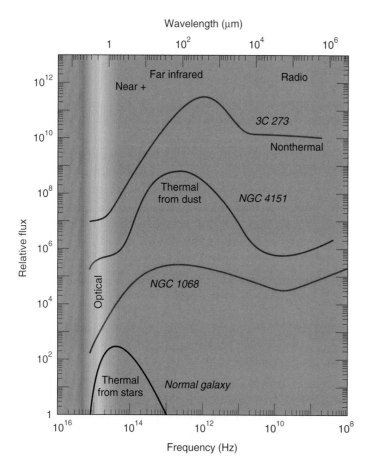

Figure 19.3 Comparison of the continuous spectra in the radio, infrared, and optical regions for a quasar (3C 273), two active galaxies (NGC 4151 and NGC 1068), and a typical normal (spiral) galaxy. Whereas the quasar and the active galaxies have a broad peak in their emission at far-infrared wavelengths (thermal emission from dust), the normal galaxy shows a Planck curve peak in the visible region from the combined emission of all its stars. The spectrum in the radio region is nonthermal. (Adapted from a diagram by R. Weymann.)

Compact radio galaxies often display very small radio sources, typically no more than a few light years in size, and coincident with the visible nucleus. Extended radio sources, in contrast, spread far beyond the optically visible galaxy – sometimes in two giant lobes of emission, up to millions of light years in extent, evenly balanced on opposite sides of the nucleus. A galaxy may show a compact source at its nucleus and also have the extended lobes. It then usually has **radio jets** connecting the nucleus to the extended radio emission.

Two High-Profile Active Galaxies

In a universe of stunning objects, active galaxies take top honors. Let's look at two examples to see what properties they have in common so that we can construct a model of their unusual activity.

Only 65 Mly away, Messier 87 is a fine example of a strong radio source and an active galaxy. A giant elliptical galaxy, M87 dominates the Virgo cluster of galaxies. Poking out from the core, a remarkable, optically visible jet fires out into space over a length of some 5000 ly (Fig. 19.4). The jet has a luminosity of roughly 10^7 solar luminosities; its emission is polarized. The jet contains blobs of material, each no more than a few tens of light years in size. Over two decades, the blobs have changed somewhat in flux and polarization.

M87 also emits x-rays with fifty times more energy than its optical emission – about 10^9 solar luminosities in x-rays from the whole galaxy. The jet itself also emits x-rays in the form of a line of knots. Radio maps confirm that the jet's radio emission coincides with the optical and x-ray emissions. The radio knots line up with the optical ones. So the jet overall emits over a wide range of frequencies, from radio to x-rays, and each knot of the jet generates this spectrum of energies. The synchrotron process within each knot produces this wide range of emission (Fig. 19.5).

Cygnus A, one of the strongest radio sources in the sky and one of the first discovered, provides an excellent example of the typical double structure of a luminous, extended radio galaxy. Its radio output, some 10^{11} solar luminosities, comes from two giant lobes set on opposite sides of the optical galaxy (Fig. 19.6). Each lobe has a diameter of 55,000 ly – about half the size of our Galaxy! The lobes hang

Figure 19.4 A jet shooting out from the nucleus of the giant elliptical galaxy Messier 87 (65 Mly distant), as imaged by the HST in the near-infrared. This false-color image shows he bright spot at center that marks the nucleus, which emits synchrotron radiation from high-speed electrons. The jet's emission is also synchrotron radiation and stretches about 5000 ly.

roughly 160,000 ly away from the central galaxy and contain a cloud of energetic electrons and vast magnetic fields that store more than 10^{52} J. A needlelike jet extends from the nucleus to one of the lobes. The radio lobes have a wispy appearance, with delicate swirls in which hot spots of emission are embedded.

The optical galaxy of Cygnus A is a supergiant elliptical galaxy with a dust lane down its middle. It has an active nuclear region, with a spectrum showing emission lines and a synchrotron continuous emission. Beyond 25,000 ly from the center, the spectrum is just that of a mix of stars; hence, the emission lines come from the nucleus alone. One key feature of these lines is that they are narrow in width (explanation to come in Section 19.3). Cygnus A is an intense source of x-rays (Fig. 19.7).

What powers the jets of M87 and Cygnus A? Observed at a wide range of wavelengths, the spectra of both jets are nonthermal. This property points to synchrotron emission as the source.

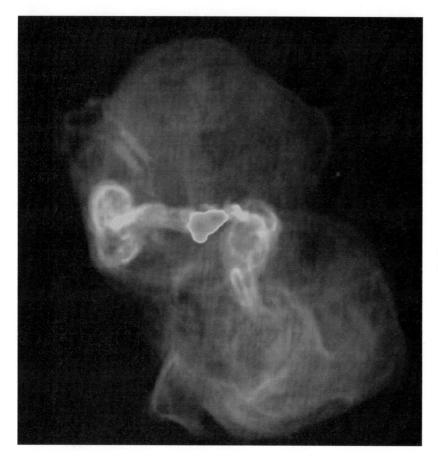

Figure 19.5 The active galaxy Messier 87 in the radio region, shown in images from the VLA. Huge bubbles of hot gas, some 200 kly across, are created by the jets, which in turn are thought to emanate from a black hole at the galaxy's center.

The large polarization of the jets' light in M87 (about 25 percent) confirms this idea. The fact that the knots in the jets coincide at different wavelengths implies that the same electrons power all the emission from the knots. The nucleus provides the high-energy electrons, which are expelled either as a fairly constant beam of particles or as a sequence of ionized blobs that are thrown out along a magnetic field. The ionized stream carries magnetic fields within it, and these help to channel the flows outward. The channel is leaky, which renders the jets' emissions visible.

Figure 19.6 The radio source Cygnus A. This false-color radio map of Cygnus A was made with the VLA. Note the thin jets from the bright nucleus and the filamentary structure in the lobes.

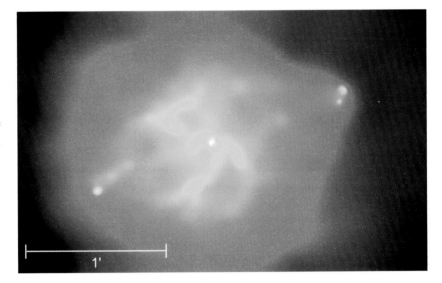

Figure 19.7 False-color x-ray image of the supergiant elliptical galaxy Cygnus A. The angular scale is at bottom left.

Other AGNs in elliptical galaxies also possess nuclear jets. In fact, radio jets are common. These jets, such as that in M87, terminate in a lobe of radio emission.

Structures of Radio Emission (Learning Outcome 19-7)

Radio astronomers have found that extended radio galaxies fall into a structural sequence, which provides clues about the environment in which galaxies are immersed. Low-luminosity tailed or distorted sources make up the majority of extended radio sources. The sequence relies on the amount of bending shown by the tails (Fig. 19.8), which ranges from 180° apart (double source – Fig. 19.8a) to 0° (head–tail source – Fig. 19.8f). The heads embrace radio jets.

What explains this structural sequence? Recall (Section 18.8) that clusters of galaxies contain a hot, ionized intracluster medium. Imagine that a galaxy, moving rapidly through this medium, shoots out material (high-speed electrons, for instance) in a jet. As the galaxy travels along, it leaves behind a radio-visible trail – a fossil record of where it has been. Here's an analogy. Imagine driving a car slowly and blowing smoke out an open window. The air stops the motion of the smoke, and it leaves a trail behind the car. Similarly, material flowing out of a galaxy is slowed by the intragalactic medium, and the moving galaxy leaves it behind.

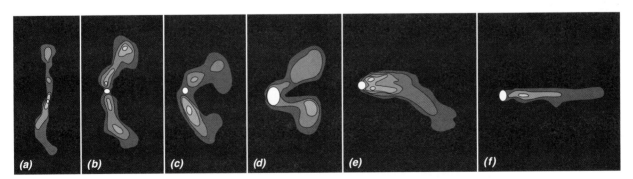

Figure 19.8 Sequence for the bending of radio tails around the nuclei of radio galaxies. These images show the extent to which the material from the jets interacts with the surrounding medium and the relative speed of the galaxy through it. The faster the speed, the greater the amount of bending. (Adapted from a diagram by G. Miley.)

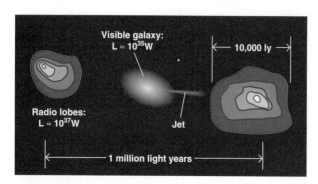

Figure 19.9 Jets, lobes, and radio galaxies. Schematic of a typical twin-lobed radio galaxy (L = luminosity). The visible galaxy is usually an elliptical one with one radio lobe on each side, separated by about 1 Mly. The lobes usually have peaks of radio emission called "hot spots"; one or both may be connected to the nucleus by a radio jet. (Adapted from a diagram by A. Bridle.)

How do the AGNs generate these lobes? We have not worked out all the details yet, but a crucial clue is the radio jets from the nucleus, which are often aligned with the lobes. These jets suggest that high-speed electrons are channeled by magnetic fields from the nucleus into the medium around the galaxy, where they pile up to form a lobe (Fig. 19.9).

Supercomputer simulations reveal the complex interactions of a supersonic jet as it tears into the medium around an active galaxy (Fig. 19.10). Because it is a plasma, the jet carries along a magnetic field that controls the flow of the jet. Still, the jet breaks up in turbulent eddies as it speeds through the medium. Such simulations, repeated under many different conditions, probe the physics and help astronomers to develop a deeper understanding of the processes involved.

TO SUM UP. Many radio galaxies emit in the form of lobes, which extend far beyond the visible galaxy and are connected to the nucleus by thin jets. The lobes are energy reservoirs that are a tangle of magnetic fields and high-speed electrons, powered by the nucleus, with material channeled to them along the jets.

19.3 SEYFERT GALAXIES AND BL LACERTAE OBJECTS
(Learning Outcomes 19-1, 19-3, 19-7)

I don't want to bog you down with all the strange members of the AGN zoo, but Seyfert galaxies and BL Lacertae objects are two galactic types that provide different insights into the AGN phenomenon at lower energy levels and with less drama. They offer nearby clues to help model AGNs.

Seyfert Galaxies

In 1943 Carl Seyfert (1911–1960) noted that some spiral galaxies showed unusual, broad emission lines. These groupings are now called **Seyfert galaxies.** About a hundred have been cataloged to date. Seyfert galaxies are close by, so they can be probed in detail. They provide good clues to a unified physics of AGNs.

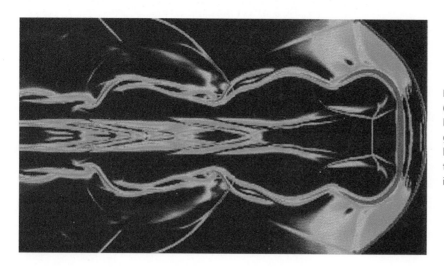

Figure 19.10 Supercomputer simulation (in false colors) of the flow of a jet (from left to right) through the medium of a galaxy. The model jet travels at 98 percent light speed and rams into and mixes with the interstellar medium. Note the breakup into turbulent eddies.

Figure 19.11 HST image of the Seyfert galaxy NGC 7742, a strong emitter of x-rays. Note the tremendously bright nucleus (about 3 kly across) and the bluish ring of star formation. This galaxy lies about 72 Mly away.

What's a Seyfert galaxy? Most galaxies have a bright nucleus, which looks fuzzy when viewed with a telescope. When imaged, a run-of-the-mill galaxy looks obviously fuzzier than a star. In contrast, a Seyfert looks like a bright star surrounded by a faint haze (Fig. 19.11).

Most Seyferts also have signature spectra: they show strong and broad emission lines. Why broad?

Consider a gas giving off emission lines. Some of this gas is moving toward us; its emission is blueshifted. Other blobs are moving away, and their emission is redshifted. When we observe a spectrum of the gas, the Doppler-shifted emission is added to the unshifted line at longer and shorter wavelengths. The emission line appears wider than it would be if the gas were stationary. If we attribute the widths of the lines to Doppler shifts produced by motion in the emitting gas, the gas in the Seyferts has very high random velocities, up to 10,000 km/s (Fig. 19.12). Some tens to thousands of solar masses of ionized gas in the nucleus moves at such speeds.

A minority of Seyferts exhibit narrow lines only, rather than a combination of broad and narrow lines. Here *broad* and *narrow* are relative terms. The gas for the region emitting the broad line churns around at about 10,000 km/s. In contrast, the region emitting the narrow lines plods along at "only" 1000 km/s. The optical emission from radio galaxies can also be split into narrow and broad categories; most fall into the broad category.

Seyferts are almost always spiral galaxies; only a few percent might be ellipticals. Compare this with the fact that most extended radio galaxies are elliptical. Overall, about 10 percent of all large spiral galaxies are Seyferts.

What ionizes the gas and stirs it up? Probably the energy source in the nucleus that generates the synchrotron emission. Also in the nucleus resides a source of high-energy electrons and gas. Observations indicate that the nuclei of Seyferts are small, only a few light years in diameter. Gas moving at 10,000 km/s would flow across such a tiny nucleus in only a century, so the gas must be replaced as it flows out.

TO SUM UP. Seyferts are (all?) spiral galaxies with active features. The most prominent are extremely small and bright nuclei; spectra that show emission lines, some of which are broad and indicate rapid motions. Also, many Seyferts have compact, low-luminosity radio sources within their nuclei.

A survey of spiral galaxies shows that Seyferts tend to be in close, binary galaxy systems. Tidal forces (Enrichment Focus 9.1) may then induce the Seyfert activity for a short period of time, making the Seyferts one form of starburst galaxy. Other Seyferts have been found to have two nuclei, a possible indication of a collision and merger from too close an interaction between a larger and a smaller galaxy (Fig. 19.13).

BL Lacertae Objects

My second AGN example consists of objects called **BL Lacertae objects** (**BL Lac** for short). BL Lac objects share the following characteristics: rapid variability (as fast as one day) at a wide range of wavelengths; extremely weak or no emission lines; nonthermal continuous radiation; strong and often rapidly varying polarization; and generally a starlike appearance.

Figure 19.13 HST false-color image of the spiral Circinus galaxy. Hot gas (pink) is ejected out of the center. Most of the gas is concentrated in two rings surrounding the central region. The outer (red) ring (700 ly from the nucleus) is actively producing new stars. The inner ring, inside the green disk, lies 130 ly from the nucleus, which surround a supermassive black hole.

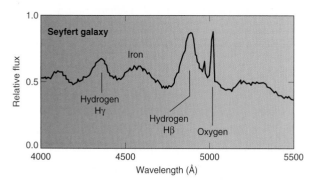

Figure 19.12 Generic emission-line spectrum of a Seyfert galaxy, showing the strongest lines. Note the especially broad lines of the hydrogen Balmer series and of iron. (Adapted from a figure by D. E. Osterbrock.)

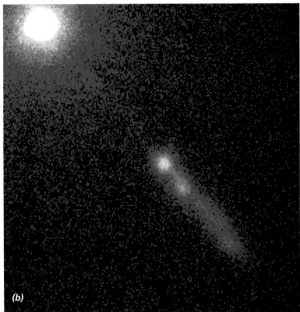

Figure 19.14 The quasar 3C 273, the first quasar to have its spectrum deciphered. *(a)* Optical image shows visible jet of energy. *(b)* X-ray image of the jet.

The BL Lac objects differ most from other active galaxies in that their emission varies frequently, erratically, and rapidly. For example, BL Lac itself fluctuates in luminosity by twenty times or so. Often BL Lac shows night-to-night variations of 10 to 50 percent in luminosity. That does not sound like much, but imagine our Galaxy changing its light output by some 20 percent in a day. That's like 10^{10} suns turning on and off simultaneously!

These rapid variations imply that the emission region of the BL Lac objects is small, no more than a light day or so in diameter. (This point was made in Section 16.6 for pulsars.) The same argument applies to active galaxies – the size of the emitting region equals the distance light can travel during the shortest variation of its brightness. A one-day variation means a size of one light day. (This argument applies only if the source emits into all directions in space.)

Most puzzling about the BL Lac objects is that they show almost *no* emission lines in their spectra! Synchrotron emission in the ultraviolet should ionize any gas near the nucleus and should produce emission lines. If the BL Lac objects are powered like other AGNs, where are their emission lines?

TO SUM UP. BL Lac objects are most peculiar because of their rapid variability and usual lack of emission lines. In contrast to Seyferts, they seem to be elliptical galaxies.

19.4 QUASARS: UNRAVELING THE MYSTERY
(Learning Outcomes 19-1, 19-2, 19-3)

During the boom of radio astronomy in the late 1950s, radio astronomers – like modern Tycho Brahes – compiled catalogs replete with radio sources that were not identified with any familiar visible objects. Hunting for possible culprits in 1960, Thomas Matthews and Allan Sandage discovered a faint starlike object – hence the name "quasi-stellar object," or **quasar** – at the position of radio object 3C 48 (object 48 from the Third Cambridge Catalogue of radio sources). This object had a spectrum of broad emission lines that could not be identified, and it emitted more ultraviolet light than an ordinary main-sequence star.

The quasar 3C 48 remained a unique object until 1963, when the strong radio source 3C 273 was identified with a faint starlike object (Fig. 19.14). The emission lines of 3C 273 were just as puzzling as the emission lines from 3C 48. Astronomers were baffled.

Quasar Redshifts
(Learning Outcome 19-5)

Maarten Schmidt finally deciphered the spectral code of 3C 273 by recognizing the prominent emission lines as those of the hydrogen Balmer series, redshifted by 16 percent compared to their normal at-rest wavelengths. After Schmidt had decoded 3C 273, Jesse Greenstein applied the same analysis to 3C 48 and found its spectrum to be redshifted by 37 percent. Other quasars fell in line with enormous redshifts (Fig. 19.15).

More than 2000 quasars have been identified, and redshifts have been measured for most of these. Many have redshifts that exceed 2.0; some even 5.0. Note that the redshift is that compared to the wavelength at rest. So a redshift of 2 means that the Lyman line of hydrogen (rest wavelength of 1216 Å) has shifted toward the red by 2 times (1216 Å × 2 = 2432 Å). If interpreted as redshifts from the expansion of the universe, the light from quasars must come from such distances that they must have originated gigayears ago. The redshifted quasars are the youngest objects we can see in the universe. Note that "youngest" here refers to objects at a point in their lives far in the past, closer to the Big Bang in time. Keep in mind that the telescope is a time machine that allows a view only of the past.

CAUTION. When the measured redshift approaches 1, the simple Doppler formula (Enrichment Focus 10.1) no longer gives the correct radial velocity. Usually a redshift of 0.16, for example, means a radial velocity of 16 percent the speed of light. But then a redshift of 2 would indicate that a quasar was moving away at 200 percent the speed of light! Not possible! A modified formula must be used instead, and all the radial velocities then come out to less than the speed of light. We measure redshifts but infer radial velocities.

General Observed Properties

The observed properties that make quasars unique and serve as identification tags are (1) starlike appearance with a large redshift, sometimes associated with a radio source (only about 10 percent are known radio sources); (2) broad emission lines in the spectrum with absorption lines sometimes present (usually the redshift of the absorption lines is less than that of the emission lines); (3) often variable luminosity; and (4) in those that are radio sources, aligned, double-lobed structures like Cygnus A.

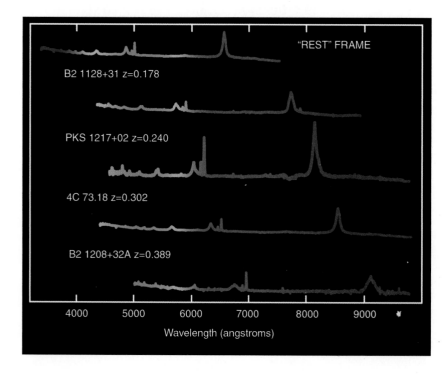

Figure 19.15 Spectra of quasars, showing its large redshifts as indicated by the emission lines of hydrogen. A comparison spectrum (top) establishes the wavelength reference scale, given for a source at rest with respect to the observer. The emission lines are the hydrogen Balmer lines at 4340 Å (blue), 4860 Å (green), and 6562 Å (red). From top to bottom, the quasars' lines are shifted more to the red end of the spectrum. These redshifts amount to 17.8, 24.0, 30.2, and 38.9 percent. The last one is moving away from the earth at about 100,000 km/s.

The main – and most remarkable! – feature of quasars is the size of their redshifts. The expansion of the universe results in the redshift of light from distant galaxies as described by Hubble's law. At first glance, the most natural explanation of the quasars' redshifts is a cosmological one: quasars participate in the expansion of the universe. If so, their enormous redshifts indicate that they are very far from us, and the light we observe was generated when the universe was very young. If they are as far away as indicated by their redshifts, however, they must expend vast amounts of energy. For example, 3C 273's redshift of 16 percent, if it is due to the expansion of the universe, implies a distance of 2.3 Gly, assuming $H = 20$ km/s/Mly. At this distance, 3C 273 emits about 10^{14} solar luminosities, or about 20 times as much as the most luminous normal galaxies. (Most of this energy is in the infrared.)

The Light from Quasars
(Learning Outcome 19-1)

Some quasars emit radio waves intensely, and all emit visible light. What produces this emission? A key clue comes from the spectrum of quasar radiation. Synchrotron radiation requires a continuous supply of clouds of energetic electrons and magnetic fields.

Optical observations of the quasars first discovered as radio sources have added more details to the general physical picture. The continuous optical emission is in part polarized, so synchrotron emission produces some of the optical radiation. Given a common magnetic field, the electrons that produce quasars' optical radiation must have higher energies than those that emit the radio radiation.

Line Spectra
(Learning Outcome 19-5)

All quasars have bright lines in their optical spectrum – the emission lines that are used to measure a quasar's redshift. Among the strongest emission lines in a quasar's spectrum are hydrogen's first Lyman series line (Lyman alpha) at 1216 Å and the second Balmer line (hydrogen beta) at 4861 Å (Fig. 19.16).

The emission-line spectra of quasars indicate

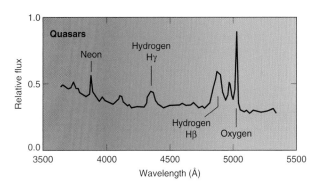

Figure 19.16 Composite spectrum of quasars showing the strongest emission lines typical for them. As for Seyfert galaxies, the lines of the hydrogen Balmer series are prominent. (Adapted from a diagram by J. S. Miller.)

that a low-density cloud of gas is radiated with photons energetic enough to ionize hydrogen. This radiation probably comes from the synchrotron emission from energetic electrons, which is a plentiful source of ultraviolet photons. So to our central synchrotron source, we add clouds or filaments of gas that convert ultraviolet radiation (and some x-rays) from the synchrotron source into visible emission lines.

The emission lines are extremely broad, which implies that the filaments or clouds from which the emission lines originate move rapidly (as argued for Seyferts in Section 19.3). The broadening of the emission lines arises from motions of the clouds or filaments themselves at radial velocities up to 20,000 km/s. This region, close to the nucleus, makes up the broad-line region in a model for a quasar. Narrow emission lines are not as obvious as the broad ones. They have widths that correspond to Doppler shifts of some 1000 km/s, so quasars have a narrow-line region farther away from the nucleus.

Many (but not all) quasars also have absorption lines in their spectra. Quasars with emission-line redshifts of less than 2.2 typically do not have absorption lines; those with greater redshifts have strong absorption lines. These lines are very narrow compared with the emission lines, so, by Kirchhoff's rules, we expect them to come from cool clouds of transparent gas. In general, the absorption-line redshifts are less than the shifts for the emission lines. And the absorption lines sometimes show more than one redshift.

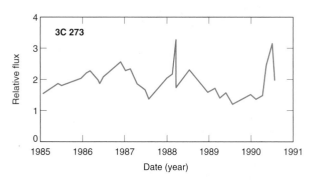

Figure 19.17 Variation of the optical flux of the quasar 3C 273. Note the large bursts and decline with some occasional smaller outbursts. (Adapted from a diagram by T J.-L. Courvoisier.)

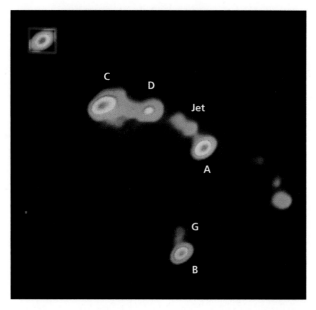

Figure 19.18 VLA map of the radio emission from the lensing of quasar 0957 + 561. The image shows up in the two elliptical regions (*A* and *B*). Blobs of emission (marked *C* and *D*) to the left of the upper quasar (*A*) are lobes at the end of a radio jet. The label *G* marks the galaxy, which causes the gravitational-lens effect.

What produces these absorption lines? Three possibilities are (1) gas clouds in or close to the quasar at temperatures cooler than the emission line regions, (2) absorption by intervening (and otherwise invisible) intergalactic gas clouds, and (3) a halo of gas around an intervening galaxy. The third model seems most likely (see the next section).

Variability in Luminosity

About 20 percent of the quasars exhibit rapid variations in light and radio output with periods on the order of days or weeks. This variability implies that the nuclear power source is very small, as was argued for BL Lac objects (Section 19.3) and any active galaxy with short-period changes in luminosity (Fig. 19.17).

19.5 GRAND ILLUSIONS: GRAVITATIONAL LENSES (Learning Outcome 19-8)

Consider two quasars only 6 arcsecs apart on the sky. Chance? Perhaps. Measure their emission-line redshifts. They are the same: 1.4. Not likely to be a coincidence! Astronomers were at first dumbfounded when they made these observations.

Then the insight came: the virtually identical emission-line redshifts of this "double quasar" forced astronomers to conclude that they were not separate twins but optical images of the *same* quasar. How so? Recall (Section 7.3) that general relativity predicts that masses will deflect the paths of light rays – in essence, acting like a lens.

If a very small, dense mass (a black hole, for instance) lay along our line of sight to a quasar, its image would be split into two, one above and one below the quasar's actual position. This phenomenon of image making by a mass is called a **gravitational-lens effect.** Such lenses tend to make imperfect, distorted images.

Radio astronomers found the "double quasar" quite puzzling at first, since the radio sources there had complex structures. There were radio blobs (*C* and *D* in Fig. 19.18) to the east of the northern quasar (*A*) that were not visible west of the southern quasar (*B*). It turns out that this emission is that of a jet flowing into two lobes!

Images of the quasars on a night of exceptionally good seeing show quasar *B* with a little bit of fuzz sticking out. This fuzz turns out to be the poorly resolved image of a faint elliptical galaxy – the gravitational lens! Since the galaxy is an extended mass, it acts like an imperfect lens and produces a complex pattern of images, as predicted by general relativity (Fig. 19.19). (The view is also complicated by the fact that the elliptical galaxy lies in a group of galaxies, and each member helps to bend the light.

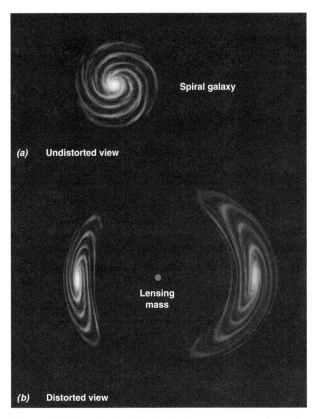

Figure 19.19 Gravitational lensing by large masses produces distorted images. In this example, you can see the distorted view of a spiral galaxy.

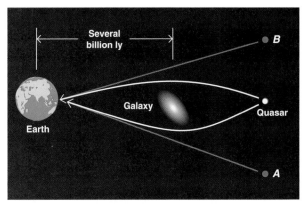

Figure 19.20 Schematic drawing of the optical illusion of the double quasar by a gravitational-lens effect. The bending of the quasar's light by the massive galaxy a few billion light years away makes two images (*A* and *B*) appear on the sky.

This cluster is closer to us than the quasar, for the cluster's redshift is only 0.4.) By a quirk of placement, we see an incomplete image.

A gravitational lens – an intervening, probably elliptical, galaxy – forms two images about halfway between us and the quasar (Fig. 19.20). The galaxy bulks up at some 10^{13} solar masses to bend the light enough to form the images.

This discovery, and others like it, have three important implications.

1. It provides another confirmation of general relativity.
2. It proves that in this case the quasar is more distant than the galaxy, so the quasar's redshift is cosmological.
3. Gas around the galaxy creates the quasar's absorption-line spectrum; this situation may be the case for other quasars as well.

We know from the observations of the bending of sunlight in the solar system that masses can, in fact, act like lenses. The light is bent just like light that is refracted when it passes through a lens. Depending on the distribution of the mass doing the lensing, the light from a distant quasar may be split into many images or distorted into rings, incomplete rings visible as arcs, or other distorted shapes.

Astronomers have found many instances of gravitational imaging with deep-seeing telescopes (Fig. 19.21). A dramatic example is the "Einstein cross" (Fig. 19.22). A quasar shows four distinct, bright images and a diffuse central object that is the bright core of the intervening galaxy. The quasar lies about 8 Gly away. Narrow absorption lines in its spectrum originate in the halo of the galaxy, which is at a distance of only 400 million ly, some twenty times closer. The mass of this galaxy causes the imperfect lensing effect. We now also have images of distant galaxies that are distorted by a closer cluster of galaxies, whose dark matter acts as an imperfect magnifying lens.

19.6 TROUBLES WITH QUASARS
(Learning Outcomes 19-4, 19-6)

If quasars are actually Gly away, then the quasar 3C 273, for example, must emit a total radiative energy in one second equivalent to all the energy produced by the sun in about 3 million years. A typical quasar produces up to 10,000 times as much power as an ordinary spiral galaxy.

Figure 19.21 HST image shows blue multiple images of the same galaxy. They are duplicated by the gravitational-lens effect of the yellow cluster of galaxies in the center, whose enormous gravitational field bends light to magnify and distort a more distant galaxy. This effect also distorts the blue galaxy's spiral shape into an arc.

Not only do quasars blast out energy at enormous rates, but the energy comes from relatively small regions of space in the centers of quasars – from possibly light hours to light months in diameter. Because quasars' light output varies on time scales of days to years, their energy-emitting regions cannot be larger than light days to light years – smaller than the distance from the sun to the nearest star.

Figure 19.22 Einstein Cross (gravitational lens G2237 + 0305), imaged by the HST. The central cloverleaf is not the nucleus of the galaxy, but light emitted by a background quasar. Gravitational lensing by the visible foreground galaxy breaks the quasar's light up into four bright images.

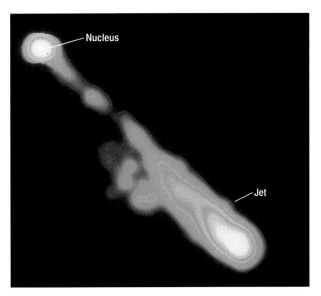

Figure 19.23 Radio jet of 3C 273: closeup view of the nucleus and the jet made at a frequency of 408 MHz. This false-color map has white as the strongest radio emission and blue the weakest.

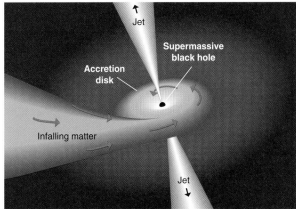

Figure 19.24 One model for a supermassive black hole powering a quasar (or active galaxy). An accretion disk, made of infalling material, surrounds the black hole; it is thick close to the hole and thins out away from it. A sharp funnel forms where material from the accretion disk streams into the black hole. The disk is hot, so gas blows off it. The funnel directs the gas up along the rotation axis of the accretion disk, creating a bipolar outflow from both sides of the accretion disk. (Adapted from a diagram by J. Burns and R. M. Price.)

Ponder that for a moment. Quasars typically emit about a hundred times the total energy output of our Galaxy from regions no more than light years in diameter!

Intercontinental radio interferometers can resolve very small angular diameters. They show that 3C 273 consists of separate radio components. The smallest parts – if the quasar is at the distance demanded by a cosmological redshift – cannot be more than a few light years across. The map also shows a radio jet (Fig. 19.23).

High-resolution studies have also revealed the radio structure of other quasars, especially those that are relatively near. These observations show that the radio structures are generally symmetrical doubles, similar to Cygnus A. And some of these show nuclear jets. In a few, the material in the jets appears to travel faster than light. This effect turns out to be another optical illusion caused by jets aligned very close to our line of sight (Enrichment Focus 19.1).

Energy Sources in Quasars (Learning Outcome 19-9)

What energy source lies in the heart of a quasar? The most developed quasar models invoke supermassive black holes. One such model relies on

a black hole of 100 million solar masses. This model builds on the model for binary x-ray sources with black holes, in which material from the normal star forms an accretion disk around the black hole before the material falls into it. A supermassive black hole in a dense galactic nucleus is fueled by the tidal disruption of passing stars (Fig. 19.24). The stellar material forms an accretion disk and radiates as it spirals into the black hole, thus powering the quasar. The model calculations show that luminosities of 10^{12} solar luminosities, about that of bright quasars, are possible. To feed the black hole requires about 1 solar mass of material a year.

The model must also deal with the nuclear jets now known for some quasars. One way is as follows. The accretion disk can restrict the flow of ionized gas from it, along its spin axis, either by magnetic fields or simply from the formation of a funnel of material around the black hole. This initiates a narrow beam that becomes visible as a jet, once it has passed beyond the main body of the quasar.

ENRICHMENT FOCUS 19.1

Faster Than Light?

Intercontinental radio astronomy picks out the finest details of distant radio sources – even quasars. Such observations have detected changes in five quasars that, if taken at face value, lead to the conclusion that parts of these objects are moving at speeds greater than that of light! Impossible! So says special relativity.

Motions that seem faster than light are called **super-luminal.** The familiar quasar 3C 273 features a jet that sticks out from its core. Observations from 1977 to 1980 (Fig. F.19) show that a knot in the jet moved away from the nucleus. If 3C 273 is at the distance implied by its redshift, the observed separation results from a motion

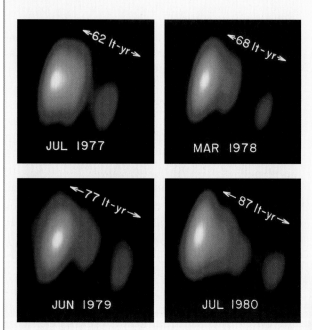

across our line of sight at ten times the speed of light! What's happening? Like the double quasar, the superluminal effect may simply be an optical illusion.

The sources that exhibit superluminal motions have a common feature: a radio jet from the nucleus. These jets are thought to contain high-energy electrons moving close to the speed of light. If quasars have such jets, and if the jets are almost (but not quite) pointing at us, a blob of material moving out along the jet will appear to travel faster than light speed.

The superluminal effect arises from an almost head-on orientation of the jet and the finite speed of light. Here's how. Suppose a jet points at us at an angle of 8° with respect to our line of sight (Fig. F.20). Imagine that a blob of electrons ejected from the nucleus (point N) gets to point A in 101 years and the light emitted from the blob at N takes 100 years to get to point B. For an 8° angle, the separation of A and B is 14 ly. But the light from B is one year ahead of that emitted by the blob when it finally reaches A. It has taken 100 years for the light from N to reach B, 101 years to reach A.

Many years later, the light that was at B reaches us. Only one year later, that emitted at A arrives. To us, it appears that the source has moved from B to A – a distance of 14 ly – in only one year. It seems to have a speed of fourteen times that of light across our line of sight. In fact, no such physical motion has occurred, only an optical illusion. The smaller the angle of the jet from our line of sight, the greater the superluminal motion will appear. So almost any faster-than-light speed seems possible because some jets will be tilted very close to our line of sight.

Figure F.19 Very-high-resolution radio maps, made at a frequency of 10.65 GHz, of the quasar 3C 273, which show the radio jet extending from the nucleus. Note how the emission moves out along the jet from 1977 to 1980. In addition, a blob of the jet has moved away from the nucleus from a distance of 62 ly in 1977 to 87 ly in 1980. At face value, the blob has moved faster than the speed of light.

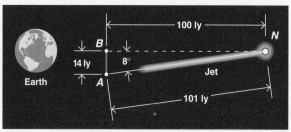

Figure F.20 Geometry for superluminal motion in a radio jet.

This model has a number of successful aspects. The supermassive black hole easily generates the level of quasar luminosity in a region of space only a few light years in size (the Schwarzschild radius of a 10^8-solar-mass black hole is only 3×10^8 km, or about 2 AU). It also does the energy conversion (from gravitational to radiative) with high efficiency.

Table 19.1 Comparison of Active Galactic Nuclei and Quasars

Property	Radio Galaxies	Seyfert Galaxies	BL Lac Objects	Quasars
Redshifts from spectra	0.01–0.3	0.003–0.06	0.05–0.6	0.2–4
Continuous spectrum	Nonthermal	Nonthermal	Nonthermal	Nonthermal
Emission lines	Broad and narrow (rare)	Broad and narrow	None or weak	Broad and narrow
Absorption lines	From stars in galaxy	None	None (?)	From intervening gas clouds
Shape (optical)	Elliptical	Spiral	Unclear	Starlike with fuzz
Shape (radio)	Jets and lobes	Weak emission from nucleus	Weak emission from nucleus	Jets and lobes

A Generic Quasar and AGN Model

You've probably noticed that quasars and AGNs share some observed characteristics (Table 19.1). Can a generic model explain them all and link them by common physical processes?

One popular idea interprets active galaxies and quasars as similar objects viewed from different directions, so that AGNs and quasars are driven by the same basic physics. This generic model tries to link together the nuclear engine (a supermassive black hole), an accretion disk, both the broad- and narrow-emission-line regions, and the jets – the key observational clues. Let's use these pieces to build a model.

A Model for Quasars and AGNs

A generic model has as the energy source a **supermassive black hole,** which is surrounded by an accretion disk a few light days in diameter (Fig. 19.25). The supermassive black hole provides the power and the central synchrotron source as it swallows material from the accretion disk. Twin jets emerge from the accretion disk; far out from this central region, they plow into gas to pile up in radio lobes. Surrounding the central engine are two zones of ionized gas. The smaller (few light months in diameter) and closer one has gas moving at high random speeds (some 10,000 km/s); this region produces the broad emission lines. Another ionized region filled with clouds that have slower random motions (only a thousand kilometers a second) extends farther out, up to a few thousand light years. Here the narrow emission lines are generated. Between these two regions lies a thick, dusty ring of molecular gas. It may have a diameter of 10 to 1000 ly.

Note that we can increase the power output from this model by simply making the supermassive black hole larger and the rate of mass infall greater. For Seyfert galaxies, a supermassive black hole of 10^6 to 10^7 solar masses would work; for quasars, we'd need some 10^8 to 10^9 solar masses. Other AGNs would fall within these ranges.

Our view determines what we see. Imagine a view almost edge-on (line A in Fig. 19.25). The dusty ring blocks our view of the broad-line region and the central source. So we see narrow emission lines, perhaps the jets, and the radio lobes, and we'd call the object a narrow-line Seyfert or a narrow-line radio galaxy. In contrast, if we observe the model more pole-on (line B in Fig. 19.25), we can see the central source, the broad-line region, and some of the narrow-line region, perhaps the jets, and the radio lobes. We would then call the object a quasar, a broad-line radio galaxy, or a Seyfert with mostly broad and some narrow lines. In either case, the dust in the ring can absorb ultraviolet and visual photons, heat up, and radiate in the infrared.

Now imagine that we observed the model very close to or right along the axis of the jets. Then we would observe apparent greater-than-light speeds and a very concentrated point of emission – a superluminal (faster than light) quasar or BL Lac object. In this orientation, we'd be looking through one radio lobe with the other on the far side of the galaxy. Then we would have a hard time detecting both lobes. At best, we might see a faint halo of radio emission (from the lobe in front) with a strong nuclear source embedded within. We might even see x-rays from the accretion disk.

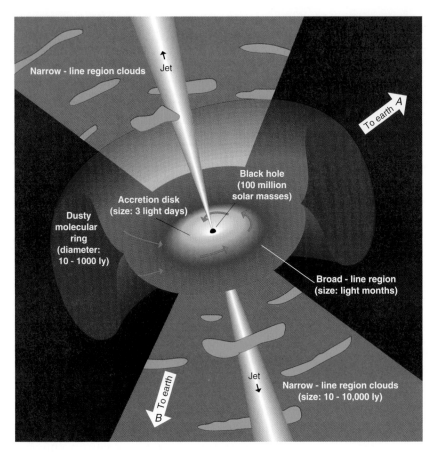

Narrow - line region clouds

Jet

Black hole
(100 million
solar masses)

Accretion disk
(size: 3 light days)

To earth A

Dusty
molecular
ring
(diameter:
10 - 1000 ly)

Broad - line region
(size: light months)

To earth
B

Jet

Narrow - line region clouds
(size: 10 - 10,000 ly)

Figure 19.25 Generic model for an AGN or quasar. The central engine may be a supermassive black hole. It is surrounded by an accretion disk made from infalling material. Close to the disk are blobs of fast-moving hot gas which make the broad emission lines in the spectrum. Much farther out is a dusty ring of molecular gas. Beyond that lie blobs of slow-moving hot gas; these make the narrow emission lines in the spectrum. If our line of sight is along A, the dusty ring blocks out the broad-emission-line region; only the narrow lines are visible. If our line of sight is along B, we see both narrow and broad emission lines. The orientation of the AGN in space determines our view.

These various orientations may explain why sometimes we see only one nuclear jet or one much stronger than the other. The relativistic beaming effect implies that one bipolar jet close to our line of sight will appear to be very strong, the other much weaker (or not at all). This generic, unified model naturally explains many aspects of AGNs and quasars: they may be essentially the same kinds of object viewed from different angles. Nature is sometimes the master of illusion!

If we accept a model for the energy output of active galaxies and quasars that imagines a black hole in the nucleus eating up huge amounts of material, the nucleus should show an intense, pointlike source of light (due to a concentration of stars around the black hole); in addition, stars orbiting the center should have high velocities, and emission lines with high Doppler shifts might be visible from the infalling matter. Observations of such effects have been reported for M87: a sharp peak of emission from M87's nucleus (such a peak is not found in the nucleus of normal elliptical galaxies),

as well as a dramatic increase in the velocities of stars in the nucleus compared to those outside the nucleus.

A model consistent with these observations is one of a 10^9 solar-mass black hole hiding in the inner 300 ly of the nucleus. This supermassive black hole would cluster the stars in the nucleus closely around it, and they would orbit at high velocities. In addition to M87, the Andromeda Galaxy needs some 10 million solar masses to explain the light distribution and motions visible in its inner 300 ly. This appears to be true for M32, a companion galaxy to M31.

By using spectroscopy, HST has confirmed the notion of a supermassive black hole in the cores of galaxies (Fig. 19.26). Recall (Section 18.5) the Doppler shift observations that result in the rotation curves for spiral galaxies. When HST observes across a galaxy's core, the blueshifts and redshifts make a telltale zigzag of a rotating disk. The total shifts imply speeds of a few hundred kilometers per second in the inner few tens of light

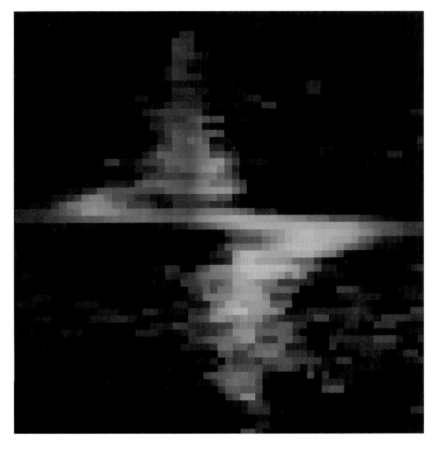

Figure 19.26 HST image of the "signature" of a supermassive black hole in galaxy M 84. The image shows the Doppler shift from the rotation of stars and gas. The sharp "zigzag" is the signature of the supermassive black hole. The Doppler shift indicates speeds of 400 km/s within 26 ly of the nucleus. That implies a black hole of at least 300 million solar masses.

years. Such results imply supermassive black holes of a few hundred million solar masses or greater!

Now that we have a good sample, it may well be the case that *every* galaxy has a supermassive black hole in its core. The AGNs simply have the most massive examples. It also appears that these core objects contain about 0.2 percent of the total mass of their host galaxy and range from 10^6 to 10^9 solar masses.

The Host Galaxies for Quasars (Learning Outcome 19-6)

Given that a quasar may be the hyperactive nucleus of a very faraway galaxy, what about the rest of the galaxy? At cosmological distances, the disk of a quasar's parent galaxy, called the *host galaxy*, would be too small and faint to be easily seen. For example, if the Andromeda Galaxy were placed at a distance of 4×10^9 ly – equivalent to a redshift of 0.2 (small for a quasar!) – its angular diameter would be a mere 4 arcsec. Remember that the earth's atmosphere limits seeing to roughly one arcsec, so a distant galaxy would be hard to distinguish from that of a faint star.

We might, however, look for a quasar in a cluster of galaxies – a good sign that we're not looking at stars. A nice example is the quasar 3C 206, which is surrounded by some twenty faint galaxies. The quasar's redshift is 0.206; that of a nearby pair of galaxies, 0.203 (and the same is presumed for the cluster). The essentially same redshifts not only show that the quasar resides in the cluster but also prove that the quasar's redshift is cosmological. In fact, almost all nearby quasars appear to have an extended structure associated with them – the underlying galaxy!

Observations suggest that the hosts are often distorted, as a consequence of tidal interactions with a nearby galaxy. About 30 percent of nearby quasars (redshifts less than 0.6) appear to be interacting with another nearby galaxy. Most of these are located in a small cluster of galaxies, such as 3C 206. Hence, the quasar-AGN phenomenon may be activated by interactions between galaxies in clusters.

Figure 19.27 HST image of a quasar merging with its companion galaxy. On the left is the quasar and the thin tidal arms of the galaxy; on the right is a higher-contrast image that shows the galaxy more clearly, just above the larger quasar. These objects are about 1.5 Gly from earth.

HST images of quasars' host galaxies drive this point home (Fig. 19.27). They encompass a variety, from normal to distorted galaxies. Many show structures of gas and dust that imply a collision and merger of two other galaxies – and some display mergers in action. If so, the mergers took place some gigayears ago within young clusters of galaxies.

KEY CONCEPTS

1. The nucleus of the Galaxy contains a small energetic source with a partly nonthermal spectrum; it emits radio, infrared, x-rays, and gamma rays.

2. Continuous spectra can be divided into two types: thermal (blackbody) and nonthermal. A common form of nonthermal emission is synchrotron radiation, produced by high-speed electrons spiraling in magnetic fields. Thermal emission can come from stars or from any hot, opaque material.

3. Active galaxies have much of their power provided by their nuclei, which have nonthermal spectra, at least in part. Some thermal emission (continuous emission) comes from stars, heated dust, and accretion disks.

4. Radio-emitting galaxies are one type of active galaxy; they fall into two classes: compact (emission from the nucleus) and extended (emission usually in two lobes widely separated from the nucleus).

5. High-resolution radio observations show that in many cases, a jet of emission extends from the galaxy's nucleus to at least one of the lobes of an extended radio galaxy. The synchrotron process may be the source of the emission; it shows that high-speed electrons are somehow channeled along the jet from AGNs out to the lobes.

6. Seyfert galaxies, another type of active galaxy, are spirals, with compact luminous nuclei and very broad emission lines in their spectra.

7. BL Lacertae objects (also active galaxies) have nonthermal spectra; they vary rapidly in luminosity and usually have no emission lines in the spectra of their nuclei.

8. Quasars have a starlike appearance, very large redshifts, and broad emission lines in their spectra (along with narrower dark lines); they sometimes have detectable radio emission with nuclear jets.

9. If the redshifts in the emission lines from quasars arise from the expansion of the cosmos, quasars are very far away – 10^9 light years. If this is true, their luminosities are huge, up to thousands of times what we find for normal galaxies. Yet the emitting region cannot be larger than a few tens of light years.

10. Quasars and AGNs share some of the same observational characteristics, so they might be the objects of the same types (galaxies) at different evolutionary stages. In particular, quasars are young galaxies with hyperactive cores, perhaps triggered by collisions and mergers.

11. We have pretty well identified the energy machine in the cores of quasars and AGNs; it is a supermassive black hole (surrounded by an accretion disk) eating up material in the nucleus. Such core black holes may be a common feature in all galaxies.

STUDY EXERCISES

1. What observational evidence do we have that the synchrotron process produces some radiation from AGNs and quasars? (Learning Outcomes 19-1, 19-2, and 19-6)
2. Contrast the evidence for violence in the Milky Way Galaxy with that for an active galaxy. (Learning Outcomes 19-2 and 19-7)
3. What evidence do we have that quasars are far away? *Hint:* How large are their redshifts? (Learning Outcomes 19-3, 19-5, and 19-6)
4. What observational evidence indicates that in many cases the light of quasars must pass through clouds of thin, cool gases? (Learning Outcome 19-4)
5. What is thought to power a quasar? (Learning Outcomes 19-4 and 19-9)
6. What connection does extended radio emission often have with the nucleus of an active galaxy? (Learning Outcomes 19-2 and 19-10)
7. Give one way in which the observed properties of radio galaxies and quasars differ; one way in which they are similar. (Learning Outcomes 19-6 and 19-7)
8. What physical conditions are required to produce synchrotron emission in AGNs and quasars? (Learning Outcome 19-1)
9. Suppose that you read an announcement in the newspaper tomorrow that astronomers have measured Hubble's constant to be twice the value used in this book. What would happen to the distances to quasars? (Learning Outcome 19-4)
10. How does the duration of variation in the light from an active galaxy give some indication of the size of the emitting region? (Learning Outcome 19-7)
11. Emission lines from quasars and some Seyfert galaxies are very broad. What physical process can account for this observation? (Learning Outcome 19-7)
12. In the generic model for an AGN or quasar, what is thought to be the source of power? (Learning Outcome 19-9)
13. Suppose that we observe the generic model for an AGN side-on so that we are looking right into the dusty ring surrounding the accretion disk and black hole. What would you expect to see? (Learning Outcome 19-9)
14. Suppose that we observe the generic model for an AGN face-on so that we are looking right onto the accretion disk and black hole. What would you expect to see? (Learning Outcome 19-9)
15. Many active galaxies show a broad peak in flux in their spectra at far-infrared wavelengths. What is a likely source of this emission? (Learning Outcome 19-1)
16. The spectra of AGN and quasars contain emission lines. What does this observation tell us about the physical conditions of at least part of the nuclear region? (Learning Outcomes 19-1 and 19-3)

PROBLEMS AND ACTIVITIES

1. The quasar 3C 273 has a redshift of 0.16. What is its radial velocity? Its distance? *Hint:* Assume the redshift is cosmological.
2. A typical quasar has a luminosity that is about 10,000 times that of the Milky Way Galaxy. If this energy were produced completely by fusion reactions, how much matter would a quasar convert to energy each second?
3. What is the Schwarzschild radius of a 10^8-solar-mass black hole?
4. What would be the angular diameter of the disk of the Milky Way Galaxy if placed at a distance of 5 Gly?

20 Cosmic History

LEARNING OUTCOMES

After studying this chapter, you should be able to:

20-1 State the basic assumptions of scientific cosmology about the physical universe.

20-2 Present, in a short paragraph, key observations that have cosmological import.

20-3 Describe briefly the Big Bang model, focusing on its assumptions, predictions, and observations.

20-4 Describe briefly the observed properties of the cosmic background radiation.

20-5 Present at least one argument for ascribing a cosmic origin to the background radiation and explain how it is a natural consequence of a Big Bang model.

20-6 Discuss the importance and the impact of cosmic radiation for the Big Bang model.

20-7 Outline the process of element formation in the standard Big Bang model and cite at least one observation that supports theoretical ideas.

20-8 Describe a physical basis for particle production from light in the young, hot universe.

20-9 Outline a history of matter and radiation from the time they stopped interacting strongly to the present.

20-10 Describe a model of galaxy formation from the Big Bang and pinpoint problems with this model, especially as influenced by the cosmic background radiation.

20-11 Identify at least two problems associated with the standard Big Bang model and explain how an inflationary model attempts to cope with them.

20-12 Specify the nature of the "dark matter problem" for the cosmic geometry and future of the universe.

CENTRAL CONCEPT

The universe originated at a finite time in the past and has evolved to its present state guided by fundamental physical laws. The beginning of the universe was branded by the Big Bang, the relic radiation that we detect today.

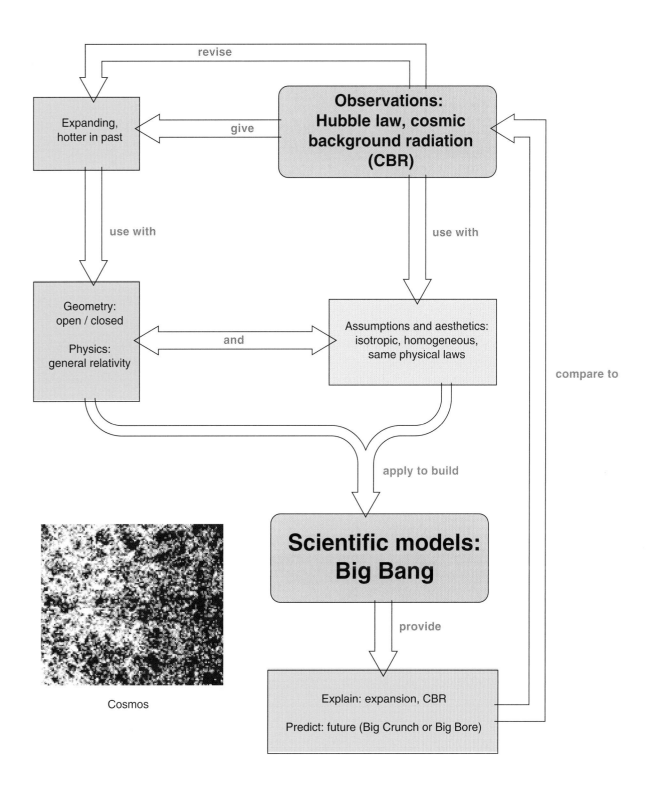

Observations:
Hubble law, cosmic
background radiation
(CBR)

revise

give

Expanding,
hotter in past

use with

use with

Geometry:
open / closed

Physics:
general relativity

and

Assumptions and aesthetics:
isotropic, homogeneous,
same physical laws

compare to

apply to build

Scientific models:
Big Bang

provide

Cosmos

Explain: expansion, CBR

Predict: future (Big Crunch or Big Bore)

Chapter 20
Celestial Navigator™

Some 20 Gy ago, the cosmic bomb exploded. Perhaps you have seen images of an H-bomb blast. Split seconds after detonation, an awesome fireball rips violently through the air (Fig. 20.1). Our universe emerged out of a similar fireball – a cosmic fireball, the Big Bang – in whose violence all that we now see was created.

That, in a nutshell, is a picture of origin accepted by many astronomers today. **Cosmology,** the subject of this chapter, is the study of the nature and evolution of the universe.

Cosmologists have been fed a meager diet of observational facts about the universe. Despite (or perhaps because of) this lack, they have been able to dream up many models. Some of the models have been quite bizarre, but only one has a substantial following now: the Big Bang model. I – and most other astronomers – believe that the present evidence indicates that the universe began in a Big Bang.

Despite its general acceptance, the Big Bang model has shortcomings. A powerful theoretical scenario, called the *inflationary universe model,* has been proposed to deal with some troublesome issues. This inflationary model complements and enhances the standard Big Bang model. Whether future observations will finally pin down the inflationary model, a main conceptual basis will linger: that the universe

is directly linked to elementary particles during its early history. The Big Bang signals the intimate connection of the very small with the very large.

20.1 COSMOLOGICAL ASSUMPTIONS AND OBSERVATIONS

Modern astronomy has arrived at models of the cosmos in which the universe evolves; it is dynamic. How to study the evolution of the universe in detail? We need a few fundamental starting assumptions, difficult-to-prove assertions about the nature of the cosmos.

Assumptions (Learning Outcome 20-1)

Assumption 1. The universality of physical laws. This assumption covers both local (the earth and solar system) and distant regions. It means that we apply the physical laws we uncover locally to all regions at all times – and to the universe as a whole. Key observations support this assumption. For example, the spectra of distant galaxies contain the same atomic spectral lines as those produced by elements found on the earth. Newton's law of gravitation correctly describes the motion of double stars and galaxies.

Figure 20.1 A terrestrial analog of the cosmic fireball. The temperature, density, and pressure in this nuclear fireball, fractions of a second after the detonation of a hydrogen bomb, correspond to the violence in the early moments of the universe.

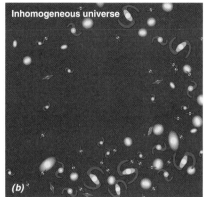

Figure 20.2 Distribution of matter in the universe. *(a)* A homogeneous universe. Matter and radiation are spread out uniformly, if you look over large enough distances. *(b)* A nonhomogeneous universe. Even at large distances, matter and radiation are not uniform.

Assumption 2. The cosmos is homogeneous (Fig. 20.2). This means that matter and radiation are spread out uniformly, with no large gaps or bunches (even for dark matter). This assumption is not strictly true, for clumps of matter, such as galaxies and superclusters, do exist, but the cosmologist assumes that the size of the clumping is much smaller than the size of the universe. It is like looking at the earth from space: bumps, such as mountains, are too small to be seen, so the globe looks smooth.

Assumption 3. The universe is isotropic (Fig. 20.3). This idea relates to a quality of space itself rather than to the matter in it. One way to think of it: space has the same properties in all directions. So, in accordance with the **cosmological principle,** no direction or place in space can be distinguished from any other by any experiment or observation. The universe has no center in space, because there is no way to tell if you are there. There is no "there,

there"; it is the same as here. No direction in space provides special properties. For example, your mass does not increase as you travel in one direction or decrease when you go another way. Another example: the expansion of the universe is the same in all directions of space. Any observer at any time would see a Hubble law.

These assumptions can be summed up in one sentence: *the universe is uniform.* All irregularities are ironed out. The result is like that of the vanishing Cheshire cat in *Alice in Wonderland* – everything gone but the grin. So cosmological models that rest on these assumptions ignore the structure and substance of planets, stars, and galaxies. Galaxies are considered, but only as tiny particles marking points in space, like a gas of atoms filling the universe. This gas is the cosmic grin. Such a mental simplification has real dangers: what lies behind the grin may be crucial. (Dark matter?) Observations must, in the end, validate our assumptions.

Figure 20.3 Isotropy and the universe. *(a)* An isotropic universe. Space has the same properties in all directions. For example, the distribution of velocities of galaxies in the expanding universe is smooth, uniform, and the same in all directions. A Hubble law is observed. *(b)* An anisotropic universe. The expansion of the universe is different at different places in space. A Hubble law is not observed.

Observations

What is the universe? If we consider all that can be seen with various telescopes, we are considering the **observable universe.** Yet this cannot be *all* of the universe. Some objects are too faint and too far away to be seen, and sections of the spectrum exist to which our instruments are, so far, blind. There is more than the observable universe: a **physical universe.** It includes directly observable matter as well as objects we detect by effects described by the laws of physics (such as dark matter).

The reality of the physical universe rests on the assumption that local physical laws apply to the rest of the universe at all other places and in all other times. That's one of our basic assumptions; without it, we could not conceive of a physical universe at all!

A Brief Review of Cosmology
(Learning Outcomes 20-2, 20-3)

Einstein's general theory of relativity underpins cosmological models today. A key aspect of these models is an expanding universe. The observed rate of expansion now is about 20 km/s/Mly. Einstein's theory shows that the rate of expansion – Hubble's constant – relates to the overall density of matter (and energy) in the universe. This average density, in turn, ordains the overall geometry of spacetime. This geometry has three possible forms:

1. Flat, an open universe with the geometry of Euclid, infinite in space and time.
2. Spherical and closed, the cosmos being finite but unbounded in space and finite in time.
3. Hyperbolic, again an open universe, infinite in space and time, but curved.

Observations have not yet settled which geometry applies. (Einstein had an aesthetic preference for a closed universe.) If enough dark matter exists, the universe may well be closed.

20.2 THE BASIC BIG BANG MODEL

Ingenious theoreticians have invented a bewildering array of cosmological models. Most fall into the arena of the **Big Bang model,** which is the standard model based on Einstein's general theory of relativity. Be warned that this model is not final; it has major problems. The inflationary universe model (to come in Section 20.8), a revision of the standard Big Bang model, represents one attempt to address deficiencies. (See the Big Bang Celestial Navigator™.)

Here's the basic idea: the universe is now expanding. Imagine time running backward. What happens? The reverse of expansion! The galaxies and all matter within and without them eventually come tightly together. The extreme squeezing heats both matter and radiation to a very high temperature – high enough to break down all structure that had been created, including all the elements fused in stars. Atoms become protons and electrons; the matter is completely ionized. In addition, the density of ionized matter is so great that photons can travel only short distances before they are absorbed. As a result, the entire universe is opaque to its own radiation.

Now, when matter is opaque to all radiation, it acts like a blackbody (Enrichment Focus 12.1). The radiation from a blackbody exhibits a specific spectral shape (Fig. 20.4). In an early dense, hot state, the whole universe acts like a blackbody radiator. If you could have been there, you would have seen a bright fog all around – like sitting in the sun's interior (but with the light at even shorter wavelengths, because of the higher temperature).

Now imagine the universe expanding from this infernal state. It is so hot that it unfolds in a violent rush; hence the name Big Bang. As the universe expands, its overall density and temperature decrease. (This is true for the expansion of any gas. Let a gas out of a container. The gas cools as it expands.) Eventually the temperature drops so low that protons can capture electrons to form neutral hydrogen and helium. This neutral gas is basically transparent to most kinds of light.

This event – the formation of neutral hydrogen from an ionized gas – marks a crucial stage in the evolution of the universe. No longer ionized, the universe becomes transparent; light is freed from its close interaction with matter. Radiation and matter are no longer connected. The radiation speeds throughout space, and the expansion dilutes it. It is also redshifted, just like the light of distant galaxies. Indeed, having been emitted long ago, it comes from far away, and Hubble's law predicts a large

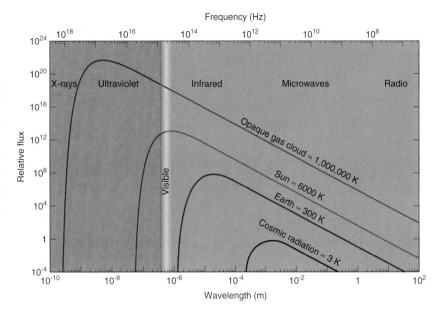

Figure 20.4 Comparison of the spectra of blackbodies at different temperatures from 3 K to 10⁶ K. Note that cosmic radiation at about 3 K peaks in intensity at infrared-microwave. Hotter blackbodies, such as the earth and sun, peak at shorter wavelengths. Also note that the overall shape of each spectrum is the same because these are all blackbody emitters whose continuous spectra are Planck curves.

redshift. The redshift lowers the temperature of the radiation.

If the universe did in fact begin in a hot Big Bang, debris (both matter and radiation) from that cosmic explosion should now lie around us. The matter is pretty obvious: you can see planets, stars, and galaxies. (Not so obvious is the dark matter.) What about the radiation produced in the Big Bang? It has been redshifted to a fairly low temperature, so its wavelength will be long compared with that of light. If the universe expanded uniformly, the radiation's spectrum should show the telltale blackbody shape, because uniform expansion preserves the blackbody properties. Have we seen such cosmic radiation? Yes! Its discovery verifies a hot Big Bang model.

20.3 THE COSMIC BACKGROUND RADIATION (Learning Outcomes 20-4, 20-5)

In 1964 Robert H. Dicke, P. James E. Peebles, Peter G. Roll, and David T. Wilkinson at Princeton University in the U.S.A. were pursuing the existence of radiation left over from a Big Bang. The Princeton group attacked this problem: What would happen if the universe went through a hot stage, so that high temperatures decomposed any heavy nuclei into elementary particles? If a primeval fireball occurred, cosmic blackbody radiation should survive today. Peebles calculated that this fossil radiation should

have a blackbody temperature of roughly 10 K. (The American physicist George Gamov and his colleagues Ralph Alpher and Robert Herman had developed such ideas in the 1940s.)

Discovery!

Just as Roll and Wilkinson were building apparatus to detect the remnants of radiation from the Big Bang, Arno Penzias and Robert Wilson, scientists with Bell Laboratories in New Jersey, detected an annoying excess radiation by means of a special low-noise radio antenna (Fig. 20.5). They intended to do a sensitive study of the radio emission from the Milky Way. The excess noise they found would affect their results, so they set about to eliminate it.

They tuned their radio receiver to 7.35 cm (4080 MHz), where the radio noise from our Galaxy is very small. Still, they picked up the static. Penzias and Wilson also discovered that the noise did not change in intensity with the direction in the sky, the time of day, or the season. Perplexed, they examined the antenna again and found a pair of pigeons roosting inside. These pigeons, oblivious to radio astronomy, had coated part of the antenna with their droppings. Perhaps the coating generated the excess noise? The birds were moved, their droppings cleaned out, but still the excess noise persisted, with a flux at 7.35 cm equivalent to that of a blackbody at 3.5 K.

Figure 20.5 Arno Penzias (right) and Robert Wilson (left) standing in front of the horn antenna with which they discovered the cosmic background radiation.

Penzias eventually called the Princeton group. When he made contact, the Princeton group concluded that the excess noise came from the cosmos – radiation left over from the Big Bang. This mental leap needed verification. After all, the measurement by Penzias and Wilson was at a single wavelength. More observations at different wavelengths were needed to confirm the blackbody shape of its spectrum and its uniform flux around the sky.

Confirmation!

Soon Roll and Wilkinson added another point at another wavelength (3.2 cm) corresponding to a temperature of 2.8 K, close to that observed at Bell Labs. Other experimental groups later contributed additional evidence to the blackbody nature of the radiation (Fig. 20.6). These clinched the case. Ground-based efforts have set the blackbody temperature of the background radiation quite well: 2.73 K for wavelengths from 10^{-1} to 10^2 cm (frequencies between a few to 100 GHz). For convenience, use 2.7 K as the radiation's temperature.

The Cosmic Background Explorer (COBE) satellite confirmed the pristine nature of the background radiation. It showed that the spectrum is extremely smooth, with less than one percent deviation from that of a blackbody (Fig. 20.7).

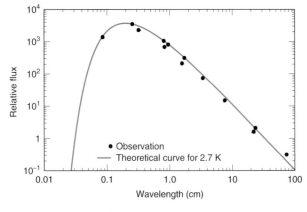

Figure 20.6 Some early observations of the cosmic microwave radiation. The points show measured fluxes at radio and infrared wavelengths; the solid curve corresponds to a theoretical blackbody of 2.7 K. (Adapted from a diagram by P. J. E. Peebles.)

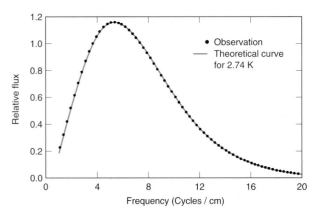

Figure 20.7 COBE measurements of the cosmic background radiation. The solid points are the data; the line is the spectrum from a blackbody at a temperature of 2.74 K. (Adapted from a NASA diagram.)

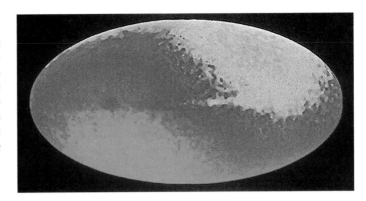

Figure 20.8 False-color, all-sky map of the cosmic microwave background radiation from the Cosmic Background Explorer (COBE). At a frequency of 53 GHz, the radiation is very smooth; the variations in the pink and blue regions are very, very small. The colors show the earth's motion relative to the radiation, which results in a Doppler shift and so a change in temperature. The pink color indicates a blueshift (higher temperature); the blue color indicates a redshift (lower temperatures).

In addition, the spatial extent is very smooth, with no sign of any early variations that might explain the clustering of matter today (Fig. 20.8). Analysis nailed down the temperature to 2.726 K.

To validate its cosmic origin, the radiation should be isotropic: that is, it should have the same measured flux from all directions in the sky. Why? Because if the radiation were from a hot primeval state, the isotropy of the universe at that time would set the isotropy of the radiation. The COBE observations support that the radiation is very isotropic, with very slight deviations.

Recent observations by a balloon-borne telescope (named Boomerang) have confirmed and extended COBE's results. They too show minuscule variations, only about a hundredth of a percent, in the background temperature (Fig. 20.9). Better yet, these observations had much better resolution than COBE (about ⅙° versus 7°). They showed the background to be blobbiest at an angular scale of a bit less than 1°. This result implies that the universe has a flat geometry.

That is the evidence that the radiation originated from the early universe and that the newborn universe was very nearly isotropic. The radiation is cosmic, existing everywhere in space, so that it arrives at the earth uniformly from every direction. It fills all space at all times.

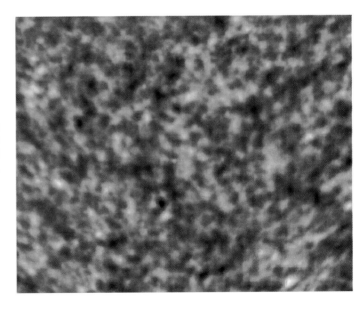

Figure 20.9 Map of the cosmic background radiation by the Boomerang telescope. This false-color map shows warm (bright) and cool (black) regions in the background radiation across a 30° patch of sky. The size of the fluctuations support a flat geometry for the cosmos.

Properties
(Learning Outcomes 20-5, 20-6)

The background radiation is usually given the long-winded name of **cosmic blackbody microwave radiation:** "cosmic" because it comes from all directions in space, "blackbody" because of its spectral shape, and "microwave" because its spectrum peaks at centimeter to millimeter wavelengths. The discovery of this radiation affirms Big Bang models of the universe. How so?

First, the radiation enables astronomers to glimpse the raw, young universe. We can conclude that the universe then was indeed quite homogeneous and isotropic.

Second, the present temperature (about 2.7 K) and the isotropy of the radiation allow us to lay out the thermal history of the universe: that is, the change of the temperature of matter and radiation with time.

Third, the radiation's presence establishes an important marker for galaxy formation. Until the radiation and matter stopped their interaction, matter could not form large clumps. Only after the ionized gas had recombined could matter form clumps that eventually became stars and galaxies (to come in Section 20.6).

Fourth, the radiation provides a reference for measuring the motion of the Galaxy and the Local Group. In the direction of any such motion, the cosmic radiation is blueshifted and so appears to be hotter in that location in the sky. In the opposite direction of the sky, the radiation is redshifted and so appears to be cooler. The observations indicate that the Local Group and Local Supercluster have a combined motion of about 600 km/s in the direction of a nearby supercluster in the constellations of Hydra and Centaurus.

TO SUM UP. The discovery of cosmic blackbody microwave radiation boosts cosmology as much as that of the discovery of the expansion of the universe. Its measured uniformity backs up our assumptions that the universe is isotropic and homogeneous. Its present measured temperature, combined with Einstein's equation of general relativity, allows us to work out the evolution of the universe as its temperature changed.

20.4 THE PRIMEVAL FIREBALL

With the addition of experimental and theoretical knowledge on how matter behaves under hot, dense conditions, we have been able to develop step-by-step details of what may happen in a Big Bang.

CAUTION. Do *not* picture the Big Bang as happening in the "center" of the universe and expanding to fill it. The Big Bang involved the entire universe; every place in it was at creation, which marked the beginning of time and space.

The Hot Start
(Learning Outcomes 20-3, 20-6)

Although the present temperature of the cosmic radiation is low, the amount of energy it contributes to the universe is large. If the energy in the radiation could be used to heat up the matter, the temperature would be greater than 10^{12} K – one clue that the early universe must have been very hot.

This hot beginning is often called the **primeval fireball,** but we don't really know what happened at this time. No one understands how matter behaves at temperatures greater than 10^{12} K. From Einstein's general relativity, we find that a temperature of 10^{12} K corresponds to a time of about 10^{-24} s after creation (time "zero"). By "creation," I mean the beginning of the expansion.

The primeval fireball produced such a fast expansion that the temperature and density dropped swiftly. The contents of the universe (Table 20.1) – elementary particles and their antiparticles – change with temperature and time. I will sketch a scenario of the young universe in which temperature plays a crucial, controlling role. Each period of time matches a specific temperature.

Roughly, the universe's thermal history can be divided into four eras: a **heavy-particle era,** when massive particles and antiparticles dominated; a **light-particle era,** when particles with less mass were made; a **radiation era,** when most particles had vanished and radiation was the main form of energy; and a **matter era,** in which we now live, when the energy of matter dominates that of radiation. (In a cubic meter of space, matter now averages about a thousand times the density of photons, considering their energy in the form of mass.)

Table 20.1 Particles in Cosmic Nucleosynthesis

Particle and Antiparticle	Symbol	Charge	Comments
Neutrino, antineutrino	ν, $\bar{\nu}$	0, 0	Massless (?) particles that travel at light speed; stable (?)
Proton, antiproton	p, $\bar{p}$	+1, −1	Nucleus of hydrogen; stable
Electron, positron	e^-, e^+	−1, +1	Particles surrounding the nucleus of an atom; stable
Neutron, antineutron	n, $\bar{n}$	0, 0	Neutron decays to a proton and an electron with half-life of 10^3 s
Photon	γ	0	Packet of radiation, electromagnetic energy
Deuteron	^{2}H	+1	Nucleus of deuterium, or "heavy hydrogen"; contains 1 proton, 1 neutron; stable
Helium-3	^{3}He	+2	Nucleus of an unusual type of helium; contains 2 protons, 1 neutron; stable
Helium-4	^{4}He	+2	Nucleus of ordinary helium; contains 2 protons, 2 neutrons; stable
Lithium-7	^{7}Li	+3	Nucleus of most abundant type of lithium; contains 3 protons, 4 neutrons; stable
Beryllium-7	^{7}Be	+4	Nucleus of most abundant type of beryllium; contains 4 protons, 3 neutrons; unstable, half-life 53 days

I will consider a proton and the nuclei of elements (such as helium) to be "heavy" particles and electrons to be "light" particles (Table 20.1). As the universe expands, its temperature and density fall during an era. Values will be representative ones.

CAUTION. Whenever I say that the universe was "smaller" or "larger," I mean that the *distance* between a pair of objects was smaller or larger. The universe itself may be finite or infinite. That overall geometry does not affect details about the Big Bang.

Creation of Matter from Photons (Learning Outcome 20-8)

Before the story unfolds, you need a little preparation about one key part: the creation of matter and antimatter from photons. (**Antimatter** has the same properties of regular matter except that it has the opposite electrical charge.)

At some time in the primeval fireball, the energy of photons was so high that their collisions produced particles. This process occurs when the energy in the colliding photons equals or exceeds the mass of the particles produced. Sounds bizarre? The result comes directly from Einstein's relation between matter and energy ($E = mc^2$). It does not restrict the *direction* of the transformation: matter can become energy, or energy can become matter.

The creation of matter from light involves both matter and antimatter. When matter and antimatter collide, they are annihilated and become converted to photons (Fig. 20.10). In reverse, two photons (if they have enough energy) create a matter–antimatter pair when they hit each other. Note that this process always results in *pairs* of particles.

How much energy must the photons have? At least the energy equivalent of the masses of the pair they produce. Making protons and antiprotons takes more energy than is needed to make electrons and positrons, because protons have more mass than electrons (about 1800 times more). Now we can talk about the energy of photons in the primeval fireball in terms of their temperature. To have enough energy to make electrons requires a temperature of at least 1.2×10^{10} K; for protons, at least 2.2×10^{13} K.

Particle production from photons plays the center ring in the very young universe. Let's see what the model tells about those times.

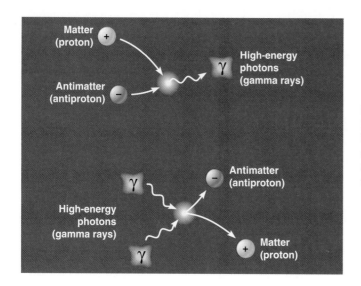

Figure 20.10 Matter and antimatter annihilation to make photons; particle and antiparticle production from photons. Here, the particles are protons; the antiparticles, antiprotons; and the photons, gamma rays. Each gamma ray must have an energy *at least* equal to that of the mass of a proton for the conversion to work.

Temperature Greater than 10^{12} K (Learning Outcomes 20-6, 20-7)

Photons create pairs of particles and antiparticles, even the most massive elementary particles. Annihilation also takes place. This balance between the production and destruction of massive particles marks the heavy-particle era. It does not last long, because the cosmos expands rapidly and the temperature declines quickly.

So at the earliest times that we can calculate, the universe is a smooth soup of high-energy light and massive, elementary particles (perhaps of unknown kinds). The nuclei of atoms have yet to be made.

Temperature about 10^{12} K (Learning Outcomes 20-6, 20-7)

Annihilation of heavy particles with their antiparticles continues (Fig. 20.11a). The remaining photons, however, lack the energy to create new heavy particles. Only light particles – electrons – can be made, because less energy is required for their production from photons. The universe enters the light-particle era. (Keep referring to Table 20.1 through this sequence.)

Protons and electrons interact to generate neutrons (Fig. 20.11b). When the temperature falls to 6×10^9 K, photons can no longer make electron–positron pairs. This temperature marks the end of the light-particle era and the beginning of the radiation era. Neutrons are no longer produced by the interaction of protons with electrons or of antiprotons with positrons (Fig. 20.11c). The neutrons now decay into protons and electrons but no new ones are made. The ones that survive are crucial to the next step.

Temperature about 10^9 K (Learning Outcome 20-7)

Now come the key nuclear reactions! The remaining neutrons and protons react to form nuclei of simple elements. The most important reaction involves the combination of a neutron and a proton to form deuterium, ^{2}H (Fig. 20.11d). All neutrons, except those that have decayed, end up in deuterium. When the deuterium has been produced, further reactions create ^{4}He, normal helium (Fig. 20.11e). The net result is about 25 percent helium by mass compared to all forms of matter. A little bit of beryllium and lithium is created (Fig. 20.11f). Some deuterium is left over.

Helium, once formed, is tough to destroy, so the present helium abundance in the universe rests on a cosmological base. The helium abundance serves as a test of the Big Bang model. It predicts that no celestial object can have a helium abundance of less than 25 percent. (Because stars form helium, the abundance now is greater than this.)

After the first few minutes, nucleosynthesis ceases. Cosmic fusion ends; the temperature is too low.

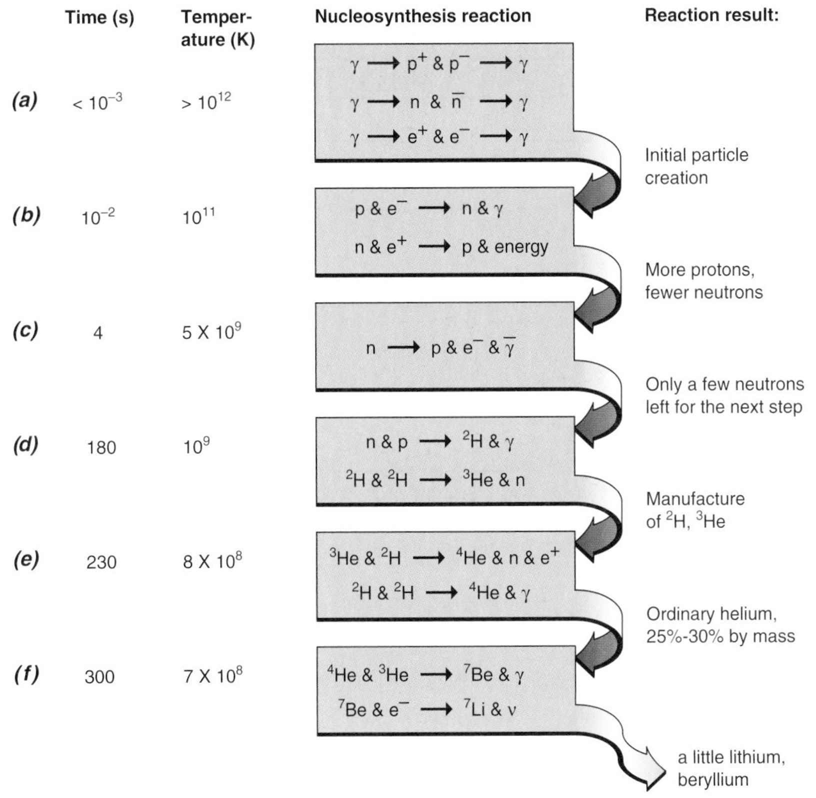

	Time (s)	Temperature (K)	Nucleosynthesis reaction	Reaction result:
(a)	$< 10^{-3}$	$> 10^{12}$	$\gamma \longrightarrow p^+ \,\&\, p^- \longrightarrow \gamma$ $\gamma \longrightarrow n \,\&\, \bar{n} \longrightarrow \gamma$ $\gamma \longrightarrow e^+ \,\&\, e^- \longrightarrow \gamma$	Initial particle creation
(b)	10^{-2}	10^{11}	$p \,\&\, e^- \longrightarrow n \,\&\, \gamma$ $n \,\&\, e^+ \longrightarrow p \,\&\, energy$	More protons, fewer neutrons
(c)	4	5×10^9	$n \longrightarrow p \,\&\, e^- \,\&\, \bar{\gamma}$	Only a few neutrons left for the next step
(d)	180	10^9	$n \,\&\, p \longrightarrow {}^2H \,\&\, \gamma$ ${}^2H \,\&\, {}^2H \longrightarrow {}^3He \,\&\, n$	Manufacture of ${}^2H$, ${}^3He$
(e)	230	8×10^8	${}^3He \,\&\, {}^2H \longrightarrow {}^4He \,\&\, n \,\&\, e^+$ ${}^2H \,\&\, {}^2H \longrightarrow {}^4He \,\&\, \gamma$	Ordinary helium, 25%–30% by mass
(f)	300	7×10^8	${}^4He \,\&\, {}^3He \longrightarrow {}^7Be \,\&\, \gamma$ ${}^7Be \,\&\, e^- \longrightarrow {}^7Li \,\&\, \nu$	a little lithium, beryllium

Figure 20.11 One possible sequence for nucleosynthesis at specific times in the Big Bang. *(a)* Particles and antiparticles are made from photons. *(b)* Neutrons are made and destroyed. *(c)* Neutrons decay; only 16 percent remain from this stage. *(d)* Neutrons and protons make deuterium and tritium. *(e)* Tritium and deuterium make regular helium. *(f)* Helium and tritium form beryllium and lithium. Note that lithium and beryllium are the heaviest elements that can be made in the first few minutes.

Temperature about 3000 K
(Learning Outcome 20-9)

So far the temperature has been so high that all atoms in the universe have been ionized. At about 1 million years after creation, the radiation's temperature has plunged to 3000 K, too low to keep matter ionized. The nuclei begin to capture electrons to form neutral atoms – a process called **recombination.** The neutral matter becomes transparent to the radiation. The matter and radiation are no longer locked together.

The radiation merrily expands with the universe. As the universe expands, the radiation cools down to become the 2.7 K cosmic radiation today. The matter, however, follows a different course because of little bumps in the generally smooth distribution of matter. Clouds of matter condense out of the primeval fireball. The radiation era ends, and the matter era begins. Material clumping could not happen until the radiation and matter had decoupled, as explained in Section 20.6. (Why not? Because the radiation exerted a pressure on the matter that kept it from gathering into clumps.) This event flags the time when galaxy formation could begin (Section 20.7).

TO SUM UP. In the first few minutes, the universe expanded and reached temperatures and densities suitable for the formation of deuterium and helium (Table 20.2). The Big Bang model predicts a helium abundance of 25 percent (by mass) and very little formation of heavier elements. Much later, the nuclei captured electrons to form neutral atoms (almost all hydrogen and helium). The universe became transparent. The radiation became the cosmic background radiation and ceased to control the universe's evolution.

Evidence for the Big Bang
(Learning Outcome 20-7)

How to judge whether the basic Big Bang model is correct? What observational evidence supports it?

Table 20.2 Sequence of Events in the Big Bang

Event	Time	Density (kg/m³)	Temperature (K)	Comments
Origin	0	?	?	Not the province of present science; relativity fails
Heavy-particle era	10^{-43} s	10^{97}	10^{33}	Photons make massive particles (such as protons) and antiparticles
Light-particle era	10^{-4} s	10^{17}	10^{12}	Photons have only enough energy to make light particles and antiparticles; protons and electrons combine to make neutrons
Radiation era	10 s	10^{7}	10^{10}	Few particles left in a sea of radiation; these partake in nucleosynthesis of deuterium, helium, lithium, and beryllium
Matter era	10^{6} y	10^{-18}	3000	Ionized hydrogen recombines; cosmos becomes transparent
Now	10^{10} y	10^{-31} (radiation)	3 (radiation)	Astronomers puzzle about origin

Note: Times are the approximate midrange of the eras.

First and foremost, observations indicate that the background radiation pervades the universe. Within the limits of observational errors, measurements confirm a blackbody spectrum with a temperature of 2.7 K. Measured in many directions of space, the radiation comes to the earth with a very uniform intensity. The background radiation has the attributes expected of a cosmic, hot origin.

Second, the model predicts that primeval helium was formed in the first 5 minutes of the universe's history and that the helium abundance should be about 25 percent by mass. The Big Bang model sets this helium abundance as the basement level; any observation of a substantially lower amount would call the model into question. Because helium is formed in stars later, the present helium abundance must be larger than 25 percent – that is, the helium made in the Big Bang plus the helium formed in stars.

Unfortunately, we can assess the present cosmic helium abundance only indirectly. As Chapter 12 pointed out, the solar photosphere is not hot enough to excite helium lines for direct viewing with a spectroscope, so it's not possible to measure the helium abundance in the solar photosphere. The corona appears to have an abundance of 28 percent. Theoretical models of stellar evolution place the initial helium abundance in the sun's interior at 25 percent.

Hot, massive stars from Population I exhibit helium absorption lines that imply a helium abundance of approximately 35 percent. H II (ionized hydrogen) regions surrounding hot stars have helium emission lines at optical and radio wavelengths that give abundances of 28 percent. Planetary nebulas also have strong helium emission lines that imply a helium abundance of greater than 30 percent. In our Galaxy, the oldest Population II stars are the best candidates. They contain about 30 percent helium.

Although the helium values have a considerable range, the helium abundance for a variety of celestial objects falls close to that predicted by the Big Bang model. The match of observations with the model is good. The accumulation of evidence for the hot Big Bang is strong, because it gives a simple, unified picture of the early universe.

20.5 THE END OF TIME?
(Learning Outcomes 20-7, 20-12)

So much for the cosmic past. What about its future? Recall (Section 7.4) that Einstein's general theory of relativity allows the cosmos to have one of two general geometries: open or closed. If open, the universe will expand forever, and time will never end. If closed, however, the universe must eventually collapse, running backward through the history outlined in the preceding section.

Which fate will be ours? The current measured

value of Hubble's constant ($H = 20$ km/s/Mly) and Einstein's theory give a critical density for a closed universe – it's about 5×10^{-27} kg/m³. If the actual density is less, the universe is open; if more, it's closed. The evidence to date is inconclusive, although it does weakly support an open geometry.

The standard hot Big Bang model provides another test. The test rests on the observed present abundance of deuterium.

How can the abundance of deuterium reveal whether the universe is open or closed? For Big Bang model calculations, you need to put in the present value of Hubble's constant, the present temperature of the cosmic radiation, and the present average density of the universe. The first two items are reasonably well known, but the third is not. The amount of helium that comes out of the Big Bang calculations does not depend very much on the value used for the present density of the universe, but the amount of deuterium does (Fig. 20.12).

So if we can measure the present cosmic abundance of deuterium, we can check this number against that predicted by the Big Bang model for various densities. The model gives the present average density of the universe, and we can compare that number with the critical density.

Ultraviolet observations of the interstellar medium show that the amount of deuterium (relative to hydrogen) in the interstellar medium is about 2.0×10^{-5} by mass. (Remember that most of the mass in the interstellar medium is hydrogen.) Now perhaps half Big Bang deuterium has been burned up in stars. So the original deuterium abundance was about twice that observed now, or 4×10^{-5} by mass relative to hydrogen. A deuterium abundance of 4×10^{-5} implies a present cosmic density of 4×10^{-28} kg/m³, considerably less than that of the critical density (5×10^{-27} kg/m³).

Conclusion from this test: the universe is open. It will expand forever. Time will not end. But enough dark matter would change that conclusion! What could that stealth matter be? Not ordinary protons, neutrons, and electrons – these make up no more than 20 percent of cosmic matter. Even neutrinos with a tiny mass may not tip the mass scales.

The leading contender among dark matter candidates is weakly interacting massive particles, fondly called *WIMPs*. This category of matter inter-

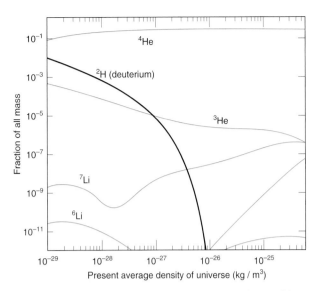

Figure 20.12 Theoretical calculations of the abundance of low-mass elements formed in the Big Bang as related to the average density of the universe now. Note how sharply the abundance of deuterium varies with different (present) average densities of the universe, while the abundance of ordinary helium changes little. The vertical axis gives the fraction of the element, by mass, compared to all other elements.

acts with ordinary matter only by gravity and the weak force. In theory, stupendous numbers of WIMPs were fabricated in the Big Bang. Because they react so seldom with ordinary matter, many WIMPs should persist today, but how do astronomers troll for WIMPs?

The preferred method involves waiting for the rare occasions when a WIMP smacks into the nucleus of an atom and generates flashes of light. Fortunately, there should be lots of WIMPs – some 10 trillion should shoot though a kilogram of matter every second. Unfortunately, they whack a nucleus only once a day per kilogram. Well, that's enough for patient scientists to give it a shot, and two experiments are now running to hook a WIMP or two. Results? Inconclusive so far – but a decisive payoff may come in a few years.

20.6 FROM BIG BANG TO GALAXIES (Learning Outcome 20-10)

The isotropy of the cosmic background radiation implies that the matter and radiation in the universe had a very uniform distribution. But that's not the situation now. Even the most spread-out of these systems of matter – clusters of galaxies – have average densities about a hundred times greater than the average density of the universe. The crucial question: how did an originally very smooth universe become clumpy?

A model of galaxy formation must leap a critical hurdle; it must operate *fast.* To date, astronomers have seen objects (the quasars) as far as some 12 Gly away, so the matter from the Big Bang formed into large clumps well before this time.

The discovery of the cosmic radiation forced us to consider what happens to disturbances in a hot, dense universe filled with matter and radiation. The radiation played a powerful role in inhibiting the growth of disturbances. A dense patch of gas in the early universe would have a high internal pressure because the radiation added to the total force that pushes the patch apart. (You can consider photons as particles like those in a gas.) Only very large disturbances would contract. So disturbances that contained roughly the mass of a galaxy could not contract to become young galaxies.

These disturbances made the early cosmos lumpy. You can think of them as sound waves, coursing through the plasma, compressing and thinning it just slightly. So some parts were warmer and some cooler than the average. When the cosmos became transparent, the sound waves disappeared. But lumps were imprinted on the photons, and these slight variations show up in the background radiation today.

Just after the universe became transparent, it was a new show. The radiation and the gas no longer interacted, so radiation pressure no longer resisted gravity. Small disturbances, amounting to only 10^5 solar masses, could condense out of the gas, along with disturbances of greater mass. This result corresponds to the range of masses we observe: for large, gravitationally bound masses range from 10^5 (globular clusters and dwarf galaxies) to 10^{16} solar masses (superclusters of galaxies). Disturbances of this size range can condense just after decoupling. So roughly 1 million years after the Big Bang marks the time at which the galaxy formation could have taken place in the young universe.

There's one real weakness in these ideas: even though disturbances can be unstable, they grow slowly – so slowly that the galaxies we see could hardly have formed by now unless the disturbances were already fairly large in the beginning. Unfortu-nately, the smoothness of the cosmic background radiation implies that any lumps were small.

> **TO SUM UP.** Galaxies probably formed early in the history of the universe, but we don't yet know when and how.

20.7 ELEMENTARY PARTICLES AND THE COSMOS (Learning Outcome 20-11)

The hot Big Bang model has a number of weaknesses, such as the great dominance of matter over antimatter in the universe today, and the formation of galaxies. Both these problems and others are being tackled now in imaginative theoretical ways that unite the universe of the small – elementary particles – with the universe itself. The connection occurs early in the Big Bang, before the first second had elapsed. To explain this connection, I will digress a bit into the world of elementary particles and their interactions.

The Forces of Nature

Let's focus on how particles are related – their inter-actions. We generally think of these relations in terms of forces between particles. You are familiar with two of these: gravitation and electromagnet-ism. These forces have one property in common: they work over large distances. According to Newton's law of gravitation, the most distant galaxies exert a force on you (and you on them). The same is true of objects that have a net electrical charge. These forces differ in their relative strengths. Electromagnetic forces are much stronger than gravity – as you know if you've lifted a nail with a magnet. The gravitational attraction of the entire earth on the nail is weaker than the force of the magnet.

Two other forces operate in nature in the sub-atomic domain. One, called the **strong force,** holds the nuclei of atoms together. Recall that an atom's nucleus has protons tightly packed together. The electric force of each proton repels the others strongly, especially when so close – only 10^{-15} m apart. The strong force overwhelms this electric repulsion and keeps the nucleus together. The other, called the **weak force,** crops up in radioac-tive decay. Without it, fission would not take place. Like the strong force, the weak force operates over very short distances, 10^{-17} m and less.

Table 20.3 Properties of the Fundamental Forces

Force	Strength (relative to strong)	Range (m)
Strong	1	10^{-15}
Electromagnetic	1/137	Infinite
Weak	10^{-5}	10^{-17}
Gravity	6×10^{-39}	Infinite

These four forces are all that are known. The strong force is the strongest of them (Table 20.3), electromagnetism second, the weak force next, and gravity takes the bottom as the weakest. Now, although these forces appear to operate very differently, might they have an underlying unity? That quest for a unified theory has tempted physicists for a century. Recently they have had some success in struggling to find a **grand unification theory,** or *GUT.* The development of GUTs (there are more than one) has, curiously enough, modified Big Bang cosmology and helped to shore up some weaknesses.

How is that connection made? First, in the 1970s, theoreticians unified the weak force and the electromagnetic force, introducing the term *electroweak force.* Buoyed by their success, they next took aim at unifying the electroweak and strong forces. As part of that work, they developed the concept of a new elementary particle, called a **quark,** that comprises other so-called elementary particles. For example, a proton consists of three quarks. Since the proton is a composite particle, it can decay (as the neutron, also a composite particle, decays) with a predicted lifetime of some 10^{30} years. (Experiments are now under way to check this prediction.)

GUTs also predict that the unification of strong and electroweak forces won't be apparent until energies of greater than some 10^5 J – the equivalent of converting about 10^{15} protons completely into energy. This unification may occur when the distances between particles are less than 10^{-31} m. Such energies cannot be made in large particle accelerators on earth, but they do occur in the Big Bang model at a time of 10^{-35} s when the temperature was 10^{26} K and the distance between particles was 10^{-40} m. The conditions then may have been just right for a unification of forces. The Big Bang model

serves as a way to test GUTs – the whole universe as a historical high-energy particle machine!

NOTE. One result of GUTs has been to simplify our view of elementary particles. We now believe that matter is composed of two classes of elementary particles: quarks and leptons. Only six particles make up the **lepton** group; they include electrons and neutrinos. (These are low-mass, "light" elementary particles.) Quarks make up all the other particles, such as protons and neutrons and some hundred others. (These are high-mass, "heavy" elementary particles.) Six quarks are known to date; each has three subtypes, for a total of eighteen. Each of these has an antiparticle, so the total of quarks of all types is thought to be thirty-six. The four basic forces act on these particles. The aim of GUTs is to reduce all particles to one kind, interacting through one force – truly a *grand* unification!

GUTs and the Cosmos

We now look at the temperature history of the Big Bang model in a new way: we attempt to determine the temperatures at which specific aspects of particles and their forces freeze into existence. Here I use the word "freeze" in the sense that when water freezes into ice (at 0° C or 273 K), its state changes abruptly. Water in the form of ice behaves differently from water as a liquid or gas.

I have already discussed two freezings in the early cosmos (Section 20.4): the formation of simple nuclei in the nucleosynthesis era (first few minutes) and of atoms during the recombination time (1 million years). GUTs predict a very special kind of freezing at 10^{-35} s, when the strong nuclear force froze out from its unification with the electroweak force. Before this time, from 10^{-43} s to 10^{-35} s, gravity had frozen out from the other three forces. Before that time, all four forces are presumed to be unified in a single force.

The final freezing took place at about one second, when the electromagnetic and weak forces split. Now, quarks could combine to form particles such as protons, which could then react in the nucleosynthesis era. Before this freezing, the universe contained a hot gas of quarks and electrons. From this time on, the standard Big Bang model applies.

A special event occurs at the separation of the electroweak force from the strong force at 10^{-35} s. Just as water releases energy in the form of heat when it freezes, the freezing of the strong force releases energy, but in enormous amounts. This event may have pumped energy into the expansion of the universe so that it grew in size by many powers of 10 in just 10^{-32} s. This marks the era of inflation, which is the hallmark of the inflationary universe model.

GUTs and Galaxy Formation
(Learning Outcomes 20-10, 20-11, 20-12)

Another aspect of freezing is related to the galaxy-formation problem. Consider the freezing of a pond: the ice sheet does not form all at once but in patches. That is, the freezing process has defects. In cosmological models, these defects have mass and survive for a period so long that at the decoupling time, when matter can clump, the defects serve as the cores – the lumps – on which gravitational instability occurs.

Another way to help make galaxies calls for neutrinos. Some GUTs predict that neutrinos should have a very small mass (and a few experiments, yet unconfirmed, support this claim) – perhaps 0.001 percent that of an electron. According to these theo-ries, neutrinos froze out at a time of 1 second, and those relic neutrinos should be with us today. If massless, they would have little effect. If they have even a slight mass, they change the universe. (Neutrinos with mass might solve the solar neutrino problem outlined in Section 12.6.)

First, neutrinos may contain enough mass in total to close the cosmos. Second, they can aid galaxy formation. Neutrinos with mass can start clumping by gravity well before other particles can. After recombination, atoms would gather around these neutrino clumps and so speed up the gravitational instability process. Third, massive neutrinos may congregate in clusters of galaxies and bind them gravitationally. Fourth, neutrinos can also gather in the halos of galaxies to make up the so-far invisible matter there.

Supercomputer models of clustering show what happens to large disturbances in a mostly smooth distribution of material. These early ripples in the universe may have fostered the formation of clusters and superclusters. The models depict the clusters in elongated structures similar to those observed in the Local Supercluster. Later stages in the simulation show bridges of material still linking the clumps of matter. The final result is a sponge-like structure of voids and matter (Fig. 20.13). Most of the matter is assumed to be in the form of

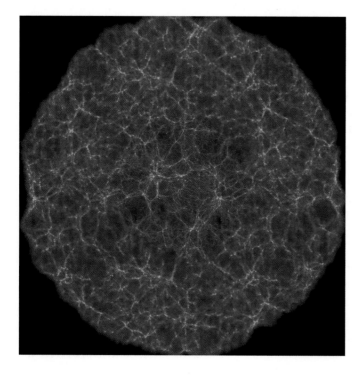

Figure 20.13 Supercomputer simulation of the gravitation clumping in the local universe. The model is constrained by an observed distribution of galaxies, so it mimics known large-scale structures well.

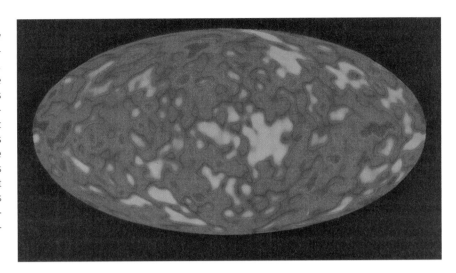

Figure 20.14 COBE false-color, all-sky map, made at microwave wavelengths. The Milky Way, if visible, would extend horizontally across the middle of the image. This map is coded so that blue indicates 0.01 percent colder than the average cosmic microwave temperature; red indicates 0.01 percent warmer than the average (see scale at bottom). This radiation is from the universe when it was about 300,000 years old. The red patches represent regions of higher-than-average density; blue, regions of lower-than-average density.

cold, dark matter (still invisible today). Small masses form first and then collide and coalesce into galaxies. Such results closely imitate the patterns found in the largest redshift maps. Hence, we are sensing that large-scale complexity grew from simple beginnings: random fluctuations in the early universe, amplified by the action of gravity.

An additional COBE result provides a solution to many of these sticky issues. After processing some 70 million observations, the COBE team carefully created a map of the sky (Fig. 20.14) that shows some of the cosmic ripples from the Big Bang – wrinkles in the cosmic tapestry that were the likely seeds for the formation of large-scale structure. The map shows minute fluctuations in the cosmic background radiation, variations that amount to only a few parts per million at a time when the cosmos was less than one million years old. In regions of higher temperature, the density of matter at the time was a little higher than average; in the regions of lower temperature, the density was a bit lower. These higher-density variations were large enough to create enough gravity to attract more matter into denser clumps, which eventually contracted into the first-born galaxies. The lower-density regions resulted in the voids. Because the young universe was very smooth but not perfectly so, we have a lumpy universe today.

What kind of matter created these clumps? The COBE results do not tell us that key piece of the large-scale puzzle. We will have to rely on models yet to be developed to interpret these startling observations, which reveal the most ancient and largest structures in the universe.

Particle physics can shore up some problems with the simple primeval fireball concept. The key point: the connection of very small particles to the Big Bang is essential to our understanding of the universe's history. The information from GUTs enhances the Big Bang model, not only in details but also in the overall picture. The model becomes more aesthetically pleasing (Section 2.1).

20.8 THE INFLATIONARY UNIVERSE (Learning Outcome 20-11)

The drive to integrate particle physics with cosmology had a new success with a variation of the Big Bang model called the **inflationary universe model.** It copes with serious flaws in the Big Bang picture and so improves it. I will focus on just two, called the *flatness problem* and the *horizon problem.*

The Flatness Problem

Recall that the universe, in terms of general relativity, can have one of two basic geometries: open or closed. We can evaluate which one applies to the cosmos by examining the *ratio* of the measured density (of matter and energy) to the critical density predicted from Einstein's general relativity and the value of Hubble's constant.

A cosmos whose actual density is exactly the critical density is flat; the ratio is 1. Now, if at the

Big Bang, the ratio were 1, it would remain 1 forever. If, however, the value differed from 1 ever so slightly, the ratio would be much different from 1 now. Surprisingly, the ratio is believed to have a value between 0.1 and 2, very close to 1. It must have started very close to 1; otherwise, as the universe evolved, the ratio would have acquired a value much different from 1. The standard Big Bang model has a special starting condition (the geometry of the universe had to be very, very close to flat) but does not give an explanation for it. That is the essence of the flatness problem.

Enter the inflationary model, which deals with times from 10^{-43} to 10^{-32} s, when massive elementary particles dominated the universe. During that interval, the universe underwent a tremendous spurt of growth, becoming perhaps as much as 10^{50} times larger. Remember, that means that the distance between two particles increased by 10^{50}. To put that increase in perspective, if we inflated the distance from the proton to the electron in a hydrogen atom (about 10^{-10} m) by the same amount as estimated for the early universe, the distance would be 10^{40} times greater – or 10^{24} ly!

This inflationary period neatly and naturally solves the flatness problem (after which the model melds into the standard hot Big Bang model). Imagine that before inflation, spacetime contained strongly curved regions. Inflation would cause these to achieve flatness automatically. As an analogy, consider the curved surface of a partially inflated balloon. The surface is clearly curved when compared to the overall size of the balloon. Now rapidly blow up the balloon, keeping a close eye on the curvature of the surface. It becomes distinctly flatter (less curved). Similarly, when the universe inflated, curved regions would have become flat, so the ratio of the actual to critical density naturally reached a value very close to 1, without the need of special assumptions.

Note that the inflationary model *requires* a flat universe. Most observations imply an open universe. But recent balloon-borne results (Fig. 20.9) support a just-flat geometry and so uphold the inflationary model.

The Horizon Problem

I have already emphasized the uniformity of the cosmic background radiation (Section 20.3), which tells us that the universe was extremely isotropic (Section 20.1) at the time of decoupling. The background radiation deviates from complete uniformity by only a few parts in a million! How did this happen? The standard Big Bang model just assumes that it started that way and stayed that way, and with this assumption, the uniformity of the cosmic radiation poses a problem.

Consider a gas in a box. If you add energy to one side of the container, the temperature goes up. To induce this effect, the particles in the gas must carry information about the addition of energy by moving around at a greater average speed and knocking into each other harder. A finite time must elapse before these collisions can carry throughout the box the information that energy has been added to it. Now imagine the box expanding much faster than the particles in it, on the average, were moving around. Then only a small region of the box would find out that energy had been added, and this part would have a temperature different from the rest.

Information cannot be communicated faster than the speed of light, yet the very early universe expanded so fast that regions of it were rapidly and widely separated. Now, in a given time, a light signal can travel some maximum distance, called the *horizon distance*. For example, after one second had elapsed, light could have gone only one second of light-travel time, for a horizon distance of about 300,000 km. As a result of rapid expansion, however, regions of the universe were separated by almost a hundred times this distance. How could these regions have evolved to the same temperature when they could not communicate with each other? That is the horizon problem.

The inflationary universe model solves the horizon problem by altering the setting of the rapid expansion of the universe: namely, the universe is said to have evolved from a region much smaller (by 10^{50} times or more) than in the standard Big Bang model. Before the inflationary era began, the universe was much smaller than its horizon distance. Thus all of it could reach the same temperature. Then inflation made it much larger, preserving the uniform temperature. That way we can account for the great uniformity of the cosmic background radiation in the past and today.

TO SUM UP. The inflationary universe model (Fig. 20.15) deals with the earliest times in the universe's history, before a time of 1 second down to a time of 10^{-43} s. Earlier than

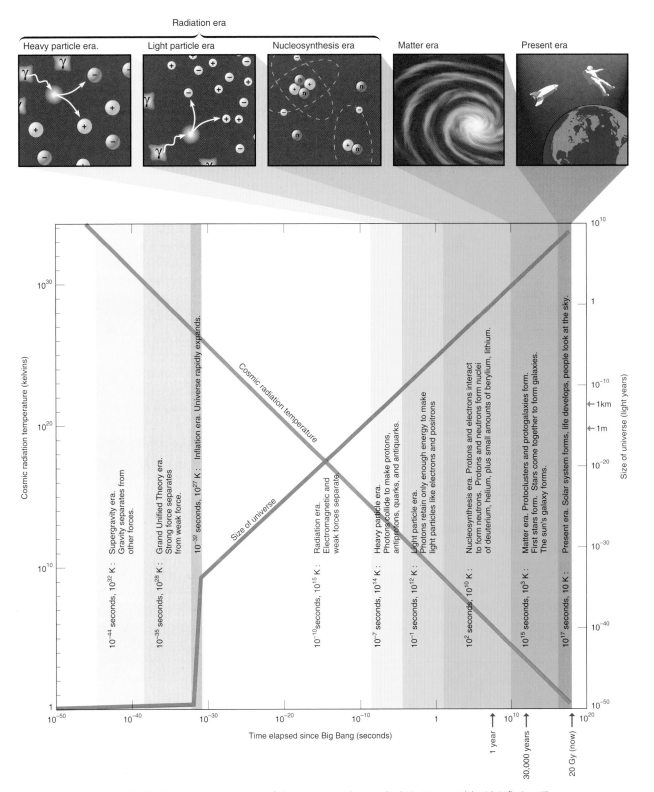

Figure 20.15 Schematic, visual history of the universe in the standard Big Bang model with inflation. The horizontal axis is time; the vertical one represents the relative size of the universe (right) and the temperature of the background radiation (left). Just after the Big Bang, gravity on the smallest scales (supergravity) may describe the universe until a time of 10^{-43} s, when gravity freezes and GUTs describe the physical processes of particle interactions. At 10^{-35} s, inflation begins, prompted by the freezing of the strong force. Electroweak interactions control a soup of quarks and electrons until the quarks form protons, which participate in nucleosynthesis. After matter and radiation have decoupled, the first galaxies form in superclusters. In particular, the cosmic background radiation comes to us from the time of recombination and decoupling. The horizontal axis at top gives the cosmic eras.

that, we have no physical model to describe the cosmos. At 10^{-43} s, gravity froze out from the other forces. At 10^{-35} s, the strong force froze out from the electroweak force and promoted an era of inflation. Then from 10^{-32} s onward, the model unfolds as the standard Big Bang.

The inflationary universe model grapples with other flaws in the Big Bang model and does so modestly well, in my opinion. It is the best model we have to date that unites the physics of the very large and the very small.

KEY CONCEPTS

1. Cosmological models assume the universality of physical laws, a homogeneous universe, and an isotropic universe.

2. Cosmological models must explain the evolution of the universe, the grouping of matter, and the expansion of the universe.

3. The main modern cosmological model is the Big Bang model (based on Einstein's theory of general relativity).

4. According to the Big Bang model, the universe began in a hot, dense state at a finite time in the past. The model predicts that radiation from the past still pervades the cosmos at a low temperature; this radiation has been observed to be isotropic and to have a temperature of about 2.7 K.

5. The properties of the cosmic blackbody radiation, when combined with Einstein's general relativity and our present knowledge of matter, permit a detailed description of the Big Bang. In it, matter is made from photons and then interacts to form light elements in the first few minutes of the universe's history.

6. The hot Big Bang model predicts a cosmos having no less than 25 percent helium by mass; observations tend to support this prediction.

7. How the universe ends depends on whether it is open or closed. Present observations support an open universe, a conclusion that can be inferred from the cosmological abundance of deuterium.

8. Galaxies could not form in the young universe until the temperature had dropped below a few thousand kelvins (about a million years after the expansion started). Galaxies did form quickly and early on, but they could not simply grow gravitationally from small disturbances; rather, they needed the help of turbulence from the Big Bang. Observations have revealed such ripples.

9. Grand unified theories (GUTs) of elementary particles and forces provide new insights into the Big Bang model and help to solve some of its major problems, such as the matter–antimatter imbalance.

10. According to the inflationary universe model, at the time of the strong force freezing out, the universe went through a stage of rapid expansion; this suggested sequence helps to solve the flatness and horizon problems of the Big Bang model.

STUDY EXERCISES

1. State in a few short sentences the assumptions of the Big Bang model. (Learning Outcomes 20-1 and 20-3)
2. Make a short list of the fundamental cosmological observations. (Learning Outcome 20-2)
3. Interpret the observations in Exercise 2 for the standard Big Bang model. (Learning Outcome 20-3)
4. Give one observational argument for asserting that the microwave background radiation is cosmic in origin. (Learning Outcomes 20-4 and 20-5)
5. How does the discovery of the cosmic microwave background radiation confirm the Big Bang model? (Learning Outcome 20-6)
6. List the elements that can be made in a hot Big Bang, and give one reason why no elements heavier than lithium and beryllium are manufactured. (Learning Outcome 20-9)
7. What observational evidence do we have to back up the standard Big Bang model? (Learning Outcomes 20-3, 20-5, 20-6, 20-7, and 20-9)
8. How can light produce particles? In the Big Bang model, how does it happen that heavy particles are not formed out of light after a certain time? (Learning Outcome 20-8)
9. In what sense does the present abundance of antimatter in the universe challenge the validity of the Big Bang model? (Learning Outcome 20-10)
10. In what sense does the Big Bang model have a flatness problem? How does the inflationary universe model cope with this problem? (Learning Outcomes 20-10 and 20-11)
11. The cosmic background radiation has a Planck curve for its continuous spectrum. What does this observation tell us about the conditions in the early universe? (Learning Outcomes 20-4 and 20-5)
12. An analysis of the COBE observations shows small fluctuations in the temperature of the cosmic background radiation. How do these variations assist in the formation of the first galaxies? (Learning Outcome 20-10)
13. In order for the universe to be closed, there must exist a large fraction of dark matter relative to the luminous matter. How is the existence (or not!) of dark matter a problem for Big Bang cosmology? (Learning Outcome 20-12)
14. We know the temperature of the cosmic background radiation now. What is needed in order to determine its temperature at specific times in the past or future? (Learning Outcome 20-6)

PROBLEMS AND ACTIVITIES

1. What is the peak wavelength of emission of the 2.7 K cosmic background radiation?
2. Gamma rays have frequencies on the order of 10^{22} Hz. In making matter from such photons, what minimum mass of elementary particles could be created by combining the gamma rays?
3. What is the energy equivalent of the mass of an electron? Of a proton?
4. When the universe was last opaque, it had a temperature of about 3000 K. At what wavelength did its emission peak then? How does that compare to the peak now?

POWERS OF TEN

Astronomers deal with quantities that range from the microcosmic to the macrocosmic. To avoid having to read and write numbers such as 149,597,870,600 meters (the average earth–sun distance), we use powers-of-ten notation. That simply means that we write a number as a fraction between one and ten times the appropriate power of ten rather than with a string of leading or trailing zeros. A positive exponent tells how many times to multiply by ten. For example:

$10^1 = 10$

$10^2 = 100$

$10^3 = 1000$

and so on. A negative exponent, in contrast, tells how many times to *divide* by 10. For instance:

$10^{-1} = 0.1$

$10^{-2} = 0.01$

$10^{-3} = 0.001$

in analogy to the positive powers. The only trick is to remember that 10 to the zero power is 1:

$10^0 = 1$

In powers-of-ten notation, the earth–sun distance is $1.495978706 \times 10^{11}$ meters. A millimeter is only 10^{-3} meter.

SIGNIFICANT FIGURES

Significant figures are the meaningful digits in a number. In experimental or observational results, the accuracy of the procedure or technique limits the number of significant figures. No more than that can be stated as the true value of the result (which should always have a range of error attached to it). In arithmetical calculations, you are sometimes interested in an approximate result to some specified degree of accuracy. Approximate ("back of the envelope") calculations using just one or two significant figures prove useful in science, especially in astronomy, because they are sufficient to illustrate a point, or because the quantities are simply not well known.

Here are some examples:

0.0045676	five significant figures
4.5676×10^{-1}	five significant figures
4.5×10^{-4}	two significant numbers
2.17	three significant figures

What about the number 2,170,000? It is not clear whether there are three or seven significant figures. But when we use powers-of-ten notation, the ambiguity disappears: 2.17×10^6 has three significant figures; 2.170000×10^6 has seven significant figures.

The general rule in doing calculations is that the result cannot have more significant figures than the smallest number of significant figures in the quantities that are used in the computation. It may have even fewer – for example, if the calculation involves subtraction of nearly equal numbers.

ROUNDING

To round off numbers, look at the last digit. If it is greater than 5, round up; round down if it is less than 5. If it is 5, then round up if the next-to-the-last digit is even, and down if it is odd. Most of the numbers in the main text of this book have been rounded off to two significant figures.

THE ENGLISH AND INTERNATIONAL SYSTEMS

The fundamental units of length, mass, and time in the English system are the yard, the slug, and the second. Other common units in this system are often strange multiples of these fundamental units. For example: the ton is 2000 pounds, the mile is 1760 yards, and the inch is 1/36 of a yard. Most countries use the much more rational metric system, which uses powers-of-ten relationships between units, making them easy to multiply and divide.

The fundamental metric system units are:

Length: 1 meter (m)
Mass: 1 kilogram (kg)
Time: 1 second (s)

This is often called the meter-kilogram-second, or mks system. An older system, often still used by astronomers, is the centimeter-gram-second, or *cgs* system, with the following fundamental units:

Length: 1 centimeter (cm)
Mass: 1 gram (g)
Time: 1 second (s)

The current international standard for physical units is called the *Système International* (in French) or the *International System of Units,* abbreviated SI units. This system is related to the mks system with many modifications. Some astronomers do not yet use SI units, though they should, because it is now the international standard. I have tried to stick with SI units in this book but have occasionally used units

of convenience, such as atmospheres for pressure. Table A.1 gives most of the fundamental SI units, and Table A.2 some conversions.

Table A.1 Selected SI Basic Units

Length: meter (m)
Time: second (s)
Mass: kilogram (kg)
Current: ampere (A)
Temperature: kelvin (K)
Force: newton (N)
Work and energy: joule (J)
Power: watt (W)
Frequency: hertz (Hz)
Charge: coulomb (C)
Magnetic field: tesla (T)
Pressure: pascal (Pa)

Some multiples of the metric system and their associated prefixes and abbreviations are:

10^{-15}	femto-	(f)
10^{-12}	pico-	(p)
10^{-9}	nano-	(n)
10^{-6}	micro-	(μ)
10^{-3}	milli-	(m)
10^{-2}	centi-	(c)
10^{-1}	deci-	(d)
10	deka-	(da)
10^2	hecto-	(h)
10^3	kilo-	(k)
10^6	mega-	(M)
10^9	giga-	(G)
10^{12}	tera-	(T)
10^{15}	peta-	(P)

Some relationships between the SI and English system are:

Length

1 kilometer (km) = 1000 m = 0.6214 mile
1 meter (m) = 1.094 yd = 39.37 inches
1 centimeter (cm) = 0.01 m = 0.3937 inch
1 millimeter (mm) = 0.001 m = 0.03937 inch
1 mile = 1.6093 km
1 inch = 2.5400 cm

Mass

> 1 metric ton = 10^6 g = 1000 kg = 2.2046×10^3 pounds
>
> 1 kilogram (kg) = 10^3 g = 2.2046 pounds
>
> 1 gram (g) = 0.0353 oz = 0.0022046 pound
>
> 1 milligram (mg) = 0.0001 g = 2.2046×10^{-6} pound
>
> 453.6 g = 1 pound
>
> 28.3495 g = 1 ounce

Table A.2 Conversion to SI Units

Length
  1 ft = 12 in. = 30.48 cm
  1 yd = 3 ft = 91.44 cm
  1 light year = 9.461×10^{15} m
  1 Å = 0.1 nm

Area
  1 m^2 = 10^4 cm^2
  1 km^2 = 0.3861 mi^2
  1 $in.^2$ = 6.4516 cm^2
  1 ft^2 = 9.29×10^{-2} m^2
  1 m^2 = 10.75 ft^2

Volume
  1 m^3 = 10^6 cm^3
  1 liter (L) = 1000 cm^3 = 10^{-3} m^3
  1 gal = 3.786 L = 231 $in.^3$

Time
  1 h = 60 min = 3.6 ks
  1 day = 24 h = 1440 min = 86.4 ks
  1 year = 365.24 days = 31.56 Ms

Speed
  1 km/h = 0.2778 m/s = 0.6125 mi/h
  1 mi/h = 0.4470 m/s = 1.609 km/h

Density
  1 g/cm^3 = 1000 kg/m^3 = 1 kg/L

Force
  1 N = 0.2248 lb = 10^5 dynes
  1 lb = 4.4482 N

Pressure
  1 Pa = 1 N/m
  1 atm = 101.325 kPa = 14.7 $lb/in.^2$ = 760 mm Hg

Energy
  1 kW · h = 36 MJ
  1 eV = 1.602×10^{-19} J
  1 erg = 10^{-7} J

Power
  1 hp = 550 ft · lb/s = 745.7 W
  1 W = 1.341×10^{-3} hp

Magnetic field
  1 G = 10^{-4} T
  1 T = 10^4 G

TEMPERATURE SCALES

The freezing and boiling points of water (at sea level) set the fundamental points on common temperature scales. In the United States, the Fahrenheit (F) system dominates. Water freezes at 32° F and boils at 212° F. Most countries use the Celsius (C; formerly centigrade) system. On it, water freezes at 0° C and boils at 100° C.

Astronomers mostly employ the Kelvin (K) system, which uses the same size degrees as the Celsius system. The Kelvin system starts from absolute zero, the temperature at which random molecular motion reaches a minimum (0 K = –273.15° C). At absolute zero, no more thermal energy can be extracted from a body. Water freezes at about 273 K and boils at about 373 K. Note that the degree mark is *not* used with temperatures on the Kelvin scale and that temperatures are often referred to as *kelvins*, not "degrees kelvin."

To convert between systems, recognize that 0 K = –273° C = –460° F and that Celsius and Kelvin degrees are larger than Fahrenheit degrees by 180/100 = 9/5. The relationships between systems are:

$$K = C + 273$$
$$C = \frac{5}{9}(°F - 32)$$

Kelvins are never negative, because absolute zero is the lowest point on any temperature scale.

ASTRONOMICAL DISTANCES

Although astronomers may use the SI system, they encounter distances so large that other measures are often used as well. In the solar system, the natural scale is the average distance from the earth to the sun, the astronomical unit (AU), which is about 1.5×10^8 km.

Beyond the solar system, the AU is too small compared to stellar distances. Here astronomers use the parsec and the light year. One parsec (pc) equals 206,265 AU or about 3.1×10^{13} km. A light year (ly) is the distance that light travels in one year, about 9.5×10^{12} km. Note that 1 pc is about 3.3 ly.

For even larger distances, astronomers most often talk in large multiples of parsecs: a kiloparsec (kpc) is 10^3 pc, a megaparsec (Mpc) is 10^6 pc, and a gigaparsec (Gpc) is 10^9 pc. Equivalent units in light years are kly, Mly, and Gly.

OTHER PHYSICAL UNITS

The SI unit of force is the *newton* (N). It is the force needed to accelerate an object having a mass of one kilogram by one meter per second squared.

The SI unit of energy is the *joule* (J). It takes about one joule to lift an apple off the floor to over your head. Table A.3 compares the energy of some familiar events and some astronomical ones.

Power is the rate of energy transfer, so it has units of joules per second. Astronomers often use the term *luminosity* in place of power. A convenient unit for luminosity or power is the *watt* (W), defined as one joule per second. Table A.3 includes some power comparisons.

When talking about stars, astronomers often measure quantities relative to the sun, for example in units of a solar mass, a solar luminosity, or a solar radius. For example, the mass of Sirius is 2.3 solar masses, its radius is 1.8 solar radii, and its luminosity is 23 solar luminosities.

Astronomers use both the *gauss* (G) and *tesla* (T) as the unit of magnetic field strength (1 T = 10^4 G); the tesla is the SI unit. The earth's magnetic field is about 5×10^{-3} T, or 0.5 G.

The SI unit of pressure is the *pascal* (Pa), which is one newton per meter squared. You can get a feel for a pascal next time you check the air in your car's tires: a typical pressure of 28 pounds per square inch (psi) equals 196 kPa (kilopascals). Astronomers also use the atmosphere (atm) as a pressure unit; one atmosphere is about 100 kPa.

Finally, astronomers face long times on the universe. To avoid confusion about "billions" (an American billion is a hundred million, 10^9; a British billion is a million million, 10^{12}), I have decided to use the SI prefix: giga – (Gy) = 10^9 years.

Table A.3 Comparative Energy and Power Outputs

Energy or Power Source	Approximate Total Energy (J) or Power (W)
ENERGY	
Earth's rotational kinetic energy	10^{29} J
Sunlight (1 y)	10^{34} J
Earth's daily input of solar energy	10^{22} J
H-bomb	10^{17} J
100% conversion of 1 gram of matter	10^{14} J
1 barrel of oil	10^{10} J
1 medium pizza	10^7 J
1 flashlight battery (D cell)	10^4 J
1 mosquito pushup	10^{-7} J
Fission of one atom of uranium-235	10^{-11} J
POWER	
Sunlight	10^{26} W
World's current energy consumption	10^{13} W
1 power plant	10^9 W
1 small person (MZ)	100 W

Note: Adapted in part from a table in the *American Journal of Physics*, volume 60, p. 596, June 1992.

Planetary Data

Table B.1 Planetary Rotation Rates and Inclinations

Planet	Rotation Period (equatorial)	Inclination of Equator to Orbital Plane
Mercury	58.65 days	0°
Venus	243.02 days (retrograde)	177° 18′
Earth	23 h 56 min 4.1 s	23° 27′
Mars	24 h 37 min 22.6 s	25° 11′
Jupiter	9 h 50.5 min	3° 07′
Saturn	10 h 14 min	26° 44′
Uranus	17 h 14 min (retrograde)	97° 52′
Neptune	16 h 3 min	29° 35′
Pluto	6.39 days (retrograde)	119° 37′

Table B.2 Distances and Periods of the Planets

Planet	Semimajor Axis (AU)	Semimajor Axis ($\times 10^6$ km)	Sidereal Period (tropical years)	Synodic Period (days)	Orbital Eccentricity
Mercury	0.387	57.9	0.241	115.9	0.206
Venus	0.723	108.2	0.615	583.9	0.007
Earth	1.000	149.6	1.000	–	0.017
Mars	1.524	227.9	1.881	779.9	0.093
Jupiter	5.203	778.4	11.86	398.9	0.048
Saturn	9.537	1427	29.42	378.1	0.054
Uranus	19.19	2871	83.75	369.7	0.047
Neptune	30.07	4498	163.7	367.5	0.009
Pluto	39.48	5906	248.0	366.7	0.249

Note: A tropical year is the year of seasons.

Table B.3 Physical Data of the Planets

Planet	Equatorial Radius (km)	Radius (earth radii)	Albedo	Surface Temperature (K)	Escape Speed (km/s)
Mercury	2440	0.38	0.11	100–700	4.25
Venus	6052	0.95	0.65	730	10.4
Earth	6378	1.00	0.37	288–293	11.2
Moon	1738	0.27	0.07	120–380	2.4
Mars	3397	0.53	0.15	183–268	5.02
Jupiter	71,492	11.2	0.52	124	59.5
Saturn	60,268	9.45	0.47	95	35.5
Uranus	25,559	4.01	0.51	59	21.2
Neptune	24,764	3.88	0.41	59	23.7
Pluto	1195	0.18	~0.50	50–70	1.3

Table B.4 Masses and Densities of the Planets

Planet	Mass (earth masses)	Mass (kg)	Density (kg/m³)	Surface Gravity (earth = 1)
Mercury	0.0562	3.30×10^{23}	5430	0.38
Venus	0.815	4.87×10^{24}	5240	0.91
Earth	1.000	5.97×10^{24}	5520	1.00
Moon	0.012	7.35×10^{22}	3340	0.16
Mars	0.1074	6.42×10^{23}	3940	0.39
Jupiter	317.9	1.90×10^{27}	1330	2.54
Saturn	95.1	5.69×10^{26}	700	1.07
Uranus	14.56	8.68×10^{25}	1300	0.90
Neptune	17.24	1.02×10^{26}	1760	1.14
Pluto	0.0018	1.3×10^{22}	1100	0.06

Table B.5 Selected Atmospheric Gases of the Major Planets

Planet	Gases (in order of relative abundance)
Mercury	Sodium, potassium, helium, hydrogen
Venus	Carbon dioxide, carbon monoxide, hydrogen chloride, hydrogen fluoride, water, argon, nitrogen, oxygen, hydrogen sulfide, sulfur dioxide, helium
Moon	Neon, helium
Earth	Nitrogen, oxygen, water, argon, carbon dioxide, neon, helium, methane, krypton, nitrous oxide, ozone, xenon, hydrogen, radon
Mars	Carbon dioxide, carbon monoxide, water, oxygen, ozone, argon, nitrogen
Jupiter	Hydrogen, helium, methane, ammonia, water, carbon monoxide, acetylene, ethane, phosphine
Saturn	Hydrogen, helium, methane, ammonia, acetylene, ethane, phosphine, propane
Titan	Nitrogen, methane, ethane, acetylene, ethylene, hydrogen cyanide
Uranus	Hydrogen, helium, methane
Neptune	Hydrogen, helium, methane
Pluto	Methane

Table B.6 Satellites of the Terrestrial Planets

Planet	Satellite	Distance ($\times 10^3$ km)	Orbital Period (days)	Radius (km)	Mass (planet = 1)	Bulk Density (kg/m^3)
Earth	Moon	384	27.32	1738	0.013	3340
Mars	Phobos	9.37	0.3189	$13 \times 11 \times 9$	1.5×10^{-8}	1900
	Deimos	23.52	1.262	$8 \times 6 \times 5$	3.1×10^{-9}	2100

Table B.7 Major Satellites of Jupiter

Name	Distance (× 10³ km)	Distance (Jupiter radii)	Orbital Period (days)	Radius (km)	Mass (planet = 1)	Bulk Density (kg/m³)
Metis	128	1.79	0.29	20	5×10^{-11}	
Andrastea	129	1.80	0.30	20	1×10^{-11}	
Amalthea	181	2.55	0.50	130×80	2×10^{-9}	3000
Thebe	222	3.11	0.67	40	4×10^{-10}	
Io	422	5.95	1.77	1820	4.7×10^{-5}	3530
Europa	671	9.47	3.55	1565	2.5×10^{-5}	3030
Ganymede	1,070	15.10	7.15	2634	7.8×10^{-5}	1930
Callisto	1,883	26.60	16.69	2403	5.7×10^{-5}	1790
Leda	11,094	156	239	~4	3×10^{-12}	
Himalia	11,480	161	251	85	5.0×10^{-9}	~1000
Lysithea	11,720	164	259	~10	4×10^{-11}	
Elara	11,737	165	260	~30	4×10^{-10}	
Ananke	21,200	291	631	8(?)	2×10^{-115}	
Carme	22,600	314	692	12(?)	5×10^{-11}	
Pasiphae	23,500	327	735	14(?)	1×10^{-10}	
Sinope	23,700	333	758	10(?)	4×10^{-11}	

Note: Distances are from the center of Jupiter.

Table B.8 Major Satellites of Saturn

Name	Distance (× 10³ km)	Distance (Saturn radii)	Orbital Period (days)	Radius (km)	Mass (planet = 1)	Bulk Density (kg/m³)
Metis	128	1.79	0.29	20	5×10^{-11}	
Atlas	137.67	2.28	0.602	20×10		
Prometheus	139.35	2.31	0.613	$70 \times 50 \times 40$		
Pandora	141.70	2.35	0.629	$55 \times 45 \times 35$		
Epimetheus	151.42	2.51	0.694	$70 \times 60 \times 50$		
Janus	151.47	2.51	0.695	$110 \times 100 \times 80$		
Mimas	185.54	3.08	0.942	195	8.0×10^{-8}	1200
Enceladus	238.04	3.95	1.370	250	1.3×10^{-7}	1200
Tethys	294.66	4.88	1.888	530	1.3×10^{-6}	1200
Telesto	297.67	4.88	1.888	$17 \times 14 \times 13$		
Calypso	294.67	4.88	1.888	$17 \times 11 \times 11$		
Dione	377.42	6.26	2.737	560	1.85×10^{-6}	1400
Helene	377.42	6.26	2.737	$18 \times 16 \times 15$		
Rhea	527.07	8.74	4.518	765	4.4×10^{-6}	1300
Titan	1221.86	20.25	15.945	2575	2.36×10^{-4}	1880
Hyperion	1481.00	24.55	21.277	$205 \times 130 \times 110$	3×10^{-8}	
Iapetus	3560.80	59.02	79.33	730	3.3×10^{-6}	1200
Phoebe	12,954	214.7	550.45	~110	$7 \times 10-10$	

Note: Distances are from the center of Saturn.

Table B.9 Major Satellites of Uranus, an.d Pluto

Planet	Satellite	Distance (× 10³ km)	Orbital Period (days)	Radius (km)	Mass (planet = 1)	Bulk Density (kg/m³)
Uranus	Ariel	191.8	2.52038	580	1.8×10^{-5}	
	Umbriel	267.3	4.14418	595	1.2×10^{-5}	
	Titania	438.7	8.70588	805	6.8×10^{-5}	~1500
	Oberon	586.6	13.46326	775	6.9×10^{-5}	~1500
	Miranda	130.1	1.414	160	0.2×10^{-5}	
Neptune	Triton	653.6	5.87683	1,400	1.3×10^{-3}	
	Nereid	5570	365	470	2.0×10^{-7}	
Pluto	Charon	19.5	6.39	600	0.1	2100

Physical Constants and Astronomical Data

PHYSICAL CONSTANTS

Gravitational constant

$$G = 6.673 \times 10^{-11} \text{ N} \cdot \text{m}^2/\text{kg}^2$$

Speed of light in a vacuum

$$c = 2.9979 \times 10^8 \text{ m/s}$$

Planck's constant

$$h = 6.62608 \times 10^{-34} \text{ J} \cdot \text{s}$$

Wien's constant

$$\sigma_{w} = 0.02898 \text{ m} \cdot \text{K}$$

Boltzmann's constant

$$k = 1.3806 \times 10^{-23} \text{ J/K}$$

Stefan–Boltzmann constant

$$\sigma = 5.6697 \times 10^{-8} \text{ W/m}^2 \cdot \text{K}^4$$

Electron mass

$$m_e = 9.10939 \times 10^{-31} \text{ kg}$$

Proton mass

$$m_p = 1.6726 \times 10^{-27} \text{ kg} = 1836.1 \, m_e$$

Neutron mass

$$m_n = 1.6749 \times 10^{-27} \text{ kg}$$

Mass of hydrogen atom

$$m_H = 1.6735 \times 10^{-27} \text{ kg}$$

ASTRONOMICAL DATA

Astronomical unit

$$\text{AU} = 1.4959789 \times 10^{11} \text{ m}$$

Parsec

$$\text{pc} = 206{,}264.806 \text{ AU}$$

$$= 3.2616 \text{ ly}$$

$$= 3.0856 \times 10^{16} \text{ m}$$

Light year

$$\text{ly} = 9.46053 \times 10^{15} \text{ m}$$

$$= 6.324 \times 10^4 \text{ AU}$$

Sidereal year

$$y = 3.155815 \times 10^7 \text{ s}$$

Mass of sun

$$M_{sun} = 1.989 \times 10^{30} \text{ kg}$$

Luminosity of sun

$$L_{sun} = 3.90 \times 10^{26} \text{ W}$$

Solar constant

$$S = 1370 \text{ W/m}^2$$

Radius of sun

$$R_{sun} = 6.96 \times 10^5 \text{ km}$$

Mass of earth

$$M_{earth} = 5.9742 \times 10^{24} \text{ kg}$$

Equatorial radius of earth

$$R_{earth} = 6.37814 \times 10^3 \text{ km}$$

Mass of moon

$$M_{moon} = 7.348 \times 10^{22} \text{ kg}$$

Radius of moon

$$R_{moon} = 1.738 \times 10^3 \text{ km}$$

APPENDIX D
Nearby Stars
in the *Hipparcos Catalogue*

Name or Designation	Absolute Visual Magnitude	Visual Luminosity (sun = 1)	Parallax (arcsec)	Distance (pc)	Distance (ly)
Alpha Centauri C (Proxima)	15.45	5.6×10^{-5}	0.7723	1.295	4.224
Beta Centauri	5.70	0.44	0.7421	1.348	4.497
Alpha Centauri	4.34	1.6	0.7421	1.348	4.497
Barnard's star	13.24	4.3×10^{-4}	0.5490	1.821	5.939
HIP 54035		5.5×10^{-3}	0.3924	2.548	8.311
Alpha Canis Majoris (Sirius)	1.45	22	0.3792	2.637	8.601
HIP 92403	13.00	5.3×10^{-4}	0.3365	2.972	9.693
Epsilon Eridani	6.18	0.29	0.3108	3.218	10.50
HIP 114046	9.76	0.011	0.3039	3.291	10.73
HIP 57548	13.50	3.4×10^{-4}	0.2996	3.338	10.89
61 Cygni A	7.49	0.086	0.2871	3.483	11.36
Alpha Canis Minoris (Procyon)	2.68	7.2	0.2859	3.498	11.41
62 Cygni B	8.33	0.039	0.2854	3.504	11.42
HIP 91772	11.97	1.4×10^{-3}	0.2845	3.515	11.46
HIP 91768	11.18	2.9×10^{-3}	0.2803	3.568	11.64
HIP 1475	10.33	6.3×10^{-3}	0.2803	3.568	11.64
Epsilon Indi	6.89	0.15	0.2758	3.626	11.83
Tau Ceti	5.68	0.45	0.2742	3.647	11.90
HIP 5643	14.25	1.7×10^{-4}	0.2690	3.717	12.12
Luyten's star	11.94	1.4×10^{-3}	0.2634	3.796	12.38

Note: Most nearby stars are so faint that they do not have proper names. Many are identified just by their *Hipparcos Catalogue* (HIP) number.

The Most Luminous Stars in the *Hipparcos Catalogue*

Name or Designation	Absolute Visual Magnitude	Visual Luminosity (sun = 1)	Parallax (arcsec)	Distance (pc)	Distance (ly)
Alpha Centauri C (Proxima)	15.45	5.6×10^{-5}	0.7723	1.295	4.224
Beta Orionis (Betelgeuse)	−6.69	4.1×10^4	.00422	237	773
Gamma Cygni	−6.12	2.4×10^4	.00214	467	1520
Zeta Puppis	−5.95	2.0×10^4	.00233	429	1400
Upsilon Carinae	−5.56	1.4×10^4	.00201	498	1620
Alpha Carinae (Canopus)	−5.53	1.4×10^4	.0104	96.2	314
Beta Centauri	−5.42	1.2×10^4	.00621	161	525
Alpha Leporis	−5.40	1.2×10^4	.00254	394	1280
Phi Velorum	−5.34	1.2×10^4	.00169	592	1930
Gamma Velorum	−5.31	1.1×10^4	.00388	258	841
Zeta Orionis	−5.26	1.1×104	.00399	251	818
Alpha Orionis (Rigel)	−5.14	9.6×10^3	.00763	131	427
Lambda Scorpii	−5.05	8.9×10^3	.00464	216	705
Delta Orionis	−4.99	8.4×10^3	.00356	281	917
Pi Puppis	−4.92	7.9×10^3	.00298	336	1100
h Carinae	−4.83	7.2×10^3	.00165	606	1980
Kappa Orionis	−4.65	6.1×10^3	.00452	221	721
Epsilon Carinae	−4.58	5.8×10^3	.00516	194	633
Zeta Persei	−4.55	5.6×10^3	.00332	301	982

Note: Most of the most luminous stars are some distance from the sun. Visual luminosities are calculated with an absolute visual magnitude of the sun equal to 4.82.

Periodic Table of the Elements

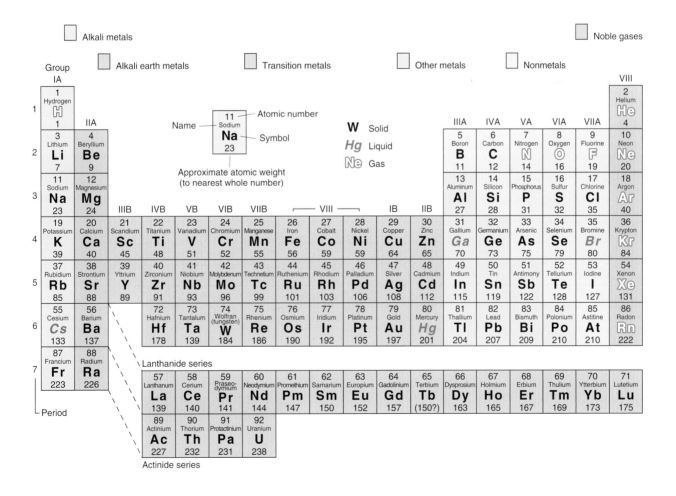

Alkali metals

Alkali earth metals   **Transition metals**   **Other metals**   **Nonmetals**

Noble gases

																	VIII
Group IA																	
1 Hydrogen **H** 1	IIA											IIIA	IVA	VA	VIA	VIIA	2 Helium **He** 4
3 Lithium **Li** 7	4 Beryllium **Be** 9											5 Boron **B** 11	6 Carbon **C** 12	7 Nitrogen **N** 14	8 Oxygen **O** 16	9 Fluorine **F** 19	10 Neon **Ne** 20
11 Sodium **Na** 23	12 Magnesium **Mg** 24	IIIB	IVB	VB	VIB	VIIB		VIII		IB	IIB	13 Aluminum **Al** 27	14 Silicon **Si** 28	15 Phosphorus **P** 31	16 Sulfur **S** 32	17 Chlorine **Cl** 35	18 Argon **Ar** 40
19 Potassium **K** 39	20 Calcium **Ca** 40	21 Scandium **Sc** 45	22 Titanium **Ti** 48	23 Vanadium **V** 51	24 Chromium **Cr** 52	25 Manganese **Mn** 55	26 Iron **Fe** 56	27 Cobalt **Co** 59	28 Nickel **Ni** 59	29 Copper **Cu** 64	30 Zinc **Zn** 65	31 Gallium **Ga** 70	32 Germanium **Ge** 73	33 Arsenic **As** 75	34 Selenium **Se** 79	35 Bromine **Br** 80	36 Krypton **Kr** 84
37 Rubidium **Rb** 85	38 Strontium **Sr** 88	39 Yttrium **Y** 89	40 Zirconium **Zr** 91	41 Niobium **Nb** 93	42 Molybdenum **Mo** 96	43 Technetium **Tc** 99	44 Ruthenium **Ru** 101	45 Rhodium **Rh** 103	46 Palladium **Pd** 106	47 Silver **Ag** 108	48 Cadmium **Cd** 112	49 Indium **In** 115	50 Tin **Sn** 119	51 Antimony **Sb** 122	52 Tellurium **Te** 128	53 Iodine **I** 127	54 Xenon **Xe** 131
55 Cesium **Cs** 133	56 Barium **Ba** 137		72 Hafnium **Hf** 178	73 Tantalum **Ta** 139	74 Wolfran (tungsten) **W** 184	75 Rhenium **Re** 186	76 Osmium **Os** 190	77 Iridium **Ir** 192	78 Platinum **Pt** 195	79 Gold **Au** 197	80 Mercury **Hg** 201	81 Thallium **Tl** 204	82 Lead **Pb** 207	83 Bismuth **Bi** 209	84 Polonium **Po** 210	85 Astitine **At** 210	86 Radon **Rn** 222
87 Francium **Fr** 223	88 Radium **Ra** 226																

Period

Name [box] 11 — Atomic number / Sodium / **Na** — Symbol / 23

Approximate atomic weight (to nearest whole number)

W Solid
Hg Liquid
Ne Gas

Lanthanide series

57 Lanthanum **La** 139	58 Cerium **Ce** 140	59 Praseo-dymium **Pr** 141	60 Neodymium **Nd** 144	61 Promethium **Pm** 147	62 Samarium **Sm** 150	63 Europium **Eu** 152	64 Gadolinium **Gd** 157	65 Terbium **Tb** (150?)	66 Dysprosium **Dy** 163	67 Holmium **Ho** 165	68 Erbium **Er** 167	69 Thulium **Tm** 169	70 Ytterbium **Yb** 173	71 Lutetium **Lu** 175
89 Actinium **Ac** 227	90 Thorium **Th** 232	91 Protactinium **Pa** 231	92 Uranium **U** 238											

Actinide series

Expanded Glossary

This glossary includes terms not used in this book.

aberration of starlight The apparent displacement in the position of a star that results from the finite speed of light and the earth's orbital speed.

absolute magnitude A measure of the brightness a star would have if it were to be placed at a standard distance of 10 parsecs (32.6 light years) from the sun.

absorption The decrease of flux of radiation as it passes through a medium.

absorption (dark) lines Discrete colors missing in a continuous spectrum because of the absorption of those colors by atoms.

absorption-line spectrum Dark lines superimposed on a continuous spectrum.

abundance The relative number of atoms or isotopes in an object.

acceleration The rate of change of velocity with respect to time.

accretion The colliding and sticking together of small particles to make larger masses.

accretion disk A disk made by infalling material around a mass; the conservation of angular momentum results in the disk shape.

achondrite A type of stony meteorite, which crystallized from molten rock so that no chondrules formed.

active galactic nuclei (AGNs) The nuclei of galaxies that have a nonthermal continuous spectrum over a wide range of wavelengths and signs of unusual, energetic activity such as radio jets; these include Seyferts, BL Lac objects, and quasars.

active galaxies Galaxies characterized by a nonthermal spectrum and a large energy output compared to a normal galaxy.

active optics Process of feedback to correct the shape of a mirror to maintain a sharp image.

active regions Areas on the sun (and other stars) in which magnetic fields are concentrated; these generate sunspots and flares.

adaptive optics Feedback process for improving optical images to compensate for seeing.

aerobe An organism that is dependent on free oxygen for its metabolism.

age the time interval from the fermentation of a celestial object up to the present time.

albedo A measure of an object's reflecting power; the ratio of reflected light to incoming light for a solid surface, where complete reflection gives an albedo of 1.0 or 100 percent.

almanac A set of tables giving positions of the sun, moon, and planets at various times, as well as other astronomical information; an ephemeris.

Alpha Centauri The closest star system to the sun; a

499

triple-star system; the component Alpha Centauri A has almost the same luminosity, mass, and surface temperature as the sun. Alpha Centauri C, also called Proxima, is the closest to the sun.

alpha particle A helium nucleus emitted in the radioactive decay of heavy elements.

altitude In the horizon system, the angle along a vertical circle from the horizon to a point on the celestial sphere.

Amalthea A small asteroidal moon close to Jupiter; elongated and irregular in shape with a reddish, cratered surface.

amino acids The building blocks of proteins, consisting mostly of carbon, nitrogen, oxygen, and hydrogen atoms.

anaerobe An organism that does not depend on free oxygen for its metabolism.

Andromeda Galaxy (M31) The closest spiral galaxy to the Milky Way Galaxy at a distance of 2.2 Mly (680 kpc); it has a diameter of about 160 kly (50 kpc).

angstrom (Å) A unit of length, equal to 10^{-10} m, often used to measure the wavelength of visible light.

angular diameter The apparent diameter of an object in angular measure; the angular separation of two points on opposite sides of the object.

angular distance The apparent angular spacing between two objects in the sky. See **angular separation.**

angular momentum The tendency for bodies, because of their inertia, to keep spinning or orbiting.

angular separation The observed angular distance between two celestial objects, measured in degrees, minutes, and seconds of angular measure.

angular speed The rate of change of angular position of a celestial object viewed in the sky.

anorthosite A basaltic mineral composed of calcium and sodium with aluminum silicate; the predominant mineral of the lunar highlands.

antapex The direction in the sky from which the sun appears to be moving relative to local stars; located in the constellation Columbia.

antimatter Elementary particles having their electric charge or other property reversed compared to ordinary matter.

aperture synthesis Interferometric techniques in which an array of radio antennas are used to attain high resolutions characteristic of their widest spacings.

apex The direction in the sky toward which the sun appears to be moving relative to local stars; located in the constellation Hercules. See **antapex.**

aphelion The point on the orbit of a body orbiting the sun that is farthest from the sun.

Aphrodite Terra A large highland region on Venus.

apogee The point in its orbit at which an earth satellite is farthest from the earth.

apparent magnitude The brightness of a star (or any other celestial object) as seen from the earth; an astronomical measure of the object's flux.

apparent solar time Time measure based on the actual daily motion of the sun.

arroyo A dry channel carved in the ground by sporadic water flows.

asterism A conspicuous pattern of stars with a popular name that does not make up a constellation; the Big Dipper is an example.

asteroid Minor planet; one of several thousand very small members of the solar system that revolve around the sun, generally between the orbits of Mars and Jupiter.

asteroid belt The region lying between the orbits of Mars and Jupiter, containing the majority of asteroids.

astronomical unit (AU) The average distance between the earth and the sun, also the semimajor axis of the earth's orbit; 149.6×10^6 km or 8.3 light minutes.

astrophysical jets Collimated beams of material (usually ions and electrons) expelled from astrophysical objects, such as the nuclei of active galaxies.

astrophysics The study of physical concepts applied to astronomical objects.

asymptotic giant branch (AGB) The path on the H–R diagram that a star traverses during the phase of evolution when it has both a hydrogen-burning and a helium-burning shell, after its main-sequence phase.

atmosphere A gaseous envelope surrounding a planet, or the visible layers of a star; also a unit of pressure (abbreviated atm) equal to the pressure of air at sea level on the earth's surface.

atmospheric absorption or extinction The decrease in light caused by passage through the atmosphere.

atmospheric escape The process by which particles in the upper atmosphere, with greater than escape speed and unhindered by collisions, leave a planet for space.

atmospheric reddening Preferential scattering of blue light over red by air particles, which results in an object, such as the setting sun, appearing to be redder than it actually is.

atom The smallest particle of an element that exhibits the chemical properties of the element.

AU Abbreviation for astronomical unit(s).

aurora Light emission from atmospheric atoms and molecules excited by collisions with energetic, charged particles from the magnetosphere.

autumnal equinox The fall equinox; see **equinox.**

axis One of two or more reference lines in a coordinate system; also, the straight line, through the poles, about which a body rotates.

azimuth Angular position along the horizon, measured clockwise from north.

Balmer jump The large change in opacity at 3650 Å due to bound–free transitions from the second energy level of hydrogen atoms.

Balmer series The set of transitions of electrons in a hydrogen atom between the second energy level and higher levels; also the set of absorption or emission lines corresponding to these transitions that lies in the visible part of the spectrum, the first of which is the hydrogen-alpha (H-alpha) line.

bandwidth The range of frequencies detected simultaneously by a radio receiver (or other electromagnetic sensor).

barred spirals A subclass of spiral galaxies that have a bar across the nuclear region.

barycenter The center of mass of a system of masses moving under their mutual gravitation.

baryon Name for group of elementary particles, usually of large mass, such as protons and neutrons.

basalt An igneous rock, composed of olivine and feldspar, that makes up much of the earth's lower crust.

basins Large, shallow lowland areas in the crusts of terrestrial planets created by asteroidal impact or plate tectonics.

Becklin–Neugebauer object An infrared source associated with the Orion Nebula; probably a very young, massive star.

belts Regions of downflow, hence low pressure, in the atmosphere of a Jovian planet.

beta decay A process of radioactive decay in which a neutron disintegrates into a proton, an electron, and a neutrino.

beta particle An electron or positron emitted by a nucleus in radioactive decay.

Beta Regio A highland region on Venus containing at least two large shield volcanoes.

Betelgeuse A red supergiant star having a luminosity roughly 40,000 times that of the sun.

Big Bang model A model of the evolution of the universe that postulates its origin, in an event called the Big Bang, from a hot, dense state that rapidly expanded to cooler, less dense states.

Big Crunch In a universe with a closed geometry, its end in a singularity.

binary accretion model A model for the origin of the moon in which the moon and the earth formed by accretion of material from the same cloud of gas and dust.

binary galaxies Two galaxies bound by gravity and orbiting a common center of mass.

binary stars Two stars bound together by gravity that revolve around a common center of mass.

binary system Any two masses bound by gravity and orbiting a common center of mass following Kepler's laws.

binary x-ray source A binary system containing an x-ray emitter, which is usually a collapsed object surrounded by a hot accretion disk giving off x-rays.

biological evolution The natural development over time from simple to complex organisms, generated by mutations that change the gene structure and directed by natural selection of those individuals best-adapted to the environment.

bipolar outflows High-speed outflows of gas in opposite directions from a young stellar object (YSO); the result of a magnetized accretion disk around the YSO, collimating its stellar wind.

blackbody A (hypothetical) perfect radiator of light that absorbs and reemits all radiation incident upon it; its light output depends only on its temperature.

blackbody spectrum The continuous spectrum emitted by a blackbody; the flux at each wavelength is given by the formula known as Planck's law.

black dwarf The cold remains of a white dwarf after all its thermal energy has been exhausted.

black hole A mass that has collapsed to such a small radius that the escape speed from its surface is greater than the speed of light; thus light is trapped by the intense gravitational field.

BL Lacertae (BL Lac) object A type of active galaxy whose nonthermal emission from the nucleus varies rapidly (one day or so) and is highly polarized.

blueshift A decrease in the wavelength of the radiation emitted by an approaching celestial body as a consequence of the Doppler effect; a shift toward the short-wavelength, high-frequency (blue) end of the spectrum.

Bohr model of the atom A simple model of atomic structure in which electrons have well-defined orbits (energy levels) about the nucleus of the atom.

bolometer A detector of infrared radiation, usually a small chip of semiconductor material cooled to a few kelvins; absorption of infrared radiation causes a change in its resistance, which can be measured in an electronic circuit.

bolometric luminosity The total energy output per second at all wavelengths emitted by an astronomical object.

bolometric magnitude The magnitude of an object measured over all wavelengths of the electromagnetic spectrum.

Boltzmann's constant The number that relates pressure and temperature, or kinetic energy and temperature in a gas; the gas constant per molecule (see Appendix C).

bound–bound transition A transition of an electron between two bound energy states of an atom or ion.

bound–free transition A transition of an electron between a bound and an unbound (free) state.

bow shock wave The shock wave created by the interaction of the solar wind with a planetary magnetosphere.

breccias Rock and mineral fragments cemented together; a common part of the lunar surface.

bremsstrahlung Electromagnetic radiation generated when an electron slows down as it passes close to an atomic nucleus.

bright-line spectrum See **emission-line spectrum.**

bright nebula See **diffuse nebula.**

brightness An ambiguous term, usually meaning the energy per unit area received from or emitted by an object – its flux – but sometimes used to refer to an object's luminosity.

broad absorption line quasars High redshift quasars in which very broad troughs of absorption are found.

broad line region The gas immediately surrounding an AGN that contributes very broad emission line radiation to the galaxy's spectrum.

brown dwarf A very-low-mass object (roughly 0.05

solar mass) of low temperature and luminosity; its core never becomes hot enough to ignite thermonuclear reactions.

caldera A large volcanic crater made by collapse or explosive eruption.

Callisto Second largest of the Galilean moons of Jupiter and the farthest out; has a density about twice that of water; a heavily cratered, icy surface with multiringed basins.

Caloris Basin A large multiringed basin on Mercury, surrounded by a mountain range, located near a "hot" longitude of 180E.

canali Italian term for channels used by Giovanni Schiaparelli to describe dark linear features seen on the surface of Mars.

capture model A model of the moon's origin proposed about 1955 that pictures the moon as having been captured by the earth's gravity, after which it spiraled in toward the earth, reversed orbital direction, and spiraled outward.

carbonaceous chondrites A class of meteorites that contain chondrules embedded in a matrix material with a large percentage of carbon (about 4 percent).

carbon–nitrogen–oxygen (CNO) cycle A series of thermonuclear reactions taking place in a star's core in which carbon, nitrogen, and oxygen aid the fusion of hydrogen into helium; in the sun it is a secondary process of energy production, but in high-mass, main-sequence stars it is the major process.

Carina arm A local segment of the spiral arms of the Milky Way Galaxy, located inward from the sun toward the constellation Carina.

Cassegrain reflector The design for reflecting telescopes in which the secondary mirror directs the beam to a focus through a hole in the center of the primary mirror.

Cassini's division A gap about 2000 km wide in Saturn's rings, discovered in 1675 by Giovanni Cassini; now known to contain many small ringlets.

cataclysmic variable A variable star whose luminosity suddenly increases.

catastrophic models Models for the origin of the solar system in which an improbable event involving a large mass (usually collision with another star) led to the expulsion and then to the collection of gaseous materials that became the planets.

causally connected A description of events in spacetime

that can be affected by a given event; all the events within the light cone of a given event are causally connected to it.

cD galaxies See **supergiant elliptical galaxies.**

celestial coordinate system A system for specifying positions on the celestial sphere, similar to the system of latitude and longitude used to specify positions on the earth.

celestial equator An imaginary projection of the earth's equator onto the celestial sphere; declination is zero along the celestial equator.

celestial meridian See **meridian.**

celestial pole An imaginary projection of the earth's pole (north or south) onto the celestial sphere; a point about which the apparent daily rotation of the stars takes place.

celestial sphere An imaginary sphere of very large radius, centered on the earth, on which the celestial bodies appear fastened and against which their motions are charted.

cell Most common form of a living organism on the earth.

Centaurus arm A spiral arm of the Milky Way Galaxy located about 12,000 ly away in the direction of the constellation Centaurus.

Centaurus X-3 A binary x-ray source in the constellation Centaurus; contains a low-mass neutron star (the x-ray source) orbiting a blue giant star.

center of mass The balance point of a set of interacting or connected bodies.

central force A force directed along a line connecting the centers of two objects.

centripetal acceleration The acceleration of a body toward the center of a circular path.

centripetal force A force required to divert a body from a straight path into a curved one, directed toward the center of the curve.

cepheid variables (cepheids) Stars that vary in brightness as a result of a regular variation in size and temperature; a class of variable stars for which the star Delta Cephei is the prototype.

Ceres The first observed asteroid, discovered by Father Giuseppe Piazzi in 1801.

CETI Acronym for "communication with extraterrestrial intelligence."

Chandrasekhar limit The maximum mass for a white dwarf star (about 1.4 solar masses); this amount leads to the highest density and smallest radius for a star made of a degenerate electron gas; if more than this mass is present, the star cannot be supported by the electron gas pressure and so collapses further gravitationally.

charge-coupled device (CCD) A small chip of semiconductor material that emits electrons when it absorbs light; the electrons are trapped in small regions called pixels; the pattern of charges is read out in such a way as to preserve the image striking the chip.

Charon Moon of Pluto, with a diameter of about 1200 km.

chemical condensation sequence The sequence of chemical reactions and condensation of solids that occurs in a low-density gas as it cools at specific densities and temperatures.

chemical evolution The natural development from simple to complex molecules, such as proteins and nucleic acids, which are the building blocks of life.

chondrite A stony meteorite characterized by the presence of small, round silicate granules (chondrules).

chondrules Round silicate granules lacking volatile elements; found in chondritic meteorites, or chondrites, they are believed to be primitive solar system materials.

chromosphere The part of the sun's atmosphere just above the photosphere; hotter and less dense than the photosphere; it creates the flash spectrum seen during eclipses.

circular velocity The speed at which an object must travel to maintain uniform circular motion around a gravitating body.

circumpolar stars For an observer north of the equator, the stars that are continually above the northern horizon and never set; for a southern observer, the stars that never set below the southern horizon.

circumstellar matter Gas and/or dust surrounding a star.

closed geometry See **spherical geometry.**

cluster of galaxies A grouping of galaxies containing a few to thousands of galaxies.

CNO cycle See **carbon–nitrogen–oxygen cycle.**

collisional de-excitation Loss of energy by an electron of an atom in a collision so that the electron drops to a lower energy level.

collisional excitation Forcing an electron of an atom to a higher energy level by a collision.

color excess The difference between the actual color of a star and its observed color; usually redder because of interstellar dust along the line of sight.

color index The difference in the magnitudes of an object measured at two different wavelengths; a measure of the color, hence the temperature, of a star.

color-magnitude diagram An H–R diagram using on the x-axis.

color temperature Temperature inferred from color, usually by fitting a Planck curve to the continuous spectrum of a star at two wavelengths.

coma The bright, visible head of a comet.

Coma cluster A cluster of galaxies located in the sky in the direction of the constellation Coma Berenices; it contains thousands of galaxies spread out over millions of light years.

comets Bodies of small mass that revolve around the sun, usually in highly elliptical orbits; in the dirty snowball model, comets consist of small, solid particles (probably of rocky material) embedded in frozen gases.

compact (radio) galaxy An active galaxy whose nucleus contains a small, strong radio source.

compound A substance composed of the atoms of two or more elements bound together by chemical forces.

condensation The growth of small particles by the sticking together of atoms and molecules.

condensation sequence See **chemical condensation sequence.**

conduction Transfer of thermal energy by particles colliding into one another.

conjunction The time at which two celestial objects appear closest together in the sky.

conservation of angular momentum The fundamental principle stating that with no externally applied torques, the total angular momentum of an isolated system is constant.

conservation of energy A fundamental principle in physics that states that the total energy of an isolated system remains constant regardless of internal changes that may occur.

conservation of magnetic flux The physical principle stating that under certain circumstances, the number of magnetic field lines passing through an area remains constant even as the area changes in size.

conservation of momentum The physical principle stating that with no external net forces, the total momentum of a system is constant.

constellation An apparent arrangement of stars on the celestial sphere, usually named after an ancient god, hero, animal, or mythological being; now an agreed-upon region of the sky containing a group of stars.

contact binary A binary system in which the two stars are in contact via their Roche lobes or share a common envelope.

continental drift The theory that the present continents were at one time a unified landmass that broke up.

continuous spectrum A spectrum showing emission at all wavelengths, unbroken by either absorption lines or emission lines.

contour map A diagram showing how the intensity of some kind of radiation varies over a region of the sky; lines on such a map connect points of equal flux; closely spaced lines mean that the flux changes rapidly over a small distance, widely spaced lines mean it changes more slowly.

convection The transfer of energy by the moving currents of a fluid.

coplanar Lying in the same plane.

core–mantle grains Interstellar dust particles with cores of dense materials (such as silicates) surrounded by a mantle of icy materials (such as water ice).

core (of the earth) The central region of the earth; it has a high density, is probably liquid, and is believed to be composed of iron and iron alloys.

core (of the Galaxy) The inner few light years of the nucleus, which contains a small, nonthermal radio source and fast-moving clouds of ionized gas and perhaps a supermassive black hole.

core (of the sun) The inner 25 percent of the sun's radius, where the temperature is great enough for thermonuclear reactions to take place.

corona The outermost region of the sun's atmosphere, consisting of thin, ionized gases at a temperature of about 1×10^6 K; large magnetic loops give the corona its structure.

coronal holes Regions in the sun's corona that lack a concentration of high-temperature plasma; here magnetic field lines extend out into interplanetary space and mark the source of the solar wind.

coronal interstellar gas High-temperature interstellar plasma made visible by its x-ray emission.

Cosmic Background Explorer (COBE) A satellite launched in 1989 to measure the spectrum and intensity distribution of the microwave background radiation.

cosmic blackbody microwave radiation Radiation with a blackbody spectrum at a temperature of about 2.7 K; the remains of the primeval fireball in which the universe was created; this radiation permeates the universe, offering an observational confirmation of the Big Bang model.

cosmic nucleosynthesis The production of the lightest few nuclei during the first 100 seconds of the Big Bang.

cosmic rays Charged atomic particles moving in space with very high energies (the particles travel close to the speed of light); most originate beyond the solar system, but some low-energy cosmic rays are produced in solar flares.

cosmological principle The statement that the universe, averaged over a large enough volume, appears the same from any location.

cosmology The study of the nature and evolution of the physical universe.

cosmos The universe considered as an orderly and harmonious system.

Crab Nebula A supernova remnant, located in the constellation Taurus, produced by the supernova explosion visible from earth in A.D. 1054; a pulsar in the nebula marks the corpse of the exploded star.

crater A circular depression of any size, usually caused by the impact of a solid body or by a surface eruption.

cratered terrain Landscape with an abundance of craters, which implies that it is old and unevolved.

crescent The phase during which a moon or planet has less than half of its visible surface illuminated.

critical density In cosmology, the density that marks the transition from an open to a closed universe; the critical density that provides enough gravity to just bring the expansion to a stop after infinite time.

crossing time In a cluster of galaxies, the size of the cluster compared to the average random velocities of galaxies in the cluster, which gives a measure of how long a galaxy would take to drift across the cluster.

crust The thin, outermost surface layer of a planet; on the earth, it is composed of basaltic and granitic rocks.

C-type asteroids Dark asteroids with low albedos (around 0.04), which probably contain a large percentage of carbon materials.

curvature of spacetime See **spacetime curvature.**

cyclone Spiral flows in a planet's atmosphere produced by the planet's rotation.

Cygnus arm A segment of one of the spiral arms in the outer part of our Galaxy, about 14 kpc (45 kly) from the center.

Cygnus X-1 A binary x-ray source in the constellation Cygnus; it contains a probable black hole orbiting a blue supergiant star.

dark cloud An interstellar cloud of gas and dust that contains enough dust to blot out the light of stars behind it (as seen from the earth).

dark-line spectrum See **absorption-line spectrum.**

dark matter Matter in an unknown form, detectable by its gravitation effects on luminous matter.

declination In the equatorial coordinate system, the angle measured north or south along an hour circle from the celestial equator to a point on the celestial sphere.

decoupling The time in the universe's history when the density and opacity became so low that matter and light stopped interacting.

deferent An ancient geometric device used to account for the apparent eastward motion of the planets; a large circle, usually centered on the earth, that carries around a planet's epicycle.

degenerate electron gas An ionized gas in which nuclei and electrons are packed together as much as possible, filling all possible low-energy states, so that the perfect ideal gas law relating pressure, temperature, and density no longer applies.

degenerate gas pressure A force exerted by degenerate matter that depends mostly on how dense the matter is and very little on its temperature.

degenerate neutron gas Matter made up of neutrons packed together as tightly as possible.

Deimos The smaller of the two moons of Mars; it has a dark, cratered surface.

density The amount of mass per volume in an object or region of space.

density-wave model A model for the generation of spiral structure in galaxies in which spiral density waves (similar to sound waves) are pictured as plowing through the interstellar matter and sparking star formation.

deoxyribonucleic acid (DNA) The basic genetic material of life as we know it; the DNA molecule is very large, consisting of subunits called nucleotides.

detector Any device, sensitive to electromagnetic radiation, placed at the focus of a telescope.

differential rotation The tendency of a fluid, spherical body to rotate faster at the equator than at the poles.

differentiated Layered interior of a planet or a moon, generally with the less dense materials atop more dense ones.

diffraction The spreading out of light waves as they pass the edge of an opaque body.

diffraction grating Optical device used to disperse light into spectra.

diffuse (bright) nebula A cloud of hot ionized gas, mostly hydrogen, with an emission-line spectrum.

dipole field A magnetic field configuration like that of a bar magnet with opposed north and south poles.

dirty snowball comet model A model for comets that pictures the nucleus as a compact solid body of frozen materials, mixed with pieces of rocky matter, that turn into gases as a comet nears the sun, creating the head and tail.

disk (of a galaxy) The flattened wheel of stars, gas, and dust outside the nucleus of a spiral galaxy.

dispersion The effect that causes pulses of radiation emitted simultaneously at different frequencies to arrive at different times after traversing the interstellar medium.

diurnal motion Daily motion of a celestial object over a sidereal day.

D lines A pair of dark lines in the yellow region of the spectrum, produced by sodium.

DNA See **deoxyribonucleic acid.**

Doppler broadening An increase in the width of a spectral line from the random, internal motions of atoms and molecules in the emitting source.

Doppler shift A change in the wavelength of waves from a source reaching an observer when the source and the observer are moving with respect to each other along the line of sight; the wavelength increases (redshift) or decreases (blueshift) according to whether the motion is away from or toward the observer.

double quasar An optical illusion, produced on the sky by a gravitational-lens effect, of two adjacent images of a single quasar.

Drake equation A relationship of the important factors for estimating the number of civilizations in the Galaxy now, devised by radio astronomer Frank Drake.

dust tail The part of a comet's tail containing dust particles, pushed out by radiation pressure from sunlight.

dwarf A star of relatively low light output and relatively small size; a main-sequence star of luminosity class V.

dynamo model A model for the generation of a planet's (or star's) magnetic field by the circulation of conducting fluids in its core (or corrective zone).

Earth Our home planet; model for the other terrestrial planets: the moon, Mercury, Venus, and Mars.

earthshine A faint illumination of the dark (night) regions of the moon facing the earth; an effect especially noticeable when the moon is a thin crescent; arises from sunlight reflected from the earth onto the moon.

eccentric An ancient geometric device used to account for nonuniform planetary motion; a point offset from the center of circular motion.

eccentricity The ratio of the distance of a focus from the center of an ellipse to its semimajor axis.

eclipse The phenomenon of one body passing in front of another, cutting off its light.

eclipsing binary system Two stars that revolve around a common center of mass, the orbits lying edge-on to the line of sight, so that each star periodically passes in front of the other to cause a decrease in the amount of light.

ecliptic From the earth, the apparent yearly path on the celestial sphere of the sun with respect to the stars; also, the plane of the earth's orbit.

ecosphere The region around a star in which an orbiting planet's surface temperature is suitable for life as we know it.

Eddington luminosity For a given mass of an object, the luminosity that would just provide sufficient radiation pressure on surrounding material to prevent gravity from pulling it in.

Eddington mass For a given luminosity of an object, the mass that would produce just enough gravity to overcome radiation pressure on surrounding material and allow it to fall inward.

effective temperature The temperature a body would have if it were a blackbody of the same size radiating the same luminosity.

Einstein X-Ray Observatory An earth-orbiting x-ray telescope, with a 58-cm objective, launched in 1978; the telescope ceased functioning in 1981, but observational data are contained in archives that can be examined now.

electromagnetic force One of the four forces of nature; particles with electromagnetic charge either attract or repel each other depending upon whether the two charges are opposite or identical.

electromagnetic radiation A self-propagating electric and magnetic wave, such as light, radio, ultraviolet, or infrared radiation; all types travel at the same speed and differ in wavelength or frequency, which is a function of the energy.

electromagnetic spectrum The range of all wavelengths of electromagnetic radiation.

electron A lightweight, negatively charged subatomic particle; a lepton.

electron volt (eV) A convenient unit of energy for atomic physics; the energy gain of an electron when it is accelerated by one volt; $1 \text{ eV} = 1.60 \times 10^{-19} \text{ J}$.

electroweak force The combination of the weak and electromagnetic forces at temperatures greater than 1015 K.

element A substance that is made of atoms having the same chemical properties that cannot be decomposed chemically into simpler substances.

ellipse A plane curve drawn so that the sum of the distances from a point on the curve to two fixed points is constant.

elliptical galaxy A gravitationally bound system of stars that has rotational symmetry but no spiral structure and contains mainly old stars, with little gas or dust.

elongation The angular separation of an object from the sun as seen in the sky.

emission (bright) lines Light of specific wavelengths or colors emitted by atoms; sharp energy peaks in a spectrum caused by downward electron transitions from a discrete quantum state to another discrete state.

emission-line galaxy Any galaxy with emission lines in its spectrum.

emission-line spectrum A spectrum containing only emission lines.

emission nebula See diffuse nebula; a hot cloud of mostly hydrogen gas whose visible spectrum is dominated by emission lines.

empirical Derived from experiment or observation.

energy The ability to do work.

energy level One of the possible quantum states of an atom, with a specific value of energy.

enzyme A protein that brings about or accelerates reactions at body temperatures without itself undergoing destruction.

ephemeris (pl., ephemerides) A table that gives the positions of celestial objects at various times.

epicycle A small circle whose center lies on a larger one (the deferent), used by ancient astronomers, such as Ptolemy, to account for the westward retrograde motion and other irregular motions of the planets.

epoch The time for which the stellar positions (which change as a result of precession) given in an almanac or ephemeris are calculated.

equant An ancient geometrical device invented by Ptolemy to account for variations in planetary motion; essentially an eccentric in which the center of the circle is not the center of uniform motion.

equation of state A relationship that describes the conditions in a solid, liquid, or gas, such as an equation relating the pressure, temperature, and density of a gas.

equatorial bulge The excess diameter, about 43 km, of the earth through its equator compared with the diameter through its poles; any planet will have a large equatorial bulge if it has a fluid interior and rotates rapidly.

equatorial system A celestial coordinate system based on the celestial equator and the celestial poles, in which two angles, right ascension and declination, are used to specify the position of a point on the celestial sphere.

equilibrium A state of a physical system in which there is no overall change.

equilibrium temperature The temperature that has been achieved when a body emits the same energy per second that it absorbs.

equinox Time of year when day and night are of equal length; the two times of the year when the sun crosses the celestial equator; spring (vernal) equinox occurs about March 21, and fall (autumnal) equinox about September 22 in the Northern Hemisphere.

escape speed The speed a body must achieve to break away from the gravity of another body and never return to it.

escape temperature The temperature that particles in an atmosphere need to enter space with the escape velocity.

Euclidean (flat) geometry The geometry in which only one parallel line can be drawn through a point near another line; the sum of the angles in a triangle drawn on a flat surface is always 180°.

Europa The smallest of the Galilean moons of Jupiter;

has a density about three times that of water; its icy, cracked surface is devoid of impact craters.

event A point in four-dimensional spacetime.

event horizon See **Schwarzschild radius.**

evolution Changes in the physical properties of a body with time.

evolutionary lifetime In stellar evolution, the total length of time for a complete cycle of evolution, from birth to death.

evolutionary track On a temperature–luminosity (H–R) diagram, the path made by the points that describe how the temperature and luminosity of a star change with time.

excitation The process of raising an atom to a higher energy level.

exosphere The topmost region of a planet's atmosphere, from which particles in the atmosphere can escape into space.

expanding (3-kpc) arm A segment of spiral-arm structure encircling the center of our Galaxy at a distance of about 3 kpc (10 kly); it appears to be moving toward us and away from the Galaxy's center.

expansion of the universe The expansion of spacetime, as indicated by markers such as the distances between galaxies; believed to be a result of the Big Bang.

exponential decay A process, such as radioactive decay, for which the rate of change is directly proportional to the quantity present, so that the quantity remaining is given by an exponential function of time.

extended radio galaxies Active galaxies that show extended radio emission, usually in the form of two lobes on either side of the nucleus.

extinction The dimming of light when it passes through some medium, such as the earth's atmosphere or interstellar material.

extragalactic Outside the Milky Way Galaxy.

eyepiece A magnifying lens used to view the image produced by the main light-gathering lens of a telescope.

f-ratio The ratio of the focal length of a lens or mirror to its diameter.

filter A device used with a detector to narrow its range of spectra sensitivity.

first dredge-up Convection acting to bring to a star's surface the material processed by hydrogen burning during the star's first episode as a red giant.

first quarter The phase of the moon a quarter of the way around its orbit (eastward) from new moon, when it looks half-illuminated when viewed from the earth.

fission See **nuclear fission.**

fission model The earliest of the major models for the origin of the moon, suggesting that a young, rapidly spinning, molten earth lost a piece that spiraled out into orbit and cooled down to form the moon.

flare A sudden burst of energy in a star's chromosphere and corona. See **solar flare.**

flare star A dwarf star of spectral type M that exhibits flares, usually more energetic than the sun's.

flash spectrum The spectrum that appears immediately before the totality of a solar eclipse as the normal absorption spectrum is replaced briefly by the chromosphere's own emission spectrum.

flat geometry See **Euclidean geometry.**

flatness problem In the Big Bang model, the fact that the geometry of the universe is flat (or very close to it) and remains flat as the universe evolves.

flocculent galaxies Disk galaxies that show a puffy structure in their disks rather than well-defined spiral arms.

fluorescence The process by which a high-energy photon is absorbed by an atom and reemitted as two or more photons of lower energy.

flux The amount of energy flowing through a given area in a given time.

focal length The distance from a lens (or mirror) to the point at which it brings light to a focus for a distant object.

focus (pl., foci) The point at which light is gathered in a telescope.

forbidden line An emission line from an atom produced by a transition with a low probability of occurrence.

force A push or pull; a measure of the strength of a physical interaction.

forced motion Any motion under the action of a net force.

frame of reference A set of axes with respect to which the position or motion of something can be described or physical laws can be formulated.

Fraunhofer lines The name given to the strongest absorption lines in the spectrum of a star, especially the sun.

free–bound transition A transition of an electron between a free energy state and one bound to an atom;

results in the atom adding an electron and emitting a photon; the reverse process is a bound–free transition.

free-fall Gravitational collapse when there is no resisting internal pressure caused by collisions among the particles.

free-fall collapse time The time for a cloud to collapse gravitationally under free-fall conditions; a function of the initial density of the cloud.

free–free absorption or emission Photons absorbed or emitted when electrons and ions interact without the capture of the electron by the ion.

free–free transition A transition of an electron between two different free states; if energy is lost by the electron, the process is free–free emission; if gained, it is free–free absorption.

frequency The number of waves that pass a particular point in some time interval (usually a second); usually given in units of hertz. One hertz (Hz) is one cycle per second.

full moon The lunar phase during which the moon is opposite the sun in the sky and looks fully illuminated as seen from the earth.

fusion See **nuclear fusion.**

galactic cannibalism A model for galaxy interaction in which more massive galaxies strip material, by tidal forces, from less massive galaxies.

galactic center (core) The innermost part of the Galaxy's nuclear bulge; contains a cluster of stars, ionized gas clouds, and a rotating ring of material.

galactic (open) cluster A small group (about ten to a few hundred) of gravitationally bound stars of Population I, found in or near the plane of the Galaxy.

galactic equator The great circle along the line of the Milky Way, marking the central plane of the Galaxy.

galactic latitude The angular distance north or south of the galactic equator.

galactic longitude The angular distance along the galactic equator from a zero point in the direction of the galactic center.

galactic rotation curve A description of how fast an object some distance from the center of a galaxy revolves around it; the orbital speed as a function of distance.

galaxy A huge assembly of stars (between 10^6 and 10^{11}), plus gas and dust, that is held together by gravity; the Milky Way Galaxy, containing the sun, is designated the Galaxy.

Galilean moons The four largest satellites of Jupiter (Io, Europa, Ganymede, Callisto), discovered by Galileo with his telescope.

gamma ray A very-high-energy photon with a wavelength shorter than that of x-rays.

gamma-ray bursters Astronomical sources of short bursts of gamma rays; thought to originate from highly magnetic neutron stars.

Ganymede The largest of the Galilean moons of Jupiter, with a density about 1.9 times that of water; has both heavily cratered and faulted terrain.

gas tail The part of a comet's tail that consists of ions and molecules; it is shaped by interaction with the solar wind.

gauss (G) A physical unit measuring magnetic field strength.

G-band Absorption band in stellar spectra caused by metals.

general theory of relativity The idea developed by Albert Einstein that mass and energy determine the geometry of spacetime and that any curvature of spacetime shows itself by what we commonly call gravitational forces; Einstein's theory of gravity.

geocentric Centered on the earth.

geomagnetic axis The axis that connects the earth's magnetic poles; it is inclined about 12E from the geographic spin axis and does not pass through the earth's center.

giant Star of luminosity class III; typically has a radius between 10 and 100 times the sun's.

giant impact model A scenario for the moon's origin in which a Mars-sized object strikes the young earth with a glancing blow; materials from the colliding object and the earth form a disk around the earth out of which the moon accretes.

giant molecular clouds Large interstellar clouds, with sizes up to tens of light years and containing 100,000 solar masses of material; found in the spiral arms of the Galaxy, giant molecular clouds are the sites of massive star formation.

gibbous The phase of the moon occurring between first quarter and full moon, when the moon is more than half-illuminated as seen from the earth.

globular cluster A gravitationally bound group of about 10^5 to 10^6 Population II stars (of roughly solar mass), symmetrically shaped, found in the halo of the Galaxy and orbiting the galactic center.

globule A small dark cloud of interstellar gas and dust, usually seen with a bright nebula; possible region of star formation.

gnomon An ancient instrument for measuring time; most simply, a stick stuck vertically into the ground whose shadow is used to indicate the sun's position with respect to the horizon.

grand design spirals Spiral galaxies that have a well-defined spiral-arm structure.

grand unification theories (GUTs) Physical theories that attempt to unite the elementary particles and the four forces in nature as the actions of one particle and one force.

granite The type of rock making up the continental regions of the earth's crust.

granules Hot regions that appear briefly (3 to 10 min) as a rough texture on the solar photosphere; they mark the tops of areas of convective upflow from the convection zone.

gravitation In Newtonian terms, a force between masses that is characterized by their acceleration toward each other; the size of the force depends directly on the product of the masses and inversely on the square of the distance between them; in Einstein's terms, the curvature of spacetime.

gravitational bending of light The effect of gravity on the straight path of a photon; results in gravitational lensing.

gravitational collapse The unhindered contraction of any mass from its own gravity.

gravitation contraction Gravitational collapse in which internal pressure slows down the rate of collapse.

gravitational field The property of space having the potential for producing gravitational force on objects within it; characterized by the acceleration of free masses.

gravitational focusing The directing of the motions of small masses by a larger one so that their paths cross, which enhances their accretion onto the larger mass.

gravitational force The weakest of the four forces of nature; all particles with nonzero mass attract each other.

gravitational instability The tendency for a disturbed region in a gas to undergo gravitational collapse.

gravitational lensing The bending effect of a large mass on light rays, which causes them to form an imperfect image of the source of light.

gravitational mass The mass of an object as determined by the gravitational force it exerts on another object.

gravitational potential energy Potential energy related to a body's position in a gravitational field.

gravitational redshift The change to longer wavelengths that marks the loss of energy by a photon that moves from a stronger to a weaker gravitational field.

great circle Shortest distance between two points on a sphere.

Great Red Spot A large, long-lived, high-pressure storm in Jupiter's atmosphere.

greenhouse effect The production of an increased equilibrium temperature at the surface of a planet that occurs when the opacity of its atmosphere in the infrared results in the trapping of outgoing heat radiation; increasing the opacity is likely to increase the equilibrium temperature.

ground state The lowest energy level of an atom.

GUTs See grand unification theories.

H I region A region of neutral hydrogen in interstellar space; emits photons at a wavelength of 21 cm.

H II region A zone of hot, ionized hydrogen in interstellar space; it usually forms a bright nebula around a hot, young star or cluster of hot stars.

habitable zone See **ecosphere.**

Hadley cells Large regions of convective flow in the earth's or any planet's atmosphere.

half-life The time required for half the radioactive atoms in a sample to disintegrate by fission.

Halley's comet The periodic comet (orbital period about 76 years) whose orbit was first worked out by Edmond Halley from Newton's laws; has a small (about 10 km diameter), dark, irregular nucleus and emits jets of gas and dust.

halo (of a galaxy) The spherical region around a galaxy, not including the disk or the nucleus, containing globular clusters, some gas, a few stray stars, and an invisible form of matter.

halo star A star in the halo of a galaxy.

H-alpha line The first or alpha line of the Balmer series, the set of transitions in a hydrogen atom between the second energy level and levels with higher energy; it lies in the red part of the visible spectrum.

H and K lines Strongest lines in the visible spectrum from once-ionized calcium, at 3934 and 3968 Å.

head–tail (radio) galaxy An active galaxy whose radio

lobes have been swept back to form a tail because of interaction with the surrounding medium.

heat The total random energy of motion of particles in an object; the flow of thermal energy.

heavy-particle era In the hot Big Bang model, the time up to 0.001 second when gamma rays collided to make high-mass particles, such as protons.

heliocentric Centered on the sun.

heliocentric stellar parallax An apparent shift in the positions of nearby stars (relative to more distant ones) due to the changing position of the earth in its orbit around the sun; the size of the shift can be used to measure the distances to close stars; see trigonometric parallax.

helioseismology The study of the interior of the sun from an analysis of vibrations that appear at its surface.

helium burning Fusion of helium into carbon by the triple-alpha process; requires a temperature of at least 108 K.

helium flash The rapid burst of energy generation with which a star initiates helium burning by the triple-alpha process in the degenerate core of a low-mass red giant star.

hertz (Hz) A physical unit of frequency equal to 1 cycle per second.

Hertzsprung–Russell (H–R) diagram A graphic representation of the classification of stars according to their spectral class (or color or surface temperature) and luminosity (or absolute magnitude); the physical properties of a star are correlated with its position on the diagram, so a star's evolution can be described by its change of position on the diagram with time (see **evolutionary track**).

highlands Regions of higher than average elevation on a planet or satellite; usually refers to the older, cratered region of the moon's surface.

high-velocity clouds Clouds of gas associated with the Galaxy that move at speeds of hundreds of kilometers per second.

high-velocity stars Stars in the Galaxy having velocities greater than 60 km/s relative to the sun; these stars have orbits with high eccentricities, often at large angles with respect to the galactic plane.

homogeneous Having a consistent and even distribution of matter, the same in all parts.

Homo sapiens Humankind.

horizon The intersection with the sky of a plane tangent to the earth at the location of the observer.

horizon distance In cosmology, the maximum distance that light can travel in some epoch of the universe.

horizon problem In the Big Bang model, the problem that arises from the rapid expansion of the early universe, during which different regions could not communicate.

horizon system The celestial coordinate system based on the horizon plane and the zenith, in which altitude and azimuth are the two angles specifying a point on the celestial sphere.

horizontal branch The portion of the Hertzsprung–Russell diagram reached by Population II stars of low mass after the red giant stage and typically found in a globular cluster; it ranges from yellowish to reddish stars all having about the same luminosity (about 100 times that of the sun).

hour circle A great circle on the celestial sphere passing through the celestial poles.

H–R diagram See **Hertzsprung–Russell diagram.**

Hubble classification Hubble's classification of galaxies by shape (spiral, elliptical, irregular).

Hubble constant The proportionality constant relating radial velocity and distance in the Hubble law; the value, now around 75 km/s/Mpc [20 (km/s)/Mly], changes with time as the universe expands.

Hubble flow The uniform motion of galaxies from the expansion of the universe.

Hubble's law A description of the expansion of the universe: the more distant a galaxy lies from us, the faster it is moving away; the relation $v = Hd$, between the expansion velocity (v) and distance (d) of a galaxy, where H is the Hubble constant.

Hubble Space Telescope (HST) A space telescope launched in 1990 by the U.S. space shuttle into a low earth orbit; with a 2.4-meter mirror, HST can observe faint objects with high resolution, especially in the ultraviolet.

Hubble time Numerically the inverse of the Hubble constant; it represents, in order of magnitude, the age of the universe.

hydrogen burning Any fusion reaction that converts hydrogen (protons) to heavier elements, such as helium.

hydrogen corona A shell of hydrogen gas emitting Lyman-alpha around the earth, coming from the atmospheric escape of hydrogen produced by the dissociation of water.

hydrostatic equilibrium An equilibrium characterized

by the absence of mass motions, when pressure balances gravity.

hyperbola The shape of the orbit of a body with more than escape speed.

hyperbolic geometry An alternative to Euclidean geometry, constructed by N. I. Lobachevski on the premise that more than one parallel line can be drawn through a point near a straight line; the sum of the angles of a triangle drawn on a hyperbolic surface is always less than 180°.

hypernova Model for a gamma-ray burster, in which a massive star implodes to make a black hole.

ideal gas law The pressure of a gas is directly proportional to its temperature and density.

igneous rock A rock formed by the cooling of molten lava.

image Light rays gathered at the focus of a lens or mirror in the same relative alignment as the real object.

image processing The computer manipulation of digitized images to enhance aspects of them.

impact cratering The idea that craters are formed by the impact of solid objects onto a surface.

inertia The resistance of an object to a force acting on it because of its mass.

inertial mass Mass determined by subjecting an object to a known force (not gravity) and measuring the acceleration that results.

inferior conjunction For a planet orbiting interior to another, the alignment with the sun when the interior planet lies on the same side of the sun as the outer planet.

inferior planet Planets whose orbits lie interior to that of the earth's.

inflationary universe model A modification of the Big Bang model in which the universe undergoes an early, brief interval of rapid expansion.

Infrared Astronomical Satellite (IRAS) An infrared space telescope (joint project of the United States, the United Kingdom, and the Netherlands) with a 57-cm mirror; IRAS mapped almost all the entire sky at far-infrared wavelengths; its data are available in a computer archive.

infrared cirrus Patches of interstellar dust, emitting infrared radiation, which look like cirrus clouds on the images of the sky produced by IRAS.

infrared galaxy A galaxy that emits most of its luminosity in the infrared.

infrared telescope A telescope, optimized for use in the infrared part of the spectrum, fitted with an infrared detector.

injection velocity The speed and direction at which a satellite is placed into an orbit; a specific speed at right angles to the radius of the orbit gives a circular velocity, whereas somewhat larger or smaller speeds result in elliptical orbits.

instability strip The locus on the H–R diagram of cepheid variable stars; these stars are burning helium in their cores.

interacting galaxies Galaxies that move close enough together to produce distortions in shape by tidal forces.

interferometer See **radio interferometer.**

intergalactic medium The gas and dust found between the galaxies.

International Ultraviolet Explorer (IUE) A space telescope with a 45-cm mirror designed to record spectra in the ultraviolet region; its data are available in computer archives.

interplanetary medium The gas and dust between the planets.

interstellar dust Small, solid particles in the interstellar medium.

interstellar extinction curve The amount of extinction from interstellar dust as a function of wavelength.

interstellar gas Atoms, molecules, and ions in the interstellar medium.

interstellar medium All the gas and dust found between stars.

interstellar molecules The molecules found in the interstellar gas, especially in molecular clouds.

inverse beta decay The process in which electrons and protons are forced together to form neutrons and neutrinos; the reverse process of neutron decay.

inverse-square law for gravitation The decrease of the strength of gravitational force with the inverse square of the distance from a mass.

inverse-square law for light The decrease of the flux of light with the inverse square of the distance from the source.

Io Third largest and most dense Galilean moon of Jupiter, with a density about 3.5 times that of water; a volcanically active body with a sulfurous surface; internally heated by tidal stresses.

ion An atom that has become electrically charged by the gain or loss of one or more electrons.

ionization The process by which an atom loses or gains electrons; collisions and absorption of energetic photons are the two most common processes in astrophysics.

ionization energy The minimum energy required to ionize an atom.

ionized gas A gas that has been ionized so that it contains free electrons and positively charged ions; a plasma.

ionosphere A layer of the earth's atmosphere ranging from about 100 to 700 km above the surface in which oxygen and nitrogen are ionized by sunlight, producing free electrons.

ion tail The part of a comet's tail that contains ionized gas.

irons One of the three main types of meteorite, typically made of about 90 percent iron and 9 percent nickel, with traces of other elements.

irregular cluster (of galaxies) A cluster of galaxies lacking spherical symmetry and having no obvious single center.

irregular galaxy A galaxy without spiral structure or rotational symmetry, containing mostly Population I stars and abundant gas and dust, with active starbirth.

Ishtar Terra A large upland plateau in the northern hemisphere of Venus, bounded on the east by Maxwell Montes, a shield volcano.

isotopes Atoms with the same number of protons but different numbers of neutrons.

isotropic Having no preferred direction in space.

Jeans length The minimum size a disturbance in a gas must have to result in gravitational contraction; it depends on the pressure, temperature, and density of the medium.

Jeans mass The mass contained in the Jeans length; the minimum mass a disturbance in a gas must have to grow larger.

jet streams Latitudinal, coherent flows at high speed in the upper atmosphere of the earth or another planet.

joule (J) A physical unit of work and energy in the SI system.

Jovian planets Planets with physical characteristics similar to Jupiter: large mass and radius, low density, mostly liquid interior.

Julian date (JD) The successive interval in days since noon at Greenwich on 1 January 4713 B.C.

kelvin The unit of temperature in the SI system.

Keplerian motion Orbital motion that follows Kepler's laws.

Kepler's laws of planetary motion The three laws of planetary motion propounded by Johannes Kepler that describe the properties of elliptical orbits with an inverse-square force law.

kiloparsec (kpc) One thousand parsecs.

kinematic distance In our Galaxy, the distance to an object inferred from its Doppler shift, direction, and the Galaxy's rotation curve.

kinetic energy The ability to do work because of motion.

Kirchhoff's rules Empirical descriptions of the physical conditions under which the main types of spectra originate.

Kleinmann–Low Nebula A diffuse dust cloud with a temperature of about 70 K near the Orion Nebula, emitting strongly at far-infrared wavelengths.

KREEP A lunar material composed of potassium (K), rare-earth elements (REE), and phosphorus (P).

Large Magellanic Cloud (LMC) A small galaxy, irregular in shape, about 50 kpc (160 kly) from the Milky Way; it contains abundant gas, dust, and young stars.

last quarter The phase during which the moon is three-quarters of the way around its path from new moon, when it looks half-illuminated when viewed from the earth.

latitude Angular distance north or south of the equator.

LAWKI Acronym for "life as we know it."

laws of motion The physical descriptions of the nature of forced and natural motion; see Newton's laws of motion.

least-energy orbit The orbit connecting two points in the solar system (or any other gravitating system) that requires the least change of energy to move from one point to the other; the least-energy path connecting two circular orbits is the ellipse just tangent to both of them.

lens A curved piece of glass designed to bring light rays to a focus.

lenticular galaxies Galaxies of Hubble type S0, with a

disk like a spiral galaxy, but lacking spiral arms and without gas or dust.

lepton An elementary particle that has low mass, such as an electron.

lepton era See **light-particle era.**

light curve A graph of a star's changing brightness with time.

light-gathering power The ability of a telescope to collect light as measured by the area of its objective.

lighthouse model For a pulsar, a rapidly rotating neutron star with a strong magnetic field; the rotation provides the pulse period, and the magnetic field generates the electromagnetic radiation.

light-particle era In the hot Big Bang model, the interval from 0.0001 to 4 seconds when gamma rays can collide to make low-mass particles, such as electrons.

light pollution An increase in the background level of the night sky brightness caused by the scattering of light from artificial sources; as light pollution increases, fewer faint astronomical sources are visible.

light rays Imaginary lines in the direction of propagation of a light wave.

light year (ly) The distance light travels in a year, about 3.1×10^{13} km.

limb darkening The apparent darkening of the sun along its edge caused by an optical depth effect.

line profile The variation of a spectral line's flux as a function of wavelength; greatest at the center of the line.

local celestial meridian An imaginary line through the north and south points on the horizon and the zenith overhead.

Local Group (of galaxies) A gravitationally bound group of about thirty galaxies to which our Milky Way Galaxy belongs.

local standard of rest A frame of reference that serves to measure the average motion of the sun and nearby stars around the center of the Milky Way Galaxy.

Local Supercluster The supercluster of galaxies in which the Local Group is located; spread over 100 million light years, it contains the Virgo cluster.

local time Apparent solar time or mean solar time at a particular location.

longitude Angular distance, east or west, along the equator; on the earth, the reference longitude is that of Greenwich, England; a similar reference is used on other planets.

longitudinal wave A sound wave that moves in a push–pull motion through solids, liquids, and gases with a velocity that depends on the density of the medium.

long-period comets Comets with orbital periods greater than 200 years.

long-period variables Variable stars, usually irregular in cycles, with periods between about 100 and 1000 days.

low-ionization nuclear emission region A type of emission-line galaxy that may be the lowest level of activity of AGN galaxies.

lowlands Regions of below-average elevation on a planet or satellite; usually refers to the younger, impact basins of the moon.

low-velocity stars Stars with close to circular orbits in the plane of the Galaxy; they travel at less than 60 km/s with respect to the sun.

luminosity The total rate at which radiative energy is given off by a celestial body over all wavelengths; the sun's luminosity is about 4×10^{26} W.

luminosity class The categorization of stars that have the same surface temperatures but different sizes, resulting in different luminosities; based on the widths of dark lines in a star's spectrum, giant stars having narrower lines than dwarf stars of the same surface temperature.

luminosity function (of galaxies) The number of galaxies as a function of their luminosity in a cluster of galaxies.

luminosity function (of stars) The number of stars in a limited region of the Galaxy as a function of their luminosity.

lunar eclipse The cutoff of sunlight from the moon, when the moon lies on the same line as that formed by the earth and sun so that it passes through the earth's shadow; a lunar eclipse can occur only at full moon, and when the full moon lies very close to the ecliptic.

lunar occultation The passage of the moon in front of a star or planet.

lunar soil The fine particles created by the bombardment of the lunar surface by meteorites plus larger rock fragments.

Lyman-alpha line First line of the Lyman series.

Lyman series Transitions in a hydrogen atom to and from the lowest energy level; they involve large energy changes corresponding to wavelengths in the ultraviolet part of the spectrum; also, the set of absorption or emission lines corresponding to these transitions.

Magellanic Clouds Two neighboring galaxies, the Large Magellanic Cloud (LMC) and the Small Magellanic Cloud (SMC), visible in the Southern Hemisphere to the unaided eye; gravitationally bound to our Galaxy.

magnetic field The property of space having the potential of exerting magnetic forces on bodies within it.

magnetic field lines A graphic representation of a magnetic field showing its direction and, by the degree of packing of the lines, its intensity.

magnetic flux The number of magnetic field lines passing through an area.

magnetic reconnection The sudden connection of magnetic field lines of opposite polarity; a process that releases energy stored in the magnetic field.

magnetometer A device to measure the strength of a magnetic field.

magnetosphere The region around a planet in which particles from the solar wind are trapped by the planet's magnetic field.

magnifying power The ability of a telescope to increase the apparent angular size of a celestial object.

magnitude An astronomical measurement of an object's brightness; larger magnitudes represent fainter objects.

main sequence The principal series of stars in the Hertzsprung–Russell diagram; such stars are converting hydrogen to helium in their cores by the proton–proton process or by the carbon–nitrogen–oxygen cycle; this is the longest stage of a star's active life.

main-sequence lifetime The duration of time that a star spends fusing hydrogen to helium in its core; the longest phase of its evolution.

major axis The larger of the two axes of an ellipse.

mantle The major portion of the earth's interior below the crust, made of a plastic rock probably composed of olivine.

mare (pl., maria; Latin for "sea") A lowland area on the moon that appears darker and smoother than the highland regions, probably formed by lava that solidified into basaltic rock about 3 to 3.5 Gy ago.

mare basalts Basaltic rocks found on the surface of the lunar maria; tend to be the youngest (most recently formed) rocks on the lunar surface, with ages around 3.2 Gy.

mascons High-density concentrations of mass beneath the lunar maria; they have been detected by their effect on the orbits of moon-orbiting satellites.

maser Small regions of molecular clouds that emit microwave radio signals, produced by molecules, which are strongly amplified by natural processes similar to that of a laser.

mass A measure of an object's resistance to change in its motion (inertial mass); a measure of the strength of gravitational force an object can produce (gravitational mass).

mass extinction The sudden disappearance from the fossil record of a large number of species; thought to result from catastrophic changes in the earth's environment.

mass loss The rate at which a star loses mass, usually by a stellar wind, per year.

mass–luminosity ratio For galaxies, the ratio of the total mass to the luminosity; a rough measure of the kind of stars in the galaxy; a large ratio implies a large fraction of dark matter.

mass–luminosity relation An empirical relation, for main-sequence stars, between a star's mass and its luminosity, roughly proportional to the fourth power of the mass; for stars of other types, the numerical value of the power is different.

mass transfer The flow of material from one star to another in a binary system via a Roche lobe.

matter era In the Big Bang model, the time interval from about one million years after the Big Bang to now, in which matter dominates the universe.

maximum elongation The greatest angular distance (east or west) of an object from the sun.

mean lifetime The time it takes for the number of atoms of a radioactive substance to decrease by one factor of e (≈ 2.718).

mechanics A branch of physics that deals with forces and their effects on bodies.

megaparsec (Mpc) 1 million parsecs, or about 3.26 million light years.

megaton An explosive force equal to that of 1 million tons of TNT (about 4×10^{15} J).

meridian (celestial) An imaginary line drawn through the north and south points on the horizon and through the zenith.

mesosphere Region of the earth's atmosphere between 50 and 100 km, where the temperature falls rapidly.

Messier catalog The French astronomers Charles Messier and Pierre Méchain compiled this catalog of about 100 of the brightest galaxies, star clusters, and nebulas as hazy objects that might be mistaken for comets.

Messier 17 (M17) A nearby H II region that is the site of recent massive star formation at one end of a giant molecular cloud.

metal abundance In astronomy, the percentage of all elements except hydrogen and helium.

metallic hydrogen A state, reached at high pressures, in which hydrogen is able to conduct electricity.

metal-poor stars Stars with metal abundances much less than the sun's.

metal-rich stars Stars with metal abundances like that of the sun (about 1 to 2 percent of the total mass).

meteor The bright streak of light that occurs when a solid particle (a meteoroid) from space enters the earth's atmosphere and is heated by friction with atmospheric particles; sometimes called a falling star.

meteorite A solid body from space that survives a passage through the earth's atmosphere and hits the ground.

meteorite fall A meteorite seen in the sky and recovered on the ground.

meteorite find A recovered meteorite that was not seen to fall.

meteoroid A very small solid body moving through space in orbit around the sun.

meteor shower A rapid influx of meteors that appear to come out of a small region of the sky called the radiant.

meteor stream A uniform distribution of meteoroids along an orbit around the sun.

meteor swarm Meteoroids grouped in a localized region of an orbit around the sun; the source of meteor showers.

meteor trail The visible path of a meteor through the atmosphere, created by ionization of the air and vaporization of the meteoroid.

micrometeorites Very small meteorites (about 0.1 to 1 μm in diameter) that cool and solidify before they hit the ground.

micrometer (μm) A millionth of a meter (10^{-6}); common unit of measurement of the wavelength of light.

microwave background radiation A universal bath of low-energy photons having a blackbody spectrum with temperature of about 2.7 K.

midoceanic ridge The almost continuous submarine mountain chain that extends some 64,000 km through the earth's ocean basins.

Milky Way The band of light that encircles the sky, caused by the blending of light from the many stars lying near the plane of the Galaxy; also sometimes designates the Galaxy to which the sun belongs.

millisecond pulsar Generic name given to any pulsar with a pulse period of a few milliseconds.

minimum resolvable angle The smallest angle a telescope can clearly show.

minute of arc (arcmin) One-sixtieth of a degree (1E/60).

missing mass As yet undetected matter of an amount needed to close the universe. See **dark matter.**

molecular clouds Large, dense, massive clouds in the plane of a spiral galaxy; they contain dust and a large fraction of gas in molecular form.

molecular maser Microwave amplification by stimulated emission of radiation from a molecule.

molecule A combination of two or more atoms bound together electrically; the smallest part of a compound that has the properties of that substance.

momentum The product of an object's mass and velocity.

month The time period for the moon to complete one orbit of the earth; the exact interval depends on the reference point; the most common are synodic (phase) and sidereal (stars).

morning star Venus or Mercury visible in the predawn sky.

moving-cluster method A method for finding a distance to a cluster of stars by determining their radial velocities and the convergent point of their proper motions.

M-type asteroids Asteroids having albedos of about 10 percent and resembling metals in their reflective properties.

mutation A basic change in gene structure.

nadir The point on the celestial sphere directly below the observer, opposite the zenith.

nanometer (nm) 10^{-9} of a meter; common unit of measurement of the wavelength of light.

narrow-line region The gas surrounding an AGN which contributes relatively narrow forbidden emission lines.

narrow-tailed radio galaxies Radio galaxies that show a U-shaped tail behind the nucleus; they are fast-moving galaxies in a cluster of galaxies.

natal astrology The belief that treats the supposed influence of the stars and planets on human affairs by the use of their positions and relationships at the time of an individual's birth.

natural motion Motion without forces.

natural selection The process by which individuals with genes that produce characteristics that are best adapted to their environment have greater genetic representation in future generations.

nebula (Latin for "cloud") A cloud of interstellar gas and dust.

nebular model A model for the origin of the solar system in which an interstellar cloud of gas and dust contracted gravitationally to form a flattened disk out of which the planets formed by accretion.

neutrino An elementary particle with no (or very little) mass and no electric charge that travels at the speed of light and carries energy away during certain types of nuclear reaction.

neutron A subatomic particle with about the mass of a proton and no electric charge; one of the main constituents of an atomic nucleus; the union of a proton and an electron.

neutron star A star of extremely high density and small size (10 km) that is composed mainly of very tightly packed neutrons; cannot have a mass greater than about 3 solar masses.

new moon The phase during which the moon is in the same direction as the sun in the sky, appearing completely unilluminated as seen from the earth. The moon must be new for solar eclipses to occur.

newton (N) The SI unit of force.

Newtonian reflector A reflecting telescope designed so that a small mirror at a 45° angle in the center of the tube brings the focus outside the tube.

Newton's laws of motion The three laws describing motion from Newton's viewpoint; they are the inertial, force, and reaction laws.

node The point on the celestial sphere where an orbit intersects a coordinate reference, such as the two nodes of the moon's path that cross the ecliptic.

nonthermal radiation Emitted energy that is not characterized by a blackbody spectrum; in astronomy, usually designates synchrotron radiation.

noon Midday; the time halfway between sunrise and sunset when the sun reaches its highest point in the sky with respect to the horizon.

Norma arm A segment of a spiral arm of our Galaxy,

about 4000 pc (13,000 ly) from the sun toward the center of the Galaxy in the direction of the constellation Norma.

north magnetic pole One of the two points on a star or planet from which magnetic lines of force emanate and to which the south pole of a compass points. See **south magnetic pole.**

nova (Latin for "new") A star that has a sudden outburst of energy, temporarily increasing its brightness by hundreds to thousands of times; now believed to be the outburst of a degenerate star in a binary system; also used in the past to refer to some stellar outbursts that modern astronomers now call supernovas.

nuclear bulge The central region of a spiral galaxy, containing mostly old Population I stars.

nuclear fission A process that releases energy from matter: a heavy nucleus hit by a particle splits into two or more lighter nuclei whose combined mass is less than the original, the missing mass being converted into energy.

nuclear fusion A process that releases energy from matter by the joining of nuclei of lighter elements to make heavier ones; the combined mass is less than that of the constituents, the difference appearing as energy.

nucleic acid A huge spiral-shaped molecule, commonly found in the nucleus of cells, that is the chemical foundation of genetic material.

nucleosynthesis The chain of thermonuclear fusion processes by which hydrogen is converted to helium, helium to carbon, and so on, through all the elements of the periodic table.

nucleus (of an atom) The massive central part of an atom, containing neutrons and protons, about which the electrons orbit.

nucleus (of a comet) Small, bright, starlike point in the head of a comet; a solid, compact (diameter a few tens of kilometers) mass of frozen gases with some rocky material embedded in it as dust.

nucleus (of a galaxy) The central portion of a galaxy, composed of old Population I stars, some gas and dust, and, for many galaxies, a concentrated source of nonthermal radiation.

number density The number of particles per unit volume.

OB association Loose groupings of O and B stars in small subgroups; they are not bound by gravity and so dissipate in a few tens of millions of years.

OB subgroup A small collection of about ten O and B

stars, a few tens of light years across, within an OB association.

objective The main light-gathering lens or mirror of a telescope.

observable universe The parts of the universe that can be detected by the light they emit.

occultation The eclipse of a star or planet by the moon or another planet.

Olbers' paradox The statement that if there were an infinite number of stars distributed uniformly in an infinite space, the night sky would be as bright as the surface of a star, in obvious contrast to what is observed.

Olympus Mons A large shield volcano on the surface of Mars in the Tharsis ridge region; probably the largest volcano in our solar system.

Oort Cloud A cloud of comet nuclei in orbit around the solar system, formed at the time the solar system formed; the reservoir for new comets.

Oort's constant In Oort's equations describing differential galactic rotation, the number relating the sun's distance from the center of the Galaxy and the change in angular speed as a function of radius; current value is 15 km/s/kpc.

opacity The property of a substance that hinders (by absorption or scattering) light passing through it; opposite of transparency.

open cluster See **galactic cluster.**

open geometry See **hyperbolic geometry.**

opposition The time at which a celestial body lies opposite the sun in the sky as seen from the earth; the time at which it has an elongation of 180E.

optics The manipulation of light by reflection or refraction, usually to make an image.

orbital angular momentum The angular momentum of a revolving body; the product of a body's mass, orbital velocity (the component perpendicular to the orbital axis), and the distance from the system's center of mass.

orbital inclination The angle between the orbital plane of a body and some reference plane; in the case of a planet in the solar system, the reference plane is that of the earth's orbit; in the case of a satellite, the reference is usually the equatorial plane of the planet; for a double star, it is the plane perpendicular to the line of sight.

organic Relating to the branch of chemistry concerned with the carbon compounds of living creatures.

organism An ordered living creature.

Orion Nebula The hot cloud of ionized gas that is a nearby region of recent star formation, located in the sword of the constellation of Orion; also called Messier 42 (M42).

outgassing Release of gases from nongaseous materials; extrusion of gases from the body of a planet after its formation.

ozone layer (ozonosphere) A layer of the earth's atmosphere about 40 to 60 km above the surface, characterized by a high content of ozone (O_3).

parabola A geometric figure that describes the shape of an escape speed orbit.

parallax The change in an object's apparent position when viewed from two different locations; specifically, half the angular shift of a star's apparent position as seen from opposite ends of the earth's orbit.

parent meteor bodies Small solid bodies, a few hundreds or thousands of kilometers in size, believed to be the source of nickel–iron meteorites; these objects formed early in the history of the solar system and broke up through collisions.

parsec (pc) The distance an object would have to be from the earth to attain a heliocentric parallax of 1 second of arc; equal to 3.26 light years; a kiloparsec (kpc) is 10^3 parsecs, a megaparsec (Mpc) is 10^6 parsecs, and a gigaparsec is 10^9 parsecs.

pascal (Pa) Unit of pressure in the SI system.

Pauli exclusion principle The concept that no two electrons can be in the same quantum state at the same time.

perfect cosmological principle The statement that the universe appears the same to an observer at all locations and at all times.

perigee The point in its orbit at which an earth satellite is closest to the earth.

perihelion The point at which a body orbiting the sun is nearest to it.

period The time interval for some regular event to take place; for example, the time required for one complete revolution of a body around another.

periodic comets Comets that have relatively small elliptical orbits around the sun, with periods of less than 200 years.

periodic (regular) variables Stars whose light varies with time in a regular fashion.

period–luminosity relationship For cepheid variables,

a relation between the average luminosity and the time period over which the luminosity varies; the greater the luminosity, the longer the pulsational period.

Perseus arm A segment of a spiral arm that lies about 3 kpc (10 kly) from the sun in the direction of the constellation Perseus.

perturbations Small changes in the motions of a mass because of the gravitational effects of another mass.

phases of the moon The monthly cycle of the changes in the moon's appearance as seen from the earth; at new, the moon is in line with the sun and so not visible; at full, it is in opposition to the sun and we see a completely illuminated surface.

Phobos The larger of the two moons of Mars.

photodissociation The breakup of a molecule by the absorption of light with enough energy to break the molecular bonds.

photometer A light-sensitive detector placed at the focus of a telescope; used to make accurate measurements of small photon fluxes.

photometry Measurement of the intensity of light.

photon A discrete amount of light energy; the energy of a photon is related to the frequency f of the light by the relation E = hf, where h is Planck's constant.

photon excitation Raising an electron of an atom to a higher energy level by the absorption of a photon.

photosphere The visible surface of the sun; the region of the solar atmosphere from which visible light escapes into space.

physical universe The parts of the universe that can be seen directly plus those that can be inferred from the laws of physics.

pitch angle The angle between a spiral-arm's direction and the direction of circular motion about the Galaxy.

pixel The smallest picture element in a two-dimensional detector.

Planck curve The continuous spectrum of a blackbody radiator.

Planck's constant The number that relates the energy and frequency of light; it has a value of 6.63×10^{-34} J·s.

planet From the Greek word for "wandering"; any of the nine (so far known) large bodies that revolve around the sun; traditionally, any heavenly object that moved with respect to the stars (in this sense, the sun and the moon were also considered planets).

planetary nebula A thick shell of gas ejected from and moving out from an extremely hot star; thought to be the outer layers of a red giant star thrown out into space, the core of which eventually becomes a white dwarf.

planetesimals Asteroid-sized bodies that, in the formation of the solar system, combined to form the protoplanets.

planetology The comparative study of the planets and how they evolved.

plasma A gas consisting of equal numbers of ionized atoms and electrons.

plate tectonics A model for the evolution of the earth's surface that pictures the interaction of crustal plates driven by convection currents in the mantle.

Pluto The last major planet discovered in the solar system; actually a double-planet system with its moon, Charon, a small body of low mass with a thin atmosphere, an icy surface, and a density about twice that of water.

polar caps Icy regions at the north and south poles of a planet.

Polaris The current North Pole star; the outermost star in the handle of the Little Dipper.

polarization A lining up of the planes of vibration of light waves.

polarized light Light waves whose planes of oscillation are the same.

polodial magnetic field Magnetic field configuration with two poles and field lines running along meridians.

Population I stars Stars found in the disk of a spiral galaxy, especially in the spiral arms, including the most luminous, hot, and young stars, having a heavy-element abundance similar to that of the sun (about 2 percent of the total); an old Population I is found in the nucleus of spiral galaxies and in elliptical galaxies.

Population II stars Stars found in globular clusters and the halo of a galaxy; somewhat older than any Population I stars and containing a smaller abundance of heavy elements.

positron An antimatter electron; essentially an electron with a positive charge.

potential energy The ability to do work because of position; it is storable and can be converted into other forms of energy.

Poynting–Robertson effect The tendency for small particles in the solar system to spiral into the sun as a result of a drag produced by solar radiation pressure acting in a slightly nonradial direction; the force due to radiation appears to have a component in the direction opposite the particle's motion, just as vertical raindrops appear to come from the direction toward which a person runs.

PP chain See **proton–proton chain.**

precession of Mercury's orbit The turning, with respect to the stars, of the major axis of Mercury's orbit at a rate of 43 arcsec per century.

precession of the equinoxes The slow westward motion of the equinox points on the sky relative to the stars of the zodiac, due to the wobbling of the earth's spin axis.

pre-main-sequence star The evolutionary phase of a star just before it reaches the main sequence and starts hydrogen core burning.

pressure Force per unit area.

pressure gradient The rate of change of pressure with respect to distance along a certain direction.

primary The more luminous of the two stars in a binary system.

prime focus The direct focus of an objective without diversion by a lens or mirror.

primeval fireball The hot, dense beginning of the universe in the Big Bang model, when most of the energy was in the form of high-energy light.

principle of equivalence The fundamental idea in Einstein's general theory of relativity; the statement than you cannot distinguish between gravitational accelerations and accelerations of other kinds, or, equivalently, a statement about the equality of inertial mass and gravitational mass; a consequence is that gravitational forces can be made to vanish in a small region of spacetime by choosing an appropriate accelerated frame of reference.

prominences Cool clouds of hydrogen gas above the sun's photosphere in the corona; they are shaped by the local magnetic fields of active regions.

proper motion The angular displacement of a star on the sky from its motion through space.

protein A long chain of amino acids linked by hydrogen bonds.

protogalaxies Clouds with so much mass that they are destined to contract gravitationally into galaxies.

proton A massive, positively charged elementary particle; one of the main constituents of the nucleus of an atom; a baryon.

proton–proton (PP) chain A series of thermonuclear reactions that occur in the interiors of stars, by which four hydrogen nuclei are fused into helium; this process is believed to be the primary mode of energy production in the sun.

protoplanet A large mass formed by the accretion of planetesimals; the final stage of formation of the planets from the solar nebula.

protoplasm The fluid within a cell.

protostar A collapsing mass of gas and dust out of which a star will be born (when thermonuclear reactions turn on); its energy comes from gravitational contraction.

Ptolemaic system A complete geocentric solar system model, described by the Greek astronomer Ptolemy in the Almagest.

pulsar A radio source that emits signals in very short, regular bursts; thought to be a highly magnetic, rotating neutron star.

pulsating star A star with an unstable internal structure that causes it to expand and contract.

quadrature When a planet or the moon has an angular distance of 90E from the Sun as viewed from the earth.

quantum (pl., quanta) A discrete packet of energy.

quantum number In quantum theory, one of the four special numbers that determine the energy structure and quantum state of an electron within an atom.

quantum state The quantum description of the arrangement of electrons in an atom; allowed quantum states are filled starting with those of lowest energy first.

quark An elementary particle that makes up others, such as protons.

quasar or quasi-stellar object An intense, pointlike source of light and radio waves that is characterized by large redshifts of the emission lines in its visible spectrum.

radar mapping The surveying of the geographic features of a planet's surface by the reflection of radio waves from the surface.

radial velocity The component of relative velocity that lies along the line of sight.

radial velocity curve For a binary star system, a graph of the radial velocities of the two stars as a function of time or orbital phase.

radian (rad) A unit of angular measurement; 1 rad = 57.3°; 2π rad = 360°.

radiant The point in the sky from which a meteor shower appears to come.

radiation Usually refers to electromagnetic waves, such as light, radio, infrared, x-rays, ultraviolet; also sometimes used to refer to atomic particles of high energy, such as electrons (beta radiation) and helium nuclei (alpha radiation).

radiation belts In a planet's magnetosphere, regions with a high density of trapped solar wind particles.

radiation era In the Big Bang model, the time in the universe's history in which the energy in the universe was dominated by radiation.

radiative energy The capacity to do work that is carried by electromagnetic waves.

radiative transfer The process by which electromagnetic radiation interacts with matter.

radioactive dating A process that determines the age of an object by the rate of decay of radioactive elements within the object; for rocks, gives the date of the most recent solidification.

radioactive decay The process by which an element is converted by fission into lighter elements.

radio galaxies Galaxies that emit large amounts of radio energy by the synchrotron process, generally characterized by two giant lobes of emission situated on opposite ends of a line drawn through the nucleus; they are divided into two types, compact and extended.

radio interferometer A radio telescope that achieves high angular resolution by combining signals from at least two widely separated antennas.

radio jets Astrophysical jets that are visible to radio telescopes; generally, the jets extending from the nuclei of active galaxies.

radio recombination line Sharp energy peaks at radio wavelengths caused by low-energy transitions in atoms from one very high energy level to another nearby level following recombination of an electron with an ion.

radio telescope A telescope designed to collect and detect radio emissions from celestial objects.

rapid process (r-process) Formation of very heavy elements by the rapid addition of neutrons to a nucleus followed by beta decay.

ray On the moon or other satellite, a bright streak formed by material ejected from an impact crater.

recombination The joining of an electron to an ion; the reverse of ionization.

recombination line Emission line from an electron following the process of recombination.

reddening The preferential scattering or absorption of blue light by small particles, resulting in red light dominating the blue.

red giant A large, cool star with a high luminosity and a low surface temperature (about 2500 K), which is largely convective and has fusion reactions going on in shells; lies on the red giant branch or asymptotic giant branch in an H–R diagram.

redshift An increase in the wavelength of the radiation received from a receding celestial body described by the Doppler effect; a shift toward the long-wavelength (red) end of the spectrum.

red variables A class of cool stars, variable in light output.

reference frame A set of coordinates by which position and motion may be specified.

reflecting telescope A telescope that has a uniformly curved mirror as its primary light gatherer.

reflection The return of a light wave from the interface between two media.

reflection nebula A bright cloud of gas and dust that is visible because of the reflection of starlight by the dust.

refracting telescope A telescope that uses glass lenses to gather light.

refraction Bending of the direction of a light wave as it crosses the interface between two media, such as air and glass.

regolith The pulverized surface soil of the moon (or any airless body) caused by meteorite impacts.

regular cluster (of galaxies) A cluster of galaxies with definite symmetry and a well-defined core.

relativistic Doppler shift Wavelength shift from the radial velocity of a source as calculated in special relativity; so that very large redshifts (greater than 1) do *not* imply that the source moves faster than light.

relativistic jet A beam of particles moving at speeds close to that of light; such jets contained charged particles channeled by magnetic fields.

relativity Two theories developed by Albert Einstein; the special theory describes the motion of nonaccelerated objects, and general relativity is a theory of gravitation.

resolution The ability of a telescope to separate close stars or to pick out fine details of celestial objects.

retrograde motion The apparent anomalous westward

motion of a planet with respect to the stars, which occurs near the time of opposition (for an outer planet) or inferior conjunction (for an inner planet).

retrograde rotation Rotation from east to west.

revolution The motion of a body in orbit around another body or a common center of mass.

rift valley A depression in the surface of a planet created by the separation of crustal masses.

ring galaxy A galaxy with a ring-like structure, probably the result of an interaction with another galaxy.

ringlets The subsections of a planetary ring system, containing many small particles orbiting together and obeying Kepler's laws.

ring system The complete set of rings around a planet; these consist of smaller ringlets.

Roche limit The minimum distance from the center of a planet that a satellite can orbit and not undergo tidal disruption; roughly 2.5 times the radius of the planet.

Roche lobe In a binary star system, the region in the space around the pair in which the stars' gravitational fields on a small mass are equal.

rotation The turning of a body, such as a planet, on its axis.

rotation curve The relation between rotational speed of objects in a galaxy and their distance from its center.

RR Lyrae stars A class of giant, pulsating variable stars with periods of less than one day; they are Population II objects and commonly found in globular clusters.

Rydberg constant A number relating to the spacing of the energy levels in a hydrogen atom.

S0 galaxy A type of galaxy intermediate between ellipticals and spirals; it has a disk but no spiral arms.

Sagittarius A (Sgr A) Radio sources at the center of the Galaxy; Sgr A West is a thermal radio source (H II region), Sgr A East a nonthermal source, and Sgr A* a pointlike source that may mark the Galaxy's core.

Sagittarius arm A portion of spiral-arm structure of the Galaxy that lies about 2 kpc (6 kly) from the center of the Galaxy in the direction of the constellation Sagittarius.

satellite Any small body orbiting a more massive parent body.

scarp A long, vertical wall running across a flat plain.

scattering (of light) The change in the paths of photons without absorption or change in wavelength.

Schwarzschild radius The critical size that a mass must reach to be dense enough to trap light by its gravity, that is, to become a black hole.

scientific model A mental image of how the natural world works, based on physical, mathematical, and aesthetic ideas.

seafloor spreading The lateral motion of the ocean basins away from oceanic ridges.

secondary The less luminous of the two stars in a binary system.

second dredge-up The process by which convection brings the products of helium burning to the surface of a massive star during the second time it becomes a red giant.

second of arc (arcsec) 1/3600 of a degree (0.000278°), or 1/60 of a minute of arc (0.0167 arcmin).

secular parallax A method of determining the average distance of a group of stars by examining the components of their proper motions produced by the straight-line motion of the sun through space.

seeing The unsteadiness of the earth's atmosphere that blurs telescopic images and limits their resolving power.

seismic waves Sound waves traveling through and across the earth that are produced by earthquakes.

seismometer An instrument used to detect earthquakes and moonquakes.

semimajor axis Half of the major axis of an ellipse; distance from the center of an ellipse to its farthest point.

sequential star formation A model for the formation of massive stars from a giant molecular cloud in which the hot, ionized gas around a small cluster of OB stars creates a shock wave that initiates the collapse of another part of the molecular cloud to give birth to another small cluster of OB stars until the cloud's material is used up.

SETI Acronym for the "search for extraterrestrial intelligence."

sexagesimal system A counting system based on the number 60, such as 60 minutes in an hour or 60 minutes of arc in one degree.

Seyfert galaxy A galaxy having a bright nucleus showing broad emission lines in its spectrum; Seyferts probably are spiral galaxies; one type of AGN.

shepherd satellites Small moons that confine a planet's ring in a narrow band; one is located on the inside edge of the ring, one on the outside.

shield volcano A large volcano with gentle slopes formed by the slow outflow of magma.

shock wave A discontinuity in a medium created when an object travels through it at a speed greater than the local sound speed.

short-period comets Comets with orbital periods less than 200 years, probably captured from longer-period orbits by an encounter with a major planet.

sidereal month The period of the moon's revolution around the earth with respect to a fixed direction in space or a fixed star; about 27.3 days.

sidereal period The time interval needed by a celestial body to complete one revolution around another with respect to the background stars.

sidereal year The orbital period of the earth around the sun with respect to the stars.

signs of the zodiac The twelve angular 30° segments into which the ecliptic is divided; each corresponds to a zodiacal constellation.

silicate A compound of silicon and oxygen with other elements, very common in rocks at the earth's surface.

singularity A theoretical point of zero volume and infinite density to which any mass that becomes a black hole must collapse, according to the general theory of relativity.

Sirius B The white dwarf companion to Sirius A; Sirius A and B make up a binary star system.

slow process (s-process) Formation of very heavy elements by the slow addition of neutrons to the nucleus followed by beta decay.

Small Magellanic Cloud (SMC) The smaller of the two companion galaxies to the Milky Way; it is an irregular galaxy containing about 2×10^9 solar masses.

solar core Region of the sun's interior in which temperatures and densities are high enough for fusion reactions to take place.

solar cosmic rays Low-energy cosmic rays generated in solar flares.

solar day The interval of time from noon to noon.

solar dynamo model See **dynamo model.**

solar eclipse An eclipse of the sun by the moon, caused by the passage of the new moon in front of the sun.

solar flare Sudden burst of electromagnetic energy and particles from a magnetic loop in an active region; probably triggered by magnetic reconnection.

solar mass The amount of mass in the sun, about 2×10^{30} kg.

solar nebula The thin disk of gas and dust around the young sun, out of which the planets formed.

solar wind A stream of charged particles, mostly protons and electrons, that escapes into the sun's outer atmosphere (from coronal holes) at high speeds and streams out into the solar system.

solstice The time at which the day or the night is the longest; in the Northern Hemisphere, the summer solstice (around June 21) is the time of the longest day and the winter solstice (around December 21) the time of the shortest day; the dates are reversed in the Southern Hemisphere.

south magnetic pole A point on a star or planet from which the magnetic lines of force emanate and to which the north pole of a compass points.

space A three-dimensional region in which objects move and events occur and have relative direction and position.

space astronomy Astronomy done by instruments in orbit above the earth's atmosphere.

spacetime Space and time unified; a continuous system of one time coordinate and three space coordinates by which events can be located and described.

spacetime curvature The bending of a region of spacetime because of the presence of mass and energy.

spacetime diagram A diagram with one axis representing the three dimensions of space and the other axis representing time; it shows the relation of events and worldlines.

space velocity The total velocity of an object through space, combining the components of radial and transverse velocities.

special theory of relativity Einstein's theory describing the relations between measurements of physical phenomena as viewed by observers who are in relative motion at constant velocities.

spectral line A particular wavelength of light corresponding to an energy transition in an atom.

spectral sequence A classification scheme for stars based on the strength of various lines in their spectra; the sequence runs O-B-A-F-G-K-M, from hottest to coolest.

spectral type (or class) The designation of the type of a star based on the relative strengths of various spectral lines.

spectrograph or spectrometer An instrument used on

a telescope to obtain and record a spectrum of an astronomical object; a spectrograph records on a photographic plate; a spectrometer uses another type of spectral detector, such as a CCD.

spectroscope An instrument for examining spectra; also a spectrometer or spectrograph if the spectrum is recorded and measured.

spectroscopic binary Two stars revolving around a common center of mass that can be identified by periodic variations in the Doppler shift of the lines of their spectra.

spectroscopic distances A technique for measuring distance by comparing the brightnesses of stars with their actual luminosities, as determined by their spectra.

spectroscopy The analysis of light by separating it into wavelengths (colors).

spectrum (pl., spectra) The array of colors or wavelengths obtained when light is dispersed, as by a prism; the amount of energy given off by an object at every different wavelength.

speed The rate of change of position with respect to time.

spherical (closed) geometry An alternative to Euclidean geometry, constructed by G. F. B. Riemann on the premise that no parallel lines can be drawn through a point near a straight line; the sum of the angles of a triangle drawn on a spherical surface is always greater than 180°.

spicules Spears of hot gas that reach up from the sun's photosphere into the chromosphere.

spin angular momentum The angular momentum of a rotating body; the product of a body's mass distribution, rotational speed, and radius.

spiral arm Part of a spiral pattern in a galaxy: a structure composed of gas, dust, and young stars, which winds out from near the galaxy's center.

spiral galaxy A disk galaxy with spiral arms; the presumed shape of our Milky Way Galaxy. See barred spirals.

spiral nebula An older term for a spiral galaxy as it appeared visually through a telescope.

spiral tracers Objects that are commonly found in spiral arms and so are used to trace spiral structure; for example, Population I cepheids, H II regions, molecular clouds, and OB stars.

spontaneous emission The emission of a photon by an excited atom in which an electron falls to a lower energy level.

spontaneous generation The supposed natural origin of living things from lifeless matter.

sporadic meteor A meteor that occurs at random and so is not associated with a shower.

stadium (pl., stadia) An ancient Greek unit of length, probably about 0.2 km.

standard candle An astronomical object of known luminosity used to estimate distances to galaxies.

starburst galaxies Active galaxies that show evidence of large-scale formation of massive stars.

star counting A technique to measure the extent of the Galaxy, first used by William Herschel, in which it is assumed that the directions in space in which more stars are found (in a specific area) mark regions of greater extent of the Galaxy.

star model See **stellar interior model.**

starspots Dark, magnetic active regions in the photospheres of stars analogous to sunspots.

statistical parallax A parallax, and so a distance, determined from the average proper motions of selected groups of stars.

steady-state model A theory of the universe based on the perfect cosmological principle, in which the universe looks basically the same to all observers at all times; largely discredited by current observations of the cosmic microwave background radiation.

Stefan–Boltzmann law For a blackbody radiator, the relation between temperature and power emitted per unit area of surface; the flux is proportional to the fourth power of the temperature.

stellar corona The hot (1×10^6 K), thin outer atmosphere of a star in analogy to the sun's corona; visible by the emission of x-rays.

stellar flares Flares associated with active regions on stars, analogous to solar flares.

stellar interior model A table or plot of values of physical characteristics (such as temperature, density, and pressure) as a function of position within a star for a specified mass, chemical composition, and age, calculated from theoretical ideas of the basic physics of stars.

stellar lifetime The total duration of a star's lifetime from birth to death; the sun's lifetime is expected to be about 10 Gy.

stellar nucleosynthesis A process in which nuclear fusion builds up heavier nuclei while supplying the energy by which stars shine.

stellar parallax See **heliocentric stellar parallax.**

stellar spectral sequence See **spectral sequence.**

stellar surface temperature The temperature of the photosphere of a star.

stellar wind The outflow, both steady and sporadic, of gas at hundreds of kilometers per second from the corona of a star.

stimulated emission Radiation produced by the effect of a photon stimulating an atom in an excited state to emit another photon of the same wavelength.

stochastic star formation A model for the generation of spiral arms by the random formation of stars in a molecular cloud, triggered by a supernova explosion, which then are drawn out into a spiral pattern by differential galactic rotation.

stones Meteorites made of light silicate materials.

stony-irons A type of meteorite that is a blend of nickel–iron and silicate materials.

straight line The shortest distance between two points in any geometry.

stratosphere A layer in the earth's atmosphere in which temperature changes with altitude are small and clouds are rare.

string theory A theory of elementary particles that views them as one-dimensional entities.

strong force One of the four forces of nature; the strong force acts over very short distances to keep the nuclei of atoms together.

S-type asteroids Asteroids whose albedos indicate a surface made of silicates.

summer solstice See **solstice.**

sunspot A temporary cool region in the sun's photosphere, associated with an active region, with a magnetic field of some 0.1 tesla (a few thousand gauss).

sunspot cycle The 11-year number cycle and a 22-year magnetic polarity cycle of the formation of sunspots.

superclusters Clusters of clusters of galaxies; superclusters appear as long, filamentary chains with voids in between.

supergiant A massive star of large size and high luminosity; red supergiants are thought to be the progenitors of Type II supernovas.

supergiant elliptical (cD) galaxies The largest and most massive elliptical galaxies, sometimes with more than one nucleus; found at the core of a rich cluster of galaxies; they may have been formed by mergers.

supergravity A model that combines quantum ideas with gravity to describe the unification of all forces of nature during the first 10^{-43} second of the universe's history.

superior conjunction A planetary configuration in which an inner planet lies in the same direction as the sun but on the opposite side of the sun as viewed from the outer planet.

superluminal motion Motion apparently faster than the speed of light.

supermassive black hole A black hole with a mass of a million solar masses or more; probably powers active galaxies and quasars.

supernova A stupendous explosion of a star, which increases its brightness hundreds of millions of times in a few days; Type II results from the core implosion of a massive star at the end of its life; Type I may originate from a carbon-rich white dwarf in a binary system.

supernova remnant An expanding gas cloud from the outer layers of a star, blown off in a supernova explosion; detectable at radio wavelengths; moves through the interstellar medium at high speeds to create shock waves.

superwind A very strong stellar wind.

surface temperature Photospheric temperature of a star.

synchronous rotation A solid body that rotates and revolves at the same rate; for a satellite, this means that the body keeps the same face to its parent planet.

synchrotron radiation Radiation from an accelerating charged particle (usually an electron) in a magnetic field; the wavelength of the emitted radiation depends on the strength of the magnetic field and the energy of the charged particles.

synodic month The time interval between similar configurations of the moon and the sun (for example, between full moon and the next full moon); about 29.5 days.

synodic period The interval between successive similar lineups of a celestial body with the sun (for example, between oppositions).

temperature A measure of the average kinetic energies of the microscopic particles in a substance.

temperature gradient The change in temperature over a unit change in distance.

terminator The fairly sharp boundary between the portions of a planet illuminated and unilluminated by sunlight; easily visible on the moon.

terrestrial planets Planets similar in composition (mostly rocks and metals) and size to the earth: Mercury, Venus, Mars, and the moon.

tesla (T) In the SI system, a unit of measure of magnetic flux.

Tharsis ridge A highland region on Mars containing a cluster of volcanoes, including Olympus Mons.

theoretical resolution A telescope's resolving power based on its optics alone; ground-based optical telescopes have a resolution limited by the seeing of the atmosphere.

thermal energy The internal energy of an object from the random motions of the particles that make it up.

thermal equilibrium Steady-state situation characterized by an absence of large-scale temperature changes.

thermal pulses Bursts of energy generation from the triple-alpha process in the shell of a red giant star.

thermal radiation Electromagnetic radiation from a body that is hot; often characterized by a blackbody spectrum.

thermal speed The average speed of the random motion of particles in a gas.

thermosphere A layer in the earth's atmosphere, above the mesosphere, heated by x-rays and ultraviolet radiation from the sun.

threshold temperature The temperature at which photons have enough energy to create a given type of particle–antiparticle pair.

tidal force The difference in gravitational force between two points in a body caused by a second body, which may result in the deformation of the second body.

tidal friction Friction caused by the tidal motion of the water in the ocean basins.

time A measure of the flow of events.

Titan Saturn's largest satellite, the first satellite detected to have an atmosphere.

Titius–Bode law A nonphysical formula that gives the approximate distances of the planets from the sun in AU.

ton (metric) 1000 kg.

torodial magnetic field A magnetic field configuration in which the field lines run parallel to the equator.

torque A twisting force applied through a lever arm.

total energy The sum of all forms of energy attributed to an isolated body or a system of bodies; usually just the sum of the kinetic and potential energies.

transition (in an atom) A change in the electron arrangements in an atom, which involves a change in energy.

transition region In the sun's atmosphere, the region of rapid temperature rise lying between the chromosphere and the corona.

transverse fault A crack in a solid surface where the ground has moved sideways.

transverse velocity The component of an object's velocity that is perpendicular to the line of sight.

transverse wave A wave in which the oscillatory motion is perpendicular to the direction of propagation; such waves cannot travel through liquids.

Trapezium cluster A small cluster of young massive stars located in the Orion Nebula; the hot stars ionize the gas around them.

trigonometric parallax A method of determining distances by measuring the angular position of an object as seen from the ends of a baseline having a known length; see **heliocentric stellar parallax.**

triple-alpha reaction A thermonuclear process in which three helium atoms (alpha particles) are fused into one carbon nucleus.

Triton The largest moon of Neptune; Triton has a thin atmosphere.

tropical year The time interval for the earth to orbit the sun relative to the equinoxes.

tropopause The boundary in the earth's atmosphere between the troposphere and the stratosphere.

troposphere The lowest level of the earth's atmosphere, reaching 10 km from the surface; the area in which most of the weather takes place.

T-Tauri stars Newly formed stars of about 1 solar mass; usually associated with dark clouds; some show evidence of flares and starspots, others have circumstellar disks; one type of young stellar object.

Tully–Fisher relation The relation between the luminosity of a galaxy and the width of its 21-cm emission line; the greater the luminosity, the wider the line.

turbulence Irregular and sometimes violent convective motion.

turbulent viscosity The property of a gas (or any fluid) by which turbulent flow in one part affects the flow of a nearby part; an important effect in the transfer of angular momentum outward in the solar nebula.

turnoff point The point on the H–R diagram of a cluster at which the main sequence appears to terminate at the high-luminosity end.

21-cm line The emission line, at a wavelength of 21.11 cm, from neutral hydrogen gas; it is produced by atoms in which the directions of spin of the proton and electron change from parallel to antiparallel.

two-sphere universe The basic premise of the celestial coordinate systems that the universe is composed of two concentric spheres, the earth and the celestial sphere.

Type I, Type II supernovas Classification of supernovas by their light curves and spectral characteristics: Type I show a sharp maximum and slow decline with no hydrogen lines; Type II have a broader peak and a very sharp decline after 100 days with strong hydrogen lines in the spectrum.

ultraviolet telescope A telescope optimized for use in the ultraviolet region; must be used above the earth's atmosphere.

umbra The dark central region of a sunspot; the darkest part of an eclipse shadow.

uniformity of nature The assumption that astronomical objects of the same type are the same throughout the universe.

universality of physical laws The assumption, borne out by some evidence, that the physical laws understood locally apply throughout the universe and perhaps to the universe as a whole.

universal law of gravitation Newton's law of gravitation; see gravitation.

universe The totality of all space and time; all that is, has been, and will be.

upland plateaus Large highland masses on the surface of Venus.

UV Ceti star A flare star.

Valles Marineris Extensive canyonlands region near the equator of Mars.

Van Allen radiation belts Belts of charged particles (from the sun) concentrated and trapped in the earth's lower magnetosphere.

variable star Any star whose luminosity changes over a short period of time.

vector A quantity that expresses magnitude and direction; for example, forces and accelerations are vector quantities.

velocity The rate and direction in which distance is covered over some interval of time.

velocity dispersion The range of velocities around an average velocity for a group of objects, such as a cluster of stars or galaxies.

vernal equinox The spring equinox; see **equinox.**

vertical circle On the celestial sphere, any great circle through the zenith.

Very Large Array (VLA) A radio interferometer located in New Mexico, U.S.A.; it consists of 27 antennas arranged in a Y-shaped pattern.

Very-Large-Baseline Array (VLBA) A radio interferometer with antennas spread across the United States; the processing and control center is in New Mexico, U.S.A.

Very-Long-Baseline Interferometry (VLBI) Radio interferometry carried out by telescopes at widely separated locations (across continents and oceans!); the signals are brought together and processed by computer; the VLBA is a special form of VLBI.

Virgo cluster The nearest large cluster of galaxies containing thousands of galaxies; it appears to lie in the direction of the constellation of Virgo and makes up a large part of the Local Supercluster.

virial theorem The statement that the gravitational potential energy is twice the negative of the kinetic energy of a system of particles in equilibrium.

visual binary Two stars that revolve around a common center of mass, both of which can be seen through a telescope, allowing their orbits to be plotted.

visual flux The flux from a celestial object measured across the visual part of the electromagnetic spectrum.

visual luminosity The luminosity from a celestial object measured across the visual part of the electromagnetic spectrum.

vis viva equation A relation between orbital speed and distance from the focus for a body moving in an elliptical orbit; a form of the law of conservation of energy for a body moving in a gravitational field.

voids Regions between superclusters that contain no large concentrations of luminous matter; they have somewhat a spherical shape and extend over millions of light years.

volatiles Materials, such as helium or methane, that vaporize at low temperatures.

volcanic model The formation of craters as cones left over from lava eruptions.

volcanism All processes by which any material is expelled from a body's interior to its crust.

watt (W) The SI unit of power; one joule expended per second.

wavelength The distance between two successive peaks or troughs of a wave.

weak force A short-range force that operates in radioactive decay.

weight The total force on some mass produced by gravity.

weightlessness The condition of apparent zero weight, produced when a body is allowed to fall freely in a gravitational field; in general relativity, weightlessness signifies motion on a straight line in spacetime, that is, natural motion.

white dwarf A small, dense star that has exhausted its nuclear fuel and shines from residual heat; such stars have an upper mass limit of 1.4 solar masses, and their interior is a degenerate electron gas.

Widmanstätten figures Large crystal patterns that appear on the surfaces of iron meteorites when they are polished and etched; they are formed by slow cooling of the material.

Wien's law The relation between the wavelength of maximum emission in a blackbody's spectrum and its temperature; the higher the temperature, the shorter the wavelength at which the peak occurs.

winter solstice See **solstice**.

Wolf-Rayet star A rare type of very hot stars (20,000 to 50,000 K) that have emission lines in their spectra; some are the central stars of planetary nebulas.

worldline A series of events in spacetime.

W Virginis stars Pulsating variable stars; Population II cepheids.

x-ray burster An x-ray source that emits brief, powerful bursts of x-rays; probably occurs from accretion onto a neutron star in a binary system.

x-rays High-energy electromagnetic radiation with a wavelength of about 10^{-10} m.

year The time for the earth to orbit the sun; its exact value depends on the reference point.

young Population I stars A class of stars found in the disk of a spiral galaxy, especially in the spiral arms; they have a metal abundance similar to the sun's and are the youngest stars in the galaxy.

young stellar object (YSO) Generic name for any star prior to its main-sequence phase.

ZAMS See **zero-age main sequence**.

Zeeman effect The splitting of spectral lines because of strong magnetic fields.

Zeilik An American form of a Ukrainian family name.

zenith The point on the celestial sphere that is located directly above the observer at 90E angular distance from the horizon.

zero-age main sequence (ZAMS) The position on an H–R diagram reached by a protostar once it has come to derive most of its energy from thermonuclear reactions rather than from gravitational contraction.

zodiac The traditional twelve constellations through which the sun travels in its yearly motion, as seen from the earth; a thirteenth constellation, Ophiuchus, is actually part of the zodiac.

zodiacal light Sunlight reflected from dust in the plane of the ecliptic.

zone A region of high pressure in the atmosphere of a Jovian planet.

zone of avoidance A region near the plane of the Galaxy in which very few other galaxies are visible because of obscuration by dust.

PHOTO CREDITS

12.22 Transition Region and Coronal Explorer (TRACE)/Stanford-Lockheed Institute for Space Research and NASA

13.1 Hubble Space Telescope/NASA

13.10 Courtesy of Harvard College Observatory

13.11 Roger Bell, University of Maryland, and Michael Briley, University of Wisconsin, Oshkosh

13.18 Courtesy of Nancy Houk, from *An Atlas of Objective-Prism Spectra* (1984) by N. Houk, N. J. Irvine, and D. Rosenbush

14.1 Bill Schoening/AURA/NOAO/NSF

14.4 Jayanne English (CGPS/STSCI), using data acquired by the Canadian Galactic Plane Survey (NRC/NSERC) and produced with the support of Russ Taylor (University of Calgary)

14.7 Atlas Image obtained as part of the Two Micron All Sky Survey (2MASS), a joint project of the University of Massachusetts and the Infrared Processing and Analysis Center/California Institute of Technology, funded by the National Aeronautics and Space Administration and the National Science Foundation

14.8 N. A. Sharp/AURA/NOAO/NSF

14.9 © Anglo-Australian Telescope Board

14.10 Gemini Observatory/AURA/NOAO/NSF

14.11 K. L. Luhman, G. Schneider, E. Young, G. Rieke, A. Catera, H. Chen, M. Rieke, and R. Thompson/NASA

14.14 **(a)–(c)** © Dr. Alan P. Boss, Carnegie Institute of Washington

14.15 IPAC, courtesy of The Archives, California Institute of Technology

14.17 C. Burrows/WFPC 2 Team/NASA; J. Hester/WFPC 2 Team/NASA; and J. Morse/WFPC 2 Team/NASA

14.20 C. Burrows and J. Krist (STSCI)/WFPC2 IDT/NASA

14.21 Hubble Space Telescope/NASA

15.11 NASA and the Hubble Heritage Team (STSCI/AURA), with acknowledgments to Dr. Raghvendra Sahai (JPL) and Dr. Arsen R. Hajian (USNO)

15.13 © Anglo-Australian Observatory/Royal Observatory, Edinburgh

15.16 NASA and the Hubble Heritage Team (STSCI/AURA)

16.4 N. A. Sharp/WIYN/NOAO/NSF

16.7 WIYN/NOAO/NSF

16.8 Jay Gallagher (University of Wisconsin)/ N. A. Sharp (NOAO)/WIYN/NOAO/NSF

16.10 European Southern Observatory

16.12 P. Garnavich (Harvard Smithsonian Center for Astrophysics)/NASA

16.14 NASA/GSFC/U. Hwang et al.

16.15 © Anglo-Australian Observatory

16.16 NASA/Chandra X-Ray Observatory/SAO

16.19 N. A. Sharp/AURA/NOAO/NSF

16.28 Carl Akerlof/ROTSE Collaboration

17.2 COBE/NASA

17.3 Hubble Heritage Team (AURA/STSCI/NASA)

17.14 2MASS Project/University of Massachusetts/IPAC, California Institute of Technology/NSF/NASA

17.15 N. E. Kassim, D. S. Briggs (Naval Research Laboratory), J. Imamura (TJHSST), T. J. W. Lazio (NRL), T. N. LaRosa (Kennesaw State University), and S. D. Hyman (Sweet Briar College)

17.16 Observers: F. Yusef-Zadeh, M. R. Morris, and D. R. Chance/Courtesy NRAO/AUI

17.17 Courtesy of F. Yusef-Zadeh

17.18 NASA/Massachusetts Institute of Technology/ Pennsylvania State University

18.1 Jane C. Charlton (Penn State) et al./Hubble Space Telescope/ESA/NASA

18.4 Blair Savage, Chris Howk (University of Wisconsin)/N. A. Sharp (NOAO)/WIYN/ NOAO/NSF

18.5 AURA/NOAO/NSF

18.6 Hubble Space Telescope/NASA

18.7 Hubble Space Telescope/NASA

18.8 © David Malin, Anglo-Australian Observatory

18.9 Hubble Heritage Project/NASA

18.10 NASA, A. Fruchter and the ERO Team (STSCI, ST-ECF)

18.11 Hubble Space Telescope/NASA

18.13 Courtesy of Palomar Observatory, California Institute of Technology

18.16 **(a)–(c)** Courtesy of Vera Rubin, Carnegie Institution of Washington

18.19 AURA/NOAO/NSF

18.20 T. R. Lauer (NOAO)/NASA

18.21 AURA/NOAO/NSF

18.22 AURA/NOAO/NSF

18.23 AURA/NOAO/NSF

18.24 B. Whitmore (STSCI)/NASA

18.25 Hubble Space Telescope/NASA

18.29 NASA/Chandra X-Ray Observatory/SAO

19.1 D. Dixon (UCR), D. Hartmann (Clemson), E. Kolaczyk (U. Chicago)/NASA

19.2 © National Radio Astronomy Observatory

19.4 J. A. Biretta et al./Hubble Heritage Team (STSCI/AURA)/NASA

19.5 F. N. Owen, J. A. Eliek, and N. E. Kassim (NRAO)/AUI

19.6 © NRAO

19.7 NASA/Chandra X-Ray Observatory/SAO

19.10 P. Hughes (University of Michigan), C. Duncan (BGSU)/NASA

19.11 Hubble Heritage Team/AURA/STSCI/NASA

19.13 Andrew S. Wilson (University of Maryland) et al./WFPC2/Hubble Space Telescope/NASA

19.14 **(a)** NASA/STSCI

19.14 **(b)** NASA/CXC/SAO/R. Marshall et al.

19.15 AURA/NOAO/NSF

19.18 National Radio Astronomy Observatory

19.21 W. N. Colley and E. Turner (Princeton University), J. A. Tyson (Bell Labs, Lucent Technologies)/NASA

19.22 J. Rhoads (STSCI) et al./WIYN/AURA/NOAO/NSF

19.23 Courtesy of Richard Davis, Jodrell Bank; observations with the MERLIN array

19.26 Gary Bower, Richard Green (NOAO)/STIS Instrument Definition Team/NASA

19.27 John Bahcall (Institute for Advanced Study, Princeton)/NASA

F.19 Based on observations by T. J. Pearson, S. C. Urwin, M. H. Cohen, R. P. Linfield, A. C. S. Readhead, G. A. Seielstad, R. S. Simon, and R. C. Walker at the Owens Valley Radio Observatory and NRAO/AUI

20.1 Courtesy of Los Alamos National Laboratory

20.5 Courtesy of Bell Labs

20.8 Goddard Space Flight Center/NASA

20.9 BOOMERANG Project/NASA

20.13 © Springel, Mathis, White, Kauffmann, Lemson, & Dekel (1998)

20.14 Goddard Space Flight Center/NASA

Part Openers

1 European Southern Observatory

2 JPL/NASA

3 M. Heydari-Malayeri/STSCI/ESA/NASA

4 NASA and the Hubble Heritage Team (STSCI/AURA)

Chapter Openers

1 B. & S. Fletcher, Science & Art Products

2 (Detail) By permission of the Houghton Library, Harvard University

3 (Detail) Erich Lessing/Art Resource, New York

4 T. A. Rector, I. P. Dell'Antonio, AURA/NOAO/NSF

5 Michael Zeilik

6 U.S. Air Force photo

7 AURA/NOAO/NSF

8 (Detail) Image produced by F. Hasler, M. Jentoft-Nilsen, H. Pierce, K. Palaniappan, and M. Manyin/NASA Goddard Lab for Atmospheres. Data from National Oceanic and Atmospheric Administration (NOAA)

9 U.S. Geological Survey/NASA

10 Jet Propulsion Laboratory/NASA

11 Johns Hopkins University, Applied Physics Laboratory/NASA

12 Transition Region and Coronal Explorer (TRACE)/Stanford-Lockheed Institute for Space Research and NASA

13 Roger Bell, University of Maryland, and Michael Briley, University of Wisconsin, Oshkosh

14 Bill Schoening AURA/NOAO/NSF

15 NASA and the Hubble Heritage Team (STSCI/AURA), with acknowledgments to Dr. Raghvendra Sahai (JPL) and Dr. Arsen R. Hajian (USNO)

16 WIYN/NOAO/NSF

17 2MASS Project/University of Massachusetts/IPAC/California Institute of Technology/NSF/NASA

18 AURA/NOAO/NSF

19 © NRAO

20 BOOMERANG Project/NASA

Index